# INTRODUCTION TO BANACH SPACES: ANALYSIS AND PROBABILITY

This two-volume text provides a complete overview of the theory of Banach spaces, emphasising its interplay with classical and harmonic analysis (particularly Sidon sets) and probability. The authors give a full exposition of all results, as well as numerous exercises and comments to complement the text and aid graduate students in functional analysis. The book will also be an invaluable reference volume for researchers in analysis.

Volume 1 covers the basics of Banach space theory, operator theory in Banach spaces, harmonic analysis and probability. The authors also provide an annex devoted to compact Abelian groups.

Volume 2 focuses on applications of the tools presented in the first volume, including Dvoretzky's theorem, spaces without the approximation property, Gaussian processes and more. Four leading experts also provide surveys outlining major developments in the field since the publication of the original French edition.

**Daniel Li** is Emeritus Professor at Artois University, France. He has published over 40 papers and two textbooks.

**Hervé Queffélec** is Emeritus Professor at Lille 1 University. He has published over 60 papers, two research books and four textbooks, including *Twelve Landmarks of Twentieth-Century Analysis* (2015).

CAMBRIDGE STUDIES IN ADVANCED MATHEMATICS

All the titles listed below can be obtained from good booksellers or from Cambridge University Press. For a complete series listing visit: www.cambridge.org/mathematics.

# Introduction to Banach Spaces: Analysis and Probability

## Volume 2

DANIEL LI
*Université d'Artois, France*

HERVÉ QUEFFÉLEC
*Université de Lille I, France*

*Translated from the French by*
DANIÈLE GIBBONS and GREG GIBBONS

# CAMBRIDGE
## UNIVERSITY PRESS

University Printing House, Cambridge CB2 8BS, United Kingdom

One Liberty Plaza, 20th Floor, New York, NY 10006, USA

477 Williamstown Road, Port Melbourne, VIC 3207, Australia

314-321, 3rd Floor, Plot 3, Splendor Forum, Jasola District Centre, New Delhi - 110025, India

79 Anson Road, #06-04/06, Singapore 079906

Cambridge University Press is part of the University of Cambridge.

It furthers the University's mission by disseminating knowledge in the pursuit of education, learning and research at the highest international levels of excellence.

www.cambridge.org
Information on this title: www.cambridge.org/9781107162624
DOI: 10.1017/9781316677391

Originally published in French as *Introduction à l'étude des espaces de Banach* by Société Mathématique de France, 2004

© Société Mathématique de France 2004

First published in English by Cambridge University Press 2018
English translation © Cambridge University Press 2018

Reprinted 2018

*A catalogue record for this publication is available from the British Library*

ISBN – 2 Volume Set  978-1-107-16263-1  Hardback
ISBN – Volume 1  978-1-107-16051-4  Hardback
ISBN – Volume 2  978-1-107-16262-4  Hardback

Dedicated to the memory of
Jean-Pierre Kahane

# Contents

## Volume 2

# Contents

## Volume 1

# Preface

This book is dedicated to the study of Banach spaces.

While this is an introduction, because we trace this study back to its origins, it is indeed a "specialized course",[1] in the sense that we assume that the reader is familiar with the general notions of Functional Analysis, as taught in late undergraduate or graduate university programs. Essentially, we assume that the reader is familiar with, for example, the first ten chapters of Rudin's book, Real and Complex Analysis (RUDIN 2); QUEFFÉLEC–ZUILY would also suffice.

It is also a "specialized course" because the subjects that we have chosen to study are treated in depth.

Moreover, as this is a textbook, we have taken the position to completely prove all the results "from scratch" (i.e. without referring within the proof to a "well-known result" or admitting a difficult auxiliary result), by including proofs of theorems in Analysis, often classical, that are not usually taught in French universities (as, for example, the interpolation theorems and the Marcel Riesz theorem in Chapter 7 of Volume 1, or Rademacher's theorem in Chapter 1 of Volume 2). The exceptions are a few results at the end of the chapters, which should be considered as complementary, and are not used in what follows.

We have also included a relatively lengthy first chapter introducing the fundamental notions of Probability.

As we have chosen to illustrate our subject with applications to "thin sets" coming from Harmonic Analysis, we have also included in Volume 1 an Annex devoted to compact Abelian groups.

This makes for quite a thick book,[2] but we hope that it can therefore be used without the reader having to constantly consult other texts.

---

[1] The French version of this book appeared in the collection "Cours Spécialisés" of the Société Mathématique de France.

[2] However, divided into two parts in the English version.

We have emphasized the aspects linked to Analysis and Probability; in particular, we have not addressed the geometric aspects at all; for these we refer, for example, to the classic DAY, to BEAUZAMY or to more specialized books such as BENYAMINI–LINDENSTRAUSS, DEVILLE–GODEFROY–ZIZLER or PISIER 2.

We have hardly touched on the study of operators on Banach spaces, for which we refer to TOMCZAK-JAEGERMANN and to PISIER 2; DIESTEL–JARCHOW–TONGE and PIETSCH–WENZEL are also texts in which the part devoted to operators is more important. DUNFORD–SCHWARTZ remains a very good reference.

Even though Probability plays a large role here, this is not a text about Probability in Banach spaces, a subject perfectly covered in LEDOUX–TALAGRAND.

Probability and Banach spaces were quick to get on well together. Although the study of random variables with values in Banach spaces began as early as the 1950s (R. Fortet and E. Mourier; we also cite Beck [1962]), their contribution to the study of Banach spaces themselves only appeared later, for example, citing only a few, Bretagnolle, Dacunha-Castelle and Krivine [1966], and Rosenthal [1970] and [1973]. However, it was only with the introduction of the notions of type and cotype of Banach spaces (Hoffmann-Jørgensen [1973], Maurey [1972 b] and [1972 c], Maurey and Pisier [1973]) that they proved to be intimately linked with Banach spaces.

Moreover, Probability also arises in Banach spaces by other aspects; notably it allows the derivation of the very important Dvoretzky's theorem (Chapter 1 of Volume 2), thanks to the concentration of measure phenomenon, a subject still highly topical (see the recent book of M. Ledoux, *The Concentration of Measure Phenomenon*, Mathematical Surveys and Monographs **89**, AMS, 2001), dating back to Paul Lévy, and whose importance for Banach spaces was seen by Milman at the beginning of the 1970s.

We will also use Probability in a third manner, through the method of selectors, due to Erdös around 1955,[3] and afterwards used heavily by Bourgain, which allows us to make random constructions.

For all that, we do not limit ourselves to the probabilistic aspects; we also wish to show how the study of Banach spaces and of classical analysis interact (the construction by Davie, in Chapter 2 of Volume 2, of Banach spaces without the approximation property is typical in this regard); in particular we have concentrated on the application to thin sets in Harmonic Analysis.

Even if we have privileged these two points of view, we have nonetheless tried to give a global view of Banach spaces (with the exception of the

---

[3] Actually, this method traces back at least to Cramér [1935] and [1937].

geometric aspect, as already mentioned), with the concepts and fundamental results up through the end of the 1990s.

We point out that an interesting survey of what was known by the mid 1970s was given by Pełczyński and Bessaga [1979].

This book is divided into 14 chapters, preceded by a preliminary chapter and accompanied by an Annex. The first volume contains the first eight chapters, including the preliminary chapter and the Annex; the second volume contains the six remaining chapters. Moreover, it also contains three surveys, by G. Godefroy, O. Guédon and G. Pisier, on the major results and directions taken by Banach space theory since the publication of the French version of this book (2004), as well as an original paper of L. Rodríguez-Piazza on Sidon sets.

Each chapter is divided into sections, numbered by Roman numerals in capital letters (**I**, **II**, **III** etc.), and each section into subsections, numbered by Arabic numerals (**I.1** etc.). The theorems, propositions, corollaries, lemmas, definitions are numbered successively in the interior of each section; for example in Chapter 5 of Volume 1, Section IV they thus appear successively in the form: Proposition IV.1, Corollary IV.2, Definition IV.3, Theorem IV.4, Lemma IV.5, ignoring the subsections. If we need to refer to one chapter from another, the chapter containing the reference will be indicated.

At the end of each chapter, we have added comments. Certain of these cite complementary results; others provide a few indications of the origin of the theorems in the chapter. We have been told that "this is a good occasion to antagonize a good many colleagues, those not cited or incorrectly cited." We have done our best to correctly cite, in the proper chronological order, the authors of such and such result, of such and such proof. No doubt errors or omissions have been made; they are only due to the limits of our knowledge. When this is the case, we ask forgiveness in advance to the interested parties. We make no pretension to being exhaustive, nor to be working as historians. These indications should only be taken as incitements to the reader to refer back to the original articles and as complements to the contents of the course.

The chapters end with exercises. Many of these propose proofs of recent, and often important, results. In any case, we have attempted to decompose the proofs into a number of questions (which we hope are sufficient) so that the reader can complete all the details; just to make sure, in most cases we have indicated where to find the corresponding article or book.

The citations are presented in the following manner: if it concerns a book, the name of the author (or the authors) is given in small capitals, for example BANACH, followed by a number if there are several books by this author: RUDIN 3; if it concerns an article or contribution, it is cited by the name of the author or authors, followed in brackets by the year of publication, followed possibly by a lower-case letter: Salem and Zygmund [1954], James [1964 a].

We now come to a more precise description of what will be found in this book.

In the Preliminary Chapter, we quickly present some useful properties concerning the weak topology $w = \sigma(E, E^*)$ of a Banach space $E$ and the weak* topology $w^* = \sigma(X^*, X)$ in a dual space $X^*$. Principally, we will prove the Eberlein–Šmulian theorem about weakly compact sets and the Krein–Milman theorem on extreme points. We then provide some information about filters and countable ordinals.

Chapter 1 of Volume 1 is intended for readers who have never been exposed to Probability Theory. With the exception of Section V concerning martingales, which will not be used until Chapter 7, its contents are quite elementary and very classical; let us say that they provide "Probability for Analysts." Moreover, in this book, we use little more than (but intensively!) Gaussian random variables (occasionally stable variables), and the Bernoulli or Rademacher random variables. The reader could refer to BARBE–LEDOUX or to REVUZ.

Section III provides the theorems of Kolmogorov for the convergence of series of independent random variables, and the equivalence theorem of Paul Lévy.

In Section IV, we show Khintchine's inequalities, which, even if elementary, are of capital importance for Analysis. We also find here the majorant theorem (Theorem IV.5) which will be very useful throughout the book.

Section V, a bit delicate for a novice reader of Probability, remains quite classical; we introduce martingales and prove Doob's theorems about their convergence.

In Chapter 2 (Volume 1) we begin the actual study of Banach spaces. We treat the Schauder bases, which provide a common and very practical tool.

After having shown in Section II that the projections associated with a basis are continuous and given a few examples (canonical bases of $c_0$, $\ell_p$, Haar basis in $L^p(0, 1)$, Schauder basis of $\mathcal{C}([0, 1])$), we prove that the space $\mathcal{C}([0, 1])$ is universal for the separable spaces, i.e. any separable Banach space is isometric to a subspace of $\mathcal{C}([0, 1])$.

In Section III, we see how the use of bases, or more generally of basic sequences, allows us to obtain structural results; notably, thanks to the Bessaga–Pełczyński selection theorem, to show that any Banach space contains a subspace with a basis. We next show a few properties of the spaces $c_0$ and $\ell_p$. Finally, we see how the spaces possessing a basis behave with respect to duality; this leads to the notions of shrinking bases and boundedly complete bases and to the corresponding structure theorems of James.

In Chapter 3 (Volume 1), we study the properties of unconditional convergence (i.e. commutative convergence) of series in Banach spaces.

After having given different characterizations of this convergence (Proposition II.2) and showed the Orlicz–Pettis theorem (Theorem II.3) in Section II, we introduce in Section III the notion of unconditional basis, and show, in particular, that the sequences of centered independent random variables are basic and unconditional in the spaces $L^p(\mathbb{P})$.

In Section IV, we study in particular the canonical basis of $c_0$, and prove the theorems of Bessaga and Pełczyński which, on one hand, characterize the presence of $c_0$ within a space by the existence of a scalarly summable sequence that is not summable, and, on the other hand, state that a dual space containing $c_0$ must contain $\ell_\infty$.

In Section V, we describe the James structure theorems characterizing, among the spaces having an unconditional basis, those containing $c_0$, or $\ell_1$, or those that are reflexive.

All of the above work was done before 1960 and is now very classical.

In Section VI, we prove the Gowers dichotomy theorem, stating that every Banach space contains a subspace with an unconditional basis or a hereditarily indecomposable subspace (that is, none of its infinite-dimensional closed subspaces can be decomposed as a direct sum of infinite-dimensional closed subspaces). In addition, we provide a sketch of the proof of the homogeneous subspace theorem: every infinite-dimensional space that is isomorphic to all of its infinite-dimensional subspaces is isomorphic to $\ell_2$.

In Chapter 4 (Volume 1), we study random variables with values in Banach spaces.

Section II essentially states that the properties of convergence in probability, almost surely, and in distribution, seen in Chapter 1 in the scalar case can be generalized "as such" for the vector-valued case. Prokhorov's theorem (Theorem II.9) characterizes the families of relatively compact probabilities on a Polish space. The conditional expectation, more delicate to define than in the scalar case, is introduced, as well as martingales; the vectorial version of Doob's theorem (Theorem II.12) then easily follows from the scalar case.

In Section III we describe the important symmetry principle, also known as the Paul Lévy maximal inequality, which allows us to obtain the equivalence theorem for series of independent Banach-valued random variables between convergence in distribution, almost sure and in probability.

The contraction principle of Section IV will be of fundamental importance for all that follows; in its quantitative version, it essentially states that for a real (respectively complex) Banach space $E$, the sequences of independent centered random variables in $L^p(E)$, $1 \leqslant p < +\infty$, are unconditional basic sequences with constant 2 (respectively 4).

In Section V, we generalize the scalar Khintchine inequalities to the vectorial case (Kahane inequalities); the proof is much more difficult than for the scalar case. These inequalities will turn out to be very important when we define the type and the cotype of Banach spaces (Chapter 5). The proof of the Kahane inequalities uses probabilistic arguments; in Subsection V.3, we will see how the use of the Walsh functions allowed Latała and Oleskiewicz, thanks to a hypercontractive property of certain operators (Proposition V.6), to obtain, in the case "$L^1 - L^2$," the best constant for these inequalities (Theorem V.4).

Chapter 5 (Volume 1) introduces the fundamental notions of type and cotype of Banach spaces.

It is now common practice to define these using Rademacher variables, but it is often more interesting to use Gaussian variables, notably for their invariance under rotation. We thus begin, in Section II, by providing some complements of Probability; we first define Gaussian vectors, and show their invariance under rotation (Proposition II.8); we take advantage of this to present the vectorial version of the central limit theorem, which we will use in Chapter 4 of Volume 2. We next prove the existence of $p$-stable variables, also to be used in Chapter 4 of Volume 2, and present the classical theorems of Schönberg on the kernels of positive type, and of Bochner, which characterizes the Fourier transforms of measures.

As notions of type and cotype are local, i.e. only involving the structure of finite dimensional subspaces, we give a few words in Section III to ultraproducts and to spaces finitely representable within another; we prove the local reflexivity theorem of Lindenstrauss and Rosenthal, stating, more or less, that the finite-dimensional subspaces of the bidual are almost isometric to subspaces of the space itself.

In Section IV, we define the type and cotype, give a few examples (type and cotype of $L^p$ spaces, cotype 2 of the dual of a $C^*$-algebra), a few properties, and see how these notions behave with duality; this leads to the notion of $K$-convexity. We also show that in spaces having a non-trivial type, respectively cotype, we can, in the definition, replace the Rademacher variables by Gaussian variables (Theorem IV.8).

In Section V, we prove Kwapień's theorem, stating that a space is isomorphic to a Hilbert space if, and only if, it has at the same time type 2 and cotype 2; for this we first study the operators that factorize through a Hilbert space.

In Section VI, we present a few applications, and in particular show how to obtain the classical theorems of Paley and Carleman (Theorem VI.2).

In Chapter 6 (Volume 1), we will study a very important notion, that of a $p$-summing operator, brought out by Pietsch in 1967, and which soon afterward allowed Lindenstrauss and Pełczyński to highlight the importance of Grothendieck's theorem, which, even though proven in the mid 1950s, had not until then been properly understood.

We begin with an introduction showing that the 2-summing operators on a Hilbert space are the Hilbert–Schmidt operators.

In Section II, after having given the definition and pointed out the ideal property possessed by the space of $p$-summing operators, we prove the Pietsch factorization theorem, stating that the $p$-summing operators $T: X \rightarrow Y$ are those that factorize by the canonical injection (or rather its restriction to a subspace) of a space $C(K)$ in $L^p(K, \mu)$, where $K$ is a compact (Hausdorff) space and $\mu$ a regular probability measure on $K$; in particular the 2-summing operators factorize through a Hilbert space. It easily follows that the $p$-summing operators are weakly compact and are Dunford–Pettis operators. We next prove, thanks to Khintchine's inequalities, a theorem of Pietsch and Pełczyński stating that the Hilbert–Schmidt operators on a Hilbert space are not only 2-summing, but even 1-summing.

In Section III, we show Grothendieck's inequality (Theorem III.3), stating that scalar matrix inequalities are preserved when we replace the scalars by elements of a Hilbert space, losing at most a constant factor $K_G$, called the Grothendieck constant. We then prove Grothendieck's theorem: every operator of a space $L^1(\mu)$ into a Hilbert space is 1-summing. The proof is "local," meaning that it involves only the finite-dimensional subspaces; in passing we also show that the finite-dimensional subspaces of $L^p$ spaces can be embedded, $(1 + \varepsilon)$-isomorphically, within spaces of sequences $\ell_p^N$ of finite dimension $N$. We then give the dual form of this theorem: every operator of a space $L^\infty(\nu)$ into a space $L^1(\mu)$ is 2-summing.

In Section IV, we present a number of results, originally proven in different ways, that can easily be obtained using the properties of $p$-summing operators (note that these do not depend on Grothendieck's theorem, contrary to what might be suggested by the order of the presentation): the Dvoretzky–Rogers theorem (every infinite-dimensional space contains at least one sequence unconditionally convergent but not absolutely convergent), John's theorem (the Banach–Mazur distance of every space of dimension $n$ to the space $\ell_2^n$ is $\leqslant \sqrt{n}$), and the Kadeč–Snobar theorem (in any Banach space, there exists, on every subspace of dimension $n$, a projection of norm $\leqslant \sqrt{n}$). We then see that Grothendieck's theorem allows us to show that every normalized unconditional basis of $\ell_1$ or of $c_0$ is equivalent to their canonical basis (this is also true for $\ell_2$, but this case is easy).

Finally, Section V is devoted to Sidon sets (see Definition V.1). The fundamental example is that of Rademacher variables in the dual of the Cantor group $\Omega = \{-1, +1\}^{\mathbb{N}}$; another example is that of powers of 3 in $\mathbb{Z}$. We prove a certain number of properties, functional, arithmetical and combinatorial, demonstrating the "smallness" of Sidon sets; we show in passing the classical inequality of Bernstein. Grothendieck's theorem allows us to show that a set $\Lambda$ is Sidon if and only if the space $\mathcal{C}_\Lambda$ is isomorphic to $\ell_1$. We next present a theorem that is very important for the study of Sidon sets, Rider's theorem (Theorem V.18), which involves, instead of the uniform norm of polynomials, another norm $[\![ \, . \, ]\!]_R$, obtained by taking the expectation of random polynomials constructed by multiplying the coefficients by independent Rademacher variables. This allows us to obtain Drury's theorem (Theorem V.20), stating that the union of two Sidon sets is again a Sidon set, and the fact, due to Pisier, that $\Lambda$ is a Sidon set as soon as $\mathcal{C}_\Lambda$ is of cotype 2; for this last result, we need to replace, in the norm $[\![ \, . \, ]\!]_R$, the Rademacher variables by Gaussian variables, and are led to show a property of integrability of Gaussian vectors, due to Fernique (Theorem V.26), a Gaussian version of the Khintchine–Kahane inequalities, which will also be useful in Chapter 6 of Volume 2.

In Chapter 7 (Volume 1), we present a few properties of the spaces $L^p$. In Section II, we study the space $L^1$. After having defined the notion of uniform integrability, we give a condition for a *sequence* of functions to be uniformly integrable (the Vitali–Hahn–Saks theorem), which allows us to deduce that the spaces $L^1(m)$ are weakly sequentially complete. We then characterize the weakly compact subsets of $L^1$ as being the weakly closed and uniformly integrable subsets (the Dunford–Pettis theorem). We conclude this section by showing that $L^1$ is not a subspace of a space with an unconditional basis. We will continue the study of $L^1$ in Chapter 4 of Volume 2; more specifically, we will examine the structure of its reflexive subspaces.

In Section III, we will see that the trigonometric system forms a basis of $L^p(0, 1)$ for $p > 1$. This is in fact an immediate consequence of the Marcel Riesz theorem, stating that the Riesz projection, or the Hilbert transform, is continuous on $L^p$ for $p > 1$; most of Section III is hence devoted to the proof of this result. We have chosen not to prove it directly, but to reason by interpolation, allowing us to show in passing the Marcinkiewicz theorem, at the origin of real interpolation, as well as Kolmogorov's theorem stating that the Riesz projection is of weak type $(1, 1)$ (Theorem III.6). We conclude this section with a result of Orlicz (Corollary III.9) stating that the unconditional convergence of a series in $L^p$, for $1 \leqslant p \leqslant 2$, implies the convergence of the sum of the squares of the norms, implying that the trigonometric system is unconditional only for $L^2$.

In Section IV, we show, in contrast, that the Haar basis is unconditional in $L^p(0, 1)$, for $1 < p < +\infty$. This unconditionality is linked to the facts that the Haar basis is a martingale difference and that martingale differences are unconditional in $L^p$, $1 < p < +\infty$ (Theorem IV.7). We also present some complements on martingales, notably on the behavior in $L^p$ of the square function of a martingale (Theorem IV.6). The proof used here starts with the easy case, $p = 2$, and then passes successively, by doubling, to the cases $p = 4, 8, 16, \ldots$; we finish by interpolation, using the Riesz–Thorin theorem, previously shown in Subsection IV.1. To conclude this section, we study a particular property of the Haar basis, in a way rendering it extremal; for this, we need Lyapounov's theorem, stating that the image of vector measures with values in $\mathbb{R}^n$ is convex, and we prove this (Theorem IV.10).

Finally, the aim of Section V is to present another proof of Grothendieck's theorem, as a simple consequence of a theorem of Paley stating that $\sum_{k=1}^{+\infty} |\widehat{f}(2^k)|^2 < +\infty$ for every function $f \in H^1(\mathbb{T})$. For this, we very succinctly develop the theory of the spaces $H^p$, and prove the factorization theorem $H^1 = H^2 H^2$ (Theorem V.1) and the Frédéric and Marcel Riesz theorem. Grothendieck's theorem then follows from the fact that the operator $f \in A(\mathbb{T}) \mapsto (\widehat{f}(2^k))_{k \geqslant 1} \in \ell_2$ is 1-summing and surjective (Theorem V.6).

In the Comments, we show that there is essentially only one space $L^1(m)$, if we assume it separable and the measure $m$ atomless. We also give an alternative proof of the F. and M. Riesz theorem, due to Godefroy, using the notions of nicely placed sets and Shapiro sets.

Chapter 8 (Volume 1) is essentially devoted to Rosenthal's $\ell_1$ theorem, discovered in 1974. It provides a way to very easily detect when a Banach space contains $\ell_1$; it is a very general dichotomy theorem: in any Banach space, from every bounded sequence, we can extract either a weakly Cauchy subsequence or a subsequence equivalent to the canonical basis of $\ell_1$. The majority of proofs currently given use a Ramsey-type theorem of infinite combinatorics, the Nash–Williams theorem; we proceed differently, by first showing, in Section II, by a method due to Debs in 1987, the Rosenthal–Bourgain–Fremlin–Talagrand theorem (Theorem II.3), which is also a dichotomy theorem for the extraction of subsequences, this time for the pointwise convergence of sequences of continuous functions on a Polish space. We then derive Rosenthal's theorem for real Banach spaces. The complex case does not follow immediately; Dor was the first to show how to adapt the proof of the real case to show the complex case; we use here a method due to Pajor [1983] which uses combinatorial arguments to obtain the complex case from the real case.

In Section III, we prove the Odell–Rosenthal theorem (Theorem III.2), stating that a separable Banach space $X$ does not contain $\ell_1$ if and only if every

element of the unit ball $B_{X^{**}}$ of its bidual is the limit, for the weak* topology
$\sigma(X^{**}, X^*)$, of a *sequence* of elements of the ball $B_X$ of $X$. We next show a
result of Pełczyński (Theorem III.5), by a method due to Dilworth, Girardi and
Hagler [2000], stating that a Banach space contains $\ell_1$ if and only if its dual
contains $L^1(0, 1)$, or if and only if this dual contains the space of measures
$\mathcal{M}([0, 1])$ on $[0, 1]$.

The Annex (Volume 1) serves especially to give a general framework to
the elements of Harmonic Analysis that we use in this book, even though
we essentially use those of the group $\mathbb{T} = \mathbb{R}/\mathbb{Z}$ and the Cantor group
$\Omega = \{-1, +1\}^{\mathbb{N}^*}$ (sometimes its finite version), as well as those of finite
Abelian groups in Chapter 2 of Volume 2. In Section II, we present various
notions on Banach algebras: invertible elements, maximal ideals, spectrum of
an element, spectral radius; characters of a commutative algebra; involutive
Banach algebras and their positive linear functionals (Theorem II.12); $C^*$-
algebras. We show that every commutative $C^*$-algebra is isometric to the
algebra of continuous functions on a compact space (Theorem II.14).

Section III concerns compact Abelian groups $G$, which we assume metrizable for simplicity. We begin by proving the existence, and uniqueness, of
the Haar measure, thanks to the use of a *strictly* convex and lower semi-
continuous function on the set of probabilities on $G$ equipped with the weak*
topology (this approach requires the metrizability). We then give some results
on convolution. We next define the dual group $\Gamma = \widehat{G}$ as the set of characters
of $G$ and note that the metrizability of $G$ implies that the dual is countable;
we then determine the dual of the Cantor group (Proposition III.9), and show
that $\widehat{G}$ separates the points of $G$ (Theorem III.10; in fact shown in Theorem
III.16), and hence that the set $\mathcal{P}(G)$ of trigonometric polynomials, i.e. finite
linear combinations of characters, is dense in $\mathcal{C}(G)$ and in $L^p(G)$ for $1 \leqslant
p < +\infty$; moreover $\Gamma$ is an orthonormal basis of $L^2(G)$. We next define the
Fourier transform and show that it is injective. We conclude with results on
approximate identities and on the Fejér and de la Vallée-Poussin kernels. We
deduce that the norm of the convolution operator by a measure $\mu$ on $L^1(G)$,
and also on $\mathcal{C}(G)$, is equal to the norm of $\mu$.

The contents of Chapter 1 (Volume 2) are essentially of a local nature.
We show a fundamental structure theorem concerning the finite-dimensional
subspaces of Banach spaces, Dvoretzky's theorem, which states that every
$n$-dimensional space $E$ contains, for any $\varepsilon > 0$, "large" subspaces (of
dimension on the order of $\log n$) which are $(1 + \varepsilon)$-isomorphic to Hilbert
spaces. The proof is based on an argument of compactness, the Dvoretzky–
Rogers lemma, and, in an essential manner, on a probabilistic argument linked
to the concentration of measure phenomenon.

We thus begin in Section II with some results from Probability; after reviewing the asymptotic behavior of Gaussian variables, we examine that of the associated maximal functions of independent Gaussian variables and their absolute value. We then prove the Maurey–Pisier deviation inequality (Theorem II.3), from which we can deduce their inequality of the concentration of measure (Theorem II.4). For this, we need Rademacher's theorem (more or less classical, but rarely taught) on the almost everywhere differentiability of Lipschitz functions in $\mathbb{R}^N$. This inequality of concentration of measure allows us to prove Dvoretzky's theorem in both real and complex spaces; nonetheless we also use another approach, due to Gordon, valid only for the real case, as it can easily be adapted to prove the isomorphic version of Milman and Schechtman (Subsection IV.5).

In Section III, we prove a theorem concerning the comparison of Gaussian vectors, in a form due to Maurey (Theorem III.3). This allows us to easily obtain some important probabilistic results: Slepian's lemma (Theorem III.5) and its variant, the Slepian–Sudakov lemma (Theorem III.4), to be used in the proof of Dvoretzky's theorem for the real case, and Sudakov's minoration (Theorem III.6); these three results will again serve, in an essential manner, in Chapters 3 and 6 (Volume 2). To prove Dvoretzky's theorem, we need to be able to compare stochastically not only the max of Gaussian variables, but also their minimax; this is the purpose of Gordon's theorem (Theorem III.7).

The actual proof of Dvoretzky's theorem is in Section IV. We in fact present two proofs; in both cases the principle is the same. First, we introduce the Gaussian dimension (Pisier calls it the concentration dimension) $d(X)$ of a Gaussian vector $X$ (Definition IV.9). Dvoretzky's theorem is derived from what is known as the Gaussian version of Dvoretzky's theorem (Theorem IV.10), stating that when a Banach space $E$ contains a Gaussian vector $X$ made up of $m$ independent Gaussian variables, then $E$ contains, for any $\varepsilon > 0$, a subspace $(1 + \varepsilon)$-Hilbertian of finite dimension controlled by the Gaussian dimension $d(X)$ of $X$. The derivation from the Gaussian version is based on the Dvoretzky–Rogers lemma (Proposition IV.1), itself based on a compactness property in the spaces of operators between finite-dimensional spaces, given by Lewis' lemma (Lemma IV.3). Next we prove Theorem IV.10. For this, we construct, out of independent copies of the Gaussian vector $X$, random operators on $\ell_2^k$ with values in $E$, where $k$ is an appropriate multiple, dependent on $\varepsilon$, of $d(X)$. In the real case, the Slepian–Sudakov lemma allows us to limit from above the expectation of their norms, and Gordon's theorem to limit them from below. In the second proof (for the complex case, but for the real case as well), the two estimations are obtained at the same time by the

Maurey–Pisier concentration of measure inequality, by using the invariance of complex standard Gaussian vectors under the unitary group.

In the rest of Section IV, we examine certain examples; we see for example that the theorem is optimal for $E = \ell_\infty^n$. We also show that, with control of the cotype-2 constant of $E$, we can find, for any $\varepsilon > 0$, subspaces $(1+\varepsilon)$-Hilbertian of dimension proportional to that of $E$ (Theorem IV.14). This will be useful in Chapter 6 (Volume 2). To conclude this section, we prove the isomorphic version (Theorem IV.15), due to Milman and Schechtman. This allows, in a real Banach space $E$ of dimension $n$, to find, for any integer $k \leqslant n$, a subspace of dimension $k$, and whose distance to $\ell_2^k$ is this time no longer arbitrarily close to 1, but is instead controlled by an explicit function of $n$ and $k$. For this, we *admit* a delicate result, due to Bourgain and Szarek, that is an improvement of the Dvoretzky–Rogers lemma, and then apply Gordon's theorem.

Finally, we show in Section V the Lindenstrauss–Tzafriri theorem, for whose proof Dvoretzky's theorem, associated with Kwapień's theorem (Chapter 5 of Volume 1, Section V), is essential; it states that if in a Banach space all the closed subspaces are complemented, then this space is isomorphic to a Hilbert space.

Chapter 2 (Volume 2), quite short, is dedicated to the construction by Davie of a separable Banach space without the approximation property. The problem of the existence of such a space was posed by Grothendieck in the mid 1950s; it generalized the old problem of the existence of a basis in every Banach space, which dates back to Banach himself, and was resolved in 1972 by Enflo. The construction given soon afterward by Davie is simpler than that of Enflo. It combines a probabilistic argument (method of selectors) with an argument from Harmonic Analysis concerning finite groups. It fits particularly well with the objectives of this book.

In Section II, we give a certain number of equivalent formulations of the approximation property, and Section III contains the actual construction. We show that, for any $p > 2$, $\ell_p$ contains a closed subspace without the approximation property. This is also the case for $c_0$ and for $\ell_p$ with $1 \leqslant p < 2$ (Szankowski), but the proof is more delicate; it can be found, for example, in LINDENSTRAUSS–TZAFRIRI, Volume II, Theorem 1.g.4.

In Chapter 3 (Volume 2), we study in more detail Gaussian vectors, as well as the more general notion of Gaussian processes.

These are defined at the beginning of Section II. To each Gaussian process $X = (X_t)_{t \in T}$ we associate a (semi)-metric on $T$ by setting $d_X(s,t) = \|X_s - X_t\|_2$; we show, with the aid of Slepian's lemma, that the condition $d_Y \leqslant d_X$ is sufficient to ensure that if $X$ possesses a bounded version (respectively a continuous version), then so does $Y$ (the Marcus–Shepp theorem).

In Section III, we define Brownian motion as an example of a Gaussian process.

Sections IV and V form the heart of this chapter.

In Section IV, we define the *entropy integral* associated with a Gaussian process: this is the integral, for $\varepsilon \in [0, +\infty[$, of $\sqrt{\log N(\varepsilon)}$, where $N(\varepsilon)$ is the entropy associated with the metric $d_X$ of the process $X$, i.e. the minimum number of open $d_X$-balls necessary to cover $T$. The Dudley majoration theorem gives an upper bound for the expectation of the supremum of the modulus (absolute value) of a process with the aid of this entropy integral; a process has continuous paths as soon as the entropy integral is finite (Theorem IV.3). We then give an example showing that this condition is not necessary.

Next, in Section V, we see that, when the process is indexed by a compact metrizable Abelian group $G$ and is *stationary*, i.e. its distribution does not change under translation, then the finiteness of the entropy integral $J(d)$ becomes necessary to have continuous trajectories, and $J(d)$ is, up to a constant, equivalent to $\mathbb{E}\left(\sup_{t \in G} |X_t|\right)$: this is the Fernique minoration theorem (Theorem V.4). We conclude the section by giving an equivalent form of the entropy integral (Proposition V.5) that will be useful in Chapter 6 (Volume 2).

Section VI returns to Banach spaces; we present the Elton–Pajor theorem (Theorem VI.12), which gives Elton's theorem: in a real Banach space, if there are $N$ vectors $x_1, \ldots, x_N$ with norm $\leqslant 1$ such that the average of $\| \pm x_1 + \cdots \pm x_N \|$ over all choices of signs is $\geqslant \delta N$, then there is a subset of these vectors, of cardinality $N' \geqslant c(\delta) N$, which is equivalent to the canonical basis of $\ell_1^{N'}$, with constant $\beta(\delta)$ depending only on $\delta$ (Corollary VI.18). The proof uses probabilistic arguments: introduction of a Gaussian process and Dudley's majoration theorem, combinatorial arguments, notably Sauer's lemma (Proposition VI.3) and Chernov's inequality (Proposition VI.4), and volume arguments: Urysohn's inequality (Corollary VI.8), deduced from the Brunn–Minkowski inequality (Theorem VI.6), itself deduced from the Prékopa–Leindler inequality (Lemma VI.7). The complex version of Pajor requires several additional combinatorial lemmas (whose infinite-dimensional versions were used in Chapter 8 of Volume 1); it shows in particular that if a complex Banach space contains $\delta$-isomorphically, as a real Banach space, the space $\ell_1^N$, then it contains, in the complex sense, the complex space $\ell_1^{cN}$, where $c$ depends only on $\delta$ (Corollary VI.21).

In Chapter 4 (Volume 2), we concentrate on the reflexive subspaces of $L^1$.

In Section II, we first see that the reflexive subspaces of $L^1$ are those for which the topology of the norm coincides with that of the convergence in measure (the Kadeč–Pełczyński theorem) and that, in consequence, any non-reflexive subspace contains a *complemented* subspace isomorphic to $\ell_1$ (Corollary II.6).

We then examine their local structure. Even though *a priori*, as $L^1$ is weakly sequentially complete (Chapter 7 of Volume 1, Theorem II.6), its reflexive subspaces are those that do not contain $\ell_1$, by the Rosenthal $\ell_1$ theorem (Chapter 8 of Volume 1), in fact we have much more: the reflexive subspaces of $L^1$ are those not containing $\ell_1^n$'s uniformly (Theorem II.7). We then show that the Banach spaces that do not contain $\ell_1^n$'s uniformly are exactly those with a type $p > 1$ (Theorem II.8, of Pisier), so that the reflexive subspaces of $L^1$ have a non-trivial type $p > 1$ (Corollary II.9).

In Section III, we present some examples of reflexive subspaces. We first see that, for $1 < p \leqslant 2$, the sequences of independent $p$-stable variables generate isometrically $\ell_p$ in the real $L^1$ space (Theorem III.1). We then succinctly study the $\Lambda(q)$-sets, which are the reflexive and translation-invariant subspaces of $L^1(\mathbb{T})$. In particular we prove the Rudin transfer theorem, stating that the properties of Rademacher functions in the dual of the Cantor group are transferred to all the Sidon sets (Theorem III.10), so that, thanks to the Khintchine inequalities, every Sidon set $\Lambda$ is a $\Lambda(q)$-set for any $q < +\infty$, and, more precisely, $\|f\|_q \leqslant C S(\Lambda) \sqrt{q} \|f\|_2$ for every trigonometric polynomial $f$ with spectrum in $\Lambda$, where $C$ is a numerical constant and $S(\Lambda)$ is the Sidon constant of $\Lambda$ (Theorem III.11). The converse, due to Pisier, is shown in two different ways, first, in Chapter 5 (Volume 2), with a method of random extraction due to Bourgain, and then, in Chapter 6 (Volume 2), with the aid of Gaussian processes, which was the original proof of Pisier.

Section IV is devoted to the deep theorem of Rosenthal showing that the reflexive subspaces of $L^1$ embed in $L^p$, for some $p > 1$ (Theorem IV.1). We use in the proof the Maurey factorization theorem (Theorem IV.2), that Maurey isolated from the original proof of Rosenthal. We thus deduce that every $\Lambda(1)$-set is in fact $\Lambda(q)$ for some $q > 1$ (Corollary IV.3).

In Section V, we study the finite-dimensional subspaces of $L^1$, and, more precisely, the dimension $n$ of spaces $\ell_1^n$ that they can contain (Theorem V.2, of Talagrand). To make this statement more precise, we first need to study the $K$-convexity constant of finite-dimensional spaces (Theorem V.3), and in particular of those of $L^1$ (Theorem V.5). We also see that, up to a constant, nothing changes in the definition if the Rademacher variables are replaced by Gaussian variables (Theorem V.8). We need to prove an auxiliary result, due to Lewis (Theorem V.9). The proof of Talagrand's theorem is then based on the method of selectors, as well as Pajor's theorem from the preceding chapter to reduce to the real case.

Chapter 5 (Volume 2) contains three results of Bourgain illustrating the method of selectors.

This method was already used, in Chapters 2 and Chapter 4 of Volume 2; it involves selecting an independent sequence of Bernoulli variables $\varepsilon_1, \ldots, \varepsilon_n$, taking on the values 0 and 1 with a certain probability, and then making constructions by randomly choosing the set $\Lambda(\omega)$ of integers $k \leqslant n$ for which $\varepsilon_k(\omega)$ takes the value 1.

Section II treats the extraction of *quasi-independent sets*; these are particular Sidon sets, defined in an arithmetical manner, and whose Sidon constant is bounded by a fixed constant ($\leqslant 8$). We prove a theorem of Pisier stating that a set $\Lambda$ is Sidon if, and only if, there exists a constant $\delta$ such that every finite subset $A$ of $\Lambda$, not reduced to $\{0\}$, contains a quasi-independent subset $B$ of cardinality $|B| \geqslant \delta |A|$ (Theorem II.3). In fact, we show that from every finite subset $A$, not reduced to $\{0\}$, we can extract a quasi-independent subset $B$ of cardinality $|B| \geqslant K(|A|/\psi_A)^2$, where $K$ is a numerical constant and $\psi_A$ depends only on $A$ (Theorem II.6). As an immediate consequence we have Drury's theorem (Corollary II.4), and we easily obtain Pisier's theorem (Theorem II.13), the converse of Rudin's theorem seen in Chapter 4 of Volume 2, as well as Rider's theorem (Theorem II.14).

In Section III, we show that, for any $N \geqslant 1$, there exists a subset $\Lambda \subseteq \mathbb{N}^*$ of cardinality $N$ such that $\left\| \sum_{k \in \Lambda} \sin kx \right\|_\infty \leqslant C_0 N^{2/3}$, where $C_0$ is a numerical constant (Theorem III.1). The interest in this result is linked to the vector-valued Hilbert transform: if $E$ is a Banach space of finite dimension $N$, John's theorem immediately implies that the Hilbert transform with values in $E$ has a norm $\leqslant \sqrt{N}$ in $\mathcal{L}\big(L^2(E)\big)$; if $E = \ell_1^N$, this norm is dominated by $\log N$; the preceding result shows that for every $N \geqslant 1$, we can find a Banach space $E$ of dimension $N$ so that this norm dominates $N^{1/3}$.

In Section IV, we show that the majoration $K(X) \leqslant C \log n$ for the $K$-convexity constant of spaces of dimension $n$ seen in Chapter 4 (Volume 2) can essentially not be improved (Theorem IV.1).

Chapter 6 (Volume 2) is for the most part devoted to Pisier's space $\mathcal{C}^{as}$.

In Section II, we prove two results that will be needed in the next section. The first is the Itô–Nisio theorem, stating that, when $\sum_{n \geqslant 1} X_n$ is a series of independent symmetric random variables with values in $\mathcal{C}(K)$, where $K$ is a metrizable compact space, such that, for every $t \in K$, the series $\sum_{n=1}^{+\infty} X_n(\cdot, t)$ converges almost surely to $X_t$, and in addition we assume that the process $(X_t)_{t \in K}$ has a continuous version, then the series is almost surely uniformly convergent (Theorem II.2). We then show a Tauberian theorem (the Marcinkiewicz–Zygmund–Kahane theorem): if $\sum_{n \geqslant 1} X_n$ is a series of independent symmetric random variables with values in a Banach space $E$, then the fact that it is almost surely bounded (respectively almost surely convergent) according to a summation procedure implies that this holds in the usual sense (Theorem II.4).

In Section III, the space $\mathcal{C}^{as}$ is defined: let $G$ be a compact metrizable Abelian group and $\Gamma = \{\gamma_n \, ; \; n \geqslant 1\}$ its dual group; let $(Z_n)_{n\geqslant 1}$ be a standard sequence of independent complex Gaussian variables; then $\mathcal{C}^{as}(G)$ is the space of all the functions $f \in L^2(G)$ for which, almost surely in $\omega$, the sum of the series $\sum_{n\geqslant 1} Z_n(\omega)\widehat{f}(\gamma_n)\gamma_n$ is a continuous function $f^\omega \in \mathcal{C}(G)$. Theorem III.1 gives several equivalent formulations (one of these being Billard's theorem). Equipped with the norm defined by $[\![f]\!] = \sup_{N\geqslant 1} \mathbb{E} \big\| \sum_{n=1}^N Z_n \widehat{f}(\gamma_n)\gamma_n \big\|_\infty$, which is $\geqslant \|f\|_2$, $\mathcal{C}^{as}(G)$ is a Banach space for which the characters $\gamma_n \in \Gamma$ form a 1-unconditional basis (Theorem III.4). The Marcus–Pisier theorem (Theorem III.5) allows the Gaussian variables $Z_n$ in the definition to be replaced by Rademacher variables; the proof uses the Dudley majoration theorem and the Fernique minoration theorem. The fundamental result concerning $\mathcal{C}^{as}$ is Theorem III.9. It establishes a duality between $\mathcal{C}^{as}$ and the space of multipliers $M_{2,\Psi_2}$ from $L^2(G)$ to $L^{\Psi_2}(G)$, where $\Psi_2$ is the Orlicz function $\Psi_2(x) = e^{x^2} - 1$, and shows that with this duality $M_{2,\Psi_2}$ can be identified, isomorphically, with the dual of $\mathcal{C}^{as}$. The first part of the theorem again uses the Fernique minoration theorem; the second part is more delicate, and in addition to the Marcus–Pisier theorem, requires several auxiliary results. Thanks to this duality, we easily establish a result of Salem and Zygmund that gives upper and lower bounds of the norm $[\![\,.\,]\!]$ of a sum of exponentials (Proposition III.13).

In Section IV we present two more applications of $\mathcal{C}^{as}$. First we prove a theorem due to Pisier, a converse to Rudin's theorem (Chapter 4 of Volume 2), that characterizes Sidon sets $\Lambda$ as those for which $\|f\|_q \leqslant C \sqrt{q} \, \|f\|_2$ for every trigonometric polynomial $f$ with spectrum in $\Lambda$ (Theorem IV.1); note that this uses only the existence of a duality between $\mathcal{C}^{as}$ and $M_{2,\Psi_2}$, and not the fact that $M_{2,\Psi_2}$ is the dual of $\mathcal{C}^{as}$, and the Gaussian Rider theorem seen in Chapter 6 of Volume 1. Next, this space provides a response to the Katznelson dichotomy problem. Katznelson showed that only the real-analytic functions operate on the Wiener algebra $A(\mathbb{T})$, while it is clear that all continuous functions operate on $\mathcal{C}(\mathbb{T})$; the problem was to know if, for every Banach algebra $B$ possessing certain "nice" properties, and such that $A(\mathbb{T}) \subseteq B \subseteq \mathcal{C}(\mathbb{T})$, either all continuous functions operate on $B$ or only the analytic functions operate on $B$. Zafran found a counterexample to this conjecture; Theorem IV.2 (Pisier) reinforces the result of Zafran: $\mathfrak{P} = \mathcal{C}^{as}(\mathbb{T}) \cap \mathcal{C}(\mathbb{T})$, equipped with the norm $\|f\|_{\mathfrak{P}} = 8 \, \|f\|_\infty + [\![f]\!]$, is a Banach algebra possessing the required qualities, but in which all the Lipschitz functions operate.

To conclude Chapter 5, we prove the Bourgain–Milman theorem (Theorem V.1): $\Lambda$ is a Sidon set as soon as $C_\Lambda$ has a finite cotype (we have already seen in Chapter 6, Volume 1, that this is the case if the cotype is 2). The proof uses the notions of *Banach diameter* $n(E)$ of a finite-dimensional Banach space $E$ (Definition V.2) and *arithmetic diameter* for the finite subsets of the dual of a compact metrizable Abelian group $G$, where the latter is the entropy number $\overline{N_A}(1/2)$ for a pseudo-metric $\overline{d_A}$ on $G$, associated with the finite subset $A$ of the dual of $G$ (Definition V.3). Using Dvoretzky's theorem for cotype-2 spaces (in fact for $\ell_1$), the proof combines Theorem V.4 (of Maurey), which gives a lower bound for $n(E)$ as a fonction of the cotype constant of $E$, and Theorem V.5 (of Pisier): if $\Lambda$ is a finite subset in the dual of $G$ and if $\overline{N_A}(\delta) \geqslant e^{\delta |A|}$ for every $A \subseteq \Lambda$, then the Sidon constant of $\Lambda$ is bounded above by $a\,\delta^{-b}$, where $a, b > 0$ are numerical constants.

In the Comments, Section VI, as an application of random Fourier series, we prove two more results: one concerning functions of the Nevanlinna class (Theorem VI.1), and the other about random Dirichlet series (Theorem VI.2).

For the reader who would like to dig a bit deeper, we refer to the works cited in the bibliography, and in particular to the recent HANDBOOK OF THE GEOMETRY OF BANACH SPACES, Vols. I and II.

We kindly thank everyone that has assisted us in the preparation of this text; in particular Gabriel Li, who created the figures, and B. Calado, D. Choimet, M. Déchamps-Gondim, G. Godefroy, P. Lefèvre, F. Lust-Piquard, G. Pisier and Martine Queffélec, who proofread all or parts of the manuscript, for their comments and the improvements they have helped us to bring. We are especially thankful to G. Godefroy, J.-P. Kahane, B. Maurey and G. Pisier, from whom we have learned a great part of what is in this book. We also warmly thank the referees for their very precise and pertinent remarks, which we found extremely useful.

## Acknowledgements for the English Edition

We have not updated the French edition in this translation, but we have taken the opportunity to correct some mistakes and add some missing arguments.

We warmly thank G. Godefroy, O. Guédon, G. Pisier and L. Rodríguez-Piazza, who were kind enough to write, especially for this English version, three surveys and an original paper (see Appendices A, B, C and D in Volume 2).

Danièle and Greg Gibbons did a beautiful job with the translation of a very long, and at times highly specialized, mathematical text. Let them be warmly thanked for this achievement.

## Conventions

(1) In this book, the set $\mathbb{N}$ of natural numbers is $\mathbb{N} = \{0, 1, 2, \ldots\}$, and $\mathbb{N}^* = \{1, 2, \ldots\}$.
(2) Compact spaces are always assumed to be Hausdorff.

# 1

# Euclidean Sections

## I Introduction

In this chapter (Section IV), a fundamental theorem of structure of Banach spaces is proved: *Dvoretzky's theorem*. This result is local in nature, stating that every space $E$ of dimension $n$ contains *large* subspaces $F$ (of dimension of the order of $\log n$) that are *almost Hilbertian*: the Banach–Mazur distance $d_F$ between $F$ and $\ell_2^{\dim F}$ is close to 1. The proof uses both compactness (the Dvoretzky–Rogers and Lewis lemmas) and a probabilistic argument.

Two proofs of this theorem are presented. The first, due to Gordon [1985], is valid only in the real case, but can be well adapted to give an isomorphic version to be presented later; it is based on the comparison of Gaussian vectors (the Slepian–Gordon theorem). The second is valid in both cases, and relies on an inequality of concentration of measure due to Maurey and Pisier (see Pisier [1986 b] or PISIER 2). Probability plays an important role in these proofs, so, in Sections II and III, the required probabilistic tools are developed.

The chapter finishes with another theorem of structure, the Lindenstrauss–Tzafriri theorem (Section V), of global nature this time, stating that every Banach space, with all its closed subspaces complemented, is isomorphic to a Hilbert space. The proof essentially relies on Dvoretzky's theorem, plus an argument of compactness (ultraproducts) to pass from local to global nature.

## II An Inequality of Concentration of Measure

The phenomenon of *concentration of measure*, dear to V. Milman, turns out to be crucial in the proof of Dvoretzky's theorem, as Milman discovered (Milman [1971 b]). He used an isoperimetric inequality on the Euclidean sphere due to Paul Lévy. Here, an alternative, simpler inequality of this type is presented,

1

due to Maurey and Pisier (see Pisier [1986 b] or PISIER 2). Beforehand, a few complements on Gaussian variables are provided, to be used in Section III.

## II.1 Asymptotic Behavior of Gaussian Variables

Recall that a standard Gaussian (always denoted $g$) is a real random variable $g$ with density $(2\pi)^{-1/2}e^{-x^2/2}$; this variable is not bounded, but everything turns out as if it were "almost" bounded, as the following proposition, the analytic version of the Gaussian "bell curves", shows:

**Proposition II.1**   *Let $g$ be a standard Gaussian on a space $(\Omega, \mathcal{A}, \mathbb{P})$. Then:*

1) $\mathbb{P}(|g| > t) \leqslant e^{-t^2/2}$ *for any $t > 0$;*
2) $\mathbb{P}(|g| > t) \sim \sqrt{\frac{2}{\pi}}\, t^{-1}e^{-t^2/2}$ *when $t \to +\infty$;*
3) *for any $\delta > 0$, there exists $\alpha = \alpha(\delta) > 0$ such that $\mathbb{P}(g > t) \geqslant \dfrac{\alpha}{t}\, e^{-t^2/2}$ when $t \geqslant \delta$.*

*Proof*
1) We have:

$$\mathbb{P}(g > t) = \frac{1}{\sqrt{2\pi}} \int_t^{+\infty} e^{-x^2/2}\, \mathrm{d}x = \frac{1}{\sqrt{2\pi}} \int_0^{+\infty} e^{-(x+t)^2/2}\, \mathrm{d}x$$

$$= \frac{e^{-t^2/2}}{\sqrt{2\pi}} \int_0^{+\infty} e^{-tx}\, e^{-x^2/2}\, \mathrm{d}x \leqslant \frac{e^{-t^2/2}}{\sqrt{2\pi}} \int_0^{+\infty} e^{-x^2/2}\, \mathrm{d}x$$

$$= \frac{1}{2}\, e^{-t^2/2};$$

then $\mathbb{P}(|g| > t) = 2\,\mathbb{P}(g > t) \leqslant e^{-t^2/2}$.

2) It follows from the calculation in 1) that:

$$\mathbb{P}(g > t) = \frac{e^{-t^2/2}}{\sqrt{2\pi}\, t} \int_0^{+\infty} e^{-y}\, e^{-y^2/2t^2}\, \mathrm{d}y,$$

a quantity equivalent to:

$$\frac{e^{-t^2/2}}{\sqrt{2\pi}\, t} \int_0^{+\infty} e^{-y}\, \mathrm{d}y = \frac{e^{-t^2/2}}{\sqrt{2\pi}\, t} \qquad \text{when } t \to +\infty.$$

3) The function $t e^{t^2/2}\, \mathbb{P}(g > t)$ is continuous, $> 0$ on $[\delta, +\infty[$, and tends to $(2\pi)^{-1/2}$ as $t \to +\infty$ by 2), hence the result.    $\square$

When we consider $n$ independent copies $g_1, \ldots, g_n$ of a standard Gaussian $g$, Proposition II.1 leads to the following bounds:

**Proposition II.2**  *Let $g_1, \ldots, g_n$ be $n$ independent copies of a standard real Gaussian $g$, and for $n = 1, 2, \ldots$, consider the two maximal functions $M_n = \max(g_1, \ldots, g_n)$ and $M_n^* = \max(|g_1|, \ldots, |g_n|)$. Then:*

1) $\mathbb{P}\big(M_n > \sqrt{\log n}\big) \geqslant 1 - \varepsilon_n$, *with* $0 < \varepsilon_n < 1$ *and* $\sum_{n=1}^{+\infty} \varepsilon_n < +\infty$;
2) $C_1 \sqrt{\log n} \leqslant \mathbb{E}(M_n) \leqslant \mathbb{E}(M_n^*) \leqslant C_2 \sqrt{\log(n+1)}$, *with $C_1$ and $C_2$ positive numerical constants.*

*Proof*

1) If $n = 1$, the inequality holds with $\varepsilon_1 = 1/2$. If $n \geqslant 2$, we apply 3) of Proposition II.1, with $\delta = \sqrt{\log 2}$ and $t = \sqrt{\log n}$, to obtain:

$$\mathbb{P}(M_n < t) = \big(\mathbb{P}(g_1 < t)\big)^n = \big(1 - \mathbb{P}(g_1 > t)\big)^n \leqslant e^{-n\mathbb{P}(g_1 > t)}$$
$$\leqslant \exp\left(-n \frac{\alpha}{\sqrt{\log n}} n^{-1/2}\right) = \exp\left(-\alpha \sqrt{\frac{n}{\log n}}\right) = \varepsilon_n,$$

and the sequence $(\varepsilon_n)_{n \geqslant 1}$ thus defined works.

2) The upper bound was proved in Chapter 1 of Volume 1, Corollary IV.4, with $C_2 = \sqrt{8/3}$. For the lower bound, first note that, for $n$ large enough:

$$(*) \qquad \mathbb{E}(M_n^+) \geqslant \delta \sqrt{\log n} \quad \text{and} \quad \mathbb{E}(M_n^-) \leqslant \frac{c}{\sqrt{n}},$$

with $\delta$ and $c$ positive constants. In fact,

$$\sum_{n=1}^{+\infty} \mathbb{P}(M_n^+ \leqslant \sqrt{\log n}) \leqslant \sum_{n=1}^{+\infty} \varepsilon_n < +\infty,$$

hence the Borel–Cantelli lemma gives $\liminf_{n \to +\infty} \frac{M_n^+}{\sqrt{\log n}} \geqslant 1$ almost surely, and thus $\liminf_{n \to +\infty} \frac{\mathbb{E}(M_n^+)}{\sqrt{\log n}} \geqslant 1$ by Fatou's lemma. Moreover, by 1) of Proposition II.1:

$$\mathbb{E}(M_n^-) = \int_0^{+\infty} \mathbb{P}(M_n < -t)\, dt = \int_0^{+\infty} \big(\mathbb{P}(g_1 < -t)\big)^n dt$$
$$\leqslant \int_0^{+\infty} \big(\mathbb{P}(|g| > t)\big)^n dt \leqslant \int_0^{+\infty} e^{-nt^2/2}\, dt = \frac{c}{\sqrt{n}} \cdot$$

It follows from $(*)$ that:

$$\mathbb{E}(M_n) = \mathbb{E}(M_n^+) - \mathbb{E}(M_n^-) \geqslant \frac{\delta}{2} \sqrt{\log n}$$

for $n \geqslant n_0$. To obtain the lower bound of 2), it thus suffices to see that, for $2 \leqslant n < n_0$, $\mathbb{E}(M_n) > 0$. However, for $n \geqslant 2$:

$$\mathbb{E}(M_n) \geqslant \mathbb{E}(M_2) = \frac{1}{2\pi} 2 \int_{\mathbb{R}} \left[ \int_{x_1 \leqslant x_2} x_2 e^{-\frac{1}{2}(x_1^2 + x_2^2)} \, dx_2 \right] dx_1$$

$$= \frac{1}{\pi} \int_{-\infty}^{+\infty} e^{-x_1^2/2} \, e^{-x_1^2/2} \, dx_1 = \frac{1}{\sqrt{\pi}} > 0. \qquad \square$$

## II.2 The Maurey–Pisier Inequality

This inequality is needed for the proof of Dvoretzky's theorem in the complex case. Hence it is stated in this framework, even though $\mathbb{C}$ does not play any particular role.

First, we equip $\mathbb{C}^m$ with its standard Gaussian measure $\gamma$, of density:

$$\gamma(z) = \frac{1}{(2\pi)^m} \exp\left( -\frac{1}{2} \sum_{j=1}^{m} |z_j|^2 \right),$$

where $z = (z_1, \ldots, z_m)$. In other words, if we write $z_j = x_j + i y_j$, with $x_j$, $y_j \in \mathbb{R}$, we have:

$$\int_{\mathbb{C}^m} f(z) \, d\gamma(z) =$$

$$\frac{1}{(2\pi)^m} \int_{\mathbb{R}^{2m}} f(x_1, \ldots, x_m, y_1, \ldots, y_m) \exp\left( -\frac{1}{2} \sum_{j=1}^{m} (x_j^2 + y_j^2) \right)$$

$$dx_1 \ldots dx_m dy_1 \ldots dy_m$$

for every function $f \colon \mathbb{C}^m \to \mathbb{C}$ for which this makes sense. The usual Hermitian norm of $\mathbb{C}^m = \ell_2^m$ is denoted by $\| \, . \, \|_2$.

**Theorem II.3** (The Maurey–Pisier Deviation Inequality)  *Let $\Phi \colon \mathbb{C}^m \to \mathbb{R}$ be a $\sigma$-Lipschitz function:*

$$|\Phi(z) - \Phi(w)| \leqslant \sigma \, \|z - w\|_2, \qquad \forall \, z, w \in \mathbb{C}^m.$$

*If $M = \int_{\mathbb{C}^m} \Phi(z) \, d\gamma(z)$, then:*

$$\gamma\big(|\Phi - M| > t\big) \leqslant 2 \exp\left( -K \frac{t^2}{\sigma^2} \right)$$

*for any $t > 0$, with $K > 0$ a numerical constant ($K = 2/\pi^2$ is suitable).*

In particular, this leads to the following corollary:

**Theorem II.4** (The Maurey–Pisier Concentration of Measure Inequality)  *Let $E$ be a (complex) Banach space, $v_1, \ldots, v_m \in E$, $Z_1, \ldots, Z_m$ independent standard (complex) Gaussians, and $Z = \sum_{j=1}^m Z_j v_j$. Set:*

$$\sigma_Z = \sup_{\varphi \in B_{E^*}} \left( \mathbb{E} |\varphi(Z)|^2 \right)^{1/2};$$

*then:*

$$\mathbb{P} \left( \big| \, \|Z\| - \mathbb{E}\|Z\| \, \big| > t \right) \leqslant 2 \exp \left( -K \frac{t^2}{\sigma_Z^2} \right)$$

*for any $t > 0$.*

*Proof*  Take $\Phi(z) = \big\| \sum_{j=1}^m z_j v_j \big\|$, for $z = (z_1, \ldots, z_m) \in \mathbb{C}^m$. We have:

$$|\Phi(z) - \Phi(w)| \leqslant \left\| \sum_{j=1}^m (z_j - w_j) v_j \right\| \leqslant \sigma_Z \|z - w\|_2,$$

since $\sigma_Z = \sup_{a \in B_{\ell_2^m}} \big\| \sum_{j=1}^m a_j v_j \big\|$, and the result ensues from Theorem II.3.
$\qquad \square$

To prove Theorem II.3, we need to establish that Lipschitz functions are differentiable almost everywhere. This is the aim of the following theorem:

**Theorem II.5** (The Rademacher Theorem)  *Every Lipschitz function $\Phi \colon \mathbb{R}^N \to \mathbb{R}$ is differentiable almost everywhere.*

*Proof*  The case $N = 1$ is assumed well known: every absolutely continuous function $\phi \colon \mathbb{R} \to \mathbb{R}$, in particular every Lipschitz function, is differentiable almost everywhere and $\int_a^b \phi'(t)\,dt = \phi(b) - \phi(a)$ for any $a, b \in \mathbb{R}$; for this, we refer to RUDIN 2, Chapter 7 (Theorem 7.18).

For any $u \in \mathbb{R}^N$ with norm 1, denote by $\partial_u \Phi(x)$ the derivative in the direction $u$ of $\Phi$ at $x \in \mathbb{R}^N$, when it exists. Let $\mathcal{N}_u$ be the set of $x \in \mathbb{R}^N$ for which $\partial_u \Phi(x)$ does not exist. We can easily verify that this set is measurable. By applying the single-dimensional case to the function $t \mapsto \Phi(x + tu)$, we obtain, for any $x \in \mathbb{R}^N$, the negligibility of $\mathcal{N}_u \cap (x + \mathbb{R}u)$. Then, by Fubini's theorem, the measure of $\mathcal{N}_u$ is null. Hence, for every unitary vector $u$, $\partial_u \Phi(x)$ exists for almost any $x \in \mathbb{R}^N$.

Now consider the gradient $\nabla \Phi(x) = \big( \partial_1 \Phi(x), \ldots, \partial_N \Phi(x) \big)$, where the $\partial_j \Phi(x)$, $1 \leqslant j \leqslant N$, are the usual partial derivatives: the derivatives in the direction of the vectors $e_j$ of the canonical basis of $\mathbb{R}^N$. We have:

$$\partial_u \Phi(x) = \langle u, \nabla \Phi(x) \rangle$$

for almost any $x \in \mathbb{R}^N$. In fact, this is well known to be the case for any continuously differentiable function; then, if $\psi : \mathbb{R}^N \to \mathbb{R}$ is $C^1$ smooth and compactly supported, we have, via a change of variables:

$$
\int_{\mathbb{R}^N} \frac{\Phi(x + hu) - \Phi(x)}{h} \psi(x)\, dx = -\int_{\mathbb{R}^N} \frac{\psi(x) - \psi(x - hu)}{h} \Phi(x)\, dx
$$

$$
\xrightarrow[h \to 0]{} -\int_{\mathbb{R}^N} \partial_u \psi(x)\, \Phi(x)\, dx
$$

$$
= -\int_{\mathbb{R}^N} \langle u, \nabla \psi(x)\rangle \Phi(x)\, dx
$$

$$
= -\sum_{j=1}^{N} u_j \int_{\mathbb{R}^N} \partial_j \psi(x)\, \Phi(x)\, dx
$$

$$
= \sum_{j=1}^{N} u_j \int_{\mathbb{R}^N} \psi(x)\, \partial_j \Phi(x)\, dx,
$$

by integrating by parts with respect to the $j$-th variable,

$$
= \int_{\mathbb{R}^N} \psi(x) \langle u, \nabla \Phi(x)\rangle dx,
$$

thus the result, since, by the dominated convergence theorem (applicable as $\Phi$ is Lipschitz), the first integral tends to $\int_{\mathbb{R}^N} \partial_u \Phi(x)\, \psi(x)\, dx$.

Now let $\Delta$ be a countable dense subset in the unit sphere $S$ of $\mathbb{R}^N$. For each $u \in \Delta$, let $A_u$ be the set of $x \in \mathbb{R}^N$ such that $\nabla \Phi(x)$ and $\partial_u \Phi(x)$ exist and satisfy $\partial_u \Phi(x) = \langle u, \nabla \Phi(x)\rangle$, and let $A = \bigcap_{u \in \Delta} A_u$. By the above, $\mathbb{R}^N \setminus A$ has measure zero. Let us show that $\Phi$ is differentiable for every $x \in A$.

Fix $x \in A$, and, for $u \in S$ and $h \neq 0$, set:

$$
L_h(u) = \frac{\Phi(x + hu) - \Phi(x)}{h} - \langle u, \nabla \Phi(x)\rangle.
$$

It suffices to show that $\lim_{h \to 0} L_h(u) = 0$ uniformly for $u \in S$. Indeed, if $C$ is the Lipschitz constant of $\Phi$, then, for every $u, u' \in S$:

$$
|L_h(u) - L_h(u')| \leqslant (N + 1)\, C \, \|u - u'\|_2.
$$

The set of functions $L_h$, for $h > 0$, is hence equicontinuous on the compact set $S$. As it converges to 0 on the dense set $\Delta$, it converges uniformly on $S$ to 0, by Ascoli's theorem, and the proof is thus complete.                                     $\square$

*Proof of Theorem II.3*   Write $z = x + iy \in \mathbb{C}^m$ with $x, y \in \mathbb{R}^m$, and denote $x = \operatorname{Re} z$ and $y = \operatorname{Im} z$. Also denote:

$$
\Phi'_x = \left( \frac{\partial \Phi}{\partial x_1}, \dots, \frac{\partial \Phi}{\partial x_m} \right) \quad \text{and} \quad \Phi'_y = \left( \frac{\partial \Phi}{\partial y_1}, \dots, \frac{\partial \Phi}{\partial y_m} \right).
$$

Let $z = x + iy$ and $w = u + iv \in \mathbb{C}^m$; for $0 \leqslant \theta \leqslant 2\pi$, set:

$$z(\theta) = z \sin \theta + w \cos \theta,$$

so that $z'(\theta) = z \cos \theta - w \sin \theta$. Since $z(\pi/2) = z$ and $z(0) = w$, we obtain:

$$
\begin{aligned}
\Phi(z) - \Phi(w) &= \Phi\big[z(\pi/2)\big] - \Phi\big[z(0)\big] = \int_0^{\pi/2} \frac{\mathrm{d}}{\mathrm{d}\theta}\big(\Phi[z(\theta)]\big)\,\mathrm{d}\theta \\
&= \int_0^{\pi/2} \big[\langle\Phi_x'\big(z(\theta)\big), \operatorname{Re} z'(\theta)\rangle + \langle\Phi_y'\big(z(\theta)\big), \operatorname{Im} z'(\theta)\rangle\big]\,\mathrm{d}\theta,
\end{aligned}
$$

where $\langle\,\cdot\,,\,\cdot\,\rangle$ is the scalar product in $\mathbb{R}^m$.

For $\lambda \in \mathbb{R}$, we introduce the convex function $\phi_\lambda : t \in \mathbb{R} \mapsto e^{\lambda t}$. By Jensen's inequality:

$$
\begin{aligned}
&\phi_\lambda\big(\Phi(z) - \Phi(w)\big) \\
&\leqslant \frac{2}{\pi}\int_0^{\pi/2} \phi_\lambda\Big[\frac{\pi}{2}\big(\langle\Phi_x'\big(z(\theta)\big), \operatorname{Re} z'(\theta)\rangle + \langle\Phi_y'\big(z(\theta)\big), \operatorname{Im} z'(\theta)\rangle\big)\Big]\,\mathrm{d}\theta.
\end{aligned}
$$

The crucial point now is that, for each $\theta$, the Gaussian measure $d\gamma(z)d\gamma(w)$ is invariant under the unitary map $(z, w) \mapsto \big(z(\theta), z'(\theta)\big)$. Integrating the preceding inequality, and using Fubini's theorem, we thus obtain:

$$
\begin{aligned}
&\iint_{\mathbb{C}^m \times \mathbb{C}^m} \phi_\lambda\big(\Phi(z) - \Phi(w)\big)\,\mathrm{d}\gamma(z)d\gamma(w) \\
&\leqslant \iint_{\mathbb{C}^m \times \mathbb{C}^m} \phi_\lambda\Big[\frac{\pi}{2}\big(\langle\Phi_x'(z), \operatorname{Re} w\rangle + \langle\Phi_y'(z), \operatorname{Im} w\rangle\big)\Big]\,\mathrm{d}\gamma(z)d\gamma(w).
\end{aligned}
$$

The equality, in which $c \in \mathbb{C}^m$ and $\alpha \in \mathbb{R}$:

$$\int_{\mathbb{C}^m} \exp \alpha\big(\langle\operatorname{Re} c, \operatorname{Re} w\rangle + \langle\operatorname{Im} c, \operatorname{Im} w\rangle\big)\,\mathrm{d}\gamma(w) = \exp\left(\frac{\alpha^2}{2}\|c\|_2^2\right),$$

is used here, with $c = \Phi_x'(z) + i\,\Phi_y'(z)$ and $\alpha = \lambda\,\pi/2$, to obtain:

$$
\begin{aligned}
&\iint_{\mathbb{C}^m \times \mathbb{C}^m} \phi_\lambda\big(\Phi(z) - \Phi(w)\big)\,\mathrm{d}\gamma(z)d\gamma(w) \\
&\leqslant \int_{\mathbb{C}^m} \exp\left(\frac{\pi^2}{8}\lambda^2\big(\|\Phi_x'(z)\|_2^2 + \|\Phi_y'(z)\|_2^2\big)\right)\,\mathrm{d}\gamma(z).
\end{aligned}
$$

However $\|\Phi_x'(z)\|_2^2 + \|\Phi_y'(z)\|_2^2 \leqslant \sigma^2$ for almost all $z$, since $\Phi: \mathbb{C}^m \to \mathbb{R}$ is $\sigma$-Lipschitz; consequently:

$$\iint_{\mathbb{C}^m \times \mathbb{C}^m} \phi_\lambda\big(\Phi(z) - \Phi(w)\big)\,\mathrm{d}\gamma(z)d\gamma(w) \leqslant \exp\left(\frac{\pi^2}{8}\lambda^2\sigma^2\right).$$

Again using Jensen's inequality, we obtain, if $M = \int_{\mathbb{C}^m} \Phi(z) \, d\gamma(z)$:

$$\int_{\mathbb{C}^m} \phi_\lambda(\Phi(z) - M) \, d\gamma(z) \leqslant \exp\left(\frac{\pi^2}{8} \lambda^2 \sigma^2\right).$$

The rest of the proof is routine: for $\lambda > 0$, Markov's inequality gives:

$$\gamma(\Phi - M > t) = \gamma\left(e^{\lambda(\Phi - M)} > e^{\lambda t}\right) \leqslant e^{-\lambda t} \int_{\mathbb{C}^m} e^{\lambda(\Phi(z) - M)} \, d\gamma(z)$$

$$\leqslant \exp\left(-\lambda t + \frac{\pi^2}{8} \lambda^2 \sigma^2\right).$$

We optimize in $\lambda$, by taking $\lambda = \dfrac{4}{\pi^2} \dfrac{t}{\sigma^2}$, to obtain:

$$\gamma(\Phi - M > t) \leqslant \exp\left(-Kt^2/\sigma^2\right),$$

with $K = 2/\pi^2$. Applying this inequality to $(-\Phi)$, we also have:

$$\gamma(\Phi - M < -t) \leqslant \exp\left(-Kt^2/\sigma^2\right),$$

so finally, by addition:

$$\gamma(|\Phi - M| > t) \leqslant 2 \exp\left(-Kt^2/\sigma^2\right),$$

as claimed.                                                                                      $\square$

**Remark**   By replacing the integration over $[0, \pi/2]$ by Itô's formula for Brownian motion, we obtain the best constant $K = 1/2$ (see Pisier [1986 b]).

## III  Comparison of Gaussian Vectors

### III.1  Statement of the Problem

Let $X = (X_1, \ldots, X_n)$ and $Y = (Y_1, \ldots, Y_n)$ be two centered Gaussian vectors. They can be considered as processes $i \mapsto X_i$ and $i \mapsto Y_i$ indexed by the instants $i = 1, \ldots, n$, and (with no loss of generality) will always be assumed non-degenerate, i.e. possessing a density. The vectors $X$ and $Y$ are determined by their respective covariance matrices $\left(c_{ij}^X\right)_{i,j=1,\ldots,n}$ and $\left(c_{ij}^Y\right)_{i,j=1,\ldots,n}$, where $c_{ij}^X = \mathbb{E}(X_i X_j)$ and $c_{ij}^Y = \mathbb{E}(Y_i Y_j)$).

Our goal is the comparison of the expectations $\mathbb{E}[\varphi(X)]$ and $\mathbb{E}[\varphi(Y)]$, where $\varphi : \mathbb{R}^n \to \mathbb{R}$ is a measurable map of moderate growth, meaning that: $|\varphi(x)| \leqslant a \, e^{b|x|}$, where $|x|$ is the Euclidean norm of $x \in \mathbb{R}^n$. We will use a variational method for this. If:

$$Z_t = \sqrt{1 - t}\, X + \sqrt{t}\, Y, \qquad 0 \leqslant t \leqslant 1$$

and $h(t) = \mathbb{E}[\varphi(Z_t)]$, we study the sign of $h'(t)$ (note that $Z_0 = X$ and $Z_1 = Y$). The following notation is useful:

$$\begin{cases} M_{ij} = c_{ij}^Y - c_{ij}^X \\ N_{ij} = \mathbb{E}(Y_i - Y_j)^2 - \mathbb{E}(X_i - X_j)^2 \end{cases}$$

and:

$$\partial_{ij} = \frac{\partial^2}{\partial x_i \partial x_j}, \qquad \partial_t = \frac{\partial}{\partial t}.$$

Note that $N_{ii} = 0$ and that the $M_{ij}$, $N_{ij}$ are linked by the trivial identity:

$$(*) \qquad M_{ij} = -\frac{1}{2}N_{ij} + \frac{1}{2}(M_{ii} + M_{jj}).$$

## III.2  The Comparison Theorem. Applications

Independently of any hypothesis on the variation of $\varphi$ and the sign of the $N_{ij}$'s, the following lemma gives a nice expression for $h'(t)$ when $X$ and $Y$ are assumed independent, which can always be done, since this affects neither the hypotheses nor the conclusions of the theorems to follow.

**Lemma III.1**  *Let $X$ and $Y$ be two independent centered Gaussian vectors in $\mathbb{R}^n$ and $Z_t = \sqrt{1-t}\,X + \sqrt{t}\,Y$ ($0 \leqslant t \leqslant 1$). For every $\varphi \in C^2(\mathbb{R}^n, \mathbb{R})$ of moderate growth, set $h(t) = \mathbb{E}[\varphi(Z_t)]$. Then, for $0 < t < 1$:*

$$h'(t) = -\frac{1}{4}\mathbb{E}\left(\sum_{i \neq j} N_{ij}\partial_{ij}\varphi(Z_t)\right) + \frac{1}{2}\sum_i \mathbb{E}\left(M_{ii}\sum_j \partial_{ij}\varphi(Z_t)\right).$$

*Proof*  This expression could be obtained as an application of Itô's formula, but we prefer a direct proof. Let $f(t, x)$ be the density of $Z_t$, and $F(t, u)$ its characteristic function. The function $f$ satisfies the heat equation:

$$(1) \qquad \partial_t f = \frac{1}{2}\sum_{i,j} M_{ij}\partial_{ij}f, \qquad 0 < t < 1.$$

Indeed, the Fourier inversion formula gives:

$$f(t, x) = (2\pi)^{-n}\int_{\mathbb{R}^n} e^{i\langle x, u\rangle}F(t, u)\,\mathrm{d}u,$$

so that:

$$\partial_t f(t, x) = (2\pi)^{-n}\int_{\mathbb{R}^n} e^{i\langle x, u\rangle}\partial_t F(t, u)\,\mathrm{d}u.$$

Calculate $F(t, u)$:

$$F(t, u) = \exp\left(-\frac{1}{2}\sum_{i,j} u_i u_j c_{ij}^{Z_t}\right),$$

with ($X$ and $Y$ being independent):

$$c_{ij}^{Z_t} = \mathbb{E}\left[(\sqrt{1-t}\,X_i + \sqrt{t}\,Y_i)(\sqrt{1-t}\,X_j + \sqrt{t}\,Y_j)\right] = (1-t)\,c_{ij}^X + t\,c_{ij}^Y;$$

hence:

$$F(t, u) = \exp\left[-\frac{1}{2}\sum_{i,j} u_i u_j \left((1-t)\,c_{ij}^X + t\,c_{ij}^Y\right)\right],$$

and:

$$\partial_t F = -\frac{1}{2}\sum_{i,j} u_i u_j M_{ij} F.$$

We thus obtain the relation:

$$\partial_t f(t, x) = -\frac{1}{2}\sum_{i,j} M_{ij} (2\pi)^{-n} \int_{\mathbb{R}^n} e^{i\langle x, u\rangle} u_i u_j\, F(t, u)\, du.$$

However, by differentiating under the integral, we also have:

$$\partial_{ij} f(t, x) = -(2\pi)^{-n} \int_{\mathbb{R}^n} u_i u_j\, e^{i\langle x, u\rangle}\, F(t, u)\, du,$$

and the comparison of these two formulas leads to the relation (1) as announced.

It is now easy to obtain the expression for $h'(t)$.

In fact, $h(t) = \int_{\mathbb{R}^n} f(t, x)\varphi(x)\, dx$, hence, via (1):

$$h'(t) = \int_{\mathbb{R}^n} \partial_t f(t, x)\varphi(x)\, dx = \frac{1}{2}\int_{\mathbb{R}^n} \sum_{i,j} M_{ij}\partial_{ij} f(t, x)\varphi(x)\, dx$$

$$= \frac{1}{2}\sum_{i,j} M_{ij}\int_{\mathbb{R}^n} f(t, x)\,\partial_{ij}\varphi(x)\, dx = \frac{1}{2}\sum_{i,j} M_{ij}\mathbb{E}[\partial_{ij}\varphi(Z_t)],$$

after two integrations by parts (we note that $f(t, x)$ and $\partial_t f(t, x)$ are $O\left(e^{-\varepsilon|x|^2}\right)$, while $\varphi(x)$ is $O(e^{b|x|})$, which validates the preceding formal calculations). Finally, the formula $(*)$ used on $h'(t)$ above leads to the expression in the statement.  $\square$

Lemma III.1 will be used with functions $\varphi$ that are not always $\mathcal{C}^2$; then we must either consider the derivatives as distributions, or regularize and make them discrete. It is this second option that we choose, with the operators $\Delta_h$ and $\Delta_{hk}$ of first and second differences defined, for $x, h, k \in \mathbb{R}^n$, as:

$$\begin{cases} \Delta_h\varphi(x) = \varphi(x+h) - \varphi(x) \\ \Delta_{hk}\varphi(x) = [(\Delta_h \circ \Delta_k)\varphi](x) \\ \qquad\qquad = \varphi(x+h+k) - \varphi(x+h) - \varphi(x+k) + \varphi(x). \end{cases}$$

The following trivial lemma shows their link with derivatives:

**Lemma III.2**  *Let* $\varphi \in C^2(\mathbb{R}^n, \mathbb{R})$, *and* $a = (a_1, \ldots, a_n) \in \mathbb{R}^n$ *and* $b = (b_1, \ldots, b_n) \in \mathbb{R}^n$. *Then*:

$$\lim_{\substack{r \to 0 \\ r \in \mathbb{R}^*}} r^{-2} \Delta_{ra,rb}\varphi(x) = \sum_{i,j} a_i b_j \, \partial_{ij}\varphi(x).$$

Denote $e = e_1 + \cdots + e_n = (1, \ldots, 1)$, where $e_1, \ldots, e_n$ is the canonical basis of $\mathbb{R}^n$. Lemmas III.1 and III.2 lead to the fundamental result of this section:

**Theorem III.3** (The Comparison Theorem)  *Let X and Y be two centered Gaussian vectors in* $\mathbb{R}^n$; *set, for* $1 \leqslant i, j \leqslant n$:

$$N_{ij} = \mathbb{E}(Y_i - Y_j)^2 - \mathbb{E}(X_i - X_j)^2 \quad and \quad M_{ij} = \mathbb{E}(Y_iY_j) - \mathbb{E}(X_iX_j).$$

*Let* $\varphi \colon \mathbb{R}^n \to \mathbb{R}$ *be a measurable function of moderate growth. Suppose that additionally it satisfies the two following hypotheses:*

(a) *For* $i \neq j$:

$$N_{ij} \, \Delta_{re_i, re_j}\varphi(x) \leqslant 0$$

*for every* $x \in \mathbb{R}^n$ *and any* $r > 0$, *or for every* $x \in \mathbb{R}^n$ *and any* $r < 0$.

(b) *For* $1 \leqslant i \leqslant n$:

$$M_{ii} \, \Delta_{re_i, re}\varphi(x) \geqslant 0$$

*for every* $x \in \mathbb{R}^n$ *and any* $r > 0$, *or for every* $x \in \mathbb{R}^n$ *and any* $r < 0$.

*Then we have the inequality*:

$$\mathbb{E}[\varphi(X)] \leqslant \mathbb{E}[\varphi(Y)].$$

*Proof*  As previously mentioned, the statement of the theorem only involves the distributions of $X$ and $Y$, and not the variables themselves; we can thus assume them independent, and therefore use Lemma III.1.

First assume $\varphi$ $C^2$ smooth. By the hypotheses, $N_{ij}\Delta_{re_i, re_j}\varphi(x) \leqslant 0$ for $r > 0$ (and the same for $r < 0$); hence, for $r > 0$, $r^{-2}N_{ij}\Delta_{re_i, re_j}\varphi(x) \leqslant 0$, and, thanks to Lemma III.2, a passage to the limit provides the inequality:

$$N_{ij} \, \partial_{ij}\varphi(x) \leqslant 0 \quad \text{if } i \neq j \text{ and } x \in \mathbb{R}^n.$$

We similarly obtain:

$$M_{ii} \sum_j \partial_{ij}\varphi(x) \geqslant 0 \quad \text{for } 1 \leqslant i \leqslant n \ \text{ and } \ x \in \mathbb{R}^n.$$

Replacing $x$ by $Z_t = Z_t(\omega)$ and taking the expectation, we obtain:

$$\mathbb{E}\left(\sum_{i \neq j} N_{ij}\, \partial_{ij}\varphi(Z_t)\right) \leqslant 0 \quad \text{and} \quad \sum_i \mathbb{E}\left(M_{ii} \sum_j \partial_{ij}\varphi(Z_t)\right) \geqslant 0.$$

Then the expression of $h'$ in Lemma III.1 shows that $h'(t) \geqslant 0$ when $0 < t < 1$. In particular: $h(0) \leqslant h(1)$, i.e. $\mathbb{E}[\varphi(X)] \leqslant \mathbb{E}[\varphi(Y)]$.

In the general case, we regularize. Let $\rho \in C^{\infty}(\mathbb{R}^n, \mathbb{R}^+)$, with integral 1 and support in the unit ball, and let $\rho_\varepsilon(x) = \varepsilon^{-n}\rho(x/\varepsilon)$ and $\varphi_\varepsilon = \varphi * \rho_\varepsilon$. The operators $\Delta_{h,k}$ commute with convolution, so $\varphi_\varepsilon$, which is $C^2$, inherits the hypotheses (a) and (b) on $\varphi$; thus $\mathbb{E}[\varphi_\varepsilon(X)] \leqslant \mathbb{E}[\varphi_\varepsilon(Y)]$. Let us prove:

$$(1) \qquad\qquad \mathbb{E}[\varphi_\varepsilon(X)] \underset{\varepsilon \to 0^+}{\longrightarrow} \mathbb{E}[\varphi(X)].$$

For this, first observe that $\mathbb{E}[\varphi(X - \varepsilon y)] \underset{\varepsilon \to 0^+}{\longrightarrow} \mathbb{E}[\varphi(X)]$ for every $y \in \mathbb{R}^n$; indeed, with $\gamma$ the density of $X$, we have:

$$\mathbb{E}[\varphi(X - \varepsilon y)] = \int_{\mathbb{R}^n} \varphi(x - \varepsilon y)\, \gamma(x)\, dx = \int_{\mathbb{R}^n} \varphi(x)\, \gamma(x + \varepsilon y)\, dx$$

$$\underset{\varepsilon \to 0^+}{\longrightarrow} \int_{\mathbb{R}^n} \varphi(x)\, \gamma(x)\, dx = \mathbb{E}[\varphi(X)]$$

by the dominated convergence theorem (recall that $\varphi$ is of moderate growth and that $\gamma(x) = O(e^{-a|x|^2})$). Next, by Lebesgue's theorem again:

$$\mathbb{E}[\varphi_\varepsilon(X)] = \int_{\mathbb{R}^n} \mathbb{E}[\varphi(X - \varepsilon y)]\, \rho(y)\, dy \underset{\varepsilon \to 0^+}{\longrightarrow} \mathbb{E}[\varphi(X)].$$

This proves (1). An analogous result holds for $Y$. A passage to the limit in $\mathbb{E}[\varphi_\varepsilon(X)] \leqslant \mathbb{E}[\varphi_\varepsilon(Y)]$ then provides the desired result, $\mathbb{E}[\varphi(X)] \leqslant \mathbb{E}[\varphi(Y)]$. $\qquad\square$

Here are two applications of the comparison theorem:

**Theorem III.4** (The Slepian–Sudakov Lemma) *Let* $X = (X_1, \ldots, X_n)$ *and* $Y = (Y_1, \ldots, Y_n)$ *be centered Gaussian vectors such that, for any* $i, j = 1, \ldots, n$:

$$\|X_i - X_j\|_2 \leqslant \|Y_i - Y_j\|_2.$$

*Then:*

$$\mathbb{E}\left(\sup_i X_i\right) \leqslant \mathbb{E}(\sup_i Y_i).$$

*Proof*  Let $\varphi(x) = \sup(x_1, \ldots, x_n)$; then $\varphi$ is of moderate growth and, for $r \in \mathbb{R}$, satisfies the following two relations:

$$(1) \qquad\qquad \varphi(x + re) = \varphi(x) + r$$

and

$$(2) \quad \varphi(x + re_i + re_j) = \max\left[\varphi(x + re_i), \varphi(x + re_j)\right], \text{ for } i \neq j, \, r > 0.$$

In fact (1) is evident; let us prove (2). Clearly:

$$\varphi(x + re_i) = \max\left(\max_{k \neq i} x_k, x_i + r\right)$$

and

$$\varphi(x + re_j) = \max\left(\max_{k \neq j} x_k, x_j + r\right);$$

then:

$$\max\left[\varphi(x + re_i), \varphi(x + re_j)\right] = \max\left(\max_{k \neq i,j} x_k, x_i + r, x_j + r\right)$$
$$= \varphi(x + re_i + re_j).$$

The first relation implies:

$$(3) \qquad\qquad \Delta_{re_i, re} \varphi(x) = 0 \quad \text{for any } x;$$

and the second:

$$(4) \qquad\qquad \Delta_{re_i, re_j} \varphi(x) \leqslant 0 \quad \text{if } i \neq j, \, x \in \mathbb{R}^n, \, r > 0;$$

indeed, if we assume, for example, $\varphi(x + re_i) \geqslant \varphi(x + re_j)$, (2) gives:

$$\Delta_{re_i, re_j} \varphi(x) = \varphi(x + re_i + re_j) - \varphi(x + re_i) - \varphi(x + re_j) + \varphi(x)$$
$$= \varphi(x + re_i) - \varphi(x + re_i) - \varphi(x + re_j) + \varphi(x)$$
$$= \varphi(x) - \varphi(x + re_j) \leqslant 0,$$

since $r > 0$ and $\varphi$ is separately non-decreasing. Relations (3) and (4) show that the hypotheses of the comparison theorem are satisfied if $N_{ij} \geqslant 0$ for any $i, j$, whatever the signs of the $M_{ii}$'s may be; the Slepian–Sudakov lemma ensues.  □

**Remark**  If $I$ is an infinite set of indices, and $(X_i)_{i \in I}$ a process indexed by $I$, define:

$$\mathbb{E}\left(\sup_{i \in I} X_i\right) = \sup_{J \subseteq I, \, J \text{ finite}} \mathbb{E}\left(\sup_{i \in J} X_i\right).$$

With this definition, the Slepian–Sudakov lemma is again applicable. In particular, if $T$ is a separable topological space (for example, a compact metric space), and if the processes $(X_t)_{t \in T}$ and $(Y_t)_{t \in T}$ are continuous and satisfy the conditions of theorem, the conclusion again holds.

**Theorem III.5** (Slepian's Lemma)  *Let $X = (X_1, \ldots, X_n)$ and $Y = (Y_1, \ldots, Y_n)$ be centered Gaussian vectors satisfying:*

$$\|X_i - X_j\|_2 \leqslant \|Y_i - Y_j\|_2 \quad \text{for } 1 \leqslant i, j \leqslant n.$$

*Assume additionally that:*

$$\|X_i\|_2 = \|Y_i\|_2 \quad \text{for } 1 \leqslant i \leqslant n.$$

*Then:*

$$\mathbb{P}\left( \sup_i X_i > t \right) \leqslant \mathbb{P}\left( \sup_i Y_i > t \right) \quad \text{for any } t \in \mathbb{R}.$$

*Proof*  Let $\varphi$ be as in Theorem III.4, and $f = \mathbb{1}_{]t,+\infty[}$. Then $\psi = f \circ \varphi$ still satisfies (2) and hence (4), since $f$ is increasing. However, there is no reason for it to satisfy (1) or (3); this is where the additional condition $\|X_i\|_2 = \|Y_i\|_2$ (meaning $M_{ii} = 0$) intervenes. The hypotheses (a) and (b) of the comparison theorem remain valid, and we obtain $\mathbb{E}[\psi(X)] \leqslant \mathbb{E}[\psi(Y)]$: the desired conclusion is reached; it is stronger than the conclusion of the Slepian–Sudakov lemma, since not only the expectations, but also the tail functions of $\sup_i X_i$ and $\sup_i Y_i$, are compared. □

Note that the same method, applied to

$$\varphi(x_1, \ldots, x_n) = \sup(x_1 + a_1, \ldots, x_n + a_n),$$

$f = \mathbb{1}_{]t,+\infty[}$ and $\psi = f \circ \varphi$ would lead to the stronger conclusion:

$$\mathbb{P}\left( \sup_i (X_i + a_i) > t \right) \leqslant \mathbb{P}\left( \sup_i (Y_i + a_i) > t \right) \quad \text{for any } t \in \mathbb{R}.$$

An important application of the Slepian–Sudakov lemma is the following:

**Theorem III.6** (Sudakov Minoration)  *Let $X = (X_1, \ldots, X_n)$ be a centered Gaussian vector, and $a = \inf_{i \neq j} \|X_i - X_j\|_2 > 0$. Then, we have the lower bound:*

$$\mathbb{E}(\sup_i X_i) \geqslant c \, a \, \sqrt{\log n},$$

*where $c > 0$ is a numerical constant.*

*Proof*  We construct a process $(Y_1, \ldots, Y_n) = Y$ which is a minorant of $X$ in the terms of the Slepian–Sudakov lemma, and for which $\mathbb{E}(\sup Y_i)$ is easy to estimate. For this, take $Y_i = (a/\sqrt{2})\, \gamma_i$, where $\gamma_1, \ldots, \gamma_n$ are independent variables with distribution $\mathcal{N}(0, 1)$; we have $\|Y_i - Y_j\|_2 = a \leqslant \|X_i - X_j\|_2$; therefore:

$$\mathbb{E}(\sup X_i) \geqslant \mathbb{E}(\sup Y_i) = (a/\sqrt{2})\mathbb{E}[\sup (\gamma_1, \ldots, \gamma_n)] \geqslant c\,a\sqrt{\log n},$$

by Proposition II.2.                                                                                    □

## III.3  The Slepian–Gordon Theorem

The Slepian–Sudakov lemma and Slepian's lemma enable us to compare stochastically the max of Gaussian variables, but this is insufficient for Dvoretzky's theorem: a comparison of their minimax is necessary, treated in the following theorem:

**Theorem III.7** (Gordon's Theorem)

1) *Let $X = (X_{ij})$ and $Y = (Y_{ij})$, $1 \leqslant i \leqslant m$, $1 \leqslant j \leqslant n$, be two centered Gaussian vectors such that*:

(1.$a$)     $\mathbb{E}(X_{ij} - X_{ik})^2 \leqslant \mathbb{E}(Y_{ij} - Y_{ik})^2$     *for any $i, j, k$*;

(1.$b$)     $\mathbb{E}(X_{ij} - X_{i'j'})^2 \geqslant E(Y_{ij} - Y_{i'j'})^2$     *for $i \neq i'$, $1 \leqslant j, j' \leqslant n$.*

*Then we have the inequality*:

(1.$c$)     $$\mathbb{E}\left(\inf_i \sup_j X_{ij}\right) \leqslant \mathbb{E}\left(\inf_i \sup_j Y_{ij}\right).$$

2) *Let $(X_{st})$, $(Y_{st})$ be two centered Gaussian processes with continuous trajectories, indexed by a product $S \times T$ of compact metric spaces, dominated (i.e. $|X_{st}| \leqslant M$, $|Y_{st}| \leqslant M$, with $M$ an integrable random variable), and such that:*

(2.$a$)     $\|X_{st} - X_{st'}\|_2 \leqslant \|Y_{st} - Y_{st'}\|_2$     *for any $s, t, t'$*;

(2.$b$)     $\|X_{st} - X_{s't'}\|_2 \geqslant \|Y_{st} - Y_{s't'}\|_2$     *for $s \neq s'$ and $t, t' \in T$.*

*Then we have the inequality:*

(2.$c$)     $$\mathbb{E}\left(\inf_s \sup_t X_{st}\right) \leqslant \mathbb{E}\left(\inf_s \sup_t Y_{st}\right).$$

*Proof*  2) follows from 1) by a passage to the limit, since $S$ and $T$ are separable. 1) is quite delicate and the hypotheses call for several comments.
  Let $A_i = \{(i, j)\,;\, 1 \leqslant j \leqslant n\}$, for $1 \leqslant i \leqslant m$.
  If $u = (i, j)$ and $v = (i', j')$, $u$ and $v$ are said to be *relatives* if they belong to the same "block" $A_k$, i.e. if $i = i'$; and $u$ and $v$ are said to be *strangers* if they belong to two distinct blocks, i.e. if $i \neq i'$.
  With the general notation of this Section III, the hypotheses can be reformulated as follows:

(a) *if u and v are relatives, then $N_{uv} \geqslant 0$;*
(b) *if u and v are strangers, then $N_{uv} \leqslant 0$.*

Let $N = mn$, $\mathbb{R}^N = \{(x_{ij})\,;\ 1 \leqslant i \leqslant m,\ 1 \leqslant j \leqslant n\}$ and $\varphi \colon \mathbb{R}^N \to \mathbb{R}$ defined by $\varphi(x) = \inf_i \sup_j x_{ij}$, we have to show that $\varphi$ satisfies the hypotheses of the comparison theorem, which requires some preliminary results.

Let $\psi \colon \mathbb{R}^N \to \mathbb{R}$ be a separately non-decreasing function, $u, v \in [1, N]$, $u \neq v$, and let $e_u$, $1 \leqslant u \leqslant N$, be the canonical basis of $\mathbb{R}^N$.

The couple $(u, v)$ is said to be *of type max for $\psi$* if:

$$\psi(x + re_u + re_v) = \max[\psi(x + re_u), \psi(x + re_v)] \quad \forall x \in \mathbb{R}^N, \ \forall r > 0,$$

and it is said to be *of type min for $\psi$* if:

$$\psi(x + re_u + re_v) = \min[\psi(x + re_u), \psi(x + re_v)] \quad \forall x \in \mathbb{R}^N, \ \forall r < 0.$$

Then the following simple lemma holds:

**Lemma III.8**   *Let $\psi \colon \mathbb{R}^N \to \mathbb{R}$ be a separately non-decreasing function. Then:*

1) *If $(u, v)$ is of type max for $\psi$, then $\Delta_{re_u, re_v} \psi(x) \leqslant 0$ for any $x \in \mathbb{R}^N$ and any $r > 0$.*
2) *If $(u, v)$ is of type min for $\psi$, then $\Delta_{re_u, re_v} \psi(x) \geqslant 0$ for any $x \in \mathbb{R}^N$ and any $r < 0$.*
3) *For $\psi(x) = \max(x_1, \ldots, x_N)$, all the couples are of type max.*
4) *For $\psi(x) = \min(x_1, \ldots, x_N)$, all the couples are of type min.*

*Proof*

1) Let $x \in \mathbb{R}^N$, and $r > 0$. If, for example, $\psi(x + re_u) \geqslant \psi(x + re_v)$, we have:

$$\begin{aligned} \Delta_{re_u, re_v} \psi(x) &= \psi(x + re_u + re_v) - \psi(x + re_u) - \psi(x + re_v) + \psi(x) \\ &= \psi(x) - \psi(x + re_v) \leqslant 0. \end{aligned}$$

2) Let $x \in \mathbb{R}^N$ and $r < 0$. If, for example, $\psi(x + re_u) \geqslant \psi(x + re_v)$, this time we have:

$$\Delta_{re_u, re_v} \psi(x) = \psi(x) - \psi(x + re_u) \geqslant 0 \quad \text{as } r < 0.$$

3) was established within the proof of the Slepian–Sudakov lemma and
4) is shown similarly.                                                      $\square$

The function $\varphi$ used to prove the theorem is defined as follows: $[1, N]$ is divided into disjoint blocks $A_1, \ldots, A_m$. Then $\varphi_i(x) = \max_{u \in A_i} x_u$ depends only on the coordinates of $x$ in $A_i$, and $\varphi(x) = \min_i \varphi_i(x)$. The following lemma describes the nature of the couples $(u, v)$ for $\varphi$:

**Lemma III.9** *The function $\varphi$ has the following properties*:

1) *If $u$ and $v$ are relatives, $(u, v)$ is of type max for $\varphi$.*
2) *If $u$ and $v$ are strangers, $(u, v)$ is of type min for $\varphi$.*

*Proof*

1) Assume for example $u, v \in A_1$ and $\varphi_1(x + re_u) \geqslant \varphi_1(x + re_v)$ with $x \in \mathbb{R}^N$ and $r > 0$. Then, by 3) of Lemma III.8:

$$\varphi(x + re_u + re_v) = \min\big(\varphi_1(x + re_u + re_v), \varphi_2(x), \ldots, \varphi_m(x)\big)$$
$$= \min\big(\varphi_1(x + re_u), \varphi_2(x), \ldots, \varphi_m(x)\big).$$

However:

$$\varphi(x + re_u) = \min\big(\varphi_1(x + re_u), \varphi_2(x), \ldots, \varphi_m(x)\big)$$

and

$$\varphi(x + re_v) = \min\big(\varphi_1(x + re_v), \varphi_2(x), \ldots, \varphi_m(x)\big) \leqslant \varphi(x + re_u);$$

hence the result.

2) Assume for example $u \in A_1$, $v \in A_2$, and this time take $r < 0$. Then, for $x \in \mathbb{R}^N$:

$$\varphi(x + re_u + re_v) = \min\big(\varphi_1(x + re_u), \varphi_2(x + re_v), \varphi_3(x), \ldots, \varphi_m(x)\big),$$

whereas:

$$\min\big(\varphi(x + re_u), \varphi(x + re_v)\big)$$
$$= \min\big(\varphi_1(x), \ldots, \varphi_m(x), \varphi_1(x + re_u), \varphi_2(x + re_v)\big)$$
$$= \min\big(\varphi_3(x), \ldots, \varphi_m(x), \varphi_1(x + re_u), \varphi_2(x + re_v)\big),$$

since $r < 0$ and $\varphi_1, \varphi_2$ are separately non-decreasing. □

We can now easily complete the proof of Gordon's theorem.

Let $1 \leqslant u, v \leqslant N$, $u \neq v$; if $u$ and $v$ are relatives, $N_{uv} \Delta_{re_u, re_v} \varphi(x) \leqslant 0$ for $x \in \mathbb{R}^N$ and $r > 0$ (by (a) and Lemmas III.8 and III.9); if $u$ and $v$ are strangers, $N_{uv} \Delta_{re_u, re_v} \varphi(x) \leqslant 0$ for $x \in \mathbb{R}^N$ and $r < 0$ (by (b) and Lemmas III.8 and III.9).

Moreover, $\varphi(x + re) = \varphi(x) + r$, hence $\Delta_{re_u, re} \varphi(x) = 0$. The hypotheses of the comparison theorem are indeed satisfied, and the conclusion (1.c) ensues. □

**Remark** This proof of Gordon's theorem, notably simpler than that of Gordon [1985], is due to B. Maurey (personal communication; see Maurey [1990, unpublished seminar]); a short proof of this theorem can be found in Kahane [1986], but the "combinatorics" of the minimax function $\varphi$ is not developed there. The hypotheses (1.*a*) and (1.*b*) are obviously *ad hoc* for the application of the comparison theorem; however they turn out to be (miraculously?) satisfied for the application to Dvoretzky's theorem. See also Latała [1997].

# IV Dvoretzky's Theorem

## IV.1 Preliminary Remarks

A Banach space $E$ can be non-Hilbertian and contain subspaces $F$ on which the induced norm becomes (almost) Hilbertian. Here are a few examples:

1) If $E = C_b(\mathbb{R})$ and $F$ is the space generated by the functions cos and sin, $F$ is isometric to $\ell_2^2$; in fact, $\|a\cos + b\sin\|_\infty = \sqrt{a^2 + b^2}$ for any $a, b \in \mathbb{R}$.

2) If $E = M_n(\mathbb{R})$ with the usual operator norm, and if $F$ is the space generated by $E_{11}, \ldots, E_{n1}$ (where all the coefficients of the matrix $E_{ij}$ are null, except the $(i,j)$-th which is 1), this "first column" subspace is isometric to $\ell_2^n$; indeed, if $A = \sum_{i=1}^n a_i E_{i1}$, then $\|A\| = \left(\sum_{i=1}^n a_i^2\right)^{1/2}$ for any $a_1, \ldots, a_n \in \mathbb{R}$.

3) The space $E = c_0$ does not contain any subspace of dimension 2 isometric to $\ell_2^2$: indeed, otherwise there would exist $u, v \in c_0$ such that $\|u + tv\|_\infty = \sqrt{1 + t^2}$ for any $t \in \mathbb{R}$. But then, with:

$$N_i = \left\{ t \in \mathbb{R} \,;\, |u_i + tv_i| = \sqrt{1 + t^2} \right\},$$

$N_i$ has at most two elements, and $N = \bigcup_{i=1}^\infty N_i$ is countable; since $\|u + tv\|_\infty \neq \sqrt{1 + t^2}$ for $t \notin N$, this contradicts the hypothesis.

However, for $\varepsilon > 0$ and $q \in \mathbb{N}^*$, $E$ contains a subspace $F$ of dimension $q$, for which $d_F = d(F, \ell_2^q) \leqslant 1 + \varepsilon$. Indeed, let $\delta > 0$ to be adjusted, $e_1, \ldots, e_q$ the canonical basis of $\ell_2^q$, $y_1, \ldots, y_p$ a $\delta$-net of the unit sphere $S_2^q$ of $\ell_2^q$, and $f_1, \ldots, f_q \in c_0$ defined by

$$f_j(n) = \begin{cases} \langle e_j, y_n \rangle & \text{if} \quad n \leqslant p \\ 0 & \text{if} \quad n > p. \end{cases}$$

Let $F$ be the space generated by $f_1, \ldots, f_q$. If $\lambda = (\lambda_1, \ldots, \lambda_q) \in \mathbb{R}^q$, then:

$$\left| \sum_{j=1}^q \lambda_j f_j(n) \right| = \left| \sum_{j=1}^q \lambda_j \langle y_n, e_j \rangle \right| \leqslant \left( \sum_{j=1}^q \lambda_j^2 \right)^{1/2} \left( \sum_{j=1}^q |\langle y_n, e_j \rangle|^2 \right)^{1/2}$$

$$\leqslant \|\lambda\|_2 \|y_n\|_2 = \|\lambda\|_2, \quad \text{if } n \leqslant p,$$

and $\sum_{j=1}^{q} \lambda_j f_j(n) = 0$ if $n > p$; hence:

(1)
$$\left\| \sum_{j=1}^{q} \lambda_j f_j \right\|_{\infty} \leqslant \|\lambda\|_2.$$

On the other hand, for $\lambda \in \mathbb{R}^q$, select $y \in S_2^q$ such that $\|\lambda\|_2 = \langle \lambda, y \rangle$ and let $n \leqslant p$ be such that $\|y - y_n\|_2 \leqslant \delta$. We have:

$$\|\lambda\|_2 = \langle \lambda, y_n \rangle + \langle \lambda, y - y_n \rangle$$

$$= \sum_{j=1}^{q} \lambda_j f_j(n) + \langle \lambda, y - y_n \rangle \leqslant \left\| \sum_{j=1}^{q} \lambda_j f_j \right\|_{\infty} + \delta \|\lambda\|_2;$$

therefore:

$$\left\| \sum_{j=1}^{q} \lambda_j f_j \right\|_{\infty} \geqslant (1 - \delta) \|\lambda\|_2.$$

With $\delta = \varepsilon/(1 + \varepsilon)$, we get:

(2)
$$\left\| \sum_{j=1}^{q} \lambda_j f_j \right\|_{\infty} \geqslant \frac{1}{1 + \varepsilon} \|\lambda\|_2;$$

then, by (1) and (2), we obtain $d_F \leqslant 1 + \varepsilon$.

4) Example 3) is typical of the general case: as soon as $\dim F > 1$, it is well known that $F$ cannot be expected to be isometric to a Hilbert space; but if we allow a "relaxation" of the isometry by $\varepsilon > 0$, we can find such an $F$, of finite but large dimension. Of course we cannot continue up until $\dim F = \infty$: the spaces $\ell_p$ ($p \neq 2$) or $c_0$ do not contain any subspace isomorphic to $\ell_2$.

In the general case, the construction of $F$ is much more difficult than in Example 3) above: if $n = \dim E$ and if $k \in [1, n]$, $F$ is selected "at random" in the Grassmannian of all $k$-dimensional subspaces of $E$; this was the initial method of Dvoretzky [1961]. A more recent point of view chooses $F$ as a "Gaussian" subspace (PISIER 2); this proof is based on the Maurey–Pisier deviation inequality proved in Subsection II.2; its great advantage is that it works equally well for both cases, complex and real. We will also present another recent point of view (Milman and Schechtman [1995]; see also Guédon [1997] and [1998]), which uses *random Gaussian operators* $G: \ell_2^k \to E$. For a good choice of $G$ (and for $k$ not too large), the image $F = G(\ell_2^k)$ is suitable. However, this method is valid only for the real case, but it has the advantage of being adaptable to the isomorphic version of Dvoretzky's theorem, essentially developed in Subsection IV.5.

As always in Analysis, any probabilistic method must be paired with a deterministic method: this is the goal of the following subsection.

## IV.2 The Dvoretzky–Rogers and Lewis Lemmas

Dvoretzky and Rogers [1950] brought to light a totally unexpected phenomenon: every Banach space $E$ contains large subspaces $F$ that "resemble" Hilbert spaces; precision on this resemblance was provided ten years later by Dvoretzky's theorem [1961]. Initially, however, a more or less trivial corollary of this resemblance was sufficient to give a positive answer to the then-open question: "Does every infinite-dimensional Banach space contain summable sequences that are not absolutely summable?" We present here a precise form of the discovery of Dvoretzky and Rogers:

**Proposition IV.1** (The Dvoretzky–Rogers Lemma)   *Let $E$ be a Banach space of dimension $n$, and let $m = \left[\frac{n}{2}\right] + 1$. Then there exist vectors $v_1, \ldots, v_m \in E$ such that $\|v_j\| = 1$, $1 \leqslant j \leqslant m$, and such that ($\mathbb{K} = \mathbb{R}$ or $\mathbb{C}$):*

$$\left\| \sum_{j=1}^{m} \alpha_j v_j \right\| \leqslant 2 \left( \sum_{j=1}^{m} |\alpha_j|^2 \right)^{1/2}, \quad \forall \alpha_1, \ldots, \alpha_m \in \mathbb{K}.$$

This local statement leads to the following almost immediate corollary:

**Theorem IV.2** (The Dvoretzky–Rogers Theorem)   *Let $E$ be an infinite-dimensional Banach space, and $(\lambda_n)_{n \geqslant 1}$ a sequence of square summable positive real numbers: $\sum_{n=1}^{+\infty} \lambda_n^2 < +\infty$. Then, there exists a summable sequence $(x_n)_{n \geqslant 1}$ of $E$ such that $\|x_n\| = \lambda_n$ for any $n = 1, 2, \ldots$*

*In particular, $E$ contains summable sequences that are not absolutely summable.*

*Proof of the theorem*   Set $C = 2 \left( \sum_{n=1}^{+\infty} \lambda_n^2 \right)^{1/2}$, and let $n_1 < n_2 < \ldots < n_k < \ldots$ be an increasing sequence of integers such that $n_1 = 1$ and $\sum_{n \geqslant n_k} \lambda_n^2 \leqslant C^2 4^{-k}$. The intervals $I_k = [n_k, n_{k+1}[$ form a partition of $\mathbb{N}^*$, and the Dvoretzky–Rogers lemma enables us to find a sequence $(v_n)_{n \geqslant 1}$ of unitary vectors such that:

$$\left\| \sum_{n \in I_k} \alpha_n v_n \right\| \leqslant 2 \left( \sum_{n \in I_k} |\alpha_n|^2 \right)^{1/2}, \quad \text{for } k = 1, 2, \ldots$$

Let us show that $x_n = \lambda_n v_n$ answers the question: the inequality above leads to:

$$I \subseteq I_k \implies \left\| \sum_{n \in I} x_n \right\| \leqslant C 2^{-k+1} ;$$

let $J$ be a finite subset of $\mathbb{N}^*$ such that $\min J \in I_k$ and $\max J \in I_l$, $l \geqslant k$. Then:

$$\left\| \sum_{n \in J} x_n \right\| \leqslant \left\| \sum_{J \cap I_k} x_n \right\| + \sum_{p=k+1}^{l-1} \left\| \sum_{I_p} x_n \right\| + \left\| \sum_{J \cap I_l} x_n \right\|$$

$$\leqslant C 2^{-k+1} + \sum_{p=k+1}^{l-1} C 2^{-p+1} + C 2^{-l+1} \leqslant C 2^{-k+2},$$

and hence $\left\| \sum_J x_n \right\| \to 0$ when $\min J \to +\infty$. The Cauchy criterion shows that $(x_n)_n$ is summable, and thus completes the proof. $\qquad \square$

The proof of the Dvoretzky–Rogers lemma appears as a consequence of the following compactness result:

**Lemma IV.3** (Lewis' Lemma)  *Let $E$ be a Banach space of dimension $n$. Then there exists an invertible operator $u_0 \in \mathcal{L}(\ell_2^n, E)$ such that $\|u_0\| = 1$ and such that:*

$$|\operatorname{tr}(u_0^{-1} v)| \leqslant n \|v\|, \qquad \forall v \in \mathcal{L}(\ell_2^n, E).$$

*Proof of Proposition IV.1*    To deduce Proposition IV.1 from Lemma IV.3, we use an inductive method (see PISIER 2), by showing the existence of an orthonormal basis $\varepsilon_1, \ldots, \varepsilon_n$ of $\ell_2^n$ such that:

$$(1) \qquad \|u_0(\varepsilon_j)\| \geqslant \frac{n - j + 1}{n}, \qquad 1 \leqslant j \leqslant n$$

(this is an analytic way to express that the ellipsoid $u_0(B_2^n)$ has a large number of contact points with $B_E$).

First select $\varepsilon_1$ such that $\|\varepsilon_1\| = 1$ and $\|u_0(\varepsilon_1)\| = \|u_0\| = 1 = \frac{n-1+1}{n}$.

Having constructed $\varepsilon_1, \ldots, \varepsilon_j$, we apply Lewis' lemma with $v = u_0 P$, where $P: \ell_2^n \to \ell_2^n$ is the orthogonal projection on $(\varepsilon_1, \ldots, \varepsilon_j)^\perp$; this gives:

$$n - j = \operatorname{tr} P \leqslant n \|u_0 P\|,$$

and we can find $x \in \ell_2^n$, with $\|x\| = 1$ and $\|u_0 P x\| \geqslant \frac{n-j}{n}$. The vector $\varepsilon_{j+1} = \frac{Px}{\|Px\|}$ is orthogonal to $\varepsilon_1, \ldots, \varepsilon_j$, and additionally:

$$\|u_0(\varepsilon_{j+1})\| \geqslant \|u_0 P x\| \geqslant \frac{n-j}{n},$$

which proves (1) by induction.

To conclude, note that $\frac{n-j+1}{n} \geq \frac{1}{2}$ for $j \leq m = \left[\frac{n}{2}\right] + 1$, and take $v_j = \frac{u_0 \varepsilon_j}{\|u_0 \varepsilon_j\|}$, $1 \leq j \leq m$. Then:

$$\left\| \sum_{j=1}^{m} \alpha_j v_j \right\| = \left\| u_0 \left( \sum_{j=1}^{m} \frac{\alpha_j \varepsilon_j}{\|u_0(\varepsilon_j)\|} \right) \right\| \leq \left\| \sum_{j=1}^{m} \frac{\alpha_j \varepsilon_j}{\|u_0(\varepsilon_j)\|} \right\|$$

$$= \left( \sum_{j=1}^{m} \frac{|\alpha_j|^2}{\|u_0(\varepsilon_j)\|^2} \right)^{1/2} \leq 2 \left( \sum_{j=1}^{m} |\alpha_j|^2 \right)^{1/2} ,$$

for any scalars $\alpha_1, \ldots, \alpha_m$.      $\square$

Note that, since $u_0$ is invertible, the vectors $v_1, \ldots, v_m$ are furthermore linearly independent.

*Proof of Lemma IV.3*   Let $K = \{u \in \mathcal{L}(\ell_2^n, E); \|u\| \leq 1\}$ be the unit ball of $\mathcal{L}(\ell_2^n, E)$. On $K$, consider the functional $\phi$ defined by $\phi(u) = |\det u|$, and select an element $u_0 \in K$ that maximizes this functional.

Here is the geometric interpretation of this: $u(B_2^n)$ is an ellipsoid inscribed in the unit ball $B_E$ of $E$, and, up to a constant, $\phi(u)$ is its volume; we thus seek to inscribe an ellipsoid of maximum volume within $B_E$. This method was used for the first time by F. John [1948].

Clearly, $|\det u_0| > 0$ and $u_0^{-1}$ exists. Now let $v \in \mathcal{L}(\ell_2^n, E)$ and $\varepsilon > 0$; the Kuhn–Tucker method of numerical analysis is used here, illustrated by Figure 1.1 (where $\Delta u_0$ illustrates an admissible increase). We have $|\det(u_0 + \Delta u_0)| \leq |\det u_0|$, meaning:

$$|\det(u_0 + \varepsilon v)| \leq \|u_0 + \varepsilon v\|^n |\det u_0| \leq (1 + \varepsilon\|v\|)^n |\det u_0|,$$

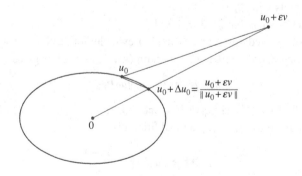

Figure 1.1

and, by dividing by $|\det u_0|$:

$$|\det(I + \varepsilon u_0^{-1}v)| \leqslant (1 + \varepsilon \|v\|)^n.$$

With a Taylor expansion, as $\varepsilon \overset{>}{\to} 0$, we obtain:

$$|1 + \varepsilon \, \text{tr}(u_0^{-1}v) + O(\varepsilon^2)| \leqslant 1 + n\varepsilon\|v\| + O(\varepsilon^2);$$

therefore:

$$\text{Re}\left[\, \text{tr}(u_0^{-1}v) \right] \leqslant n\|v\| + O(\varepsilon),$$

and hence $\text{Re}\left[\, \text{tr}(u_0^{-1}v) \right] \leqslant n\|v\|$. By then replacing $v$ by $e^{-i\theta}v$, where $\theta$ is the argument of $\text{tr}(u_0^{-1}v)$ (when the latter is not null), we obtain $|\, \text{tr}(u_0^{-1}v)| \leqslant n\|v\|$, which completes the proof. $\qquad\qquad\qquad\qquad\qquad\qquad\qquad\qquad\qquad\qquad\quad\square$

**Remark** We have maximized the functional $\phi(u) = |\det u|$ on the unit ball, for the operator norm of $\mathcal{L}(\ell_2^n, E)$; with the same proof, we can replace this operator norm by another; we thus obtain:

**Lemma IV.4** (Generalized Lewis' Lemma) *Let $E$ be a Banach space of dimension $n$, and let $\alpha$ be a norm on $\mathcal{L}(\mathbb{K}^n, E)$. Then, there exists an isomorphism $u_0 \colon \mathbb{K}^n \to E$ such that:*

$$\alpha(u_0) = 1 \quad \text{and} \quad \alpha^*(u_0^{-1}) = \sup_{\alpha(v) \leqslant 1} |\, \text{tr}(u_0^{-1}v)| = n.$$

Since the correspondence $u \mapsto \big(u(e_1), \ldots, u(e_n)\big)$, where $(e_1, \ldots, e_n)$ is the canonical basis of $\mathbb{K}^n$, identifies $\mathcal{L}(\mathbb{K}^n, E)$ and $E^n$, this lemma can be reformulated as follows:

**Theorem IV.5** (Lewis' Theorem) *Let $E$ be a Banach space of dimension $n$, $\alpha$ a norm on $E^n$ and $\alpha^*$ its dual norm on $(E^*)^n$. Then there exists a basis $(x_1, \ldots, x_n)$ of $E$, with dual basis $(x_1^*, \ldots, x_n^*)$ in $E^*$, such that:*

$$\alpha(x_1, \ldots, x_n) = 1 \quad \text{and} \quad \alpha^*(x_1^*, \ldots, x_n^*) = n.$$

From this, we deduce an important corollary (see Exercise VII.7 for another proof, and some applications), to be used later.

**Proposition IV.6** (Auerbach's Lemma) *Let $E$ be a Banach space of dimension $n$. Then there exists a normalized basis $(x_1, \ldots, x_n)$ of $E$ whose dual basis $(x_1^*, \ldots, x_n^*)$ is also normalized: $\|x_k\| = \|x_k^*\| = 1$.*

*Proof* In Theorem IV.5, take $\alpha(y_1, \ldots, y_n) = \sup_{1 \leqslant k \leqslant n} \|y_k\|$. We thus have $\alpha^*(y_1^*, \ldots, y_n^*) = \sum_{k=1}^{n} \|y_k^*\|$, and we can find a basis $(x_1, \ldots, x_n)$ of $E$ such that $\sup_{1 \leqslant k \leqslant n} \|x_k\| = 1$ and $\sum_{k=1}^n \|x_k^*\| = n$. Moreover, as $1 = x_k^*(x_k) \leqslant \|x_k^*\| \|x_k\| \leqslant \|x_k^*\|$, it must be that $\|x_k^*\| = 1$ for any $k$. The relation $1 \leqslant \|x_k^*\| \|x_k\|$ can then be read $1 \leqslant \|x_k\|$, thus $\|x_k\| = 1$ for any $k$. $\qquad\square$

We can easily verify that, when $E$ is a Hilbert space, the *Auerbach bases* are exactly the orthonormal bases.

**Corollary IV.7** *Let $X$ be a Banach space, and $E$ a subspace of $X$ of finite dimension $n$. Then there exists a projection from $X$ onto $E$ of norm $\leqslant n$.*

*Proof* Let $(x_1, \ldots, x_n)$ be an Auerbach basis of $E$; we extend the associated coordinate forms $x_1^*, \ldots, x_n^*$ to continuous linear functionals on $X$ $\widetilde{x}_1^*, \ldots, \widetilde{x}_n^*$ with norm 1; then the projection defined by $Px = \sum_{k=1}^n \widetilde{x}_k^*(x) x_k$ works. $\qquad\square$

**Remark** In fact, from Lewis' lemma, we can show that $\pi_2(u_0^{-1}) = \sqrt{n}$, which gives a projection from $X$ onto $E$ of norm $\leqslant \sqrt{n}$ (which we have already seen in another context, with Pietsch's theorem; see Chapter 6 of Volume 1, Theorem IV.4).

## IV.3 The Proof of Dvoretzky's Theorem

Now, the principal result of this chapter is stated and is proved:

**Theorem IV.8** (Dvoretzky's Theorem)

1) *For every $\varepsilon > 0$, $\delta = \delta(\varepsilon) > 0$ can be found such that, for every Banach space $E$ of dimension $n$, there exists a subspace $F$ of $E$ with $\dim F = k \geqslant \delta \log n$ and $d_F \leqslant 1 + \varepsilon$.*

   *Equivalently, $w_1, \ldots, w_k \in E$ can be found such that:*

$$(1 + \varepsilon)^{-1/2} \left( \sum_{j=1}^k |\alpha_j|^2 \right)^{1/2} \leqslant \left\| \sum_{j=1}^k \alpha_j w_j \right\| \leqslant (1 + \varepsilon)^{1/2} \left( \sum_{j=1}^k |\alpha_j|^2 \right)^{1/2}$$

   *for every $\alpha_1, \ldots, \alpha_k \in \mathbb{K} = \mathbb{R}$ or $\mathbb{C}$.*
2) *If $\dim E = +\infty$, then, for any $k \geqslant 1$ and every $\varepsilon > 0$, $E$ contains a subspace $F$ of dimension $k$ such that $d_F \leqslant 1 + \varepsilon$.*

Note that it suffices to show 1), as 2) is an immediate consequence of 1). This will be a fairly simple consequence of the following theorem. First, we give a definition:

**Definition IV.9** Let $E$ be a Banach space, and $X = \sum_{j=1}^m g_j v_j$ a non-zero Gaussian vector, where $v_1, \ldots, v_m \in E$ and $g_1, \ldots, g_m$ are standard independent Gaussians, real or complex, according to whether the space $E$ is itself real or complex. The *weak moment* $\sigma_X$ of $X$ is the number:

$$\sigma_X = \sup_{\xi \in B_{E^*}} \left( \mathbb{E}|\xi(X)|^2 \right)^{1/2},$$

and we define the *Gaussian dimension* of $X$ by:

$$d(X) = \left( \frac{\mathbb{E}\|X\|}{\sigma_X} \right)^2.$$

With this definition, we can state:

**Theorem IV.10** (Gaussian Version of Dvoretzky's Theorem) *Let $E$ be a Banach space, $X = \sum_{j=1}^{m} g_j v_j$ a Gaussian vector with values in $E$, with $g_1, \ldots, g_m$ standard independent Gaussians (real or complex), and $v_1, \ldots, v_m$ vectors of $E$. Let $d(X)$ be the Gaussian dimension of $X$. Then, for every $\varepsilon > 0$, there exists $\alpha(\varepsilon) \in ]0, 1[$ such that $E$ contains a subspace $F$ of dimension:*

$$\dim F = [\alpha(\varepsilon)\, d(X)],$$

*and such that $d_F \leqslant 1 + \varepsilon$.*

Theorem IV.10 is only interesting when $d(X)$ is large, and is then the tool used to prove Dvoretzky's theorem.

*Proof of Theorem IV.8* We prove that, when $\dim E = n$ and when the $v_j$'s of Theorem IV.10 (for $1 \leqslant j \leqslant m = \left[\frac{n}{2}\right] + 1$) are the vectors of the (deterministic) Dvoretzky–Rogers lemma, then the Gaussian dimension of $X$ is controlled. In fact, in this case, if $\xi \in B_{E^*}$, we have:

$$\left( \mathbb{E}|\xi(X)|^2 \right)^{1/2} = \left( \sum_{j=1}^{m} |\xi(v_j)|^2 \right)^{1/2} \leqslant \sup_{x \in B_{\ell_2^m}} \left\| \sum_{j=1}^{m} x_j v_j \right\| \leqslant 2,$$

and hence:

$$\sigma_X \leqslant 2.$$

Moreover, for $u > 0$, the Paul Lévy principle of symmetry gives:

$$\mathbb{P}(\|X\| > u) \geqslant \frac{1}{2}\mathbb{P}\left( \sup_{j \leqslant m} \|g_j v_j\| > u \right) = \frac{1}{2}\mathbb{P}\left( \sup_{j \leqslant m} |g_j| > u \right);$$

therefore:

$$\mathbb{E}(\|X\|) = \int_0^{+\infty} \mathbb{P}(\|X\| > u)\, du$$

$$\geqslant \frac{1}{2} \int_0^{+\infty} \mathbb{P}\left( \sup_{j \leqslant m} |g_j| > u \right) du = \frac{1}{2}\mathbb{E}\left( \sup_{j \leqslant m} |g_j| \right)$$

$$\geqslant c\,\sqrt{\log m} \geqslant c'\,\sqrt{\log n},$$

where $c$ and $c'$ are numerical constants. Hence:

$$d(X) \geqslant \frac{c'^2}{4} \log n = c'' \log n,$$

which implies the "classical" version of 1) of Dvoretzky's theorem, along with the Gaussian version.                                                                 □

*Proof of Theorem IV.10 (real case)* Let $k$ be an integer (to be adjusted), and $X^1, \ldots, X^k$ independent copies of $X = \sum_{j=1}^{m} g_j v_j$. Let $(e_1, \ldots, e_k)$ be the canonical basis of $\ell_2^k$. If we write $X^i = \sum_{j=1}^{m} g_{ij} v_j$, $1 \leqslant i \leqslant k$, the entries of the matrix $(g_{ij})_{\substack{1 \leqslant i \leqslant k \\ 1 \leqslant j \leqslant m}}$ are independent Gaussian variables with distribution $\mathcal{N}(0,1)$. Consider the random operator $G_\omega \colon \ell_2^k \to E$ defined by:

$$G_\omega = \sum_{i,j} g_{ij}(\omega)\, e_i \otimes v_j = \sum_{i,j} g_{ij}\, e_i \otimes v_j.$$

This tensor notation simply means that if $x = \sum_i \langle x, e_i \rangle e_i$, then $G_\omega(x) = \sum_{i,j} g_{ij} \langle x, e_i \rangle v_j$, or, in other words, that $G_\omega$ maps the vector $e_i$ to $\sum_j g_{ij} v_j$, $1 \leqslant i \leqslant k$, i.e. that $G_\omega$ is defined by the random matrix $(g_{ij})_{ij}$ on the "bases" $(e_i)_i$ and $(v_j)_j$ (in general, the $v_j$'s are not linearly independent).

We have to show that, for a suitable choice of $\omega$, $G_\omega$ is "almost" an isometry, i.e. that we "almost" have $\|G_\omega x\| \leqslant 1$ and $\|G_\omega x\| \geqslant 1$ for $x$ in the unit sphere of $\ell_2^k$. We proceed in two steps.

*An upper bound for $G_\omega$*

Let $S$ be the unit sphere of $\ell_2^k$, and $T$ that of $E^*$. We have $\|G_\omega\| = \sup_{x \in S, \xi \in T} X_{x,\xi}$, where:

$$X_{x,\xi} = \sum_i x_i \xi(X^i), \quad \text{with} \quad X^i = \sum_j g_{ij} v_j.$$

We are going to bound $\mathbb{E}_\omega \|G_\omega\| = \mathbb{E}(\sup X_{x,\xi})$ thanks to the Slepian–Sudakov lemma. But there is not yet a dominating process $Y_{x,\xi}$! So we use a "trick" due to S. Chevet: it consists of introducing $Y_{x,\xi}$ (while attempting to bound the $L^2$-differences of $X_{x,\xi}$), via a separation of the variables $x$ and $\xi$.

Let $(x, \xi)$ and $(x', \xi') \in S \times T$; we have:

$$\mathbb{E}(X_{x,\xi} - X_{x',\xi'})^2 = \sum_{1 \leqslant i \leqslant k} \mathbb{E}\big(x_i \xi(X^i) - x_i' \xi'(X^i)\big)^2$$

$$= \sum_{1 \leqslant i \leqslant k} x_i^2 \mathbb{E}[\xi(X)^2] + \sum_{1 \leqslant i \leqslant k} x_i'^2 \mathbb{E}[\xi'(X)^2]$$

$$- 2 \sum_{1 \leqslant i \leqslant k} x_i x_i' \,\mathbb{E}[\xi(X)\,\xi'(X)],$$

since $X$ has the same distribution as $X^i$; hence, as $\|x\|_2 = \|x'\|_2 = 1$:

$$\mathbb{E}(X_{x,\xi} - X_{x',\xi'})^2 = \mathbb{E}\big[\xi(X) - \xi'(X)\big]^2 + \sum_{1 \leqslant i \leqslant k} (x_i - x_i')^2 \mathbb{E}[\xi(X)\,\xi'(X)]$$

$$\leqslant \mathbb{E}\big[\xi(X) - \xi'(X)\big]^2 + \sum_{1 \leqslant i \leqslant k} (x_i - x_i')^2 \sigma_X^2,$$

by the Cauchy–Schwarz inequality, with $\sigma_X$ the weak moment of $X$.

Having separated the variables $x$ and $\xi$, we can set:

$$(1) \qquad Y_{x,\xi} = \xi(X) + \sigma_X \sum_{i=1}^{k} x_i \gamma_i,$$

where $(\gamma_i)_{1 \leqslant i \leqslant k}$ is a sequence of independent Gaussian variables, with distribution $\mathcal{N}(0, 1)$, and independent of $X$. The inequality thus obtained reads:

$$\mathbb{E}(X_{x,\xi} - X_{x',\xi'})^2 \leqslant \mathbb{E}(Y_{x,\xi} - Y_{x',\xi'})^2.$$

The Slepian–Sudakov lemma (with the Remark following it) leads to:

$$\mathbb{E}\,\|G_\omega\| \leqslant \mathbb{E}\left( \sup_{x,\xi} Y_{x,\xi} \right).$$

Now:

$$\sup_{x,\xi} Y_{x,\xi} = \|X\| + \sigma_X \left( \sum_{i=1}^{k} \gamma_i^2 \right)^{1/2},$$

and:

$$\mathbb{E}\left( \sum_{i=1}^{k} \gamma_i^2 \right)^{1/2} \leqslant \left( \mathbb{E} \sum_{i=1}^{k} \gamma_i^2 \right)^{1/2} = \sqrt{k};$$

hence the bound:

$$(2) \qquad \mathbb{E}\|G_\omega\| \leqslant \mathbb{E}\|X\| + \sqrt{k}\,\sigma_X.$$

*A lower bound for $G_\omega$*

Very often in Probability, lower bounds are more difficult to obtain than upper bounds. Whereas the Slepian–Sudakov lemma was sufficient for (2), the more complicated theorem of Gordon is now required. We will show:

$$(3) \qquad \mathbb{E}\left( \inf_{x \in S} \|G_\omega x\| \right) \geqslant \mathbb{E}\,\|X\| - \sqrt{k}\,\sigma_X.$$

In light of the inequality $\dfrac{1}{\|G_\omega^{-1}\|} \leqslant \|G_\omega x\| \leqslant \|G_\omega\|$ if $x \in S$, this boils down to finding an upper bound for $\mathbb{E}\|G_\omega^{-1}\|$. Surprise (and pleasant at that!): the process $(Y_{x,\xi})$ defined in (1), which gave an upper bound for $(X_{x,\xi})$ "à la" Slepian–Sudakov, provides a lower bound "à la" Gordon; indeed, as we have seen for the upper bound:

$$\mathbb{E}(Y_{x,\xi} - Y_{x,\xi'})^2 = \mathbb{E}\big(\xi(X) - \xi'(X)\big)^2 = \mathbb{E}(X_{x,\xi} - X_{x,\xi'})^2,$$

whereas, when $x \neq x'$, the calculation made above shows that:

$$\mathbb{E}(Y_{x,\xi} - Y_{x',\xi'})^2 \geqslant \mathbb{E}(X_{x,\xi} - X_{x',\xi'})^2.$$

Using 2) in Gordon's theorem, and switching there the roles of $X_{st}$ and $Y_{st}$, we obtain:

$$\mathbb{E}\left(\inf_{x \in S}\|G_\omega x\|\right) = \mathbb{E}\left(\inf_x \sup_\xi X_{x,\xi}\right) \geqslant \mathbb{E}\left(\inf_x \sup_\xi Y_{x,\xi}\right)$$

$$= \mathbb{E}\left(\|X\| - \sigma_X \left(\sum \gamma_i^2\right)^{1/2}\right) \geqslant \mathbb{E}\|X\| - \sqrt{k}\,\sigma_X,$$

which is the claimed inequality (3).

We set:

$$\begin{cases} U = U_\omega = \dfrac{1}{\|G_\omega^{-1}\|} = \inf_{x \in S}\|G_\omega x\| \\[2mm] V = V_\omega = \|G_\omega\| = \sup_{x \in S}\|G_\omega x\|. \end{cases}$$

Now, we use the Gaussian dimension of $X$: given $\varepsilon > 0$, we select $t > 0$ such that $(1+t)/(1-t) \leqslant 1 + \varepsilon$ (for example, $t = \varepsilon/(2+\varepsilon)$); we then set $k = [t^2 d(X)]$ and, thanks to (2) and (3):

$$\frac{\mathbb{E}\,V}{\mathbb{E}\,U} \leqslant \frac{\mathbb{E}\|X\| + \sqrt{k}\,\sigma_X}{\mathbb{E}\|X\| - \sqrt{k}\,\sigma_X} = \frac{\sqrt{d(X)} + \sqrt{k}}{\sqrt{d(X)} - \sqrt{k}} = \frac{1 + \sqrt{\dfrac{k}{d(X)}}}{1 - \sqrt{\dfrac{k}{d(X)}}} \leqslant 1 + \varepsilon.$$

We use the following simple remark:

**Lemma IV.11**  *Let $U$ and $V$ be two integrable positive random variables such that $\dfrac{\mathbb{E}\,V}{\mathbb{E}\,U} \leqslant \lambda$. Then there exists $\omega_0$ such that $\dfrac{V(\omega_0)}{U(\omega_0)} \leqslant \lambda$.*

The proof is immediate: since $\mathbb{E}(V - \lambda U) \leqslant 0$, there indeed exists an $\omega_0$ such that $(V - \lambda U)(\omega_0) \leqslant 0$.

Hence, with our choice of $k$ above, we can find $\omega_0$ such that:

$$\frac{V(\omega_0)}{U(\omega_0)} = \|G_{\omega_0}\| \, \|G_{\omega_0}^{-1}\| \leqslant 1 + \varepsilon.$$

Then, if $F = G_{\omega_0}(\ell_2^k)$, we have:

$$d_F \leqslant \|G_{\omega_0}\| \, \|G_{\omega_0}^{-1}\| \leqslant 1 + \varepsilon,$$

which completes the proof of Theorem IV.10 in the real case. $\qquad\square$

**Remark**  The proof leads to $\alpha(\varepsilon) \geqslant c\,\varepsilon^2$ in the real version of Theorem IV.10, and hence $\delta(\varepsilon) \geqslant c'\,\varepsilon^2$ in the real version of Dvoretzky's theorem.

*Proof of Theorem IV.10 (complex case)*  The approach is essentially the same, except that it makes use of the deviation inequality (concentration of measure) of Section II instead of the Slepian–Gordon theorem. We will use the notation of the complex setting. Beforehand, a simple technical result is required:

**Lemma IV.12**  *In every real normed space $E$ of finite dimension $n$ and for every $\varepsilon > 0$:*

$$\left(\frac{1}{\varepsilon}\right)^n \leqslant N(\varepsilon) \leqslant \left(1 + \frac{2}{\varepsilon}\right)^n,$$

*where $N(\varepsilon)$ is the* entropy *of the unit ball $B_E$ of $E$, i.e. the minimum number of closed balls of radius $\varepsilon$ necessary to cover $B_E$.*

*When the space is complex, this still holds, if the exponent $n$ is replaced by $2n$.*

Recall that a finite subset $R$ of a totally bounded metric space is an $\varepsilon$-*net* if it can be covered by closed balls of radius $\varepsilon$ centered at points of $R$.

*Proof of Lemma IV.12*

1) Let $r = N(\varepsilon)$, and let $y_1, \ldots, y_r \in B_E$ such that:

$$B_E \subseteq \bigcup_{j=1}^{r} (y_j + \varepsilon B_E).$$

We have:

$$\mathrm{Vol}(B_E) \leqslant \sum_{j=1}^{r} \mathrm{Vol}(y_j + \varepsilon B_E) = r\varepsilon^n \, \mathrm{Vol}(B_E);$$

thus $r \geqslant (1/\varepsilon)^n$.

2) For the right-hand inequality, we use the following:

**Sub-Lemma IV.13**  *Let $(X, d)$ be a totally bounded metric space. Denote by $K(\varepsilon)$ the maximal number of points $x_1, \ldots, x_k \in X$ such that $d(x_j, x_{j'}) > \varepsilon$ for $j \neq j'$. Then:*

$$N(\varepsilon) \leqslant K(\varepsilon) \leqslant N\left(\frac{\varepsilon}{2}\right),$$

*where $N(\varepsilon)$ is the entropy of $X$, i.e. the minimal number of closed balls of radius $\varepsilon$ necessary to cover $X$.*

Let then $y_1, \ldots, y_s \in B_E$ be such that $\|y_i - y_j\| > \varepsilon$ if $1 \leqslant i < j \leqslant s$, where the number $s$ of points is maximal. The balls $y_j + \frac{\varepsilon}{2} B_E$ are pairwise disjoint, and contained in $\left(1 + \frac{\varepsilon}{2}\right) B_E$ (as $y_j \in B_E$). We thus have:

$$s \left(\frac{\varepsilon}{2}\right)^n \mathrm{Vol}(B_E) = \mathrm{Vol}\left(\bigcup_{j=1}^{s} \left(y_j + \frac{\varepsilon}{2} B_E\right)\right)$$

$$\leqslant \mathrm{Vol}\left(\left(1 + \frac{\varepsilon}{2}\right) B_E\right) = \left(1 + \frac{\varepsilon}{2}\right)^n \mathrm{Vol}(B_E),$$

therefore $s \leqslant \left(1 + \frac{2}{\varepsilon}\right)^n$; hence the result, since $N(\varepsilon) \leqslant s$, according to the sub-lemma. $\qquad\qquad\square$

*Proof of the sub-lemma*  Let $R$ be an $(\varepsilon/2)$-net of $X$ with cardinality $N(\varepsilon/2)$. For every subset $A$ of $X$, with cardinality $\geqslant N(\varepsilon/2) + 1$, there are at least two distinct points $a, b \in A$ belonging to a same ball $B(r, \varepsilon/2)$, with $r \in R$. Then $d(a, b) \leqslant \mathrm{diam}\, B(r, \varepsilon/2) \leqslant \varepsilon$ and it follows that $K(\varepsilon) < |A|$; hence $K(\varepsilon) \leqslant N(\varepsilon/2)$.

For the left-hand inequality, let $m = K(\varepsilon)$. Note first that $m < +\infty$, as $X$ is a totally bounded metric space ($m \leqslant N(\varepsilon/2)$ by the above). Let $A = \{x_1, \ldots, x_m\}$ be a set of points whose mutual distances are $> \varepsilon$. Then $X = \bigcup_{j=1}^{m} B(x_j, \varepsilon)$: indeed, otherwise we could find $x \in X$ such that $d(x, x_j) > \varepsilon$ for any $j = 1, \ldots, m$, and then the set $A \cup \{x\}$ would have $m + 1$ points with mutual distances $> \varepsilon/2$, contrary to the definition of $m$. This shows that $N(\varepsilon) \leqslant m$, which completes the proof of the sub-lemma. $\qquad\qquad\square$

Let us now go back to the proof of Theorem IV.10.

Let $k \geqslant 1$ be an integer to be adjusted, and $(e_1, \ldots, e_k)$ the canonical basis of the complex space $\ell_2^k$. Let $X_1, \ldots, X_k$ be independent copies of $X$, i.e. :

$$X_l = \sum_{j=1}^{m} Z_{j,l}\, v_j,$$

where the standard complex Gaussians $Z_{j,l}$, for $1 \leqslant j \leqslant m, 1 \leqslant l \leqslant k$, are independent. A crucial point is the invariance of standard complex Gaussian vectors $(Z_{j,1}, \ldots, Z_{j,k})$ under the unitary group $\mathbb{U}(k)$ (see Chapter 5 of Volume 1, Proposition II.8, and the Remark that follows it), which implies the complex $\ell_2$-stability:

$\sum_{l=1}^{k} a_l X_l$ *has the same distribution as* $X$, *for every* $(a_1, \ldots, a_k)$ *in the unit sphere* $S$ *of the complex space* $\ell_2^k$.

For $0 < \rho < 1$, let $R$ be a $\rho$-net of the unit sphere $S$ of $\ell_2^k$, with cardinality $\leqslant (1 + 2/\delta)^{2k}$. By the complex $\ell_2$-stability and the Maurey–Pisier inequality of concentration of measure (Theorem II.4), applied to $Z = \sum_{l=1}^{k} a_l X_l = \sum_{l=1}^{k} \sum_{j=1}^{m} Z_{j,l} a_l v_j$, we obtain:

$$\mathbb{P}\left( \left| \left\| \sum_{l=1}^{k} a_l X_l \right\| - \mathbb{E}\|X\| \right| > \rho \, \mathbb{E}\|X\| \right) \leqslant 2 \exp\left( -K \frac{\rho^2 \, (\mathbb{E}\|X\|)^2}{\sigma_X^2} \right)$$

for every $a = (a_1, \ldots, a_k) \in S$. Therefore:

$$\mathbb{P}\left( \sup_{a \in R} \left| \left\| \sum_{l=1}^{k} a_l X_l \right\| - \mathbb{E}\|X\| \right| > \rho \, \mathbb{E}\|X\| \right)$$

$$\leqslant 2 \left( 1 + \frac{2}{\delta} \right)^{2k} \exp\left( -K \frac{\rho^2 \, (\mathbb{E}\|X\|)^2}{\sigma_X^2} \right)$$

$$\leqslant 2 \exp\left( \frac{4k}{\rho} - K \frac{\rho^2 \, (\mathbb{E}\|X\|)^2}{\sigma_X^2} \right)$$

$$= 2 \exp\left( \frac{4k}{\rho} - K \rho^2 \, d(X) \right).$$

Now let $k$ be the largest integer such that $\dfrac{4k}{\rho} \leqslant \dfrac{K}{2} \rho^2 \, d(X)$; then:

$$\mathbb{P}\left( \sup_{a \in R} \left| \left\| \sum_{l=1}^{k} a_l X_l \right\| - \mathbb{E}\|X\| \right| > \rho \, \mathbb{E}\|X\| \right) \leqslant 2 \exp\left( -\frac{K}{2} \rho^2 \, d(X) \right).$$

This last quantity can be assumed $< 1$: indeed, otherwise, $d(X)$ is small, and there is nothing to prove ($F$ can be taken of dimension 1). Thus we can find an $\omega_0 \in \Omega$, the underlying probability space, such that:

$$\left| \left\| \sum_{l=1}^{k} a_l X_l(\omega_0) \right\| - \mathbb{E}\|X\| \right| \leqslant \rho \, \mathbb{E}\|X\|$$

for every $a = (a_1, \ldots, a_k) \in R$. Hence:

$$(1 - \rho)\, \mathbb{E}\|X\| \leqslant \left\| \sum_{l=1}^{k} a_l X_l(\omega_0) \right\| \leqslant (1 + \rho)\, \mathbb{E}\|X\|$$

for any $a \in R$.

To conclude, for every $\varepsilon > 0$, we can choose a $\rho = \rho(\varepsilon) > 0$, small enough, so that (see Chapter 5 of Volume 1, proof of Theorem III.2):

$$(1 - \varepsilon)\, \mathbb{E}\|X\| \leqslant \left\| \sum_{l=1}^{k} a_l X_l(\omega_0) \right\| \leqslant (1 + \varepsilon)\, \mathbb{E}\|X\|$$

for any $a \in S$ (with $\omega_0$ dependent on $\varepsilon$). This estimation shows that if:

$$F = \mathrm{span}\, \big( X_1(\omega_0), \ldots, X_k(\omega_0) \big),$$

then $d_F \leqslant \dfrac{1 + \varepsilon}{1 - \varepsilon}$ and $\dim F = k$. As $k \geqslant \dfrac{1}{16}\, \rho^3\, d(X)$, this completes the proof of Theorem IV.10.                                                                           $\square$

## IV.4  Examples

1) Let $E = \ell_\infty^n$; in this case, Dvoretzky's theorem is shown to be optimal; more precisely, for $n \geqslant 2$:

*if* $\dim F = k$ *and* $d_F \leqslant 2$, *then* $k \leqslant c \log n$.

Indeed let $(x_1, \ldots, x_k)$ be a basis of $F$ such that:

$$\left( \sum_{j=1}^{k} \lambda_j^2 \right)^{1/2} \leqslant \left\| \sum_{j=1}^{k} \lambda_j x_j \right\|_\infty \leqslant 2 \left( \sum_{j=1}^{k} \lambda_j^2 \right)^{1/2}, \quad \forall \lambda_1, \ldots, \lambda_k \in \mathbb{R}.$$

We test the left-hand inequality on $\lambda_j = \varepsilon_j(\omega)$, where $(\varepsilon_j)_j$ is a Bernoulli sequence, and integrate over $\omega$ to obtain:

$$\sqrt{k} \leqslant \mathbb{E} \left\| \sum_{j=1}^{k} \varepsilon_j(\omega) x_j \right\| = \mathbb{E} \left( \sup_{\varphi \in A} \left| \sum_{j=1}^{k} \varepsilon_j(\omega) \varphi(x_j) \right| \right),$$

where $A = \{\pm e_1, \pm e_2, \ldots, \pm e_n\}$ is the set of extreme points of the unit ball of the dual $\ell_1^n$ of $E = \ell_\infty^n$. As $|A| = 2n$, the boundedness theorem for Bernoulli sequences gives:

$$\sqrt{k} \leqslant c_1 \sqrt{\log 2n} \, \sup_{\varphi \in A} \left( \sum_{j=1}^{k} |\varphi(x_j)|^2 \right)^{1/2}.$$

Now, if we test the right-hand inequality on $\lambda_j = \varphi(x_j)$, $\varphi \in A$, we obtain:

$$\sum_{j=1}^{k} \varphi(x_j)^2 \leqslant 2 \left( \sum_{j=1}^{k} \varphi(x_j)^2 \right)^{1/2} ;$$

therefore:

$$\left( \sum_{j=1}^{k} (\varphi(x_j))^2 \right)^{1/2} \leqslant 2.$$

Finally:

$$\sqrt{k} \leqslant 2c_1 \sqrt{\log 2n} \leqslant 2c_1 \sqrt{2} \sqrt{\log n},$$

hence the result with $c = 8\, c_1^2$. $\qquad\square$

2) If the cotype-$q$ constant of $E$ ($2 \leqslant q \leqslant \infty$) is well controlled (i.e. if $C_q(E) \leqslant C_q$, where $C_q$ does not depend on $n = \dim E$), the estimation of the general case can be substantially improved; we have:

$$(\forall\, \varepsilon > 0)\ (\exists\, \delta = \delta(\varepsilon, q) > 0)\ (\exists\, F \subseteq E) :$$
$$\dim F = k \geqslant \delta n^{2/q} \quad \text{and} \quad d_F \leqslant 1 + \varepsilon.$$

To see this, we apply the Gaussian version of Dvoretzky's theorem, this time with a better lower bound for $\mathbb{E}\|X\|$:

$$\mathbb{E}\|X\| = \mathbb{E}\left\|\sum_{j=1}^{m} g_j v_j\right\| \geqslant \sqrt{\frac{2}{\pi}}\, \mathbb{E}\left\|\sum_{j=1}^{m} \varepsilon_j v_j\right\| \geqslant \frac{1}{\sqrt{\pi}} \left( \mathbb{E}\left\|\sum_{j=1}^{m} \varepsilon_j v_j\right\|^2 \right)^{1/2}$$

$$\geqslant \frac{1}{\sqrt{\pi}\, C_q} \left( \sum_{j=1}^{m} \|v_j\|^q \right)^{1/q} = \frac{m^{1/q}}{\sqrt{\pi}\, C_q}$$

$$\geqslant \frac{1}{\sqrt{\pi}\, C_q\, 2^{1/q}}\, n^{1/q} = \delta_0\, n^{1/q},$$

where $(\varepsilon_j)_j$ is a Bernoulli sequence, and where we have used the comparison principle and the Khintchine–Kahane inequalities. Thus $X$ has a large Gaussian dimension:

$$d(X) = \left( \frac{\mathbb{E}\|X\|}{\sigma_X} \right)^2 \geqslant \left( \frac{\delta_0}{2}\, n^{1/q} \right)^2,$$

and an application of Theorem IV.10 provides the result, for example with $\delta = \dfrac{\delta_0^2}{4}$. $\qquad\square$

In particular, $\dim F \geqslant \delta n$ when $q = 2$, which leads to the following spectacular result:

**Theorem IV.14**    *If the cotype-2 constant of E does not depend on the dimension of E, then E contains Hilbertian subspaces of dimension proportional to that of E.*

This applies in particular to $E = \ell_1^n$, the dual of example 1).

3) For any Banach space $B$, set:

$$k_0(B) = \sup\{k \, ; \, \dim F = k, \, F \subseteq B, \text{ and } d_F \leqslant 2\}.$$

If $\dim E = n$, it can be proved (Figiel, Lindenstrauss and Milman [1977]) that:

$$k_0(E) \, k_0(E^*) \geqslant n.$$

In other words, a space, or its dual, always contains Hilbertian subspaces of dimension "truly" large, and it is thus unsurprising that the examples 1) and 2) imply, with $E = \ell_\infty^n$, $E^* = \ell_1^n$:

$$k_0(\ell_\infty^n) \, k_0(\ell_1^n) \geqslant (\delta \log n) \, (\delta' n) = \delta'' n \, \log n.$$

4) Let $E = \mathcal{L}(\ell_2^n)$; we have $\dim E = n^2$. Example 2) of Subsection IV.1 shows that $E$ trivially contains a Hilbertian subspace of dimension $n$; this result is more or less optimal, since with the notation of 3), we are going to show that:

$$k_0\big[\mathcal{L}(\ell_2^n)\big] \leqslant C \, n,$$

where $C$ is a numerical constant.

Indeed, let $F$ be a subspace of $E$ such that $\dim F = k$, $d_F \leqslant 2$, and let $T_1, \ldots, T_k$ be a basis of $F$ such that:

$$(*)\qquad \left(\sum_{j=1}^{k} \lambda_j^2\right)^{1/2} \leqslant \left\| \sum_{j=1}^{k} \lambda_j T_j \right\| \leqslant 2 \left(\sum_{j=1}^{k} \lambda_j^2\right)^{1/2}, \quad \forall \lambda_1, \ldots, \lambda_k \in \mathbb{R}.$$

We proceed as in example 1). Let $R$ be a $1/2$-net of the unit sphere $S$ of $\ell_2^n$ with cardinality $\leqslant 5^n$ (Lemma IV.12). Then:

$$(**)\qquad \|T\| \leqslant 4 \sup_{(a,b)\in R\times R} |\langle Ta, b\rangle|, \quad \forall \, T \in E.$$

Indeed, let $x \in S$, and $a \in R$ such that $\|x - a\| \leqslant 1/2$. Then:

$$\|Tx\| \leqslant \|T(x-a)\| + \|Ta\| \leqslant \frac{1}{2}\|T\| + \sup_{a\in R}\|Ta\|;$$

hence:

$$\|T\| \leqslant 2 \sup_{a \in R} \|Ta\|.$$

If $a \in R$ and $\|Ta\| = |\langle Ta, y \rangle|$, with $y \in S$, let $b \in R$ such that $\|b - y\| \leqslant 1/2$. Then:

$$\|Ta\| \leqslant |\langle Ta, y - b \rangle| + |\langle Ta, b \rangle| \leqslant \frac{1}{2} \|Ta\| + \sup_{(a,b) \in R \times R} |\langle Ta, b \rangle|;$$

hence:

$$\sup_{a \in R} \|Ta\| \leqslant 2 \sup_{(a,b) \in R \times R} |\langle Ta, b \rangle|,$$

which implies the result stated in $(**)$.

If now $(\varepsilon_j)_j$ is a Bernoulli sequence, the left-hand inequality of $(*)$, $(**)$ and the boundedness theorem lead to:

$$\sqrt{k} \leqslant \mathbb{E} \left\| \sum_{j=1}^{k} \varepsilon_j T_j \right\| \leqslant 4 \, \mathbb{E} \left( \sup_{(a,b) \in R \times R} \left| \sum_{j=1}^{k} \varepsilon_j \langle T_j a, b \rangle \right| \right)$$

$$\leqslant 4 \, C_0 \sqrt{\log |R \times R|} \, \sup_{(a,b) \in R \times R} \left( \sum_{j=1}^{k} |\langle T_j a, b \rangle|^2 \right)^{1/2}$$

$$\leqslant C_1 \sqrt{n} \, \sigma;$$

hence:

$$\sigma = \sup_{(a,b) \in R \times R} \left( \sum_{j=1}^{k} |\langle T_j a, b \rangle|^2 \right)^{1/2}.$$

However, if $(a, b) \in R \times R$, the right-hand inequality of $(*)$ leads to:

$$\left| \sum_{j=1}^{k} \lambda_j \langle T_j a, b \rangle \right| \leqslant \left\| \sum_{j=1}^{k} \lambda_j T_j \right\| \leqslant 2 \left( \sum_{j=1}^{k} \lambda_j^2 \right)^{1/2};$$

hence $\left( \sum_{j=1}^{k} |\langle T_j a, b \rangle|^2 \right)^{1/2} \leqslant 2$, and therefore $\sigma \leqslant 2$. It ensues that $\sqrt{k} \leqslant 2 \, C_1 \sqrt{n}$, and then that $k \leqslant 4 \, C_1^2 \, n = 4 \, C_1^2 \sqrt{\dim E}$. $\qquad \square$

## IV.5 The Milman–Schechtman Theorem

When $\dim E = n$, the hope of obtaining a subspace $F$ of $E$ whose distance $d_F$ to a Hilbert space is uniformly bounded is blocked at $\dim F = O(\log n)$, as the example $E = \ell_\infty^n$ shows. The point of view can be reversed: if $1 \leqslant k \leqslant n$, is there a subspace $F$ of $E$ such that $\dim F = k$ and $d_F \leqslant \varphi(n,k)$, where $\varphi(n,k)$ is much smaller than the $\sqrt{k}$ of F. John? That would lead to a new version of Dvoretzky's theorem, no longer quasi-isometric, but isomorphic. Such a version was obtained by Milman and Schechtman [1995].

**Theorem IV.15** (The Milman–Schechtman Theorem)  *There exists a numerical constant $c > 0$ such that, for every real Banach space $E$ of dimension $n$, and any integer $k = 1, \ldots, n$, $E$ contains a subspace $F$ with:*

$$\dim F = k \quad \text{and} \quad d_F \leqslant c \left[ 1 + \sqrt{\frac{k}{\log\left(1 + \frac{n}{k}\right)}} \right].$$

**Preliminary remarks** For $k \approx \log n$, this gives $d_F \leqslant c'$: we are back to Dvoretzky's theorem. The value $k = n$ leads to $d_F \leqslant c'\sqrt{n}$: we are back to F. John's theorem. An intermediate value such as $k \approx (\log n)^2$ leads to $d_F \leqslant c'\sqrt{\log n} = c' k^{1/4}$, which is much better than the general upper bound of F. John. The method used in the proof is very similar to that of Dvoretzky's theorem, but in the deterministic part, the Dvoretzky–Rogers lemma must be replaced by an improved version due to Bourgain and Szarek, whose proof is delicate and will be *admitted* (see Bourgain and Szarek [1988]).

**Lemma IV.16** (The Bourgain–Szarek Lemma)  *For any $\delta \in \,]0,1[$, there exists a constant $C_0 = C_0(\delta) > 0$ such that every Banach space $E$ of dimension $n$ contains vectors $v_1, \ldots, v_m$, with $m \geqslant \delta n$, such that, for all real numbers $\alpha_1, \ldots, \alpha_m$, we have:*

$$C_0 \sup_{j \leqslant m} |\alpha_j| \leqslant \left\| \sum_{j=1}^m \alpha_j v_j \right\| \leqslant \left( \sum_{j=1}^m \alpha_j^2 \right)^{1/2}.$$

*Proof of Theorem IV.15*  Let $(\varepsilon_1, \ldots, \varepsilon_m)$ be the canonical basis of $\ell_\infty^m$, and let $G_\omega : \ell_2^k \to E$ be again the random Gaussian operator defined by $G_\omega = \sum_{i \leqslant k, j \leqslant m} g_{ij}\, e_i \otimes v_j$. We know that there exists a subspace $F$ of $E$ with $\dim F = k$ and

$$d_F \leqslant \frac{\mathbb{E}(\sup_{x \in S} \|G_\omega x\|)}{\mathbb{E}(\inf_{x \in S} \|G_\omega x\|)}.$$

Again, an estimation of $\mathbb{E}(\sup_{x \in S} \|G_\omega x\|)$ is provided by the inequality:

(2) $$\mathbb{E}\|G_\omega\| \leqslant \mathbb{E}\|X\| + \sqrt{k}\, \sigma_X$$

seen in the first proof of Dvoretzky's theorem (the real case). However, for a lower bound of $\mathbb{E}(\inf_{x \in S} \|G_\omega x\|)$, we use the inequality:

(3) $$\mathbb{E}\left( \inf_{x \in S} \|G_\omega x\| \right) \geqslant \mathbb{E}\|X\| - \sqrt{k}\, \sigma_X$$

in two steps. First, taking advantage of the left-hand inequality of Lemma IV.16 (the novelty compared to the Dvoretzky–Rogers lemma), we bound $\|G_\omega x\|$ below by $\|\Gamma_\omega x\|_{(k)}$, where $\Gamma_\omega : \ell_2^k \to \mathbb{R}^m$ is defined by:

$$\Gamma_\omega = \sum_{i \leqslant k, j \leqslant m} g_{ij}\, e_i \otimes \varepsilon_j,$$

and where $\| \cdot \|_{(k)}$ is a norm on $\mathbb{R}^m$ (not to be confused with the usual norm on $\ell_k^n$), adapted to $k$, and dominated by $\| \cdot \|_\infty$. Next, we apply the inequality (3) to $\Gamma_\omega$, to obtain an effective lower bound of $\mathbb{E}(\inf_{x \in S} \|G_\omega x\|)$ (recall that $S$ is the unit sphere of $\ell_2^k$).

*Step 1. Description of an auxiliary norm on $\mathbb{R}^m$.*

If $\alpha = (\alpha_1, \ldots, \alpha_m) \in \mathbb{R}^m$, denote by $\alpha_1^*, \ldots, \alpha_m^*$ the non-increasing rearrangement of $|\alpha_1|, \ldots, |\alpha_m|$: $\alpha_1^* = \|\alpha\|_\infty$; and set:

$$\|\alpha\|_{(k)} = \frac{\alpha_1^* + \cdots + \alpha_k^*}{k} \leqslant \|\alpha\|_\infty.$$

This is indeed a norm; in fact, if $\alpha, \beta \in \mathbb{R}^n$ and $\gamma = \alpha + \beta$, there is a permutation $\sigma$ of $\{1, \ldots, m\}$ such that:

$$\begin{aligned}
\gamma_1^* + \cdots + \gamma_k^* &= |\gamma_{\sigma(1)}| + \cdots + |\gamma_{\sigma(k)}| \\
&\leqslant (|\alpha_{\sigma(1)}| + \cdots + |\alpha_{\sigma(k)}|) + (|\beta_{\sigma(1)}| + \cdots + |\beta_{\sigma(k)}|) \\
&\leqslant (\alpha_1^* + \cdots + \alpha_k^*) + (\beta_1^* + \cdots + \beta_k^*),
\end{aligned}$$

since, for example, $\alpha_1^* + \cdots + \alpha_k^*$ is the sum of the $k$ largest $|\alpha_i|$'s. Of course we assume $k \leqslant m$; as we require $m \geqslant \delta n$, $k$ is limited to a portion of $n$.

We now define a Gaussian variable $Y = \sum_{j=1}^m g_j\, \varepsilon_j$ with values in $\ell_\infty^k$, so that that $\Gamma_\omega x = \sum_{i=1}^k x_i\, Y^i$, where $Y^i = \sum_{j=1}^m g_{ij}\, \varepsilon_j$ is an independent copy of $Y$. The passage from $X$ to $Y$ only slightly reduces the mean of the norm, as shown by the following lemma:

**Lemma IV.17** *We have:*

$$\mathbb{E}(\|Y\|_{(k)}) \geqslant a \sqrt{\log\left(1 + \frac{n}{k}\right)},$$

*where $a$ is a positive constant.*

*Proof of the lemma* Let $l = \left[\frac{m}{k}\right]$ and $\mu = k\,l$; we divide $\{1, \ldots, \mu\}$ into $k$ consecutive blocks $A_1, \ldots, A_k$ of length $l$:

$$A_j = \{(j-1)l + 1, (j-1)l + 2, \ldots, jl\};$$

clearly:

$$g_1^* + \cdots + g_k^* \geqslant \max_{i \in A_1} |g_i| + \max_{i \in A_2} |g_i| + \cdots + \max_{i \in A_k} |g_i|.$$

Taking the expectation and using the Gaussian minoration of Proposition II.2, we obtain:

$$\mathbb{E}(g_1^* + \cdots + g_k^*) \geqslant \sum_{j=1}^k \mathbb{E}\left(\max_{i \in A_j} |g_i|\right) = k\,\mathbb{E}\left(\max_{i \in A_1} |g_i|\right)$$

$$\geqslant a_0 k \sqrt{\log(1+l)} \geqslant a_1 k \sqrt{\log\left(1 + \frac{m}{k}\right)}$$

$$\geqslant ak \sqrt{\log\left(1 + \frac{n}{k}\right)},$$

where $a$ is a constant depending only on $\delta$; hence:

$$\mathbb{E}(\|Y\|_{(k)}) = \mathbb{E}\left(\frac{g_1^* + \cdots + g_k^*}{k}\right) \geqslant a \sqrt{\log\left(1 + \frac{n}{k}\right)}. \qquad \square$$

In contrast, passing from $X$ to $Y$ greatly reduces the weak moment: here the weak moment $(\sigma_Y)_{(k)}$ of $Y$ is the quantity:

$$(\sigma_Y)_{(k)} = \sup_{\xi} \left(\mathbb{E}\,\xi(Y)^2\right)^{1/2},$$

where $\xi$ runs over the unit ball of the dual of $(\mathbb{R}^n, \|\,.\,\|_{(k)})$. In fact, Lemma IV.16 gives the inequality:

$$\sigma_X \leqslant 1,$$

and here it is replaced by:

$$(\sigma_Y)_{(k)} \leqslant \frac{1}{\sqrt{k}},$$

as indeed:

$$(\sigma_Y)_{(k)} = \sup_{\alpha \in S} \left\|\sum_{j=1}^m \alpha_j \varepsilon_j\right\|_{(k)} = \sup_{\alpha \in S} \frac{\alpha_1^* + \cdots + \alpha_k^*}{k}$$

$$\leqslant \sup_{\alpha \in S} \frac{1}{\sqrt{k}}(\alpha_1^{*2} + \cdots + \alpha_k^{*2})^{1/2} \leqslant \frac{1}{\sqrt{k}}.$$

Now we minorize $\|G_\omega x\|$ in terms of $\|\Gamma_\omega x\|_{(k)}$, thanks to the left-hand inequality of Lemma IV.16: as $G_\omega x = \sum_j(\sum_i g_{ij}x_i)\,v_j$ and $\Gamma_\omega x = \sum_j(\sum_i g_{ij}x_i)\,\varepsilon_j$, we have:

$$\|G_\omega x\| \geqslant C_0 \sup_j \left| \sum_i g_{ij} x_i \right| = C_0 \|\Gamma_\omega x\|_\infty,$$

and, *a fortiori*:

(4) $$\|G_\omega x\| \geqslant C_0 \|\Gamma_\omega x\|_{(k)}$$

for every $x \in S$.

*Step 2. Minoration of* $\|\Gamma_\omega x\|_{(k)}$.

When applying the minoration (3) to $\Gamma_\omega$, we obtain:

$$\mathbb{E} \left( \inf_{x \in S} \|\Gamma_\omega x\|_{(k)} \right) \geqslant \mathbb{E} \|Y\|_{(k)} - \sqrt{k} \, (\sigma_Y)_{(k)};$$

thus by Lemma IV.17 and the inequality $(\sigma_Y)_{(k)} \leqslant 1/\sqrt{k}$:

(5) $$\mathbb{E} \left( \inf_{x \in S} \|\Gamma_\omega x\|_{(k)} \right) \geqslant a \sqrt{\log \left( 1 + \frac{n}{k} \right)} - 1.$$

Henceforth, we take $\delta = 1/2$ in Lemma IV.16; the preceding constants $a$ and $C_0$ are hence numerical.

Let us define the constant $A$ by $a \sqrt{\log(1 + A)} = 2$. To complete the proof of the Milman–Schechtman theorem, we distinguish two cases:

– *Case 1*: $n/A \leqslant k \leqslant n$.

Then $n/k \leqslant A$, and hence:

$$1 + \sqrt{\frac{k}{\log \left( 1 + \frac{n}{k} \right)}} \geqslant \sqrt{\frac{k}{\log(1 + A)}}.$$

As $d_F \leqslant \sqrt{k}$ for every $k$-dimensional subspace $F$ of $E$, by John's theorem (Chapter 6 of Volume 1, Theorem IV.4), any constant $c \geqslant \sqrt{\log(1 + A)} = 2/a$ is suitable.

– *Case 2*: $1 \leqslant k \leqslant n/A$.

This is divided into two subcases.

a) If $\mathbb{E}\|X\| \geqslant 2\sqrt{k}$, Gordon's estimation suffices; indeed we can find $F$ with dimension $k$ such that:

$$d_F \leqslant \frac{\mathbb{E}\|X\| + \sqrt{k} \, \sigma_X}{\mathbb{E}\|X\| - \sqrt{k} \, \sigma_X},$$

and hence:

$$d_F \leqslant \frac{\mathbb{E}\|X\| + \sqrt{k}}{\mathbb{E}\|X\| - \sqrt{k}} \leqslant 3,$$

since $\sigma_X \leqslant 1$ and $\mathbb{E}\|X\| \geqslant 2\sqrt{k}$. It thus suffices to take $c \geqslant 3$.

b) If $\mathbb{E}\|X\| \leqslant 2\sqrt{k}$, we have:

$$\mathbb{E}\left(\sup_{x \in S}\|G_\omega x\|\right) \leqslant \mathbb{E}\|X\| + \sqrt{k}\,\sigma_X \leqslant 3\sqrt{k},$$

whereas the minorations (4) and (5) imply:

$$\mathbb{E}\left(\inf_{x \in S}\|G_\omega x\|\right) \geqslant C_0\,\mathbb{E}\left(\inf_{x \in S}\|\Gamma_\omega x\|_{(k)}\right) \geqslant C_0\left[a\sqrt{\log\left(1 + \frac{n}{k}\right)} - 1\right].$$

However, as $k \leqslant n/A$, we have:

$$1 \leqslant \frac{\sqrt{\log\left(1 + \frac{n}{k}\right)}}{\sqrt{\log(1 + A)}} = \frac{a}{2}\sqrt{\log\left(1 + \frac{n}{k}\right)},$$

and hence:

$$\mathbb{E}\left(\inf_{x \in S}\|G_\omega x\|\right) \geqslant \frac{a\,C_0}{2}\sqrt{\log\left(1 + \frac{n}{k}\right)}.$$

It follows that we can find a subspace $F$ of dimension $k$ such that:

$$d_F \leqslant \frac{\mathbb{E}\left(\sup_{x \in S}\|G_\omega x\|\right)}{\mathbb{E}\left(\inf_{x \in S}\|G_\omega x\|\right)} \leqslant \frac{6}{a\,C_0}\sqrt{\frac{k}{\log\left(1 + \frac{n}{k}\right)}},$$

and any constant $c \geqslant 6/(a\,C_0)$ is suitable.

This completes the proof of Theorem III.15.                                    $\square$

## V  The Lindenstrauss–Tzafriri Theorem

The second structure theorem of this chapter is the following:

**Theorem V.1** (The Lindenstrauss–Tzafriri Theorem)  *Let $X$ be an infinite-dimensional Banach space with all its closed subspaces complemented; then $X$ is isomorphic to a Hilbert space.*

*Proof*  First note that the converse of the theorem trivially holds. The proof is broken into the following three steps, with an essential role for Dvoretzky's theorem in the second step.

– *Step 1.* There exists a constant $\lambda < \infty$ such that every finite-dimensional subspace $E$ of $X$ is $\lambda$-complemented (i.e. there exists a projection $P: X \to E$ such that $\|P\| \leqslant \lambda$).

– *Step 2.* With the conclusions of Step 1, we can find $C = C(\lambda) < \infty$ such that $\sup_E d_E \leqslant C$, where $E$ runs over the finite-dimensional subspaces of $X$, and $d_E = d(E, \ell_2^{\dim E})$.

– *Step 3.* With the conclusions of Step 2, $X$ is $C$-isomorphic to a Hilbert space $H$.

The three steps together clearly imply the result; we now detail each of them.

*Proof of Step 1*   If $X$ is a Banach space, for every finite-dimensional subspace $E$ of $X$, we define:

$$\lambda(E,X) = \inf\{\|P\|;\ P\colon X \to E \text{ projection}\}$$

and

$$\lambda(X) = \sup\{\lambda(E,X)\,;\ E \subseteq X,\ \text{and}\ \dim E < \infty\}.$$

This step is then based on the following two lemmas:

**Lemma V.2**   *Suppose that $X = E \oplus Y$ with $\dim E < +\infty$. Then:*

$$\lambda(Y) < \infty \iff \lambda(X) < \infty.$$

*Proof*   The implication from right to left is evident; let us examine "$\Rightarrow$". We have a projection $P\colon X \to Y$ and the complementary projection $Q = Id_X - P\colon X \to E$. We set $C = \max(\|P\|, \|Q\|, \dim E + 2, \lambda(Y) + 2)$. Let $F$ be a finite-dimensional subspace of $X$ and let $G = P(F)$. The following assertions will be proved:

a) $X = E \oplus Y$ with $\max(\|e\|, \|y\|) \leqslant C\|e + y\|$ if $e \in E$ and $y \in Y$.
b) $Y = G \oplus Z$ with $\max(\|g\|, \|z\|) \leqslant C\|g + z\|$ if $g \in G$ and $z \in Z$.
c) $E \oplus G = F \oplus H$ with $\max(\|f\|, \|h\|) \leqslant C\|f + h\|$ if $f \in F$ and $h \in H$.

Indeed, a) derives from $y = P(e + y)$ and $e = Q(e + y)$. As $G$ is a finite-dimensional subspace of $Y$, we can find a projection $\sigma\colon Y \to G$ of norm $\leqslant \lambda(Y) + 1$, and $Z = \ker \sigma$ satisfies b). Finally, for c), we note that $F \subseteq E \oplus G$ (indeed $f = Pf + Qf$ for $f \in F$) and that the codimension of $F$ in $E \oplus G$ is $\dim E + \dim G - \dim F \leqslant \dim E = l$. Hence $F^\perp$ is a subspace of dimension $\leqslant l$ of $(E \oplus G)^*$. Thanks to Corollary IV.7, we can thus find a projection $\tau\colon (E \oplus G)^* \to F^\perp$ with norm $\leqslant l$. Then $\|\,Id_{E\oplus G} - \tau^*\| \leqslant l + 1$, and $H = \ker(Id_{E\oplus G} - \tau^*)$ satisfies c).

Consequently:

$$X = E \oplus Y = E \oplus (G \oplus Z) = (E \oplus G) \oplus Z = (F \oplus H) \oplus Z = F \oplus (H \oplus Z),$$

and hence, with $x \in X$ and obvious notation: $x = e + y = e + g + z = f + h + z$. The relations a), b), c) then cascade to give:

$$
\begin{aligned}
\|f\| &\leqslant C\|f + h\| = C\|e + g\| \leqslant C(\|e\| + \|g\|) \\
&\leqslant C\big[C\|e + y\| + C\|g + z\|\big] = C^2(\|x\| + \|y\|) \\
&\leqslant C^2(\|x\| + C\|e + y\|) = C^2(1 + C)\|x\|.
\end{aligned}
$$

By setting $R(x) = f$, we define a projection from $X$ onto $F$, with $\|R\| \leqslant C^2$ $(1 + C)$. Hence we obtain $\lambda(X) \leqslant C^2(1 + C)$.                          $\square$

**Lemma V.3** *Let $X$ be a Banach space, $E$ a finite-dimensional subspace of $X$, and $\varepsilon > 0$. Then there exists a closed subspace $Z$ of $X$, containing $E$, such that:*

$$\operatorname{codim} Z < \infty \qquad \textit{and} \qquad \lambda(E, Z) \leqslant 1 + \varepsilon.$$

*Proof* Let $\varphi_1, \ldots, \varphi_N$ be a $(\varepsilon/2)$-net of the unit sphere of $E^*$ (with $\varepsilon < 1$); then:

$$(*) \qquad \|x\| \leqslant \left(\frac{1}{1 - \varepsilon/2}\right) \sup_i |\varphi_i(x)| \leqslant (1 + \varepsilon) \sup_i |\varphi_i(x)|$$

for every $x \in E$, and extending the $\varphi_i$'s by the Hahn–Banach theorem if necessary, we can assume them to be in the unit sphere of $X^*$. Define $Y = \bigcap_{i=1}^N \ker \varphi_i$ and $Z = E \oplus Y$, the sum being direct according to $(*)$. Then $\operatorname{codim} Z \leqslant \operatorname{codim} Y \leqslant N < \infty$, and, if $P$ is the projection from $Z$ onto $E$ along the direction of $Y$, we have, for $x \in E$ and $y \in Y$:

$$\|x\| \leqslant (1 + \varepsilon) \sup_i |\varphi_i(x)| = (1 + \varepsilon) \sup_i |\varphi_i(x + y)| \leqslant (1 + \varepsilon) \|x + y\|,$$

thus $\|P\| \leqslant 1 + \varepsilon$.                          $\square$

Let us now complete Step 1. Assume $\lambda(X) = +\infty$. By induction we construct a sequence $(E_n)_{n \geqslant 1}$ of finite-dimensional subspaces of $X$, all in a direct sum, such that if $Z_n = E_1 \oplus \cdots \oplus E_n$ and $Z = \overline{\operatorname{span}}(\bigcup_1^\infty E_n)$, we have:

a) $E_n$ is poorly complemented in $X$: $\sup_n \lambda(E_n, X) = +\infty$;
b) $E_n$ is well complemented in $Z_n$: $\lambda(E_n, Z_n)$ is bounded;
c) $Z_n$ is well complemented in $Z$: $\lambda(Z_n, Z)$ is bounded.

Then $Z$ is not complemented in $X$. Indeed, if we had a projection $\pi : X \to Z$, by composing it with the projections $\sigma_n : Z \to Z_n$ and $\tau_n : Z_n \to E_n$ such that $\|\sigma_n\| \leqslant C$ and $\|\tau_n\| \leqslant C$, we would obtain a projection $\tau_n \sigma_n \pi : X \to E_n$ with $\|\tau_n \sigma_n \pi\| \leqslant C^2 \|\pi\|$, which contradicts a).

To construct the $E_n$'s, it is useful to derive from Lemma V.2 the following remark:

$(**)$     *If $Y$ is a subspace of $X$ with finite codimension, then $Y$ contains finite-dimensional subspaces that are poorly complemented in $X$.*

Indeed, if $F \subseteq Y$ and $\dim F < \infty$, then $\lambda(F, X) \geqslant \lambda(F, Y)$ (restrict a projection from $X$ onto $F$); however, by Lemma V.2, we can find $F$ such that $\lambda(F, Y)$ is large.

Now, by induction we construct two sequences $(E_k)_{k \geqslant 1}$ and $(X_k)_{k \geqslant 1}$ of closed subspaces of $X$ such that (again setting $Z_n = E_1 \oplus \cdots \oplus E_n$):

1) $\dim E_k < \infty$ and $\lambda(E_k, X) \geqslant k$;
2) $X_k = Z_k \oplus Y_k$ with a projection $P_k \colon X_k \to Z_k$ of norm $\|P_k\| \leqslant 2$;
3) $X_k \supseteq X_{k+1}$ and $\dim X_k/X_{k+1} < \infty$, with codim $X_k < +\infty$;
4) $Y_k \supseteq E_{k+1}$.

For this, set $X_0 = X$, and let $E_1$ be a finite-dimensional subspace of $X$ such that $\lambda(E_1, X) \geqslant 1$. Lemma V.3 provides an $X_1$ such that $E_1 \subseteq X_1 \subseteq X$, with $\dim X/X_1 < +\infty$, and a projection $P_1 \colon X_1 \to E_1$ of norm $\leqslant 2$. Then $X_1 = E_1 \oplus Y_1$, where $Y_1 = \ker P_1$. Assume that we have constructed $X_1, \ldots, X_n$, $E_1, \ldots, E_n$ and $Y_1, \ldots, Y_n$ satisfying 1) and 2) for $k \leqslant n$, and 3) and 4) for $k \leqslant n-1$. Then codim $Y_n = $ codim $X_n + \dim Z_n < \infty$; hence the remark (∗∗) implies the existence of a finite-dimensional subspace $E_{n+1}$ of $Y_n$ such that $\lambda(E_{n+1}, X) \geqslant n+1$.

Moreover, $Z_{n+1} = Z_n \oplus E_{n+1} \subseteq X_n$; hence, by Lemma V.3, we can find a subspace $X_{n+1}$ such that $Z_{n+1} \subseteq X_{n+1} \subseteq X_n$ and with $\dim X_n/X_{n+1} < +\infty$, as well as a projection $P_{n+1} \colon X_{n+1} \to Z_{n+1}$ with norm $\leqslant 2$. By setting $Y_{n+1} = \ker P_{n+1}$, we obtain 1) and 2) at step $n+1$, as well as 3) and 4) at step $n$.

To finish, note that:

(i) $Z \subseteq \bigcap_{j=1}^{\infty} X_j = X_\infty$,

since the $Z_n$'s are increasing and the $X_n$'s decreasing; then, if $p, q \in \mathbb{N}^*$, we have $Z_p \subseteq Z_{p+q} \subseteq X_{p+q} \subseteq X_q$; hence $Z_p \subseteq X_\infty$ and $Z \subseteq X_\infty$.

(ii) If $n \geqslant 2$, $Q_{n-1} = \mathrm{Id}_{X_{n-1}} - P_{n-1}$ projects $Z_n$ onto $E_n$ and $\|Q_{n-1}\| \leqslant 3$.

Indeed, by the decomposition $X_{n-1} = Z_{n-1} \oplus Y_{n-1}$, $Q_{n-1}$ is null on $Z_{n-1}$ and acts as $\mathrm{Id}_{X_{n-1}}$ on $Y_{n-1}$; moreover, $E_n \subseteq Y_{n-1}$ by 4).

(iii) $P_n$ projects $Z$ onto $Z_n$, with norm $\leqslant 2$,

since $Z_n \subseteq Z \subseteq X_n$, and since $P_n$ projects $X_n$ onto $Z_n$ with norm $\leqslant 2$ according to 2).

The assertions (i), (ii) and (iii) prove a), b), c), which completes Step 1. □

*Proof of Step 2* Let $E$ be a finite-dimensional subspace of $X$ and let $Q \colon X \to E$ be a projection with norm $\leqslant \lambda$. As $X$ is infinite-dimensional, so is $\ker Q$, and hence a subspace $\Lambda$ of $\ker Q$ can be found such that $\dim \Lambda = \dim E$. Then, when $x \in E$ and $y \in \Lambda$, $x = Q(x + y)$ leads to:

$$\|x\| \leqslant \lambda \|x + y\|.$$

For $u \in \mathcal{L}(E)$, we define the "norm" $\gamma_\Lambda(u)$ of the factorization of $u$ through $\Lambda$:

$$\gamma_\Lambda(u) = \inf\big\{\|v\|\,\|w\| \,;\, u = vw,\, w \in \mathcal{L}(E, \Lambda),\, v \in \mathcal{L}(\Lambda, E)\big\}.$$

There is no reason for $\gamma_\Lambda$ to satisfy the triangle inequality; however, with $d = d(E, \Lambda)$ the Banach–Mazur distance between $E$ and $\Lambda$, we have the following lemma:

**Lemma V.4**  *There is a decomposition of the identity of E such that*:

$$\text{Id}_E = a + b,$$

*with $a, b \in \mathcal{L}(E)$ and*:

$$\max\big(\gamma_\Lambda(a), \gamma_\Lambda(b)\big) \leqslant \lambda(1 + \lambda)\sqrt{d}.$$

*Proof*  Let $T: E \to \Lambda$ be an isomorphism such that $\|T\| = \|T^{-1}\| = \sqrt{d}$, and let $F$ be the graph of $T$ in $E \oplus \Lambda \subseteq X$, i.e.: $F = \{x \oplus Tx \, ; \, x \in E\}$. By Step 1, there is a projection from $X$ onto $F$, with norm $\leqslant \lambda$; then, by restriction, there is a projection $P: E \oplus \Lambda \to F$ with $\|P\| \leqslant \lambda$. We can write:

$$Px = Ax + TAx$$

for $x \in E$, with $A \in \mathcal{L}(E)$, and similarly:

$$Py = By + TBy$$

for $y \in \Lambda$, with $B \in \mathcal{L}(\Lambda, E)$. The relation $P(x + Tx) = x + Tx$ gives $x = Ax + BTx$, for $x \in E$, meaning $\text{Id}_E = a + b$ with $a = A$ and $b = BT$. Now:

$$\|By\| \leqslant \lambda\|Py\| \leqslant \lambda^2\|y\|,$$

and

$$\|TAx\| \leqslant (1 + \lambda)\|Px\| \leqslant \lambda(1 + \lambda)\|x\|;$$

consequently:

$$\|B\| \text{ and } \|TA\| \leqslant \lambda(1 + \lambda).$$

Hence:

$$\gamma_\Lambda(a) \leqslant \|T^{-1}\| \, \|TA\| \leqslant \sqrt{d}\,\lambda(1 + \lambda)$$

and

$$\gamma_\Lambda(b) \leqslant \|B\| \, \|T\| \leqslant \lambda(1 + \lambda)\sqrt{d},$$

which completes the proof of the lemma.  $\square$

The auxiliary space $\Lambda$ and Lemma V.4 reveal their full potential when we apply Dvoretzky's theorem: let $\varepsilon > 0$; since $\dim \ker Q = +\infty$, we can chose inside $\ker Q$ a subspace $\Lambda$ close to a Hilbert space: $d_\Lambda \leqslant 1 + \varepsilon$. Then, recalling that $\gamma_2$ denotes the norm of factorization through a Hilbert space, we clearly obtain:

$$\gamma_2(u) \leqslant (1 + \varepsilon)\,\gamma_\Lambda(u),$$

for $u \in \mathcal{L}(E)$. As $\gamma_2$ is a true norm, if we set $\mu = \lambda(1 + \lambda)$, Lemma V.4 hence implies:

$$d_E = \gamma_2(\mathrm{Id}_E) = \gamma_2(a + b) \leqslant \gamma_2(a) + \gamma_2(b) \leqslant (1 + \varepsilon)\big[\gamma_\Lambda(a) + \gamma_\Lambda(b)\big]$$
$$\leqslant 2(1 + \varepsilon)\mu \sqrt{d} \leqslant 2(1 + \varepsilon)^{3/2}\mu \sqrt{d_E},$$

since $d = d(E, \Lambda) \leqslant d_\Lambda \, d_E$. As $\varepsilon$ is arbitrary, it ensues that $d_E \leqslant 2\mu\sqrt{d_E}$, and hence $d_E \leqslant 4\mu^2$, which completes Step 2. $\qquad\square$

*Proof of Step 3*　This follows from Kwapień's theorem, but in fact, here we only need the compactness argument (ultraproducts) used in its proof:

> If $u \in \mathcal{L}(X, Y)$ and $\gamma_2(u|_E) \leqslant C$ for every finite-dimensional subspace $E$ of $X$, then $\gamma_2(u) \leqslant C$.

Applying this remark to $u = \mathrm{Id}_X \colon X \to X$, with $C = 4\mu^2$, thanks to Step 2 we see that $d_X = \gamma_2(\mathrm{Id}_X) \leqslant C$. In particular $X$ is isomorphic to a Hilbert space, and this completes the proof of the Lindenstrauss–Tzafriri theorem. $\qquad\square$

# VI  Comments

1) The Maurey–Pisier concentration inequality can be found in PISIER 2 and Pisier [1986 b]. A proof, giving the best possible constant $K = 1/2$, attributed by Pisier [1986 b] (pages 180–181) to Maurey, used Itô's formula. In fact it had already appeared in Ibragimov, Sudakov and Tsirelson [1976]. For Rademacher's theorem, we have followed MATTILA, pages 101–102, who credits L. Simon for the proof. See also BENYAMINI–LINDENSTRAUSS, Proposition 6.41.

2) The statement and proof of the comparison theorem III.3 are due to B. Maurey [1990, unpublished seminar], as are the notions of "relatives" and "strangers." This "combinatorics" also leads to proofs of the Slepian–Sudakov lemma and of Gordon's theorem, much simpler than the original ones (see also Kahane [1986] and Gordon [1992]). Slepian's lemma dates back to Slepian [1962]. The Slepian–Sudakov lemma can be easily deduced from Slepian's lemma, up to a possible loss of a factor of 2. This can be found in Sudakov [1971], Theorem 2: this paper is an announcement of results, but without proofs. In his review in *Mathematical Reviews* (MR 44 6027), Dudley indicated that the details would appear in Dudley [1973]; see also SUDAKOV, Chapter 2. The Sudakov minoration is found in Sudakov [1969], Theorem 1; the proof is in Sudakov [1973], Proposition 7.

3) The proof of the Dvoretzky–Rogers lemma based on Lewis' lemma is in Pisier [1986 b]. As for the operator norm of $\mathcal{L}(\ell_2^n, E)$, it can also be shown

that $\pi_2(u_0^{-1}) = \sqrt{n}$ (PISIER 2); this leads again to the result of F. John, and has applications to the local theory of Banach spaces (Gluskin, Meyer and Pajor [1994], and Queffélec [1993] and [1995]).

4) Dvoretzky's theorem was announced in Dvoretzky [1959], and its proof was later given in Dvoretzky [1961]. The proof given here of the real case of Dvoretzky's theorem is due to Gordon [1985]; for the proof based on the concentration inequality, see Pisier [1986 b] and PISIER 2, where the notion of Gaussian dimension is introduced; see also MILMAN–SCHECHTMAN, Appendix V. Another proof, due to Milman [1971 b] (see also MILMAN–SCHECHTMAN, Theorem 5.8), is given in Beauzamy [1973], along with the proof of the Lindenstrauss–Tzafriri theorem. See also Szankowski [1974], and Figiel [1976]; this is where Milman brought out the importance of the role played by the concentration of measure.

The improved versions of Dvoretzky's theorem for spaces of cotype $q$ are due to Figiel, Lindenstrauss and Milman [1977].

A completely different proof is due to Krivine [1975] and [1977] (see GUERRE-DELABRIÈRE, page 104; see also BENYAMINI–LINDENSTRAUSS, Chapter 12, § 2). It ensues from the following result (Krivine [1977]): *for every Banach space E, there exists $p \geqslant 1$ such that $\ell_p$ or $c_0$ is finitely representable in E.* A survey on Dvoretzky's theorem can be found in Milman [1992].

5) The presentation of the isomorphic version of Dvoretzky's theorem (the Milman–Schechtman theorem) is due to O. Guédon [1998], who attributes the proof of Lemma IV.17 to Gluskin.

6) The structure Theorem V.1 is due to J. Lindenstrauss and L. Tzafriri [1971]; their paper details only Step 2, Steps 1 and 3 being considered as "well known" (Step 1 was proved by Davis, Dean and Singer [1968]); see also Beauzamy [1973]. The majoration of $d_E$ thanks to the norm of factorization $\gamma_\Lambda$ is made explicit in Pisier [1996].

# VII Exercises

**Exercise VII.1**  Provide the details of the proof of the remark following Theorem III.5: if $X$ and $Y$ are real integrable random variables, then $\mathbb{P}(X > t) \leqslant \mathbb{P}(Y > t)$ for all $t \in \mathbb{R}$ implies $\mathbb{E}(X) \leqslant \mathbb{E}(Y)$.

**Exercise VII.2**  Use the Sudakov minoration to show the inequality $\mathbb{E}\|P_\omega\|_\infty \geqslant c\sqrt{n \log n}$, where $P_\omega$ is a random Gaussian trigonometric polynomial: $P_\omega(t) = \sum_{j=1}^{n} g_j(\omega) e^{ijt}$, and $(g_j)_{j\leqslant n}$ a standard sequence of normal variables.

**Exercise VII.3** (Auerbach's Lemma and Applications)    This exercise proposes an alternative proof of Auerbach's lemma, as well as a few applications.

1) Let $E$ be a normed space of dimension $n$, and $S$ its unit sphere. Show that the function $(x_1, \ldots, x_n) \in S^n \mapsto |\det(x_1, \ldots, x_n)| \in \mathbb{R}$ attains its maximum $(> 0)$ at a point $(e_1, \ldots, e_n)$ of $S^n$, which is a basis of $E$.

2) By observing that the dual basis $(e_1^*, \ldots, e_n^*)$ is given by the formulas:

$$e_i^*(x) = \frac{\det(e_1, \ldots, e_{i-1}, x, e_{i+1}, \ldots, e_n)}{\det(e_1, \ldots, e_n)},$$

show that $\|e_i\| = \|e_i^*\| = 1$ for every $i$; the sequence $(e_i)$ is called an *Auerbach basis of E*.

3) Let $X$ be a Banach space, and $E$ a subspace of $X$ with dimension $n$; show that there exists a projection $P \colon X \to E$, of norm $\leqslant n$. (*Hint*: take an Auerbach basis $(e_i)$ of $E$; by Hahn–Banach extend each $e_i^*$ to $\varphi_i \in X^*$ with norm 1, and consider $P(x) = \sum_{k=1}^n \varphi_k(x) e_k$).

4) Now let $E$ be a closed subspace of $X$, of codimension $n$. Show that, for every $\varepsilon > 0$, there exists a projection $P \colon X \to E$, of norm $\leqslant n + 1 + \varepsilon$. (*Hint*: consider an Auerbach basis $u_1, \ldots, u_n$ of $X/E$, and raise it to a basis $e_1, \ldots, e_n$ of a subspace $F = [e_1, \ldots, e_n]$ such that $\|e_i\| \leqslant 1 + \delta$; then $X = E \oplus F$; for a suitable choice of $\delta > 0$, the projection of $X$ onto $E$, along $F$, fits the bill.)

**Remark**    It can be shown (WOJTASZCZYK pages 116–117) that:

(i) If $\dim E = n$, there exists a projection $P \colon X \to E$ such that $\|P\| \leqslant \sqrt{n}$ (see Chapter 6 of Volume 1, Theorem IV.4); Exercise VII.5 shows that the result is essentially optimal.

(ii) If $\operatorname{codim} E = n$ and $E$ is closed, for every $\varepsilon > 0$ there exists a projection $P \colon X \to E$ such that $\|P\| \leqslant \sqrt{n} + 1 + \varepsilon$. The proof of (ii) is a nice example of an application of the local reflexivity theorem, and was summarized in Exercise VIII.12 of Chapter 5 (Volume 1), by admitting (i) (the Kadeč–Snobar theorem).

**Exercise VII.4**    Let $F$ be a subspace of $\ell_q^n$ $(2 \leqslant q < \infty)$ such that $\dim F = k$ and $d_F \leqslant 2$. Show that $k \leqslant C_q n^{2/q}$; the estimate of example 2) in IV.4 is hence optimal. Formulate and prove an analogous result for subspaces of the Schatten class $S_q^n$, i.e. $\mathcal{L}(\mathbb{R}^n)$ with the norm $\|A\|_q = (\operatorname{tr} |A|^q)^{1/q}$.

**Exercise VII.5**    Let $\Lambda$ be a finite subset of $\mathbb{Z}$, of cardinality $n \geqslant 1$, and let $P \colon \mathcal{C}(\mathbb{T}) \to \mathcal{C}_\Lambda$ be the canonical projection: $P(f) = \sum_{n \in \Lambda} \widehat{f}(n) \, \mathrm{e}^{int}$.

1) Show that $\|P\| \leqslant \sqrt{n}$.

2) Let $Q \colon \mathcal{C}(\mathbb{T}) \to \mathcal{C}_\Lambda$ be another projection.
   By noting that $P = \int_{\mathbb{T}} \tau_{-\theta} Q \tau_\theta \, dm(\theta)$, show that $\|P\| \leqslant \|Q\|$ (see Chapter 6 of Volume 1, Exercise VII.11).

3) Let $(\delta_j)_{j\geqslant 0}$ be the *Rudin–Shapiro sequence*: $\delta_0 = 1$, $\delta_{2j} = \delta_j$ and $\delta_{2j+1} = (-1)^j \delta_j$, which satisfies $\left\| \sum_{j=0}^n \delta_j e^{ijt} \right\|_\infty \leqslant 4\sqrt{n+1}$, and let $\Lambda_n = \{j \in \{0, 1, \ldots, n\} \, ; \, \delta_j = +1\}$. Show that $|\Lambda_n| \geqslant \frac{n+1}{2} - 2\sqrt{n+1}$ and deduce that there exists $\delta > 0$ such that:

$$\|Q\| \geqslant \delta \sqrt{|\Lambda_n|} = \delta \sqrt{\dim \mathcal{C}_{\Lambda_n}}$$

for every projection $Q \colon \mathcal{C}(\mathbb{T}) \to \mathcal{C}_{\Lambda_n}$.

**Exercise VII.6** (Dvoretzky's Theorem for $\ell_1^n$ (Schechtman [1981]))
   This exercise is devoted to giving a direct proof of an isomorphic version of Dvoretzky's theorem for the space $\ell_1^n$.

1) Let $X$ be a real random variable, whose absolute value is bounded by 1, and conditionally centered: $\mathbb{E}^{\mathcal{B}} X = 0$, where $\mathcal{B}$ is a sub-$\sigma$-algebra of the original $\sigma$-algebra.

   a) Show that $e^{\lambda X} \leqslant \operatorname{ch} \lambda + (\operatorname{sh} \lambda) X$ for any $\lambda \in \mathbb{R}$.
   b) Show that $\mathbb{E}^{\mathcal{B}}(e^{\lambda X}) \leqslant e^{\lambda^2/2}$ for any $\lambda \in \mathbb{R}$.

2) (Azuma's Inequality, Azuma [1967]) Let $(M_k)_{0\leqslant k\leqslant n}$ be a finite martingale with $M_0 = 0$, adapted to a filtration $(\mathcal{A}_k)_{0\leqslant k\leqslant n}$, where $\mathcal{A}_0$ is the trivial $\sigma$-algebra. Let $d_k = M_k - M_{k-1}$, $1 \leqslant k \leqslant n$, be the martingale differences. Assume that $|d_k| \leqslant C_k$ for any $k$, where the $C_k$'s are positive real constants.

   a) Show that $\mathbb{E}(e^{\lambda M_k} \mid \mathcal{A}_{k-1}) \leqslant e^{\lambda M_{k-1}} e^{C_k^2 \lambda^2/2}$ for $1 \leqslant k \leqslant n$ and any $\lambda \in \mathbb{R}$.
   b) Show that $\mathbb{E}(e^{\lambda M_n}) \leqslant e^{(\lambda^2/2)\left(\sum_{k=1}^n C_k^2\right)}$ for any $\lambda \in \mathbb{R}$.
   c) Show Azuma's deviation inequality:

$$\mathbb{P}(|M_n| > t) \leqslant 2 \exp\left(-\frac{t^2}{2\sum_{k=1}^n C_k^2}\right)$$

   for any $t > 0$.

3) (Yurinskiĭ's Inequality, Yurinskiĭ [1976]) Let $E$ be a real or complex Banach space, and let $X_1, \ldots, X_n$ be independent random variables with values in $E$, such that $\|X_k\| \leqslant C_k$ for $1 \leqslant k \leqslant n$, where $C_1, \ldots, C_n$ are real constants. Set:

$$S = \left\| \sum_{k=1}^{n} X_k \right\| \quad \text{and} \quad M_k = \mathbb{E}(S \mid \mathcal{T}_k) - \mathbb{E}(S),$$

where $\mathcal{T}_k = \sigma(X_1, \ldots, X_k)$. Let $d_k = M_k - M_{k-1}$, $1 \leqslant k \leqslant n$ be the martingale differences, and let $d_0 = 0$.

a) Show that $|d_k| \leqslant 2\, C_k$ almost surely.

b) Show that:

$$\mathbb{P}\big(|S - \mathbb{E}(S)| > t\big) \leqslant 2\exp\left(-\frac{t^2}{8 \sum_{k=1}^{n} C_k^2}\right)$$

for any $t > 0$.

4) Consider the real or complex space $E = \ell_1^n$. Let $(e_1, \ldots, e_n)$ be its canonical basis. Let $m \leqslant n$ be an integer to be chosen later, and let $\big(\varepsilon_{j,k}\big)_{1 \leqslant j \leqslant m, 1 \leqslant k \leqslant n}$ be a matrix of independent random Rademacher variables. Set:

$$f_j(\omega) = \frac{1}{n} \sum_{k=1}^{n} \varepsilon_{j,k}(\omega)\, e_k.$$

Denote by $\Sigma$ the unit sphere of $\ell_2^n$ (real or complex), and, for $a = (a_1, \ldots, a_n) \in \Sigma$, set:

$$X_{j,k} = \frac{1}{n} a_j \varepsilon_{j,k} e_k$$

and

$$S_a = \left\| \sum_{j=1}^{m} a_j f_j \right\|_{\ell_1^n} = \left\| \sum_{j,k} X_{j,k} \right\|_{\ell_1^n}.$$

a) Show that $1/\sqrt{2} \leqslant \mathbb{E}(S_a) \leqslant 1$.

b) Show that $\|X_{j,k}\|_{\ell_1^n} \leqslant |a_j|/n = C_{j,k}$.

c) Using Yurinskiĭ's inequality, show that, for $a \in \Sigma$ and for any $t > 0$:

$$\mathbb{P}\big(|S_a - \mathbb{E}(S_a)| > t\big) \leqslant 2\exp\left(-\frac{nt^2}{8}\right).$$

d) Using a net in $\Sigma$, and adjusting the integer $m$, deduce the following isomorphic version of Dvoretzky's theorem for $\ell_1^n$: *For every $\varepsilon > 0$, there exists $\delta(\varepsilon) > 0$ such that, for any $n \geqslant 1$, the (real or complex) space $\ell_1^n$ contains a subspace $F$ with $\dim F \geqslant \delta(\varepsilon)n$ and $d_F = d(F, \ell_2^{\dim F}) \leqslant \sqrt{2} + \varepsilon$.*

**Exercise VII.7**   Prove the Bourgain–Szarek lemma when $E = \ell_1^n$: if $m = 2^j$, $n/2 \leqslant m \leqslant n$, take $v_i = m^{-1/2} \sum_{j=1}^m a_{ij} e_j$, $1 \leqslant i \leqslant m$, where $(a_{ij})$ is a Hadamard matrix: namely $(a_{ij}) \in \mathbb{O}(m)$ and $a_{ij} = \pm 1/\sqrt{m}$, where $(e_j)$ is the canonical basis of $\ell_1^n$.

**Exercise VII.8**   Let $X = \mathcal{C}(\mathbb{T})$ be the space of $2\pi$-periodic continuous functions equipped with the norm $\|.\|_\infty$, and let $\Lambda \subseteq \mathbb{Z}$ be such that, for every $\Lambda' \subseteq \Lambda$, $C_{\Lambda'}$ is complemented in $C_\Lambda$. Then show that $\Lambda$ is a Sidon set and that $C_\Lambda \approx \ell_1$ (see Chapter 6 of Volume 1, Exercise VII.11). Why is this not in contradiction with the Lindenstrauss–Tzafriri theorem?

# 2

# Separable Banach Spaces
# without the Approximation Property

## I Introduction and Definitions

The existence of Schauder bases for every separable Banach space is a question already appearing in BANACH, page 111: *"On ne sait pas si tout espace de type (B) séparable admet une base"* (It is not known whether every type (B) separable space possesses a basis). This question is linked to an approximation problem, *a priori* less demanding:

**Definition I.1**    A Banach space $X$ is said to have the *approximation property* (abbreviated: $X$ has *AP*) if, for every compact subset $K$ of $X$ and every $\varepsilon > 0$, there exists an operator $T \in \mathcal{L}(X)$, of finite rank, such that $\|Tx - x\| \leqslant \varepsilon$ for any $x \in K$.

If $T$ can be found such that $\|T\| \leqslant \lambda$, $X$ is said to have the $\lambda$-*bounded approximation property* (abbreviated $X$ has $\lambda$-*BAP*). $X$ is said to have *BAP* if it has $\lambda$-*BAP* for some $\lambda \geqslant 1$. If, moreover, $\lambda = 1$ is obtainable, $X$ is said to have the *metric approximation property* (abbreviated: $X$ has *MAP*).

To test $\lambda$-*BAP*, it suffices to approximate the identity map up to $\varepsilon$ on finite sets $K$, which makes clear that a space with a Schauder basis $(e_n)_{n \geqslant 1}$ of constant $\lambda$ has $\lambda$-*BAP*; indeed, the canonical projections $P_n(x) = \sum_{j=1}^n e_j^*(x)e_j$ are pointwise convergent to the identity, and $\|P_n\| \leqslant \lambda$ by definition. The approximation property is hence less demanding than the existence of a basis, and moreover has little to do with separability, as the following proposition shows:

**Proposition I.2**    *Let $S$ be a compact space, metrizable or not. Then the Banach space $X = \mathcal{C}(S)$ possesses MAP.*

*Proof*    Let $K = \{f_1, \ldots, f_p\} \subseteq X$, $\varepsilon > 0$, and $O_1, \ldots, O_q$ be a covering of $S$ by non-empty open sets such that each $f_k$ oscillates less than $\varepsilon$ on each $O_j$.

Let $x_j \in O_j$, $1 \leqslant j \leqslant q$, and let $(\varphi_j)_{1 \leqslant j \leqslant q}$ be a continuous partition of unity, subordinated to $(O_j)_{1 \leqslant j \leqslant q}$. We set $Pf(x) = \sum_{j=1}^{q} f(x_j)\varphi_j(x)$, thus defining an operator $P \colon X \to X$ of rank $\leqslant q$. We have $|Pf(x)| \leqslant \|f\|_\infty \sum_{j=1}^{q} \varphi_j(x) = \|f\|_\infty$, which shows that $\|P\| \leqslant 1$. Moreover, if $f \in K$, then

$$|f(x) - Pf(x)| = \left| \sum_{j=1}^{q} [f(x) - f(x_j)] \varphi_j(x) \right|$$

$$\leqslant \sum_{j=1}^{q} |f(x) - f(x_j)| \varphi_j(x) \leqslant \sum_{j=1}^{q} \varepsilon \, \varphi_j(x) = \varepsilon;$$

therefore $\|Pf - f\|_\infty \leqslant \varepsilon$. $\qquad\qquad\qquad\qquad\qquad\qquad\qquad\qquad\square$

**Remark**    If $S$ is not metrizable, $\mathcal{C}(S)$ is not separable; it nonetheless has *MAP*. In particular, this is the case for $\ell_\infty = \mathcal{C}(\beta\mathbb{N})$ ($\beta\mathbb{N}$ is the Stone–Čech compactification of $\mathbb{N}$).

Another way to explain the word "approximation" in Definition I.1: it is the approximation of compact operators by finite rank operators. The following notation used from now on:

**Notation I.3**    Denote by:

$\mathcal{F}(Y,X)$ the space of finite rank operators $T \colon Y \to X$;

$\mathcal{A}(Y,X)$ the closure in operator norm of $\mathcal{F}(Y,X)$, or equivalently the space of operators $T \colon Y \to X$ approximable by finite rank operators;

$\mathcal{K}(Y,X)$ the space of compact operators $T \colon Y \to X$;

$\mathcal{L}(Y,X)$ the space of all bounded operators $T \colon Y \to X$.

We write $\mathcal{F}(X)$, $\mathcal{A}(X)$, $\mathcal{K}(X)$ and $\mathcal{L}(X)$ when $Y = X$.

We always have:

$$\mathcal{F}(Y,X) \subseteq \mathcal{A}(Y,X) \subseteq \mathcal{K}(Y,X) \subseteq \mathcal{L}(Y,X).$$

The last inclusion is not always strict (consider, for example, $Y = \ell_2, X = \ell_1$), and the problem of approximation of compact operators is that of the strict inclusion of $\mathcal{A}(Y,X)$ in $\mathcal{K}(Y,X)$. For the "usual" spaces, it is always an equality, but the general problem has remained open for a very long time. The link with the approximation property was found by Grothendieck [1956], in a huge work of "clarification," where this property was reduced to a problem of "hard analysis." The problem was settled in the negative by P. Enflo [1973]. S. Davie [1973] and [1975] later gave a simplified version, to this day – in our opinion – still the most accessible. But first let us examine Grothendieck's results.

## II The Grothendieck Reductions

The following theorem of Grothendieck makes the link between the approximation properties mentioned in Section I.

**Theorem II.1** *Let X be a Banach space. The following assertions are equivalent*:

(1) *X has the approximation property.*
(2) *For every Banach space Y, $\mathcal{K}(Y,X) = \mathcal{A}(Y,X)$.*
(3) *For every separable Banach space Y, $\mathcal{K}(Y,X) = \mathcal{A}(Y,X)$.*

*Proof*

(1) $\Rightarrow$ (2). Let $T \in \mathcal{K}(Y,X)$, and $\varepsilon > 0$. Set $K = \overline{T(B_Y)}$; there exists $P \in \mathcal{F}(X)$ such that $\|Px - x\| \leqslant \varepsilon$ if $x \in K$. In particular $\|PTy - Ty\| \leqslant \varepsilon$ when $y \in B_Y$; thus $PT \in \mathcal{F}(Y,X)$ and $\|PT - T\| \leqslant \varepsilon$.

(2) $\Rightarrow$ (3) is trivial.

(3) $\Rightarrow$ (1). This is the hard part.

The idea is that each compact subset $K$ of $X$ creates its own separable Banach space $Y$ on which we can test the hypothesis. First, a description of the compact subsets of a Banach space (again due to Grothendieck) is required.

**Lemma II.2** *Let X be a Banach space, and K a compact subset of X. Then there exists a sequence $(x_n)_{n \geqslant 1}$ of elements of X such that $\lim_{n \to +\infty} \|x_n\| = 0$ and $K \subseteq \overline{\mathrm{conv}}\{x_n \, ; \, n \geqslant 1\}$, the closed convex hull of the $x_n$.*

*Proof* Denote $L = 2K$. Using a dilation if necessary, we can assume that $L \subseteq B_X$. By induction, we construct finite sets $R_j, j \geqslant 1$ such that:

a) $R_j \subseteq 2^{-j+1}B_X$;
b) $L \subseteq R_1 + 2^{-1}R_2 + \cdots + 2^{-j+1}R_j + 4^{-j}B_X$.

First, for $R_1 \subseteq B_X$, we select a $1/4$-net of $L$, so that $L \subseteq R_1 + (1/4)B_X$. Having chosen $R_1, \ldots, R_j$, note that $2^j(L - R_1 - \cdots - 2^{-j+1}R_j) \cap (2^{-j}B_X)$ is a compact subset of $2^{-j}B_X$; we can thus find a finite set, $R_{j+1} \subseteq 2^{-j}B_X$, such that:

$$2^j(L - R_1 - \cdots - 2^{-j+1}R_j) \cap (2^{-j}B) \subseteq R_{j+1} + (1/4)2^{-j}B_X .$$

Then we take $l \in L$. By b), we have $l = r_1 + \cdots + 2^{-j+1}r_j + u$, where $r_k \in R_k$ and $\|u\| \leqslant 4^{-j}$; hence:

$$2^j(l - r_1 - \cdots - 2^{-j+1}r_j) = 2^j u \in R_{j+1} + (1/4)2^{-j}B ,$$

and then:

$$2^j(l - r_1 - \cdots - 2^{-j+1}r_j) = r_{j+1} + v ,$$

where $r_{j+1} \in R_{j+1}$ and $\|v\| \leqslant (1/4)2^{-j}$.

Next $l = r_1 + \cdots + 2^{-j+1} r_j + 2^{-j} r_{j+1} + w$, with $w = 2^{-j} v$ and $\|w\| \leqslant 4^{-j-1}$, which proves a) and b).

Then let $(x_n)_{n \geqslant 1}$ be the sequence obtained by concatenating the elements of $R_1$, then those of $R_2$, etc. Condition a) shows that $\|x_n\| \xrightarrow[n \to +\infty]{} 0$, and condition b) that $K \subseteq \sum_{j=1}^{\infty} 2^{-j} R_j \subseteq \overline{\mathrm{conv}}\{x_n \, ; \, n \geqslant 1\}$.    □

Now let $K$ be a compact subset of $X$. By Lemma II.2, we can assume $K \subseteq \overline{\mathrm{conv}}\{x_n \, ; \, n \geqslant 1\}$, where $\|x_n\| \searrow 0$, $x_n \neq 0$, and $\|x_1\| \leqslant 1$. We first create a convex compact set, symmetric with respect to 0, and much larger than $K$, by setting:

$$U = \overline{\mathrm{conv}} \left\{ \pm \frac{x_n}{\|x_n\|^{1/2}} \, ; \, n \geqslant 1 \right\}.$$

Since $\left\| x_n / \|x_n\|^{1/2} \right\| = \|x_n\|^{1/2} \xrightarrow[n \to +\infty]{} 0$, $U$ is again compact; moreover: $K \subseteq U \subseteq B_X$, since, on one hand,

$$x_n = \|x_n\|^{1/2} \times \frac{x_n}{\|x_n\|^{1/2}} + (1 - \|x_n\|^{1/2}) \times 0,$$

with $\|x_n\|^{1/2} \leqslant \|x_1\|^{1/2} \leqslant 1$, and, on the other hand, $\left\| x_n / \|x_n\|^{1/2} \right\| = \|x_n\|^{1/2} \leqslant 1$.

Now let $Y = \bigcup_{\lambda > 0} (\lambda U)$ be the linear space generated by $U$; we equip $Y$ with a norm (the gauge of $U$) by setting:

$$[y] = \inf\{\lambda > 0 \, ; \, y \in \lambda U\}.$$

This is indeed a norm on $Y$ (and not only a semi-norm), since $\|y\| \leqslant [y]$ for every $y \in Y$, a consequence of the inclusion $U \subseteq B_X$. To continue, we need the following approximation lemma:

**Lemma II.3**  *The norm $[\,.\,]$ has the following properties*:

(1) $(Y, [\,.\,])$ *is a separable Banach space*;
(2) *If $y^*$ is a continuous linear functional on $(Y, [\,.\,])$, then, for every $\delta > 0$, there exists a continuous linear functional $x^*$ on $(X, \|\,.\,\|)$ such that $|y^*(x) - x^*(x)| \leqslant \delta$ for any $x \in K$.*

We first show that this lemma completes the proof of Theorem II.1. Let $T$ be the injection from $(Y, [\,.\,])$ into $(X, \|\,.\,\|)$. Then $T(B_Y) = B_Y = U$, and hence $T \in \mathcal{K}(Y, X)$. By hypothesis, $T$ can be approximated in norm by a finite rank operator, and we can find $v_1, \ldots, v_p \in X$, $y_1^*, \ldots, y_p^* \in Y^*$ such that:

$$\left\| x - \sum_{k=1}^{p} y_k^*(x) v_k \right\| \leqslant \frac{\varepsilon}{2} \qquad \text{for } x \in U = B_Y.$$

By Lemma II.3, there exist $x_1^*, \ldots, x_p^* \in X^*$ such that:

$$|y_k^*(x) - x_k^*(x)| \leqslant \frac{\varepsilon}{2p \max_{j \leqslant p} \|v_j\|}, \qquad 1 \leqslant k \leqslant p, \; x \in K.$$

As $K \subseteq U$, it ensues that, for every $x \in K$:

$$\left\| x - \sum_{k=1}^{p} x_k^*(x) v_k \right\| \leqslant \left\| x - \sum_{k=1}^{p} y_k^*(x) v_k \right\| + \sum_{k=1}^{p} |x_k^*(x) - y_k^*(x)| \, \|v_k\|$$

$$\leqslant \frac{\varepsilon}{2} + p \frac{\varepsilon}{2p} = \varepsilon.$$

The finite rank operator $P = \sum_{k=1}^{p} x_k^* \otimes v_k \in \mathcal{L}(X)$ thus approximates the identity map up to $\varepsilon$ on $K$, and consequently $X$ has the approximation property. $\square$

It remains to prove Lemma II.3.

*Proof of Lemma II.3*   Let $(y_n)_{n \geqslant 1}$ be a Cauchy sequence of $(Y, [\,.\,])$:

$$(\forall \varepsilon > 0) \quad (\exists n_0 \geqslant 1) \quad (\forall p, q \geqslant n_0 : [y_p - y_q] \leqslant \varepsilon).$$

As $\|\,.\,\| \leqslant [\,.\,]$, $(y_n)_{n \geqslant 1}$ is also Cauchy in $X$, and there exists $y \in X$ such that $\|y_n - y\| \underset{n \to +\infty}{\longrightarrow} 0$. But $y_p - y_q \in \varepsilon U$ for $p, q \geqslant n_0$; as $\varepsilon U$ is closed in $X$, a passage to the limit leads to $y_p - y \in \varepsilon U$ for $p \geqslant n_0$. In particular, $y \in Y$ and $[y_p - y] \leqslant \varepsilon$, which proves (1) since additionally $\{x_n \,;\, n \geqslant 1\}$ is a total subset of $(Y, [\,.\,])$.

To prove (2), we first observe that $[x_n] \underset{n \to +\infty}{\longrightarrow} 0$ since $[x_n] \leqslant \|x_n\|^{1/2}$. Hence, for every $y^* \in (Y, [\,.\,])^*$, we can find an integer $n_0$ such that:

$$|y^*(x_n)| \leqslant \delta/4, \qquad \forall n > n_0.$$

Now we consider:

$$A = \mathrm{conv}\{\pm x_n \,;\, n > n_0\} \quad \text{and} \quad K_0 = (2/\delta)\overline{A},$$

where $\overline{A}$ is the closure of $A$ for the norm $[\,.\,]$.

We have seen that $[x_n] \underset{n \to +\infty}{\longrightarrow} 0$; hence $\overline{A}$ is compact for the norm $[\,.\,]$. *A fortiori* $\overline{A}$ is compact for the coarser norm $\|\,.\,\|$, and in particular is $\|\,.\,\|$-closed. Therefore $\overline{A}$ is the closure of $A$ both for the norm $[\,.\,]$ and for the norm $\|\,.\,\|$.

We also consider:

$$F = \{x \in \mathrm{span}\,(x_1, \ldots, x_{n_0}) \,;\, y^*(x) = 1\}.$$

If $x \in K_0$, it is approximated by a finite sum $(2/\delta) \sum_{n > n_0} \lambda_n x_n = z$, with $\sum_{n > n_0} |\lambda_n| \leqslant 1$. Thus:

$$|y^*(z)| \leqslant (2/\delta) \sum_{n>n_0} |\lambda_n| \, |y^*(x_n)| \leqslant (2/\delta) \sum_{n>n_0} |\lambda_n| (\delta/4) \leqslant 1/2 \,,$$

and hence $|y^*(x)| \leqslant 1/2$. Therefore $K_0 \cap F = \emptyset$. However, in $(X, \|\,.\,\|)$, $K_0$ is a compact convex set, and $F$ is closed (because $y^*$ is continuous on the finite-dimensional space span $(x_1, \ldots, x_{n_0})$). The Hahn–Banach separation theorem thus provides $\varphi \in X^*$ such that:

$$0 \leqslant \sup_{K_0} \varphi < \inf_{F} \varphi \,.$$

Let us find $\inf_F \varphi$. As $F$ is a level set of a linear functional, it can be written $a + G$, where $G$ is a linear subspace of span $(x_1, \ldots, x_{n_0})$. Since $\inf_F \varphi > 0$, we have $\varphi(a) + \varphi(g) > 0$ for all $g \in G$. However, as $G$ is a vector subspace, this is only possible if $\varphi(g) = 0$. Hence $\inf_F \varphi = \varphi(a) > 0$. Dividing by $\varphi(a)$ if necessary, we thus obtain:

$$\sup_{K_0} \varphi < \inf_{F} \varphi = 1 = \varphi(a) \,.$$

We show that $x^* = \varphi$ is an answer to our problem. For this, we distinguish two cases.

a) *First case: $n \leqslant n_0$.*
   Then $y^*(x_n) = \varphi(x_n)$. In fact:

   – either $y^*(x_n) = 0$, and then $y^*(x_n + a) = 1$, implying $x_n + a \in F$ and hence $\varphi(x_n + a) = 1 = \varphi(x_n) + \varphi(a) = \varphi(x_n) + 1$, so $\varphi(x_n) = 0$;
   – or $y^*(x_n) \neq 0$; then we have $\varphi\big(x_n/y^*(x_n)\big) = 1$ since $x_n/y^*(x_n) \in F$; thus $\varphi(x_n) = y^*(x_n)$.

b) *Second case: $n > n_0$.*
   Then $\pm(2/\delta) x_n \in K_0$, and hence $\pm(2/\delta) \varphi(x_n) \leqslant \sup_{K_0} \varphi < 1$, i.e. $|\varphi(x_n)| < \delta/2$. Therefore:

$$|y^*(x_n) - \varphi(x_n)| \leqslant |y^*(x_n)| + |\varphi(x_n)| \leqslant \delta/2 + \delta/2 = \delta \,.$$

Thus, $|y^*(x_n) - \varphi(x_n)| \leqslant \delta$ for any $n \geqslant 1$. By convexity and continuity, it follows that $|y^*(x) - \varphi(x)| \leqslant \delta$ for every $x \in \overline{\mathrm{conv}}\{x_n \,;\, n \geqslant 1\}$, and *a fortiori* for every $x \in K$, where $\overline{\mathrm{conv}}\{x_n \,;\, n \geqslant 1\}$ is the closure of $\mathrm{conv}\{x_n \,;\, n \geqslant 1\}$ for both norms $[\,.\,]$ and $\|\,.\,\|$, as explained above.

This ends the proof of Lemma II.3. □

Theorem II.1 reduces the problem of approximation in norm of all compact operators $T\colon Y \to X$, for an arbitrary Banach space $Y$, to a problem of approximation of the single operator $\mathrm{Id}\colon X \to X$ on all compact sets. Hence it gets rid of the "arbitrariness" of the space $Y$ and the operator $T$, at the price of a change of topology: the uniform topology on $\mathcal{L}(X)$ is replaced by the topology $\tau$ of compact-convergence on $X$, which is the locally convex topology on $\mathcal{L}(X)$ associated with the family of semi-norms:

$$p_K(T) = \sup_{x \in K} \|Tx\|, \quad K \text{ compact subset of } X, \quad T \in \mathcal{L}(X).$$

To study the approximation property in $X$, a description of the dual of $(\mathcal{L}(X), \tau)$ is useful; this is the object of a second theorem of Grothendieck.

**Theorem II.4** *The topological dual* $(\mathcal{L}(X), \tau)^*$ *consists of linear functionals* $\varphi$ *of the form*:

$$\varphi(T) = \sum_{n=1}^{+\infty} x_n^*(Tx_n),$$

*with* $x_n \in X$, $x_n^* \in X^*$, *and* $\sum_{n=1}^{+\infty} \|x_n^*\| \, \|x_n\| < +\infty$.

*Proof* First let $\varphi$ be as in the theorem. We can assume $\|x_n\| = 1$ for every $n \geqslant 1$. Let $(\lambda_n)_{n \geqslant 1}$ with $\lambda_n > 0$, such that $\lambda_n \xrightarrow[n \to +\infty]{} +\infty$, and $C = \sum_{n=1}^{+\infty} \lambda_n \|x_n^*\| < +\infty$. Finally, let $K = \{x_n/\lambda_n \, ; \, n \geqslant 1\} \cup \{0\}$. The set $K$ is compact, and we have $\varphi(T) = \sum_{n=1}^{+\infty} \lambda_n x_n^* T(x_n/\lambda_n)$, so that $|\varphi(T)| \leqslant \sum_{n=1}^{+\infty} \lambda_n \|x_n^*\| p_K(T) = C p_K(T)$. Consequently, $\varphi \in (\mathcal{L}(X), \tau)^*$.

Conversely, suppose that there exist a constant $C > 0$ and a compact set $K$ such that $|\varphi(T)| \leqslant C p_K(T)$ for any $T \in \mathcal{L}(X)$. By Lemma II.2, we can find $x_n \in X$, norm-convergent to 0, such that $K \subseteq \overline{\text{conv}}\{x_n \, ; \, n \geqslant 1\}$. This leads to:

$$|\varphi(T)| \leqslant C \sup_{n \geqslant 1} \|Tx_n\|, \quad \forall \, T \in \mathcal{L}(X).$$

This information can be interpreted as follows: if $S \colon \mathcal{L}(X) \to c_0(X)$ is the operator defined by $S(T) = (Tx_1, \ldots, Tx_n, \ldots)$, then:

$$|\varphi(T)| \leqslant C \, \|S(T)\|.$$

Hence, a continuous linear functional $\Psi$ can be defined on $S[\mathcal{L}(X)]$ by the formula $\Psi[S(T)] = \varphi(T)$, and then $\|\Psi\| \leqslant C$. The Hahn–Banach theorem allows the extension of $\Psi$ to $\widetilde{\Psi} \in [c_0(X)]^* = \ell_1(X^*)$ with $\|\widetilde{\Psi}\| \leqslant C$. Thus there exists a sequence $(x_n^*)_{n \geqslant 1}$ in $X^*$ such that $\sum_{n=1}^{+\infty} \|x_n^*\| \leqslant C$, for which $\widetilde{\Psi}(u_1, \ldots, u_n, \ldots) = \sum_{n=1}^{+\infty} x_n^*(u_n)$, for every $(u_n)_{n \geqslant 1} \in c_0(X)$. In particular $\varphi(T) = \widetilde{\Psi}[S(T)] = \sum_{n=1}^{+\infty} x_n^*(Tx_n)$, with the condition $\sum_{n=1}^{+\infty} \|x_n^*\| \, \|x_n\| \leqslant C \sup_{n \geqslant 1} \|x_n\|$, which completes the proof. $\square$

Thanks to Theorem II.4, the following theorem characterizes the Banach spaces $X$ that possess the approximation property:

**Theorem II.5**   *Let $X$ be a Banach space. The following assertions are equivalent:*

1) *$X$ has the approximation property.*
2) *Every linear functional $\beta$, which is continuous on $\big(\mathcal{L}(X), \tau\big)$ and such that $\beta(T) = 0$ when $T \in \mathcal{F}(X)$, satisfies $\beta(\mathrm{Id}) = 0$.*
3) *If $\sum_{n=1}^{+\infty} \|x_n^*\|\,\|x_n\| < +\infty$ and $\sum_{n=1}^{+\infty} x_n^*(x)x_n = 0$ for every $x \in X$, then $\sum_{n=1}^{+\infty} x_n^*(x_n) = 0$.*

*Proof*

1) $\Rightarrow$ 2). For $X$ to have $AP$ means that the identity map Id is adherent to $\mathcal{F}(X)$ in $\big(\mathcal{L}(X), \tau\big)$.

2) $\Rightarrow$ 3). If $\beta(T) = \sum_{n=1}^{+\infty} x_n^*(Tx_n)$ and if $\beta \in \big[\mathcal{L}(X), \tau\big]^*$, then, for every rank-1 operator $T = x^* \otimes x$, we have:

$$\beta(T) = \sum_{n=1}^{+\infty} x_n^*\big(x^*(x_n)x\big) = x^* \left( \sum_{n=1}^{+\infty} x_n^*(x)x_n \right) = x^*(0) = 0\,.$$

As every finite rank operator is the sum of rank-1 operators, we have $\beta(T) = 0$ for every $T \in \mathcal{F}(X)$. Consequently, 2) leads to $\sum_{n=1}^{+\infty} x_n^*(x_n) = \beta(\mathrm{Id}) = 0$.

3) $\Rightarrow$ 1). As $\big(\mathcal{L}(X), \tau\big)$ is locally convex, the Hahn–Banach theorem is applicable: $\mathrm{Id} \in \overline{\mathcal{F}(X)}$ if for every continuous functional $\beta$ on $\big(\mathcal{L}(X), \tau\big)$ we have $\beta(\mathrm{Id}) = 0$ as soon as $\beta|_{\mathcal{F}(X)} = 0$. But such a functional is of the form $\beta(T) = \sum_{n=1}^{+\infty} x_n^*(Tx_n)$, and hence $\beta(x^* \otimes x) = x^*\big(\sum_{n=1}^{+\infty} x_n^*(x)x_n\big) = 0$ for all $x \in X$, $x^* \in X^*$. Thus $\sum_{n=1}^{+\infty} x_n^*(x)x_n = 0$ for every $x \in X$, and the hypothesis implies $\sum_{n=1}^{+\infty} x_n^*(x_n) = \beta(\mathrm{Id}) = 0$.   $\square$

**Remark**   The construction of a Banach space $X$ without the approximation property thus boils down to a problem in "hard analysis": how to construct $X$ and a continuous functional $\beta$ on $\big(\mathcal{L}(X), \tau\big)$ such that $\beta(\mathrm{Id}) = 1$ but $\beta(T) = 0$ if $T$ is of finite rank. Alternatively, how to construct $X$ and two sequences, $(x_n)_{n \geqslant 1}$ in $X$ and $(x_n^*)_{n \geqslant 1}$ in $X^*$, with $\sum_{n=1}^{+\infty} \|x_n\|\,\|x_n^*\| < +\infty$ and $\sum_{n=1}^{+\infty} x_n^*(x)x_n = 0$ for every $x \in X$, with nonetheless $\sum_{n=1}^{+\infty} x_n^*(x_n) \neq 0$. Grothendieck went even further and showed the following two facts:

a) To find a space $X$ without $AP$, it suffices to construct a continuous mapping $K: [0, 1]^2 \to \mathbb{C}$ such that $\int_0^1 K(x, t)K(t, y)\,dt = 0$ for any $x, y \in [0, 1]^2$ but $\int_0^1 K(t, t)\,dt \neq 0$.
b) To find a space $X$ without $AP$, it suffices to construct an infinite matrix $A = (a_{ij})_{ij}$ such that $\sum_i \sup_j |a_{ij}| < +\infty$, $A^2 = 0$, but with $\operatorname{tr} A = \sum_i a_{ii} \neq 0$ (the problem of the trace of nuclear operators is obviously lurking underneath).

Enflo [1973], then Davie [1973], each gave a counterexample by presenting a space $X$ and an "exotic" functional $\beta$. Next, Davie [1975] showed how to reformulate his counterexample to obtain the forms a) and b) proposed by Grothendieck.

## III The Counterexamples of Enflo and Davie

In 1972, P. Enflo [1973] constructed a separable and reflexive Banach space $X$ without the approximation property and *a fortiori* without a basis; with Grothendieck's results, this also provided the following negative response to the problem of approximation of compact operators:

**Theorem III.1** *There exist a separable Banach space $Z$ and a compact operator $u: Z \to Z$ that is not the limit in norm of finite rank operators:* $u \in \mathcal{K}(Z)$, $u \notin \mathcal{A}(Z)$.

*Proof* Let $X$ be a separable Banach space without *AP*. By Theorem II.1, we can find a separable Banach space $Y$ and a non-approximable operator $T \in \mathcal{K}(Y,X)$ :

$$\|T - B\| \geqslant \delta \quad \text{for every } B \in \mathcal{F}(Y,X).$$

Let $Z = X \oplus_1 Y$ be the $\ell_1$-direct sum of $X$ and $Y$: it is a separable Banach space. Let $u: Z \to Z$ be the operator defined by $u(x,y) = (Ty, 0)$. Clearly $u \in \mathcal{K}(Z)$. If $v \in \mathcal{L}(Z)$, it has a matrix representation $\left(\begin{smallmatrix} A & B \\ C & D \end{smallmatrix}\right)$, where $A \in \mathcal{L}(X)$, $B \in \mathcal{L}(Y,X)$, $C \in \mathcal{L}(X,Y)$ and $D \in \mathcal{L}(Y)$; so $u = \left(\begin{smallmatrix} 0 & T \\ 0 & 0 \end{smallmatrix}\right)$, or, in other words:

$$v(x,y) = (Ax + By, Cx + Dy).$$

Note that $\|B\| \leqslant \|v\|$; indeed,

$$\|By\| \leqslant \|By\| + \|Dy\| = \|v(0,y)\| \leqslant \|v\| \, \|(0,y)\| = \|v\| \, \|y\|.$$

Now if $v$ is of finite rank, so is $B$, because $By = Pv(0,y)$, with $P: Z \to X$ the canonical projection. As $u - v = \left(\begin{smallmatrix} -A & T-B \\ -C & -D \end{smallmatrix}\right)$, we obtain $\|u - v\| \geqslant \|T - B\| \geqslant \delta$; thus the operator $u$ is not approximable. $\qquad \square$

We now prove the following theorem, stronger than Enflo's counterexample.

**Theorem III.2** (Davie) *For any fixed $p > 2$, $\ell_p$ contains a closed subspace $X$ without AP. Thus there exist Banach spaces which are uniformly convex (in particular reflexive) and uniformly smooth, nonetheless without the approximation property.*

**Remark**   The proof can be adapted to show that $c_0$ possesses a subspace without *AP*. Another more difficult method makes it possible to construct $X$ in $\ell_p$, when $1 \leqslant p < 2$ (Szankowski [1978]; see also LINDENSTRAUSS–TZAFRIRI II, Theorem 1.g.4).

*Proof*   Here is the idea of the proof: $X$ is generated by a direct sum of finite-dimensional subspaces $X_1, \ldots, X_k, \ldots$, wherein the subspaces are "badly positioned" with respect to each other: certain elements of $X_k$ are close to certain elements of $X_{k+1}$, so that, for every $T \in \mathcal{F}(X)$, the normalized trace $\beta^k(T)$ of $T$ on $X_k$ is close to $\beta^{k+1}(T)$, in a way to be specified. Consequently the series $\sum_{k \geqslant 1} \left( \beta^{k+1}(T) - \beta^k(T) \right)$ converges, and hence the sequence $\left( \beta^k(T) \right)_{k \geqslant 1}$ converges to a continuous linear functional $\beta$ on $\left( \mathcal{L}(X), \tau \right)$ that satisfies the hypotheses of Theorem II.5 "backwards": $\beta(T) = 0$ if $T \in \mathcal{F}(X)$, but $\beta(\mathrm{Id}) = 1$.

We will argue as follows:

1) The construction of $\beta^k$ is based on properties of orthogonality of characters, hence on *Harmonic Analysis*.
2) The "smallness" of $\beta^{k+1}(T) - \beta^k(T)$ comes from *Probability* (majoration theorem).

The first ingredient is the following probabilistic lemma:

**Lemma III.3**   *Let $G$ be a finite Abelian group of order $3N$, and $\Gamma$ its dual. Then $\Gamma$ can be partitioned into a block $(\sigma_1, \ldots, \sigma_N)$ of $N$ characters and a block $(\tau_1, \ldots, \tau_{2N})$ of $2N$ characters so that:*

$$\left\| 2 \sum_{j=1}^{N} \sigma_j - \sum_{j=1}^{2N} \tau_j \right\|_\infty \leqslant C \sqrt{N \log(N + 1)}.$$

**Preliminary remark**   The choice of the number 3 will be justified later by the necessity to partition $\Gamma$ in one block of $N$ elements and another of $2N$ elements, so that $|G| = |\Gamma| = 3N$. The characters are then complex-valued (if they were real, they would be of order 2 and the cardinality of $G$ would be of the form $2^n$). Even if the theorems of Section II were proved for real Banach spaces, clearly the proofs still hold for complex Banach spaces.

*Proof of Lemma III.3*   The proof is a variant of the method of selectors (see Chapter 5 of Volume 2). Let $\gamma_1, \ldots, \gamma_{3N}$ be an enumeration of $\Gamma$ and $\theta_1, \ldots, \theta_{3N}$ *i.i.d.* random variables such that:

$$\mathbb{P}(\theta_j = 2) = 1/3 \quad \text{and} \quad \mathbb{P}(\theta_j = -1) = 2/3.$$

In particular, we have $\mathbb{E}(\theta_j) = 0$. Let:

$$X_g = \left| \sum_{j=1}^{3N} \theta_j \, \gamma_j(g) \right|$$

and:

$$M = \sup_{g \in G} X_g = \left\| \sum_{j=1}^{3N} \theta_j \, \gamma_j \right\|_\infty .$$

To estimate $\mathbb{E}(M)$, we use the majoration theorem of Chapter 1 (Volume 1) and a symmetrization. Let $(\varepsilon_j)_{j \geqslant 1}$ be a Rademacher sequence; by independence, and since $\mathbb{E}(\theta_j) = 0$ (see Chapter 1 of Volume 1), for every $\omega' \in \Omega$, we have:

$$\int_\Omega \left\| \sum_{j=1}^{3N} \theta_j(\omega) \, \gamma_j \right\|_\infty \, d\mathbb{P}(\omega) \leqslant 4 \int_\Omega \left\| \sum_{j=1}^{3N} \theta_j(\omega) \varepsilon_j(\omega') \, \gamma_j \right\|_\infty \, d\mathbb{P}(\omega) .$$

Fubini's theorem then gives:

$$\mathbb{E}(M) \leqslant 4 \int_\Omega \left( \int_\Omega \left\| \sum_{j=1}^{3N} \varepsilon_j(\omega') \theta_j(\omega) \, \gamma_j \right\|_\infty \, d\mathbb{P}(\omega') \right) d\mathbb{P}(\omega)$$

$$\leqslant C \sqrt{\log 3N} \int_\Omega \sup_{x \in G} \left| \sum_{j=1}^{3N} |\theta_j(\omega)|^2 |\gamma_j(x)|^2 \right|^{1/2} d\mathbb{P}(\omega) ,$$

by the boundedness theorem

$$\leqslant 2C \sqrt{3N} \sqrt{\log 3N} , \qquad \text{since } |\theta_j| \leqslant 2 .$$

This can also be written:

$$\mathbb{E}(M) \leqslant C' \sqrt{N \log(N+1)} .$$

We can thus find $\omega_0 \in \Omega$ such that $M(\omega_0) \leqslant C' \sqrt{N \log(N+1)}$ and, setting $\theta_j = \theta_j(\omega_0)$, we obtain:

$$\left\| \sum_{j=1}^{3N} \theta_j \, \gamma_j \right\|_\infty \leqslant C' \sqrt{N \log(N+1)} \quad \text{with } \theta_j = -1 \text{ or } 2 .$$

Set:

$$A = \{j \, ; \, \theta_j = 2\}, \qquad B = \{j \, ; \, \theta_j = -1\}$$

and $\alpha = |A|$, $\beta = |B|$. We have:

$$\begin{cases} \alpha + \beta = 3N \\ |2\alpha - \beta| = \left| \sum_{j=1}^{3N} \theta_j \gamma_j(0) \right| \leqslant C'\sqrt{N \log(N+1)}\,; \end{cases}$$

therefore:

$$|3\alpha - 3N| \leqslant C'\sqrt{N \log(N+1)}\,,$$

and hence $\alpha = N + O\left(\sqrt{N \log N}\right)$, $\beta = 2N + O\left(\sqrt{N \log N}\right)$. The sets $A$ and $B$ can thus be corrected, with an "error" $O\left(\sqrt{N \log N}\right)$, in order to have $|A| = N$ and $|B| = 2N$ exactly, and still:

$$\left\| 2 \sum_{\gamma \in A} \gamma - \sum_{\gamma \in B} \gamma \right\|_\infty \leqslant C\sqrt{N \log(N+1)}\,,$$

where $C$ is a numerical constant. It only remains to take for the $\sigma_j$'s an enumeration of $A$, and for the $\tau_j$'s an enumeration of $B$.    $\square$

Now, here is the construction of the subspace $X$ of $\ell_p$: for each integer $k \geqslant 0$, let $G_k$ be an Abelian group of order $3.2^k$; Lemma III.3 provides an enumeration $(\sigma_j^k)_{1 \leqslant j \leqslant 2^k}$, $(\tau_j^k)_{1 \leqslant j \leqslant 2^{k+1}}$ of its dual $\widehat{G}_k$ such that:

$$\left\| 2 \sum_{j=1}^{2^k} \sigma_j^k - \sum_{j=1}^{2^{k+1}} \tau_j^k \right\|_\infty \leqslant C\sqrt{(k+1)2^k}\,.$$

Let $G$ be the disjoint union of the $G_k$'s, considered as blocks of integers placed end-to-end. The space $\ell_p$ can be assimilated to $L^p(G)$ with the counting measure. We define elements $e_j^k$, $k \geqslant 0$, $1 \leqslant j \leqslant 2^k$, of $\ell_p$ by the relations:

$$\text{(1)} \qquad e_j^k(t) = \begin{cases} \tau_j^{k-1}(t) & \text{if } t \in G_{k-1}\,, \ k \geqslant 1 \\ \varepsilon_j^k \sigma_j^k(t) & \text{if } t \in G_k \\ 0 & \text{if } t \notin G_{k-1} \cup G_k\,, \end{cases}$$

where $(\varepsilon_j^k)_{k,j}$ is a sequence of signs $\pm 1$, appearing out of the blue for now, but turning out to be very useful in the end. The formula (1) makes sense because the number of $\tau_j^{k-1}$'s is $2^k$. Finally we set:

$$X_k = \text{span}\{e_j^k\,;\ 1 \leqslant j \leqslant 2^k\} \quad \text{and} \quad X = \overline{\text{span}}\,(X_0, X_1, \ldots) \subseteq \ell_p\,.$$

Note that $e_j^k \in X_k$ and is supported by $G_{k-1} \cup G_k$; also $e_j^{k+1} \in X_{k+1}$ is supported by $G_k \cup G_{k+1}$, and there is a substantial overlap between the supports of $e_j^k$ and $e_j^{k+1}$, corresponding to the idea that certain vectors of $X_k$ and $X_{k+1}$ must be "close".

We now study the existence of a well-controlled family $d_j^k$, $1 \leqslant j \leqslant 2^k$, $k \geqslant 0$, in the dual $X^*$ of $X$, such that $d_j^k(e_i^l) = \delta_{ij}\delta_{kl}$. The sequence $(d_j^k)_{k,j}$ is in a way the bi-orthogonal family of $(e_j^k)_{k,j}$ and its existence would be automatic if $(e_j^k)_{k,j}$ were a basis of $X$; but here we are trying to show not only that $(e_j^k)_{k,j}$ is not a basis of $X$, but that $X$ does not have any basis at all, and indeed does not have the approximation property! The existence of the $d_j^k$'s is provided by the following lemma, taken from Harmonic Analysis:

**Lemma III.4** *With $k \geqslant 0$ and $1 \leqslant j \leqslant 2^k$, set, for $f \in X$:*

(a)
$$d_j^k(f) = \frac{1}{|G_k|} \sum_{t \in G_k} f(t)\, \varepsilon_j^k\, \overline{\sigma_j^k(t)}\,.$$

*Then $d_j^k \in X^*$ and $d_j^k(e_i^l) = \delta_{ij}\delta_{kl}$, where $\delta_{kl}$ is the Kronecker delta. Moreover, with $1 \leqslant j \leqslant 2^{k+1}$:*

(b)
$$d_j^{k+1}(f) = \frac{1}{|G_k|} \sum_{t \in G_k} f(t)\, \overline{\tau_j^k(t)}\,.$$

*Proof* Let $f \in X$. Formally, $f$ has a Fourier series over the $e_j^l$'s:

$$f(t) = \sum_{l,j} d_j^l(f)\, e_j^l(t)\,,$$

for $t \in G_k$, only involving the $l$'s such that $t \in \operatorname{supp} e_j^l = G_{l-1} \cup G_l$, i.e. $l = k$ or $l = k + 1$, and hence:

$$
\begin{aligned}
f(t) &= \sum_j d_j^k(f)\, e_j^k(t) + \sum_j d_j^{k+1}(f)\, e_j^{k+1}(t) \\
&= \sum_j d_j^k(f)\, \varepsilon_j^k\, \sigma_j^k(t) + \sum_j d_j^{k+1}(f)\, \tau_j^k(t)\,, \quad \text{by (1)}\,.
\end{aligned}
$$

The formulas (a) and (b) then appear as the Fourier formulas for $G_k$. However, the existence of a Fourier series for $f$ is not guaranteed at all: thus we must proceed "backwards" to prove the lemma, by distinguishing several cases:

- If $l < k$: then $d_j^k(e_i^l) = 0$, as $\operatorname{supp} e_i^l = G_{l-1} \cup G_l$ is disjoint from $G_k$.
- If $l > k + 1$: *same argument*.
- If $l = k$: then:

$$d_j^k(e_i^l) = \frac{1}{|G_k|} \sum_{t \in G_k} \varepsilon_i^k\, \sigma_i^k(t)\, \varepsilon_j^k\, \overline{\sigma_j^k(t)} = \delta_{ij}\,,$$

since the characters $\sigma_i^k$ and $\sigma_j^k$ are orthogonal when $i \neq j$ and $\varepsilon_i^k \varepsilon_j^k = 1$ when $i = j$.

- If $l = k + 1$: then:

$$d_j^k(e_i^l) = \frac{1}{|G_k|} \sum_{t \in G_k} \tau_i^k(t) \, \varepsilon_j^k \, \overline{\sigma_j^k(t)} = 0,$$

since the characters $\tau_i^k$ and $\sigma_j^k$ are orthogonal.

This proves the bi-orthogonality of the $d_j^k$'s and $e_i^l$'s.

To prove $(b)$, for $k \geqslant 1$ and $1 \leqslant j \leqslant 2^k$, set:

$$\Delta_j^k(f) = \frac{1}{|G_{k-1}|} \sum_{t \in G_{k-1}} f(t) \, \overline{\tau_j^{k-1}(t)}.$$

Then:

$$\Delta_j^k(e_i^l) = \frac{1}{|G_{k-1}|} \sum_{t \in G_{k-1}} e_i^l(t) \, \overline{\tau_j^{k-1}(t)},$$

and hence:

- when $l \neq k$ or $l \neq k - 1$, $\Delta_j^k(e_i^l) = 0$ because $(\operatorname{supp} e_i^l) \cap G_{k-1} = \emptyset$;
- when $l = k - 1$:

$$\Delta_j^k(e_i^l) = \frac{\varepsilon_i^{k-1}}{|G_{k-1}|} \sum_{t \in G_{k-1}} \sigma_i^{k-1}(t) \, \overline{\tau_j^{k-1}(t)} = 0,$$

since $\sigma_i^{k-1} \perp \tau_j^{k-1}$; and
- when $l = k$:

$$\Delta_j^k(e_i^l) = \frac{1}{|G_{k-1}|} \sum_{t \in G_{k-1}} \tau_i^{k-1}(t) \, \overline{\tau_j^{k-1}(t)} = \delta_{ij}.$$

This shows that, for fixed $j$ and $k$ $(k \geqslant 1)$, we have $\Delta_j^k(e_i^l) = d_j^k(e_i^l)$ for all $i, l$. However, by definition, the $e_i^l$'s generate $X$, and clearly $d_j^k, \Delta_j^k \in X^*$. Hence $\Delta_j^k = d_j^k$. This proves $(b)$, and then also Lemma III.4. $\qquad\square$

In particular, for $k$ fixed, $(d_j^k)_{1 \leqslant j \leqslant 2^k}$ is the dual basis of $(e_j^k)_{1 \leqslant j \leqslant 2^k}$, and the normalized trace of an operator $T$ of $X_k$ is $2^{-k} \sum_{j=1}^{2^k} d_j^k(Te_j^k)$. This motivates the definition of the linear functionals $\beta^k$:

$$\beta^k(T) = 2^{-k} \sum_{j=1}^{2^k} d_j^k(Te_j^k), \quad \forall T \in \mathcal{L}(X).$$

Note immediately that the orthogonality relations between the $d_j^k$'s and the $e_j^k$'s lead to:

$$\beta^k(\mathrm{Id}) = 1.$$

To continue, it is convenient to note:

$$N(X) = (\mathcal{L}(X), \tau)^*$$

(beware that, notwithstanding the notation, in general $N(X)$ cannot be identified with the nuclear operators of $X$, but with the projective tensor product of $X$ and $X^*$; the theory of tensor products is not needed here). We equip $N(X)$ with the norm:

$$\|\beta\|_N = \inf \left\{ \sum_{n=1}^{\infty} \|x_n^*\| \|x_n\| \ ; \ \beta(T) = \sum_{n=1}^{\infty} x_n^*(Tx_n) \, , \ \forall T \right\}.$$

Clearly every absolutely convergent series for this norm is convergent; hence $\|\cdot\|_N$ is a Banach norm on $N(X)$. The key point of Theorem III.2 is to establish, for a proper choice of the $\varepsilon_j^k$'s:

$$(2) \qquad \sum_{k=0}^{\infty} \|\beta^{k+1} - \beta^k\|_N < +\infty.$$

We first show that this completes the proof, by noting that:

$$(3) \qquad T \text{ of finite rank} \implies \beta^k(T) \xrightarrow[k \to +\infty]{} 0.$$

Indeed, we can assume $T$ of rank one: $T(f) = \varphi(f)\, u$, where $u \in X$ and $\varphi \in X^*$. As the $e_j^k$'s are by definition total in $X$, we can assume $u = e_i^l$; then:

$$\beta^k(T) = 2^{-k} \sum_{j=1}^{2^k} d_j^k \left( \varphi(e_j^k)\, e_i^l \right) = 2^{-k} \sum_{1}^{2^k} \varphi(e_j^k)\, d_j^k(e_i^l) = 0$$

as soon as $k > l$, by Lemma III.4. With that, (2) and the inequality

$$|\gamma(T)| \leqslant \|\gamma\|_N \|T\|$$

for $T \in \mathcal{L}(X)$ and $\gamma \in N(X)$, we obtain

$$\beta^k \xrightarrow[k \to +\infty]{N} \beta \in N(X),$$

and consequently, thanks to (3), if $T$ is of finite rank:

$$\beta(T) = \lim_{k \to +\infty} \beta^k(T) = 0.$$

Moreover, since

$$\beta(\mathrm{Id}) = \lim_{k \to +\infty} \beta^k(\mathrm{Id}) = \lim_{k \to +\infty} 1 = 1,$$

$\beta$ is hence an "exotic" functional on $(\mathcal{L}(X), \tau)$, and thus $X$ does not have $AP$, according to Theorem II.5.

It remains to prove (2). Note that:

$$(4) \qquad \beta^{k+1}(T) - \beta^k(T) = \frac{1}{2^{k+1}|G_k|} \sum_{t \in G_k} \delta_t(T\varphi_t^k),$$

where $\delta_t \in X^*$ is the evaluation at $t \in G$ and where:

$$\varphi_t^k = \sum_{j=1}^{2^{k+1}} \overline{\tau_j^k(t)}\, e_j^{k+1} - 2 \sum_{j=1}^{2^k} \varepsilon_j^k \overline{\sigma_j^k(t)}\, e_j^k = v_t^k - 2u_t^k,$$

with $u_t^k \in X_k$, $v_t^k \in X_{k+1}$ and with $2u_t^k, v_t^k$ the announced "close" vectors that will make the difference $\beta^{k+1}(T) - \beta^k(T)$ small. In fact, the definition of the $\beta^k$'s and the $d_j^k$'s in Lemma III.4 gives, for $T \in \mathcal{L}(X)$:

$$\beta^k(T) = 2^{-k} \sum_{j=1}^{2^k} \frac{1}{|G_k|} \sum_{t \in G_k} (Te_j^k)(t)\, \varepsilon_j^k\, \overline{\sigma_j^k(t)}$$

$$= \frac{1}{2^k|G_k|} \sum_{t \in G_k} T \left( \sum_{j=1}^{2^k} \varepsilon_j^k\, \overline{\sigma_j^k(t)}\, e_j^k \right)(t)$$

$$= \frac{1}{2^k|G_k|} \sum_{t \in G_k} \delta_t(Tu_t^k),$$

and similarly, using formula $(b)$ of Lemma III.4 instead of $(a)$, we obtain:

$$\beta^{k+1}(T) = \frac{1}{2^{k+1}|G_k|} \sum_{t \in G_k} \delta_t(Tv_t^k),$$

where $u_t^k$ and $v_t^k$ were defined just above.

We form the difference $\beta^{k+1}(T) - \beta^k(T)$ and reduce to the same denominator: here $v_t^k - 2u_t^k$ appears, and hence there appears a block of $2.2^k$ characters $\tau_j^k$ of $G_k$ and another of $2^k$ characters $\sigma_j^k$ of $G_k$, which gives $|G_k| = 3.2^k$ and explains the choice of the number 3 in Lemma III.3 (what a coincidence! a play on numbers...).

It remains to estimate $\|\varphi_t^k\|_p$ for $t \in G_k$ by noting that $\operatorname{supp} \varphi_t^k \subseteq G_{k-1} \cup G_k \cup G_{k+1}$ and that $u_t^k$ and $v_t^k$ overlap $G_k$:

- When $y \in G_{k-1}$:

$$\varphi_t^k(y) = -2 \sum_{j=1}^{2^k} \varepsilon_j^k \, \overline{\sigma_j^k(t)} \, e_j^k(y) = -2 \sum_{j=1}^{2^k} \varepsilon_j^k \, \overline{\sigma_j^k(t)} \, \tau_j^{k-1}(y) .$$

- When $y \in G_k$:

$$\varphi_t^k(y) = \sum_{j=1}^{2^{k+1}} \overline{\tau_j^k(t)} \, \tau_j^k(y) - 2 \sum_{j=1}^{2^k} \varepsilon_j^k \, \overline{\sigma_j^k(t)} \, \varepsilon_j^k \, \sigma_j^k(y)$$

$$= \sum_{j=1}^{2^{k+1}} \tau_j^k(y - t) - 2 \sum_{j=1}^{2^k} \sigma_j^k(y - t)$$

(the group structure here proves itself useful).

- When $y \in G_{k+1}$:

$$\varphi_t^k(y) = \sum_{j=1}^{2^{k+1}} \varepsilon_j^{k+1} \, \overline{\tau_j^k(t)} \, \sigma_j^{k+1}(y) .$$

- When $y \notin G_{k-1} \cup G_k \cup G_{k+1}$, then $\varphi_t^k(y) = 0$.

For $y \in G_k$, the smallness of $\varphi_t^k(y)$ is guaranteed by Lemma III.3; for $y \in G_{k-1} \cup G_{k+1}$, we must rely on the $\varepsilon_j^k$'s, the object of the following lemma:

**Lemma III.5** *For every $k \geqslant 0$ and $1 \leqslant j \leqslant 2^k$, there is a choice of signs $\varepsilon_j^k = \pm 1$ such that:*

$$\sup_{\substack{t \in G_k \\ y \in G_{k-1}}} \left| \sum_{j=1}^{2^k} \varepsilon_j^k \, \overline{\sigma_j^k(t)} \, \tau_j^{k-1}(y) \right| \leqslant C_1 \sqrt{k+1} \, 2^{k/2}$$

*(where $C_1$ is a numerical constant).*

*Proof of Lemma III.5* If $(r_j)_{1 \leqslant j \leqslant 2^k}$ is a Rademacher sequence, the majoration theorem gives:

$$\mathbb{E} \left( \sup_{\substack{t \in G_k \\ y \in G_{k-1}}} \left| \sum_{j=1}^{2^k} r_j \, \overline{\sigma_j^k(t)} \, \tau_j^{k-1}(y) \right| \right) \leqslant C_2 2^{k/2} \sqrt{\log(|G_k| \, |G_{k-1}|)}$$

$$\leqslant C_1 \sqrt{k+1} \, 2^{k/2} ;$$

thus there exists $\omega_k \in [0, 1]$ such that $\varepsilon_j^k = r_j(\omega_k)$ works. $\qquad \square$

Such a choice of the $\varepsilon_j^k$'s is made once and for all. Then:

- When $y \in G_{k-1}$:    $|\varphi_t^k(t)| \leqslant 2C_1\sqrt{k+1}\,2^{k/2}$.
- When $y \in G_k$:    $|\varphi_t^k(t)| \leqslant C\sqrt{k+1}\,2^{k/2}$.
- When $y \in G_{k+1}$:    $|\varphi_t^k(t)| \leqslant C_1\sqrt{k+2}\,2^{k+1/2}$.

We thus obtain:

$$\|\varphi_t^k\|_p^p = \sum_{y \in G_{k-1} \cup G_k \cup G_{k+1}} |\varphi_t^k(y)|^p = O\big(2^k\,k^{p/2}\,2^{kp/2}\big)$$

and:

$$\|\varphi_t^k\|_p = O\left(\sqrt{k}\,2^{k(\frac{1}{p}+\frac{1}{2})}\right).$$

From the definition of $\|\,.\,\|_N$ and from (4) (note that $\|\delta_t\|_{X^*} = 1$), it ensues that:

$$\|\beta^{k+1} - \beta^k\|_N \leqslant \frac{1}{2^{k+1}|G_k|}\sum_{t \in G_k}\|\delta_t\|\,\|\varphi_t^k\|_p$$

$$= O\big(2^{-k}\sqrt{k}2^{k(\frac{1}{p}+\frac{1}{2})}\big) = O\big(\sqrt{k}2^{k(\frac{1}{p}-\frac{1}{2})}\big).$$

Since $p > 2$, this shows (2) and hence completes the proof of Davie's theorem.      □

## IV Comments

1) If $X$ possesses a Schauder basis, or more generally an *FDD*, then there are projections $P_n$ which commute and converge uniformly to the identity map on all compact sets. The space $X$ is said to have the *property* $(\pi)$ if we can approach the identity map by *projections* of finite rank, and it has the *commuting bounded approximation property* ($X$ has $BAP_{\mathrm{comm}}$) if there exists a sequence of commuting operators of finite rank approximating the identity map. Casazza [1989] showed that having an *FDD* is equivalent to having the property $(\pi)$ and $BAP_{\mathrm{comm}}$. However a space can have $BAP_{\mathrm{comm}}$ without having an *FDD* (Read, unpublished work; see Casazza and Kalton [1990] and Casazza [2001]). Johnson [1970] showed that, when $X$ has $\pi_1$ (i.e. the approximating projections can be taken with norm 1), then it has an *FDD*.

If $X$ has $BAP_{\mathrm{comm}}$, then it can be renormed to have $MAP_{\mathrm{comm}}$, i.e. there exists an approximation of the identity map with commuting operators of norm 1 (Johnson [1972]); see Exercise V.8.

Every space that has *MAP* in fact has $MAP_{\mathrm{comm}}$ (Casazza and Kalton [1990]); consequently if $X$ has $BAP_{\mathrm{comm}}$, it can be renormed to have *MAP*, so that the following three open problems are equivalent:

*Question a.* If $X$ has property ($\pi$), does it have an *FDD*?

*Question b.* If $X$ has *BAP*, does it have $BAP_{\text{comm}}$?

*Question c.* If $X$ has *BAP*, can it be renormed to have *MAP*?

(Casazza and Kalton [1990]; see also Li [1991] for a summary presented in French).

Finally, note that knowing if $X$ has property ($\pi$) as soon as it has *AP* is also an open problem.

2) Grothendieck [1956] gave the following dual form of Theorem II.1: $X^*$ has *AP* if and only if $\mathcal{K}(X, Y) = \mathcal{A}(X, Y)$ for every Banach space $Y$.

3) An approximation property, weaker than the *AP*, is the *compact approximation property* (abbreviated *CAP*); $X$ is said to have the *CAP* if the identity map of $X$ is the uniform limit of compact operators on all compact sets. Even with this, it was shown (Szankowski [1976]; see LINDENSTRAUSS-TZAFRIRI II, Section 1.g) after the results of Enflo and Davie that there are separable Banach spaces without *CAP*, and that for $1 \leqslant p < +\infty$ and $p \neq 2$ there are closed subspaces of $\ell_p$ without *CAP* (Szankowski [1978]). Also (Willis [1992]), there are separable Banach spaces with *CAP* but without *AP*.

4) For certain *non-separable* "concrete" spaces, we can conclude in the negative. For example, $X = \mathcal{L}(H)$, where $H$ is an infinite-dimensional Hilbert space, does not have *AP* (Szankowski [1978]).

Similarly, the Calkin algebra $\mathcal{L}(H)/\mathcal{K}(H)$ does not have the *AP* (Godefroy and Saphar [1989]).

For $H^\infty$, the space of bounded analytic functions on the unit disk $\mathbb{D}$, the question remains open.

5) If $X$ is a separable dual, it has *MAP* as soon as it has *AP* (Grothendieck [1956]); this is also the case if $X$ is reflexive.

6) The approximation property does not imply the bounded approximation property, even for separable Banach spaces (Figiel and Johnson [1973]). Separable spaces with *BAP* are exactly those isomorphic to a complemented subspace of a space with a basis (Pełczyński [1971] and Johnson, Rosenthal and Zippin [1971]). There exist separable spaces (and moreover, super-reflexive, with an unconditional *FDD*) with *BAP*, but without a basis (Szarek [1987]). Read (unpublished work) also showed that there are separable Banach spaces without *FDD* but with *BAP* (see Comment 1)).

7) There are separable Banach spaces, all of whose closed subspaces possess bases, but which are not isomorphic to Hilbert spaces: for example, the 2-convexified space $T^2$ of the Tsirelson space $T$ (Johnson [1980]); $T^2$ is nonetheless a weak Hilbert space in Pisier's sense; these spaces are of

type $2 - \varepsilon$ and of cotype $2 + \varepsilon$ for any $\varepsilon > 0$ (see PISIER 2, Chapters 12 and 13; Szankowski [1978]; see also Chapter 3 of Volume 1, Section VI).

8) An exhaustive study of the approximation property and its variants can be found in Casazza [2001].

9) The construction (in more or less detail) of an infinite matrix $A$ that provides a negative response to the problem of approximation is presented in LINDENSTRAUSS-TZAFRIRI I, as are the Grothendieck reductions.

10) Casazza and Kalton [1990] introduced the real *unconditional metric approximation property* (*UMAP*); its complex version was introduced by Godefroy, Kalton and Saphar [1993]. Godefroy and Kalton [1997] showed that every separable Banach space with *UMAP* actually has *UMAP*$_{\text{comm}}$.

Subspaces of $L^1$ with *UMAP* and with unit ball closed in measure have been characterized by Godefroy, Kalton and Li [1995] and [1996].

Translation-invariant subspaces of $\mathcal{C}(\mathbb{T})$ and with *UMAP* have been studied by Li [1996] and those of $L^p(\mathbb{T})$ by Neuwirth [1998].

# V Exercises

**Exercise V.1**  Let $S$ be a normal topological space (Hausdorff space in which two disjoint closed sets can be separated by two disjoint open sets).

1) If $O_1, \ldots, O_q$ is an open covering of $S$, show that we can find a closed covering $F_1, \ldots, F_q$ of $S$ with $F_j \subseteq O_j$, $1 \leqslant j \leqslant q$.

2) By using Urysohn's theorem in 1), show that we can find a continuous partition of unity $\varphi_1, \ldots, \varphi_q$, subordinate to the open covering $O_1, \ldots, O_q$, i.e.: $\varphi_j \in \mathcal{C}(S)$, $\varphi_j \geqslant 0$, $\varphi_j$ is null outside $O_j$, $1 \leqslant j \leqslant q$, and $\sum_{j=1}^{q} \varphi_j = 1$ (this classical property is used in the proof of Proposition I.2).

**Exercise V.2**  Let $S$ be the unit sphere of $H^\infty$ and $(f_n)_{n \geqslant 1}$ a sequence of elements of $S$.

1) Show that there exists a sequence $(z_n)_{n \geqslant 1}$ of $\mathbb{D}$ such that $|z_n| \geqslant 1 - 2^{-n}$ and $|f_n(z_n)| \geqslant 1/2$.

2) If $B \in S$ is the Blaschke product associated with the sequence $(z_n)_{n \geqslant 1}$, show that $\|f_n - B\|_\infty \geqslant 1/2$ for any $n \geqslant 1$.

3) Show that $H^\infty$ is not separable.

**Exercise V.3**

1) Show that $\ell_\infty$ is not separable.

2) By using a Carleson interpolation sequence, show that there exists a continuous linear surjection from $H^\infty$ onto $\ell_\infty$, and recover the result of Exercise V.2.

3) Show nonetheless that $H^\infty$ is not isomorphic to $\ell_\infty$ (use the fact that $H^\infty$ is not complemented in $L^\infty$).

**Exercise V.4**  Let $G$ be a metrizable compact Abelian group. Show that the spaces $L^p(G)$, for $1 \leqslant p \leqslant +\infty$, $\mathcal{C}(G)$, and $\mathcal{M}(G)$ have *MAP*.

**Exercise V.5**  Explain why the theorems of Grothendieck seen in Section II remain valid for complex spaces.

**Exercise V.6**  Show that a complemented subspace of a space with *AP* itself has *AP*.

**Exercise V.7**  Use Davie's method to construct a closed subspace of $c_0$ not having the *AP*.

**Exercise V.8**  Let $(T_n)_{n \geqslant 1}$ be a sequence of commuting operators of finite rank that approximate the identity map.

1) Show that the definition $\|\|x\|\| = \sup_n \|T_n x\|$ leads to an equivalent norm on $X$.

2) Show that, for this new norm, $X$ has $MAP_{\text{comm}}$ (show that

$$S_n = \sum_{k=\frac{1}{2}n(n-1)+1}^{\frac{1}{2}n(n+1)} T_k$$

is an approximating sequence of norm 1).

# 3

# Gaussian Processes

## I Introduction

This chapter evolves around the expectation of the supremum of a Gaussian process: how to find an upper bound for it, thanks to an entropy integral (Dudley's theorem). Also, when the process is indexed by a compact metrizable Abelian group and is stationary, the same expression, up to a constant, gives a lower bound (Fernique's theorem). This is treated in Sections II through V.

Section VI shows how Dudley's theorem, along with other ingredients, combinatorial this time, leads to the following result: in a Banach space, if the means over all choices of signs of a finite sequence of $N$ vectors are proportional to the length $N$ of this sequence, then a subsequence of length $\delta N$, proportional to $N$, can be extracted, equivalent to the basis of $\ell_1^{\delta N}$ (the Elton–Pajor theorem).

## II Gaussian Processes

### II.1 Definitions and Notation

To simplify the presentation, the study is limited to real and centered Gaussian processes.

The extension to the centered complex case is straightforward.

**Definition II.1**   Let $T$ be a set and $(\Omega, \mathcal{A}, \mathbb{P})$ a probability space. A *Gaussian process indexed by* $T$ is a collection of real random variables $X_t \colon \Omega \to \mathbb{R}$ such that $(X_{t_1}, \ldots, X_{t_p})$ is a centered Gaussian vector for any $p \geqslant 1$ and every $t_1, \ldots, t_p \in T$. If $\omega \in \Omega$, the map $t \mapsto X_t(\omega) = X^\omega(t)$ is called the *trajectory* of $\omega$.

A *version of a process* $(X_t)_{t\in T}$ is a process $(Y_t)_{t\in T}$ defined on another probability space $(\Omega', \mathcal{A}', \mathbb{P}')$, indistinguishable from $(X_t)_{t\in T}$ in law, i.e. $(Y_{t_1}, \ldots, Y_{t_p}) \sim (X_{t_1}, \ldots, X_{t_p})$ for every $t_1, \ldots, t_p \in T$.

For the definition of Gaussian vectors, refer to Chapter 5 (Volume 1), Definition II.1. The law of a process is completely determined by its covariance $K(s, t) = \mathbb{E}(X_s X_t)$, which is a kernel of positive type on $T \times T$.

In what follows, we will always assume that $T$ is a *separable* topological space, and that the covariance is *continuous*.

If $T$ is considered as time and $\Omega$ as a space of particles, $X_t(\omega)$ is the position of the particle $\omega$ at the instant $t$. An important and difficult problem is to know if the trajectories $X^\omega$ are almost surely continuous, or almost surely bounded. The continuity of the covariance is not in general sufficient. The study of sufficient or necessary conditions in terms of $K(s, t)$ (which determines the law of the process) is the purpose of the theorems of Dudley and Fernique below (Theorems IV.3 and IV.4). First, we examine an important special case, which falls (up to a linear term) within the scope of random Fourier series.

## II.2 The Marcus–Shepp Theorem

The following application of Slepian's lemma (Chapter 1 of this volume) is a gem:

**Theorem II.2** (The Marcus–Shepp Theorem) *Let $(X_t)_{t\in T}$ and $(Y_t)_{t\in T}$ be two centered Gaussian process (real or complex), indexed by a set $T$, and such that:*

$$\|Y_s - Y_t\|_2 \leqslant \|X_s - X_t\|_2$$

*for every $s, t \in T$. Then:*

(a) *If $(X_t)_{t\in T}$ has a bounded version, then so has $(Y_t)_{t\in T}$.*
(b) *If $T$ is a compact metrizable space and if $(X_t)_{t\in T}$ has a continuous version, then so has $(Y_t)_{t\in T}$.*

*Proof* We only prove (b); the proof of (a) is similar albeit simpler. Without loss of generality, we can assume that $(X_t)_{t\in T}$ has continuous trajectories. Moreover, the complex case can be derived from the real case, as follows: We write:

$$\begin{cases} X_t = A_t + i B_t \\ Y_t = C_t + i D_t, \end{cases}$$

where the variables $A_t, B_t, C_t$ and $D_t$ are real. Let $(B'_t)_{t\in T}$ be a copy of $(B_t)_{t\in T}$, independent of $(A_t)_{t\in T}$, with continuous trajectories. We consider

$$M_t = A_t + B'_t.$$

Then, by hypothesis:

$$\mathbb{E}(C_s - C_t)^2 \leqslant \mathbb{E}|Y_s - Y_t|^2 \leqslant \mathbb{E}|X_s - X_t|^2 = \mathbb{E}(M_s - M_t)^2.$$

According to the real case, $(C_t)_{t \in T}$ has a continuous version. The same method, applied to $(D_t)_{t \in T}$, shows that $(Y_t)_{t \in T}$ possesses a continuous version.

Thus the processes can be assumed real. First we fix $t_0 \in T$ and a dense countable subset $\Delta$ of $T$. From the hypothesis, we deduce the following inequality, for $u > 0$:

(1) $\qquad \mathbb{P}\big(\sup_{t \in \Delta} |Y_t - Y_{t_0}| \geqslant u\big)$
$$\leqslant 4\,\mathbb{P}\big(\sup_{t \in \Delta} |X_t - X_{t_0}| \geqslant u/2\big) + 4\,\phi(u/2\rho),$$

where $\phi$ is the "error function":

$$\phi(x) = \frac{1}{\sqrt{2\pi}} \int_x^{+\infty} e^{-t^2/2}\,dt,$$

and where:

$$\rho = \sup_{t \in T} \|X_t - X_{t_0}\|_2 < +\infty,$$

since $T$ is compact and $t \mapsto X_t$ is continuous from $T$ into $L^2(\Omega, \mathcal{A}, \mathbb{P})$ (recall that Gaussian processes are always assumed to have continuous covariances). To obtain (1), we *correct* $X_t$ and $Y_t$ in order to be able to apply Slepian's lemma. We set:

$$\overline{X}_t = X_t - X_{t_0} \quad \text{and} \quad \overline{Y}_t = Y_t - Y_{t_0}.$$

By definition of $\rho$, we have $\|\overline{X}_t\|_2 \leqslant \rho$, and we can define:

$$f(t) = \big(\rho^2 - \|\overline{X}_t\|_2^2 + \|\overline{Y}_t\|_2^2\big)^{1/2}.$$

Then $0 \leqslant f(t) \leqslant \rho$, thanks to the hypothesis on $X$ and $Y$. Now let $g$ be a Gaussian $\mathcal{N}(0,1)$, independent of $(X_t)_t$ and $(Y_t)_t$, and set:

$$\begin{cases} \widetilde{X}_t = f(t)\,g + \overline{X}_t \\ \widetilde{Y}_t = \rho\,g + \overline{Y}_t. \end{cases}$$

Then:

$$\mathbb{E}(\widetilde{X}_t^2) = f^2(t) + \mathbb{E}(\overline{X}_t^2) = \rho^2 + \mathbb{E}(\overline{Y}_t^2) = \mathbb{E}(\widetilde{Y}_t^2),$$

and

$$\mathbb{E}(\widetilde{X}_s - \widetilde{X}_t)^2 = \big(f(s) - f(t)\big)^2 + \mathbb{E}(X_s - X_t)^2$$
$$\geqslant \mathbb{E}(X_s - X_t)^2 \geqslant \mathbb{E}(Y_s - Y_t)^2 = \mathbb{E}(\widetilde{Y}_s - \widetilde{Y}_t)^2.$$

Hence Slepian's lemma is applicable and gives:

$$\mathbb{P}\big(\sup_{t\in\Delta}\widetilde{Y}_t \geq u\big) \leq \mathbb{P}\big(\sup_{t\in\Delta}\widetilde{X}_t \geq u\big).$$

Since $(g \geq 0)$ and $\big(\sup_{t\in\Delta}\overline{Y}_t \geq u\big)$ imply $\sup_{t\in\Delta}\widetilde{Y}_t \geq u$, we obtain:

$$\frac{1}{2}\,\mathbb{P}\big(\sup_{t\in\Delta}\overline{Y}_t \geq u\big) = \mathbb{P}(g \geq 0)\,\mathbb{P}\big(\sup_{t\in\Delta}\overline{Y}_t \geq u\big)$$

$$= \mathbb{P}\big(g \geq 0,\ \sup_{t\in\Delta}\overline{Y}_t \geq u\big)$$

$$\leq \mathbb{P}\big(\sup_{t\in\Delta}\widetilde{Y}_t \geq u\big) \leq \mathbb{P}\big(\sup_{t\in\Delta}\widetilde{X}_t \geq u\big)$$

$$\leq \mathbb{P}\big(\sup_{t\in\Delta}\overline{X}_t \geq u/2\big) + \mathbb{P}(g \geq u/2\rho),$$

because $\sup_{t\in\Delta}\widetilde{X}_t \geq u$ implies $\sup_{t\in\Delta}\overline{X}_t \geq u/2$ or $\rho\,g \geq u/2$. This can also be written:

$$\mathbb{P}\big(\sup_{t\in\Delta}\overline{Y}_t \geq u\big) \leq 2\mathbb{P}\big(\sup_{t\in\Delta}\overline{X}_t \geq u/2\big) + 2\,\phi(u/2\rho).$$

Applying the same inequality to $(-\overline{Y}_t)_t$, we obtain:

$$\mathbb{P}\big(\sup_{t\in\Delta}|\overline{Y}_t| \geq u\big) \leq 4\,\mathbb{P}\big(\sup_{t\in\Delta}\overline{X}_t \geq u/2\big) + 4\,\phi(u/2\rho),$$

which proves (1).

Next, we will deduce from (1) that $(Y_t)_t$ is almost surely uniformly continuous on $\Delta$; the extension by uniform continuity of $(Y_t)_{t\in\Delta}$ will thus be the sought-after continuous version. For this, we denote by $H$ the Gaussian subspace of $L^2(\Omega, \mathcal{A}, \mathbb{P})$ generated by the $Y_t$, $t \in T$, and let $(g_n)_{n\geq 1}$ be an orthonormal basis of this space; this is a standard Gaussian sequence since, in a Gaussian space, orthogonality is equivalent to independence (we can assume $\dim H = \infty$, since, if $\dim H = N < +\infty$, then $\sum_{n=1}^{N}\langle Y_t, g_n\rangle g_n$ is clearly a continuous version of $(Y_t)_t$). We can also assume that, for each $t$, $Y_t = \sum_{n=1}^{+\infty}\langle Y_t, g_n\rangle g_n$. We have to show that the random variable:

$$Z = \overline{\lim_{\substack{\text{dist}(t,t')\to 0 \\ t,t'\in\Delta}}} |Y_t - Y_{t'}|$$

is null. The variable $Z$ is measurable with respect to the $g_n$'s, and for any $N$, we have:

$$Z = \overline{\lim_{\substack{\text{dist}(t,t')\to 0 \\ t,t'\in\Delta}}} |Y_t^N - Y_{t'}^N|,$$

where:

$$Y_t^N = \sum_{N+1}^{\infty}\langle Y_t, g_n\rangle g_n,$$

as $t \mapsto \langle Y_t, g_n \rangle$ is continuous. Thus $Z$ is an asymptotic variable, and by the zero–one law, it is almost surely constant. This constant is denoted $\delta = \delta(\Delta)$. Similarly, define $\delta(\Delta')$ for $\Delta' \subseteq \Delta$; we localize the inequality (1) thanks to the following remark:

*If $(B_i)_i$ is a finite covering of $\Delta$ by open balls, and if $\tilde{B}$ denotes the ball with the same center as $B$ and of twice the radius, then:*

$$\delta(\Delta) = \max_i \delta(\Delta \cap \tilde{B}_i).$$

Indeed, if $r$ is the smallest radius of these balls, as soon as $t, t' \in \Delta$ and $\mathrm{dist}(t, t') < r$, $t$ and $t'$ belong to the same ball $\tilde{B}_i$.

Then cover $\Delta$ with a finite number of open balls of radius $\varepsilon_1 = 1/2$; one of these balls $B_1$, with center $t_1$, satisfies $\delta(\Delta \cap \tilde{B}_1) = \delta$. Next cover $\tilde{B}_1 \cap \Delta$ by a finite number of open balls of radius $\varepsilon_2 = 2^{-2}$, centered in $\tilde{B}_1 \cap \Delta$; one of these balls $B_2$, with center $t_2$, satisfies $\delta(\Delta \cap \tilde{B}_2) = \delta$ and $d(t_2, t_1) \leqslant 2\varepsilon_1$. We thus construct by induction open balls $B_j$, of radius $\varepsilon_j = 2^{-j}$, with center $t_j$, satisfying $\delta(\Delta \cap \tilde{B}_j) = \delta$ and $d(t_{j+1}, t_j) \leqslant 2\varepsilon_j$. Let $t_0$ be the limit of $t_j$ when $j \to +\infty$. Note that, by the triangle inequality, $\delta(\Delta \cap \tilde{B}_j) \leqslant 2\sup_{\Delta \cap \tilde{B}_j} |Y_t - Y_{t_0}|$. We test the inequality (1) at the point $t_0$, replacing $\Delta$ by $\Delta \cap \tilde{B}_j$ and, correspondingly, $\rho$ by $\rho_j = \sup_{t \in \Delta \cap \tilde{B}_j} \|X_t - X_{t_0}\|_2$. Thus we have $\rho_j \xrightarrow[j \to +\infty]{} 0$.

Suppose $\delta > 0$, and take $u = \delta/2$. The above inequality shows that $\sup_{\Delta \cap \tilde{B}_j} |Y_t - Y_{t_0}| \geqslant \delta/2$. Formula (1) hence gives the inequality:

$$1 \leqslant 4\,\mathbb{P}\Big( \sup_{\Delta \cap \tilde{B}_j} |X_t - X_{t_0}| \geqslant \delta/4 \Big) + 4\,\phi(\delta/4\rho_j).$$

A passage to the limit in this expression and the continuity of $t \mapsto X_t$ lead to $1 \leqslant 4\,\phi(\infty) = 0$. This contradiction shows that $\delta = 0$, and completes the proof. $\qquad\square$

**Remark** If $T$ is a metrizable compact Abelian group, and $(X_t)_t$, $(Y_t)_t$ are *stationary* Gaussian processes, the necessary and sufficient conditions of Dudley–Fernique (Sections IV and V) immediately imply Theorem II.2.

## III  Brownian Motion

### III.1  Introduction

*Brownian motion* is the Gaussian process corresponding to $T = [0, +\infty[$ (equipped with its usual topology) and $K(s, t) = \min(s, t)$. Here is a model (version) of such a process: let $(e_n)_{n \geqslant 0}$ be an orthonormal basis of the real Hilbert space $L^2(\mathbb{R}^+)$, $(g_n)_{n \geqslant 0}$ a standard Gaussian sequence, $G$ the closed

(Gaussian) space it generates in $L^2(\Omega, \mathcal{A}, \mathbb{P})$, and $\phi\colon L^2(\mathbb{R}^+) \to G$ the isometry defined by $\phi\big(\sum_{n=0}^{+\infty} a_n e_n\big) = \sum_{n=0}^{+\infty} a_n g_n$, when $\sum_{n=0}^{+\infty} |a_n|^2 < +\infty$.

Set $X_t = \phi\big(\mathbb{1}_{[0,t]}\big)$. Then, $(X_t)_{t \geqslant 0}$ is a Gaussian process and, since $\phi$ preserves the scalar product, then $K(s, t) = \mathbb{E}(X_s X_t) = \langle \mathbb{1}_{[0,s]}, \mathbb{1}_{[0,t]} \rangle = \min(s, t)$.

Depending on the choice of an orthonormal basis, different versions of Brownian motion are obtained. They are indistinguishable in law; however one version can be "preferred" to another, because, for example, it has almost surely continuous trajectories, or is easier to study.

If the study is limited to $[0, 1]$, the Haar basis is well adapted, as the functions of the $n$-th dyadic generation have disjoint supports. Here we will work with a trigonometric basis: no doubt it is more delicate to handle, but it provides an excellent example of a function in Pisier's space $\mathcal{C}^{as}$ (see Chapter 6 of this volume). First observe that $\|X_s - X_t\|_2 = |s - t|^{1/2}$, and that the square of the distance $d(s, t) = |s - t|^{1/2}$ is of negative type, which implies (Schönberg's theorem, Chapter 5 of Volume 1) that $(\mathbb{R}^+, d)$ can be isometrically embedded in a Hilbert space; here $t \mapsto X_t$ provides an "explicit" embedding into $L^2(\mathbb{R}^+)$.

## III.2 The $\mathcal{C}^{as}$ Model of Brownian Motion

The study is restricted here to $T = [0, 1]$. An orthonormal basis $(e_n)_{n \geqslant 0}$ of $L^2(0, 1)$ is obtained, with $e_0 = \mathbb{1}$, $e_n(x) = \sqrt{2}\cos n\pi x$ for $n \geqslant 1$. With the notation of Section II, we know that $X_t = \phi\big(\mathbb{1}_{[0,t]}\big)$ is a model of Brownian motion on $[0, 1]$ that we can specify. Indeed, $\mathbb{1}_{[0,t]} = \sum_{n=0}^{+\infty} a_n(t)\, e_n$, with $a_0(t) = \langle \mathbb{1}_{[0,t]}, \mathbb{1} \rangle = t$, and $a_n(t) = \langle \mathbb{1}_{[0,t]}, e_n \rangle = \sqrt{2}\,\dfrac{\sin n\pi t}{n\pi}$ for $n \geqslant 1$; hence:

$$X_t = t\,g_0 + \sum_{n=1}^{+\infty} \sqrt{2}\,\frac{\sin n\pi t}{n\pi}\,g_n = t\,g_0 + X_t'.$$

Thus $X_t$ appears as a random trigonometric series (an element of $\mathcal{C}^{as}$, as will be seen) plus a "drift" term $t\,g_0$, which is a continuous function of $t$. Hence everything depends on the trigonometric term $X_t'$, which is almost surely continuous, as is shown by the following theorem:

**Theorem III.1** *For $t \in [0, 1]$, set $X_t' = \sum_{n=1}^{+\infty} \sqrt{2}\,\dfrac{\sin n\pi t}{n\pi}\,g_n$, with $(g_n)_{n \geqslant 1}$ a standard Gaussian sequence. Then:*

1) *$(X_t')_{t \in [0,1]}$ is almost surely continuous on $[0, 1]$;*
2) *more precisely, for any $\alpha < 1/2$, $(X_t')_{t \in [0,1]} \in \mathrm{Lip}_\alpha$ almost surely.*

*Proof*   It suffices to prove 2). Let

$$f(t) = \frac{\pi}{\sqrt{2}} X'_t = \sum_{n=1}^{+\infty} \frac{\sin n\pi t}{n} g_n \quad \text{and} \quad f_k(t) = \sum_{2^{k-1} \leqslant n < 2^k} \frac{\sin n\pi t}{n} g_n \,;$$

thus $f = \sum_{k=1}^{+\infty} f_k$ (for each $t$, the series is almost surely convergent, according to the three-series theorem). Then, by the majoration theorem (Chapter 1 of Volume 1), and by Bernstein's inequality (Chapter 6 of Volume 1):

$$\mathbb{E} \|f_k\|_\infty \leqslant C \left( \sum_{2^{k-1} \leqslant n < 2^k} \frac{1}{n^2} \right)^{1/2} \sqrt{\log 2^k} = O(2^{-k/2}\sqrt{k}).$$

In particular, for $\alpha < 1/2$:

$$\mathbb{E} \left( \sum_{k=1}^{+\infty} 2^{k\alpha} \|f_k\|_\infty \right) = \sum_{k=1}^{+\infty} 2^{k\alpha} \mathbb{E} \|f_k\|_\infty = O \left( \sum_{k=1}^{+\infty} 2^{k\alpha} 2^{-k/2} \sqrt{k} \right) < +\infty \,;$$

hence $\sum_{k=1}^{+\infty} 2^{k\alpha} \|f_k\|_\infty < +\infty$ almost surely, and consequently:

$$(*) \qquad\qquad \|f_k\|_\infty = O(2^{-k\alpha}) \qquad \text{almost surely.}$$

The series $\sum_{k \geqslant 1} f_k$ is hence almost surely convergent in the Banach space $\mathcal{C}([0,1])$, thus so is the series $\sum_{n \geqslant 1} \frac{g_n}{n} \sin n\pi t$, by Theorem III.5 of Chapter 4 (Volume 1). Therefore $f$, and hence $(X'_t)_{t \in [0,1]}$, are almost surely continuous. More precisely, when $|h| \leqslant 1$, from $(*)$ and Bernstein's inequality, it ensues that almost surely, we have, uniformly in $t$:

$$|f(t+h) - f(t)| \leqslant \sum_{k=1}^{l} |f_k(t+h) - f_k(t)| + \sum_{k>l} |f_k(t+h) - f_k(t)|$$

$$= O \left( |h| \sum_{k=1}^{l} 2^k 2^{-k\alpha} \right) + O \left( \sum_{k>l} 2^{-k\alpha} \right)$$

$$= O\big(|h| 2^{l(1-\alpha)} + 2^{-l\alpha}\big) = O\big(|h|^\alpha\big),$$

after adjusting $l$ so that $2^{-l}$ is of the order of magnitude of $|h|$.   $\square$

**Remark**   If $\alpha > 1/2$, a classical result of Bernstein (see KAHANE 2) states that $\text{Lip}_\alpha$ is contained in the Wiener algebra $A(\mathbb{T})$ of absolutely convergent Fourier series. Here, $\sum_{n=1}^{+\infty} \frac{|g_n|}{n} = +\infty$ almost surely, by – for example – the three-series theorem; thus $(X_t)_{t \in [0,1]} \in \text{Lip}_\alpha$ is not possible. Nor is

$(X_t)_{t\in[0,1]} \in \mathrm{Lip}_{1/2}$ almost surely, as the following subtle result (*law of the iterated logarithm for Brownian motion*) shows:

$$\varlimsup_{t \to 0} \frac{|X_t|}{\sqrt{2t \log \log 1/t}} = 1 \qquad a.s.$$

# IV Dudley's Majoration Theorem

## IV.1 The Entropy Integral

A Gaussian process $(X_t)_{t\in T}$ indexed by a separable compact space $T$ can have a continuous covariance without having a continuous – or even bounded – version, as shown in the example where $T$ is the circle $\mathbb{T} = \mathbb{R}/2\pi\mathbb{Z}$ and $X_t = \sum_{n=1}^{+\infty} \frac{1}{n} g_n \sin(2^n t)$. The covariance $K(s,t) = \sum_{n=1}^{+\infty} \frac{\sin(2^n s)\sin(2^n t)}{n^2}$ is continuous, but $(X_t)_{t\in\mathbb{T}}$ possesses no bounded version: indeed, let $(Y_t)_{t\in\mathbb{T}}$ be a version of $(X_t)_{t\in\mathbb{T}}$, and take a dense sequence $(t_j)_{j\geqslant 1}$ in $\mathbb{T}$. For any integer $N \geqslant 1$ and any $A > 0$, setting $X_t^{(N)} = \sum_{n=1}^{N} \frac{g_n}{n} \sin(2^n t)$, we have:

$$\mathbb{P}\big(\sup_j |Y_{t_j}| > A\big) = \mathbb{P}\big(\sup_j |X_{t_j}| > A\big)$$
$$\geqslant \frac{1}{2} \mathbb{P}\big(\sup_j |X_{t_j}^{(N)}| > A\big) = \frac{1}{2} \mathbb{P}(\|X_t^{(N)}\|_\infty > A),$$

the inequality being a consequence of the symmetry of the $g_n$'s. Moreover, $\{2^n\,;\, n \geqslant 1\}$ is a Sidon set of integers (see Chapter 6 of Volume 1); thus, for some numerical constant $c$:

$$\|X_t^{(N)}\|_\infty \geqslant c \sum_{n=1}^{N} \frac{|g_n|}{n}.$$

Hence:

$$\mathbb{P}\big(\sup_j |Y_{t_j}| > A\big) \geqslant \mathbb{P}\big(\textstyle\sum_{n=1}^{N}|g_n|/n > A/c\big),$$

so $\mathbb{P}\big(\sup_j |Y_{t_j}| > A\big) = 1$, since $\sum_{n=1}^{+\infty} |g_n|/n = +\infty$ almost surely, by the three-series theorem. Letting $A$ tend to $+\infty$, we obtain

$$\mathbb{P}\big(\sup_j |Y_{t_j}| = +\infty\big) = 1;$$

hence $(Y_t)_{t\in\mathbb{T}}$ is almost surely unbounded.

Thus, to obtain a continuous version, we must add conditions on the covariance $K$ (which, as already said, completely determines the law of the process). This can be done in terms of the $L^2$-norm of the process, which induces a metric $d$ on $T$:

**Definition IV.1** If $(X_t)_{t \in T}$ is a process in $L^2(\mathbb{P})$, the $L^2$-*metric of the process* is defined by:

$$d(s,t) = \|X_s - X_t\|_2 = \sqrt{K(s,s) + K(t,t) - 2K(s,t)}.$$

The space $(T,d)$ is a semi-metric space and is (quasi-) compact, since its topology is coarser than that on $T$.

**Definition IV.2** Let $(T,d)$ be a compact semi-metric space; the *entropy metric function* $N(\varepsilon) = N(T,d,\varepsilon)$ is the minimum number of open $d$-balls of radius $\varepsilon$ necessary to cover $T$.

The *entropy integral $J(d)$* is defined by:

$$J(d) = \int_0^{+\infty} \sqrt{\log N(\varepsilon)}\, d\varepsilon = \int_0^{D} \sqrt{\log N(\varepsilon)}\, d\varepsilon,$$

where $D < +\infty$ is the diameter of $(T,d)$.

As we will see, the finiteness of this entropy integral implies the existence of a continuous version of $(X_t)_{t \in T}$. This applies to Brownian motion on $[0,a]$, $a > 0$: indeed $d(s,t) = |s - t|^{1/2}$ and, by an easy calculation, $N(\varepsilon) \approx a/\varepsilon^2$; since $D = \sqrt{a}$, we obtain:

$$J(d) \approx \int_0^{D} \sqrt{\log(a/\varepsilon^2)}\, d\varepsilon < +\infty.$$

The result of Theorem III.1 is thus recovered.

Note that this definition of entropy differs slightly from the one given in Chapter 1 of this volume, where we considered closed balls: open balls are better suited to the current situation.

## IV.2 The Dudley Majoration Theorem

Let us return to general Gaussian processes. If $\Psi_2$ is the Orlicz function $e^{x^2} - 1$ and if $X$ is a centered Gaussian variable, we know (Chapter 1 of Volume 1, Corollary IV.4) that $\|X\|_{\Psi_2} = a\,\|X\|_2$, with $a = \sqrt{8/3}$. Hence, with the process metric $d$:

$$\|X_s - X_t\|_{\Psi_2} = a\,\|X_s - X_t\|_2 = a\,d(s,t) = d'(s,t).$$

Evidently $J(d) < +\infty$ if and only if $J(d') < +\infty$. For our purposes, it is convenient to ignore the Gaussian nature of the process and only consider its Lipschitz nature: $\|X_s - X_t\|_{\Psi_2} \leqslant d'(s,t)$. The resulting processes, known as *sub-Gaussian processes* (with respect to the metric $d'$), are the object of the following theorem:

**Theorem IV.3** (The Abstract Dudley Theorem)  *Let $(T, d)$ be a semi-metric compact space, with entropy integral $J(d) < \infty$. Let $X = (X_t)_{t \in T}$ be a process indexed by $T$, with $X_t \in L^{\Psi_2}$ for any $t$, and satisfying the Lipschitz condition:*

$$\|X_s - X_t\|_{\Psi_2} \leqslant d(s, t), \qquad \forall s, t \in T.$$

*Then $(X_t)_{t \in T}$ has a version with continuous trajectories on $(T, d)$. Moreover, with $t_0$ an arbitrary point of $T$:*

$$\mathbb{E}\left( \sup_{t \in T} |X_t| \right) \leqslant b \left( J(d) + \|X_{t_0}\|_2 \right),$$

*where $b$ is a numerical constant.*

*Proof*  The problem is homogeneous: if $X_t$ is changed to $X_t/c$, $d$ is changed to $d/c$ (to retain the Lipschitz condition), $N(\varepsilon)$ to $N(c\,\varepsilon)$, $J(d)$ to $J(d)/c$. We can thus assume that the $d$-diameter $D$ of $T$ is $D = 1/2$. For $j \geqslant 0$, set $N_j = N(2^{-j})$ (hence $N_0 = 1$) and let $R_j$ be a $2^{-j}$-net of $T$: $\bigcup_{a \in R_j} B(a, 2^{-j}) = T$, where $B(a, \varepsilon)$ denotes the open $d$-ball of center $a$ and radius $\varepsilon$, with $|R_j| = N_j$. Let $(\varphi_a^j)_{a \in R_j}$ be a continuous partition of unity, subordinate to the open covering $B(a, 2^{-j})$ for $a \in R_j$. We set:

$$X^j(t) = \sum_{a \in R_j} X_a \, \varphi_a^j(t).$$

Let us show that $X^j$ converges almost surely uniformly to a continuous version of $X$. It is convenient to define:

$$E_j = \{(a, b) \in R_j \times R_{j+1} \, ; \; d(a, b) < 3.2^{-j-1}\}$$

and:

$$M_j = \sup_{(a,b) \in E_j} |X_a - X_b|.$$

For $(a, b) \in R_j \times R_{j+1}$ and $t \in T$, we have

$$|X_a - X_b| \, \varphi_a^j(t) \, \varphi_b^{j+1}(t) \leqslant M_j \, \varphi_a^j(t) \, \varphi_b^{j+1}(t).$$

Indeed, if $\varphi_a^j(t) \, \varphi_b^{j+1}(t) \neq 0$, then $d(t, a) < 2^{-j}$ and $d(t, b) < 2^{-j-1}$; therefore:

$$d(a, b) \leqslant d(a, t) + d(t, b) < 2^{-j} + 2^{-j-1} = 3.2^{-j-1},$$

hence $(a, b) \in E_j$ and $|X_a - X_b| \leqslant M_j$. But then, since:

$$X^j(t) - X^{j+1}(t) = \sum_{(a,b) \in R_j \times R_{j+1}} (X_a - X_b) \, \varphi_a^j(t) \, \varphi_b^{j+1}(t),$$

we have:

$$|X^j(t) - X^{j+1}(t)| \leqslant M_j \sum_{a,b} \varphi_a^j(t)\,\varphi_b^{j+1}(t) = M_j,$$

and hence:

$$\|X^j - X^{j+1}\|_\infty \leqslant M_j.$$

Next, by the Orlicz–Jensen inequality (Chapter 1 of Volume 1, Proposition IV.3):

$$\mathbb{E}\left(\sup_{1\leqslant n\leqslant N} |X_n|\right) \leqslant \Psi_2^{-1}(N) \sup_{1\leqslant n\leqslant N} \|X_n\|_{\Psi_2}$$
$$= \sqrt{\log(N+1)} \sup_{1\leqslant n\leqslant N} \|X_n\|_{\Psi_2}\,;$$

consequently:

$$\mathbb{E}\|X^j - X^{j+1}\|_\infty \leqslant \mathbb{E}M_j \leqslant \sqrt{\log(|E_j|+1)} \sup_{(a,b)\in E_j} \|X_a - X_b\|_{\Psi_2}$$
$$\leqslant \sqrt{\log(|E_j|+1)} \sup_{(a,b)\in E_j} d(a,b)$$
$$\leqslant \frac{3}{2} 2^{-j}\sqrt{\log(|E_j|+1)}.$$

However $|E_j| \leqslant N_j N_{j+1}$ and hence:

$$\mathbb{E}\|X^j - X^{j+1}\|_\infty \leqslant \frac{3}{2} 2^{-j}\sqrt{\log(1+N_j) + \log(1+N_{j+1})}.$$

Summing up these inequalities, we get:

$$\sum_{j=0}^{+\infty} \mathbb{E}\|X^j - X^{j+1}\|_\infty \leqslant \frac{3}{2} \sum_{j=0}^{+\infty} 2^{-j}\sqrt{\log(1+N_j)}$$
$$+ \frac{3}{2} \sum_{j=0}^{+\infty} 2^{-j}\sqrt{\log(1+N_{j+1})}$$
$$\leqslant \left(\frac{3}{2}+3\right) \sum_{j=0}^{+\infty} 2^{-j}\sqrt{\log(1+N_j)}$$
$$\leqslant 5 \sum_{j=0}^{+\infty} 2^{-j}(1+\sqrt{\log N_j})$$
$$\leqslant 10 + 5\sum_{j=0}^{+\infty} 2^{-j}\sqrt{\log N_j}.$$

Now, let us give a minoration for $J(d)$:

$$J(d) \geqslant \int_0^1 \sqrt{\log N(\varepsilon)} \, d\varepsilon = \sum_{j=0}^{+\infty} \int_{2^{-j-1}}^{2^{-j}} \sqrt{\log N(\varepsilon)} \, d\varepsilon$$

$$\geqslant \sum_{j=0}^{+\infty} \int_{2^{-j-1}}^{2^{-j}} \sqrt{\log N_j} \, d\varepsilon = \sum_{j=0}^{+\infty} 2^{-j-1} \sqrt{\log N_j}.$$

Now, since $N(\varepsilon) \geqslant 2$ when $\varepsilon < 1/4$ (if not, we could cover $T$ by a single open ball of radius $\varepsilon < 1/4$, contradicting $D = 1/2$), we have as well:

$$J(d) \geqslant \frac{1}{4} \sqrt{\log 2}.$$

It follows that:

$$\sum_{j=0}^{+\infty} \mathbb{E} \| X^j - X^{j+1} \|_\infty \leqslant 100 \, J(d).$$

The almost sure uniform convergence of the $X^j$'s ensues: indeed, $J(d) < +\infty$ by hypothesis, hence $\mathbb{E}\big( \sum_{j=0}^{+\infty} \| X^j - X^{j+1} \|_\infty \big) < +\infty$; in particular $\sum_{j=0}^{+\infty} \| X^j - X^{j+1} \|_\infty < +\infty$ almost surely, and therefore the series $\sum_{j=0}^{+\infty} (X^j - X^{j+1})$ is almost surely normally convergent (say for $\omega \notin N$, with $\mathbb{P}(N) = 0$). Then we define:

$$\begin{cases} Y_t = \displaystyle\lim_{j \to +\infty} X^j(t) & \text{if } \omega \notin N \\ Y_t = \quad 0 & \text{if } \omega \in N. \end{cases}$$

By construction, $t \mapsto Y_t(\omega)$ is continuous for every $\omega$, and *a fortiori* the trajectories of the process $Y = (Y_t)_{t \in T}$ are almost surely continuous. Let us show that it is a version of $X = (X_t)_{t \in T}$. Fix $t \in T$. We have:

$$X^j(t) - X_t = \sum_{a \in R_j} (X_a - X_t) \, \varphi_a^j(t) ;$$

hence:

$$\| X^j(t) - X_t \|_{\Psi_2} \leqslant \sum_{a \in R_j} \varphi_a^j(t) \| X_a - X_t \|_{\Psi_2} \leqslant \sum_{a \in R_j} \varphi_a^j(t) \, d(a, t) \leqslant 2^{-j}.$$

We can thus find a negligible set $N_t$ and an increasing sequence of integers $j_1 < j_2 < \cdots < j_k < j_{k+1} < \cdots$ such that:

$$X^{j_k}(t) \xrightarrow[k \to +\infty]{} X_t \quad \text{if } \omega \notin N_t.$$

Hence, $X_t(\omega) = Y_t(\omega)$ if $\omega \notin N \cup N_t$. Consequently $Y$ is a continuous version of $X$.

In what follows, by convention, we make no distinction between $Y$ and $X$, and refer to $Y$ as $X$.

To conclude, if $t_0 \in T$, we can take $R_0 = \{t_0\}$. Then $X^0(t) = X_{t_0}$ for any $t$; thus $X_t = X_{t_0} + \sum_{j=0}^{+\infty} \left( X^{j+1}(t) - X^j(t) \right)$, so that:

$$\mathbb{E} \sup_{t \in T} |X_t| \leqslant \mathbb{E}|X_{t_0}| + \sum_{j=0}^{+\infty} \mathbb{E}\|X^{j+1} - X^j\|_\infty$$

$$\leqslant \|X_{t_0}\|_2 + 100 J(d) \leqslant 100\big(\|X_{t_0}\|_2 + J(d)\big). \qquad \square$$

## IV.3  Absence of a Converse to Dudley's Theorem

Dudley's theorem provides a quite fine criterion for a Gaussian process to have continuous trajectories, as it allows gigantic growth for the entropy function $N(\varepsilon)$. But this sufficient condition is not always necessary, as shown by the following "very dispersed" process.

Let $T = \{1/n, n \geqslant 3\} \cup \{0\}$, $X_0 = 0$, and, with $t_n = 1/n$, set:

$$X_{t_n} = \frac{g_n}{\sqrt{\log n} \, \log \log n} = a_n g_n,$$

where $(g_n)_{n \geqslant 3}$ is a standard Gaussian sequence. The process $(X_t)_{t \in T}$ has continuous trajectories because $X_{t_n} \xrightarrow[n \to +\infty]{as} 0$. Indeed, by Proposition II.1 of Chapter 1 of this volume, for any $\theta > 1$, we have $\mathbb{P}\big(|g_n| > \sqrt{2\theta \log n}\big) \leqslant n^{-\theta}$; hence, by the Borel–Cantelli lemma, $\overline{\lim}_{n \to +\infty} \frac{|g_n|}{\sqrt{2\theta \log n}} \leqslant 1$ almost surely. Therefore $\overline{\lim}_{n \to +\infty} \frac{|g_n|}{\sqrt{2 \log n}} \leqslant 1$ almost surely. The Lipschitz condition of Dudley's theorem is satisfied (up to the constant $\sqrt{8/3} \ldots$) with $d(s,t) = \|X_s - X_t\|_2$.

Now let $\varepsilon < 1/e$,

$$n_0 = \left[ \exp \frac{1}{4\varepsilon^2 (\log 1/\varepsilon)^2} \right]$$

and $p, q$ be distinct integers, $p, q \leqslant n_0$. Then:

$$d(t_p, t_q) = \sqrt{a_p^2 + a_q^2} \geqslant a_p \geqslant a_{n_0} = \frac{1}{\sqrt{\log n_0} \, \log \log n_0} > \varepsilon.$$

Indeed:

$$\log n_0 \leqslant \frac{1}{4\varepsilon^2 (\log 1/\varepsilon)^2} < \frac{1}{4\varepsilon^2} < \frac{1}{\varepsilon^2};$$

thus, on one hand,

$$\sqrt{\log n_0} \leqslant \frac{1}{2\varepsilon \log 1/\varepsilon},$$

and, on the other hand,

$$\log \log n_0 < 2 \log 1/\varepsilon;$$

therefore $\sqrt{\log n_0} \log \log n_0 < 1/\varepsilon$. Hence at least $n_0 - 2$ balls of radius $\varepsilon/2$ are necessary to cover $t_3, \ldots, t_{n_0}$; thus $N(\varepsilon/2) \geqslant n_0 - 2$. This leads to:

$$\log N(\varepsilon) \geqslant \frac{c}{\varepsilon^2 (\log 1/\varepsilon)^2}$$

for $\varepsilon > 0$ small enough, $c$ being a numerical constant. Consequently:

$$J(d) = \int_0^{+\infty} \sqrt{\log N(\varepsilon)} \, d\varepsilon = +\infty.$$

We will see that, for a process with low dispersion (more precisely: *stationary*), Dudley's condition becomes necessary and gives a "minoration" of the process.

# V Fernique's Minoration Theorem for Stationary Processes

## V.1 Processes Indexed by a Cantor Tree

In this subsection, we consider Gaussian processes $(X_t)_{t \in G}$, indexed by a compact metrizable Abelian group $G$, and *stationary* (also known as *translation-invariant*):

**Definition V.1**   A Gaussian process $(X_t)_{t \in G}$, indexed by a compact metrizable Abelian group $G$, is said to be *stationary* if, for every $a \in G$, the process $(X_{t+a})_{t \in G}$ has the same distribution as $(X_t)_{t \in G}$.

In particular this implies that the $L^2$-metric of the process is translation-invariant:

$$d(s + a, t + a) = \|X_{s+a} - X_{t+a}\|_2 = \|X_s - X_t\|_2 = d(s, t).$$

This homogeneity of the metric makes it possible to prove a converse of Dudley's theorem. But beforehand, it is useful to bound below an auxiliary process $(Y_t)_{t \in K}$ indexed by a tree $K$; then we compare $(X_t)_{t \in G}$ and $(Y_t)_{t \in K}$ using *Slepian's lemma*.

Let $(A_j)_{j \geqslant 1}$ be a sequence of integers $\geqslant 1$, and $K$ the Cantor tree:

$$K = \prod_{j=1}^{+\infty} \{1, \ldots, A_j\}.$$

It is compact metrizable for the product topology of the discrete topology on each factor. An element $t$ of $K$ is a sequence $t = (\varepsilon_1(t), \ldots, \varepsilon_n(t), \ldots)$ of integers such that $1 \leqslant \varepsilon_j(t) \leqslant A_j$ for any $j \geqslant 1$. For each $j \geqslant 1$ and each integer $n_j \in [1, A_j]$, we select independent Gaussian variables $Y_{n_j}^{(j)}$, with law $\mathcal{N}(0, 4^{-j})$, and a Gaussian $g$ with law $\mathcal{N}(0, 7/8)$, independent of the preceding variables. We define the following Gaussian process, indexed by $K$:

$$Y_t = \sqrt{\frac{3}{8}} \sum_{j=1}^{+\infty} Y_{\varepsilon_j(t)}^{(j)} + g.$$

Let $d = d_Y$ be the $L^2$-metric of the process: $d(s, t) = \|Y_s - Y_t\|_2$.

The following lemma lists two technical properties of $(Y_t)_{t \in K}$, and notably explains the parameters appearing "out of the blue".

**Lemma V.2**  *The process $(Y_t)_{t \in K}$ has the following two properties*:

1) *It "lives on" the unit sphere of $L^2$ : $\|Y_t\|_2 = 1$ for every $t \in K$.*
2) *If $s = (m_j)_{j \geqslant 1}$ and $t = (n_j)_{j \geqslant 1}$ are distinct and if $l$ is the smallest integer such that $m_l \neq n_l$, then $d_Y(s, t) \leqslant 2^{-l}$.*

*Proof*

1) $\mathbb{E}(Y_t^2) = \dfrac{3}{8} \displaystyle\sum_{j=1}^{+\infty} 4^{-j} + \dfrac{7}{8} = \dfrac{1}{8} + \dfrac{7}{8} = 1.$

2) $\mathbb{E}(Y_s - Y_t)^2 = \dfrac{3}{8} \displaystyle\sum_{j=l}^{+\infty} \mathbb{E}\big(Y_{m_j}^{(j)} - Y_{n_j}^{(j)}\big)^2 \leqslant \dfrac{3}{4} \displaystyle\sum_{j=l}^{+\infty} 4^{-j} = 4^{-l}.$ $\qquad\square$

Now let $K_0$ be the countable dense subset of $K$ formed by the sequences $t = (n_j)_{j \geqslant 1}$ such that $n_j = 1$ for $j$ large enough (depending on $t$). Consider the maximal functions $M = \sup_{t \in K_0} Y_t$ and $M^+ = \sup(M, 0) \leqslant +\infty$. A first step in the proof of Fernique's theorem is the following control of $M$ and $M^+$ "from below":

**Proposition V.3**

1) *There exists a numerical constant $C_1 > 0$ such that*:

$$\mathbb{E}M^+ \geqslant C_1 \sum_{j=1}^{+\infty} 2^{-j} \sqrt{\log A_j}.$$

2) *If*

$$\sum_{j=1}^{+\infty} 2^{-j}\sqrt{\log A_j} = +\infty,$$

*then* $\mathbb{P}(M = +\infty) = 1$. *In particular,* $(Y_t)_{t \in K}$ *almost surely has unbounded trajectories.*

*Proof*

1) We define the random index $v_j$ ($1 \leqslant v_j \leqslant A_j$) as the smallest index $n_j$ for which $Y_{n_j}^{(j)}$ equals max $(Y_1^{(j)}, \ldots, Y_{A_j}^{(j)})$. For every fixed $k \geqslant 1$, we have:

$$M^+ \geqslant M \geqslant \sqrt{\frac{3}{8}} \sum_{j=1}^{k} Y_{v_j}^{(j)} + \sqrt{\frac{3}{8}} \sum_{j=k+1}^{+\infty} Y_1^{(j)} + g,$$

since $(v_1, \ldots, v_k, 1, 1, \ldots) \in K_0$. Integrating this inequality (note that $M^+$ is $\geqslant 0$) and using 2) of Proposition II.2 of Chapter 1 of this volume, we obtain:

$$\mathbb{E}M^+ \geqslant \sqrt{\frac{3}{8}} \sum_{j=1}^{k} \mathbb{E}Y_{v_j}^{(j)} \geqslant \sqrt{\frac{3}{8}} C \sum_{j=1}^{k} 2^{-j}\sqrt{\log A_j},$$

hence the result by letting $k$ tend to $+\infty$.

2) Denote by $J$ the set of indices $j \geqslant 1$ for which $A_j \geqslant j$, and let $c = \sqrt{3/8}$ and $\alpha_j = c\, 2^{-j}\sqrt{\log A_j}$. Note that $\sum_{j \in J} \alpha_j = +\infty$ since, by hypothesis, $\sum_{j=1}^{+\infty} \alpha_j = +\infty$ and

$$\sum_{j \notin J} \alpha_j \leqslant c \sum_{j=1}^{+\infty} 2^{-j}\sqrt{\log j} < +\infty.$$

Let $\sum_{j \geqslant 1} \varepsilon_j$ be the convergent series of 1) of Proposition II.2 in Chapter 1 of this volume; the proof of this proposition shows that the sequence $(\varepsilon_j)_{j \geqslant 1}$ is non-increasing. Then:

$$M \geqslant c \sum_{j \leqslant k, j \in J} Y_{v_j}^{(j)} + G,$$

where

$$G = c \sum_{j \leqslant k, j \notin J} Y_1^{(j)} + c \sum_{j > k} Y_1^{(j)} + g$$

is a symmetric Gaussian variable. However, for $j \in J$:

$$\mathbb{P}\big(c\, Y_{v_j}^{(j)} \geqslant \alpha_j\big) = \mathbb{P}\big(\max(g_1, \ldots, g_{A_j}) \geqslant \sqrt{\log A_j}\big)$$
$$\geqslant 1 - \varepsilon_{A_j} \geqslant 1 - \varepsilon_j,$$

since $(g_n)_{n \geqslant 1}$ is a standard Gaussian sequence. Consequently:

$$\mathbb{P}\left(M \geqslant \sum_{j \leqslant k, j \in J} \alpha_j\right) \geqslant \prod_{j \leqslant k, j \in J} \mathbb{P}\big(c\, Y_{v_j}^{(j)} \geqslant \alpha_j\big)\, \mathbb{P}(G \geqslant 0)$$

$$\geqslant \frac{1}{2} \prod_{j=1}^{k}(1 - \varepsilon_j) \geqslant \frac{1}{2} \prod_{j=1}^{+\infty}(1 - \varepsilon_j) = \delta > 0.$$

Letting $k$ tend to $+\infty$, we obtain $\mathbb{P}(M = +\infty) \geqslant \delta$, because $\sum_{j \in J} \alpha_j = +\infty$.

The Kolmogorov zero–one law then leads to $\mathbb{P}(M = +\infty) = 1$. □

Thus, for the process $(Y_t)_{t \in K_0}$, a converse to Dudley's theorem has been obtained, quantitative in 1) and qualitative in 2).

## V.2 Minoration of Stationary Gaussian Processes

We are now prepared to prove the following theorem:

**Theorem V.4** (The Fernique Minoration Theorem) *Let* $(X_t)_{t \in G}$ *be a stationary Gaussian process, indexed by a metrizable compact Abelian group, with continuous covariance. Let* $d = d_X$ *be its* $L^2$-*metric and* $J(d)$ *its entropy integral. Then*:

1) *If* $J(d) < +\infty$, $(X_t)_{t \in G}$ *possesses a continuous version such that*:

$$a\big(J(d) + \|X_0\|_2\big) \leqslant \mathbb{E}\sup_{t \in G} |X_t| \leqslant b\big(J(d) + \|X_0\|_2\big),$$

*where* $a$ *and* $b$ *are numerical constants* $> 0$.
2) *If* $J(d) = +\infty$, $(X_t)_{t \in G}$ *does not possess a bounded version.*
3) *If* $(X_t)_{t \in G}$ *possesses a bounded version, it also possesses a continuous version.*

*Proof* Recall that $(Y_t)_{t \in K_0}$ is a *bounded version* of $(X_t)_{t \in K_0}$ if it has the same distribution as $(X_t)_{t \in K_0}$, i.e. $(Y_{t_1}, \ldots, Y_{t_p}) \sim (X_{t_1}, \ldots, X_{t_p})$ for every choice of $t_1, \ldots, t_p \in K_0$, and if there exists a positive real random variable $Z$ such that, for every $t \in K_0$, $|Y_t(\omega)| \leqslant Z(\omega)$ whenever $\omega \notin N_t$, with $\mathbb{P}(N_t) = 0$ (equivalently: $\sup \mathrm{ess}\, |Y_t| < +\infty$ almost surely).

1) By Dudley's theorem, we already know that $(X_t)_{t \in G}$ admits a continuous version satisfying the right-hand inequality. The difficulty is to show the

inequality on the left. We can assume $\|X_0\|_2 = 1$; then, by stationarity, $\|X_t\|_2 = 1$ for every $t \in G$: the process "lives" within the unit sphere of $L^2$, and has a diameter $\leqslant 2$. Set $B_{-1} = G$ and $B_j = B_G(0, 2^{-j})$ if $j \geqslant 0$ (recall that the balls are open). For any $j \geqslant 1$, we denote $R_j = \{t_n^{(j)} ; 1 \leqslant n \leqslant \mu_j\}$, a maximal system of points of $B_{j-2}$ such that $t_1^{(j)} = 0$, and:

$$d\big(t_n^{(j)}, t_m^{(j)}\big) \geqslant 2^{-j+1}, \quad m \neq n.$$

The maximality of $R_j$ leads to:

(1) $$B_{j-2} \subseteq \bigcup_{s \in R_j} B_G(s, 2^{-j+1})$$

(i.e. we cover $B_{j-2}$ by balls of half its radius centered at points of $R_j \subseteq B_{j-2}$, the first being centered at 0).
For $1 \leqslant n_1 \leqslant \mu_1, \ldots, 1 \leqslant n_j \leqslant \mu_j$, we set:

$$\begin{cases} t(n_1, \ldots, n_j) = t_{n_1}^{(1)} + \cdots + t_{n_j}^{(j)} \\ B(n_1, \ldots, n_j) = B_G\big(t(n_1, \ldots, n_j), 2^{-j+1}\big) \\ X(n_1, \ldots, n_j) = X_{t(n_1, \ldots, n_j)}. \end{cases}$$

First, let us show:

(2) $$G \subseteq \bigcup B(n_1, \ldots, n_j)$$

for any $j \geqslant 1$, and for $1 \leqslant n_1 \leqslant \mu_1, \cdots, 1 \leqslant n_j \leqslant \mu_j$. We proceed by induction. By (1), this is true for $j = 1$. Next, if this is true for $j$, and if $t \in G$, let $t(n_1, \ldots, n_j)$ be such that

$$d\big(t, t(n_1, \ldots, n_j)\big) = d\big(t - t(n_1, \ldots, n_j), 0\big) < 2^{-j+1}.$$

Then $t - t(n_1, \ldots, n_j) \in B_{j-1}$ and there exists $t_{n_{j+1}}^{(j+1)} \in R_{j+1}$ such that:

$$d\big(t - t(n_1, \ldots, n_j), t_{n_{j+1}}^{(j+1)}\big) < 2^{-j};$$

in other words, $t \in B(n_1, \ldots, n_{j+1})$. This proves (2).

This relation leads to a majoration of the entropy number $N(\varepsilon)$: for any $j \geqslant 1$, if $\varepsilon \geqslant 2^{-j+1}$, then:

$$N(\varepsilon) \leqslant \mu_1 \ldots \mu_j.$$

This inequality in turn provides a majoration of the entropy integral as a function of the $\mu_j$'s:

$$J(d) \leqslant 4 \sum_{j=1}^{+\infty} 2^{-j} \sqrt{\log \mu_j}.$$

Indeed,

$$J(d) = \int_0^{+\infty} \sqrt{\log N(\varepsilon)}\, d\varepsilon$$

$$= \int_0^2 \sqrt{\log N(\varepsilon)}\, d\varepsilon = \sum_{k=1}^{+\infty} \int_{2^{1-k}}^{2^{2-k}} \sqrt{\log N(\varepsilon)}\, d\varepsilon$$

$$\leqslant \sum_{k=1}^{+\infty} 2^{1-k}\sqrt{\log \mu_1 \ldots \mu_k} \leqslant \sum_{k=1}^{+\infty} 2^{1-k} \sum_{j=1}^{k} \sqrt{\log \mu_j}$$

$$= \sum_{j=1}^{+\infty} \sqrt{\log \mu_j} \sum_{k=j}^{+\infty} 2^{1-k} = \sum_{j=1}^{+\infty} 2^{2-j}\sqrt{\log \mu_j}.$$

The recipe is now the following: let $Y$ be the Cantor process of Proposition V.3 with $A_j = \mu_j$. From the above, $J(d)$ is dominated by $\sum_{j=1}^{+\infty} 2^{-j}\sqrt{\log \mu_j}$; however, according to Proposition V.3, with $Y^* = \sup_{t \in K} |Y_t|$, this latter expression is itself dominated by $\mathbb{E}Y^*$. If Slepian's lemma were to enable us in turn to dominate $\mathbb{E}Y^*$ by $\mathbb{E}\sup_{t \in G} |X_t|$, the proof of the minoration would be complete. At this stage, we run into a technical difficulty: the points $t(n_1, \ldots, n_j)$ are not "far enough apart" for Slepian's lemma to be applied immediately; hence we must take the indices $j$ four-by-four.

For this, we split the series $\sum_{j=1}^{+\infty} 2^{-j}\sqrt{\log \mu_j}$ into four subseries, according to $j \equiv 0, 1, 2, 3 \,(mod.\,4)$, and let $I$ be an arithmetic progression of common difference 4 for which the sum of the subseries is maximum; then

$$(3) \qquad\qquad J(d) \leqslant 16 \sum_{j \in I} 2^{-j}\sqrt{\log \mu_j}.$$

Now let $D$ be the countable subset of $G$ consisting of the elements $t = \sum_{j=1}^{+\infty} t_{n_j}^{(j)}$, with $1 \leqslant n_j \leqslant \mu_j$, but with $n_j = 1$ for $j$ large enough or for $j \notin I$. Recall that $t_1^{(j)} = 0$, and that consequently these sums are all finite. Now, the points of $D$ are "far enough apart": this means (once $t$ and $(n_1, \ldots, n_j, \cdots)$ have been identified) that:

If $t = (m_j)_{j \geqslant 1} \in D$ and $t' = (n_j)_{j \geqslant 1} \in D$ are different, and if $l$ is the first index such that $m_l \neq n_l$, then $d(t, t') = d_X(t, t') \geqslant 2^{-l}$.

Indeed, we have:

$$t - t' = t_{m_l}^{(l)} - t_{n_l}^{(l)} - \sum_{j > l} \left( t_{m_j}^{(j)} - t_{n_j}^{(j)} \right);$$

hence:

$$d(t, t') = d(t - t', 0) \geqslant d\left(t_{m_l}^{(l)} - t_{n_l}^{(l)}, 0\right) - \sum_{j>l} d\left(t_{m_j}^{(j)} - t_{n_j}^{(j)}, 0\right)$$

$$= d\left(t_{m_l}^{(l)}, t_{n_l}^{(l)}\right) - \sum_{j>l, j\in I} d\left(t_{m_j}^{(j)} - t_{n_j}^{(j)}, 0\right)$$

$$\text{because if } j \notin I, \ t_{m_j}^{(j)} = t_{n_j}^{(j)} = 0$$

$$\geqslant 2^{-l+1} - \sum_{j>l, j\in I} \operatorname{diam} B_{j-2}.$$

Since $m_l \neq n_l$, we have $m_l \neq 1$ or $n_l \neq 1$, so $l \in I$, and consequently:

$$d(t, t') \geqslant 2^{-l+1} - \sum_{j \geqslant l+4} \operatorname{diam} B_{j-2}$$

$$\geqslant 2^{-l+1} - \sum_{j \geqslant l+4} \frac{1}{2^{j-3}} = 2^{1-l} - 2^{-l} = 2^{-l}.$$

We can now use the preceding idea, setting $A_j = 1$ if $j \notin I$, and $A_j = \mu_j$ if $j \in I$. Let $K$ be the Cantor tree $\prod_{j=1}^{+\infty}\{1, \ldots, A_j\}$ and $K_0$ the countable subset of points of $K$ with coordinates eventually 1. Clearly $D$ can be identified with $K_0$ and thus the processes $(X_t)_{t\in K_0}$ and $(Y_t)_{t\in K_0}$ can be compared, $(Y_t)_{t\in K}$ being the Cantor process of Proposition V.3. With this aim, we observe:

$$\begin{cases} \mathbb{E}(X_t^2) = \mathbb{E}(Y_t^2) = 1 \\ \mathbb{E}(Y_s - Y_t)^2 \leqslant \mathbb{E}(X_s - X_t)^2, \ \forall s, t \in K_0. \end{cases}$$

Indeed, by Lemma V.2, when $s = (m_j)_{j\geqslant 1}$ and $t = (n_j)_{j\geqslant 1}$ are distinct and belong to $K_0$, and when $l$ is the first index for which $m_l \neq n_l$, we have:

$$\mathbb{E}(Y_s - Y_t)^2 = d_Y(s, t)^2 \leqslant 4^{-l} \leqslant d_X(s, t)^2 = \mathbb{E}(X_s - X_t)^2.$$

Thus, by Slepian's lemma (Chapter 1 of this volume, Theorem III.5):

(4)     $$\mathbb{P}\left(\sup_{t\in K_0} Y_t > u\right) \leqslant \mathbb{P}\left(\sup_{t\in K_0} X_t > u\right) \leqslant \mathbb{P}\left(\sup_{t\in K_0} |X_t| > u\right)$$

for any $u \geqslant 0$.

We integrate this inequality with respect to $u$, using the majoration (3) of the entropy integral and Proposition V.3.

By setting $M = \sup_{t \in K_0} Y_t$ and $M^+ = \max(M, 0)$, it ensues that:

$$J(d) \leqslant 16 \sum_{j \in I} 2^{-j} \sqrt{\log \mu_j} = 16 \sum_{j=1}^{+\infty} 2^{-j} \sqrt{\log A_j} \leqslant \frac{16}{C_1} \mathbb{E}(M^+)$$

$$= \frac{16}{C_1} \int_0^{+\infty} \mathbb{P}(M > u) \, du \leqslant \frac{16}{C_1} \int_0^{+\infty} \mathbb{P}\left(\sup_{K_0} |X_t| > u\right) du$$

$$= \frac{16}{C_1} \mathbb{E}\left(\sup_{K_0} |X_t|\right) \leqslant \frac{16}{C_1} \mathbb{E}\left(\sup_{G} |X_t|\right).$$

We thus obtain the desired minoration, after noting that:

$$\mathbb{E}\left(\sup_{G} |X_t|\right) \geqslant \|X_0\|_1 = \sqrt{2/\pi} \, \|X_0\|_2.$$

The other two assertions are now easy.

2) Assume $J(d) = +\infty$.

Then, the majoration (3) provides $\sum_{j=1}^{+\infty} 2^{-j} \sqrt{\log A_j} = +\infty$ and Proposition V.3 shows that $P(M = +\infty) = 1$. Let $(X_t')_{t \in G}$ be a version of $(X_t)_{t \in G}$. Then, for $u \geqslant 0$:

$$\mathbb{P}\left(\sup_{t \in K_0} X_t' > u\right) = \mathbb{P}\left(\sup_{t \in K_0} X_t > u\right),$$

since $K_0$ is countable; therefore, by (4):

$$\mathbb{P}\left(\sup_{t \in K_0} X_t' > u\right) \geqslant \mathbb{P}(M > u).$$

Letting $u$ tend to $+\infty$, we obtain:

$$\mathbb{P}\left(\sup_{t \in K_0} X_t' = +\infty\right) \geqslant \mathbb{P}(M = +\infty) = 1.$$

Thus $(X_t')_{t \in G}$ is almost surely unbounded.

3) If $(X_t)_{t \in G}$ possesses a bounded version, then, by 2), $J(d) < +\infty$, and, by Dudley's theorem, $(X_t)_{t \in G}$ possesses a continuous version. $\qquad \square$

## V.3 An Equivalent Form of the Entropy Integral

First recall that a complex standard Gaussian variable $Z$ is a random variable $Z = \dfrac{g_1 + i g_2}{\sqrt{2}}$, where $g_1, g_2$ are real standard independent Gaussian variables (they were introduced in Chapter 5 of Volume 1, Definition II.7). A *complex standard Gaussian sequence* is a sequence i.i.d. $(Z_n)_{n \geqslant 1}$, with $Z_n \sim Z$. It is straightforward to see that (see Proposition II.8 of Chapter 5 in Volume 1):

1) $Z$ is centered and normalized: $\mathbb{E}(Z) = 0$, $\mathbb{E}(|Z|^2) = 1$;
2) $Z$ is rotation-invariant: $e^{i\alpha}Z \sim Z$ for any $\alpha \in \mathbb{R}$;
3) more generally, if $(a_{j,k})_{j,k \leqslant n} \in \mathbb{U}(n)$ is a unitary matrix, and if $Z'_j = \sum_{k=1}^n a_{j,k}Z_k$, then $(Z'_1, \ldots, Z'_n) \sim (Z_1, \ldots, Z_n)$;
4) if $M_n^* = \max\left(|Z_1|, \ldots, |Z_n|\right)$, then:

$$C_1\sqrt{\log(n+1)} \leqslant \mathbb{E}(M_n^*) \leqslant C_2\sqrt{\log(n+1)},$$

where $C_1$ and $C_2$ are numerical constants.

A typical example of a complex stationary Gaussian process is:

$$X_t = \sum_{n \geqslant 1} a_n Z_n \gamma_n(t),$$

where $(a_n)_{n \geqslant 1} \in \ell_2$, $(Z_n)_{n \geqslant 1}$ is a complex standard Gaussian sequence and $(\gamma_n)_{n \geqslant 1}$ a sequence of distinct characters of $G$. Theorem V.4 is applicable with some evident modifications to the complex process $(X_t)_{t \in G}$ (it suffices to modify the constants $a$ and $b$ of the statement). Then $d(s, t) = \psi(s - t)$, where $\psi : G \to \mathbb{R}^+$ is the function:

$$\psi(t) = \left(\sum_{n \geqslant 1} |a_n|^2 |\gamma_n(t) - 1|^2\right)^{1/2}.$$

It is then desirable to have an equivalent form of the entropy integral $J(d)$, more directly related to $\psi$, and hence to the $a_n$'s. For this, we consider the *non-decreasing rearrangement* $\overline{\psi} : [0, 1] \to \mathbb{R}^+$ of $\psi$, defined by:

$$\overline{\psi}(x) = \sup\{y \in [0, 1] \,;\, \mu(y) < x\},$$

where $\mu(y) = m(\{t \in G \,;\, \psi(t) < y\})$ and $m$ is the Haar measure of $G$: $\mu$ is the distribution function of $\psi$; it is non-decreasing and left-continuous. The function $\overline{\psi}$ is a generalized inverse of $\mu$; it is non-decreasing and has the same distribution as $\psi$, i.e. with $\lambda$ the Lebesgue measure on $[0, 1]$:

$$\lambda(\overline{\psi} \in A) = m(\psi \in A)$$

for every Borel set $A$ of $\mathbb{R}^+$. As $\mu(\varepsilon)$ is the Haar measure of open $d$-balls of radius $\varepsilon$, it ensues that:

$$\frac{1}{\mu(\varepsilon)} \leqslant N(\varepsilon) = N(d, \varepsilon) \leqslant \frac{1}{\mu(\varepsilon/2)}.$$

Indeed, if $G$ is covered by $N = N(\varepsilon)$ balls $B_G(t_j, \varepsilon)$, that is:

$$G = \bigcup_{j=1}^N B_G(t_j, \varepsilon),$$

then, taking the Haar measures, we obtain:

$$1 \leqslant \sum_{j=1}^{N} m\big(B_G(t_j, \varepsilon)\big) = N(\varepsilon)\,\mu(\varepsilon).$$

Moreover, let $(s_1, \ldots, s_M)$ be a system of points of maximum cardinality $M$, with mutual distances $\geqslant \varepsilon$; then $N(\varepsilon) \leqslant M$ and the open balls $B_G(s_j, \varepsilon/2)$ are pairwise disjoint; therefore:

$$M\,\mu(\varepsilon/2) = m\left(\bigcup_{j=1}^{M} B_G(s_j, \varepsilon/2)\right) \leqslant 1,$$

and hence $N(\varepsilon) \leqslant M \leqslant 1/\mu(\varepsilon/2)$, which leads to the stated inequalities.

We are now ready to prove the following proposition:

**Proposition V.5**    *Let $a = (a_n)_{n \geqslant 1} \in \ell_2$, and let*

$$X_t = \sum_{n \geqslant 1} a_n Z_n \gamma_n(t)$$

*be a stationary Gaussian process (where $\gamma_n \in \widehat{G}$). Set:*

$$\psi(t) = \left(\sum_{n \geqslant 1} |a_n|^2 |\gamma_n(t) - 1|^2\right)^{1/2}$$

*and:*

$$I(\psi) = \int_0^1 \frac{\overline{\psi}(x)}{x\sqrt{\log(e/x)}}\,dx.$$

*Then:*

$$\alpha\left(I(\psi) + \|a\|_2\right) \leqslant \mathbb{E}\left(\sup_{t \in G} |X_t|\right) \leqslant \beta\left(I(\psi) + \|a\|_2\right),$$

*where $\alpha, \beta > 0$ are numerical constants.*

*Proof*    Two quantities $A, B \geqslant 0$ are said to be *numerically equivalent*, written $A \approx B$, if there exist constants $\alpha, \beta > 0$ such that $\alpha A \leqslant B \leqslant \beta A$. By Fernique's minoration theorem:

$$\mathbb{E}\left(\sup_{t \in G} |X_t|\right) \approx J(d) + \|a\|_2.$$

Moreover, by the above inequality $1/\mu(\varepsilon) \leqslant N(\varepsilon) = N(d,\varepsilon) \leqslant 1/\mu(\varepsilon/2)$, it follows that:

$$J(d) = \int_0^{+\infty} \sqrt{\log N(\varepsilon)} \, d\varepsilon$$

$$\approx \int_0^{+\infty} \sqrt{\log\big(1/\mu(\varepsilon)\big)} \, d\varepsilon = \int_0^1 \sqrt{\log(1/x)} \, d\overline{\psi}(x) = I_0(\psi),$$

after the change of variable $\mu(\varepsilon) = x$, which gives $\overline{\psi}(x) = \varepsilon$. Now set:

$$I_1(\psi) = \int_0^1 \sqrt{\log(e/x)} \, d\overline{\psi}(x).$$

By an integration by parts, we have $I_1(\psi) = \frac{1}{2} I(\psi)$. Since $I_0(\psi) \leqslant I_1(\psi)$, and also since:

$$I_1(\psi) \leqslant \int_0^1 \big(\sqrt{\log e} + \sqrt{\log(1/x)}\big) \, d\overline{\psi}(x) = \overline{\psi}(1^-) + I_0(\psi)$$

$$= \|\psi\|_\infty + I_0(\psi) \leqslant 2\|a\|_2 + I_0(\psi),$$

the three quantities:

$$I_0(\psi) + \|a\|_2, \quad I_1(\psi) + \|a\|_2, \quad I(\psi) + \|a\|_2$$

are numerically equivalent, and the proposition ensues. □

Note that we replaced $\log(1/x)$ by $\log(e/x)$ in order to write

$$I(\psi) = \int_0^1 \overline{\psi}(x) f(x) \, dx$$

with $f$ *non-increasing*; it is easily seen that $f(x) = 1/\big(x\sqrt{\log(e/x)}\big)$ is non-increasing on $(0,1)$, whereas it is obviously not the case for the function $1/\big(x\sqrt{\log(1/x)}\big)$.

**Remark** This reformulation of Fernique's minoration theorem will be useful in Chapter 6 of this volume, in the proof of the Marcus–Pisier equivalence theorem.

# VI  The Elton–Pajor Theorem

This theorem states that if the means (taken over all choices of signs) of vectors $x_1, \ldots, x_N$ of a Banach space are on the order of $N$, then a subsequence can be extracted from these vectors, equivalent to the canonical basis of $\ell_1^{\delta N}$, with $\delta$ a numerical constant. The proof, in the case of real Banach spaces, uses

Dudley's theorem, as well as several combinatorial arguments. The passage to the complex case is then handled as in Rosenthal's $\ell_1$ theorem (Chapter 8 of Volume 1).

## VI.1 Combinatorial Preliminaries

The essential result of this section is Sauer's lemma, whose proof is based on the following notion:

**Definition VI.1**  Let $S$ be a subset of $\{-1, 1\}^N$. For each subset $I$ of $\{1, 2, \ldots, N\}$, the natural projection of $\{-1, 1\}^N$ onto $\{-1, 1\}^I$ is denoted $P_I$. A subset $I$ of $\{1, 2, \ldots, N\}$ is said to be *dense for $S$* if $P_I(S) = \{-1, 1\}^I$. The *combinatorial density* of $S$ is defined to be the largest cardinality of subsets $I$ dense for $S$; it is denoted $\Delta(S)$.

Another way to define it is as follows: if $S$ is a family of subsets of $\{1, 2, \ldots, N\}$, the set of traces on $I$ of elements of $S$ is denoted $I \cap S = \{I \cap \Sigma \; ; \; \Sigma \in S\}$; then a subset $I$ of $\{1, 2, \ldots, N\}$ is said to be *dense for*, or *pulverized by*, $S$ if $\mathcal{P}(I) = I \cap S$. The *combinatorial density* of $S$, denoted $\Delta(S)$, is the largest cardinality of subsets $I$ pulverized by $S$. The set of subsets pulverized by $S$ is denoted $\mathcal{D}(S)$.

The passage from one point of view to the other is materialized by associating the element $(2\mathbb{1}_\Sigma - \mathbb{1})$ of $S$ to each $\Sigma \in S$.

When $S$ is *hereditary* (i.e. when $A \in S$ and $B \subseteq A$ imply $B \in S$), then $\mathcal{D}(S) = S$. Indeed, in this case, clearly every element of $S$ is pulverized by $S$. Conversely, if $I$ is pulverized by $S$, there exists $\Sigma \in S$ such that $I = I \cap \Sigma$, thus $I \subseteq \Sigma$, and hence $I \in S$, since $S$ is hereditary.

**Proposition VI.2**  *Let $S \subseteq \mathcal{P}(\{1, 2, \ldots, N\})$ and let $\mathcal{D}(S)$ be the set of subsets pulverized by $S$. Then:*

$$|\mathcal{D}(S)| \geqslant |S|.$$

In other words, $S$ pulverizes more subsets than its cardinality.

*Proof*  Let $\mathfrak{T}$ be the family of sets of subsets $\mathcal{T} \subseteq \mathcal{P}(\{1, 2, \ldots, N\})$ such that:

(i) $|\mathcal{T}| \geqslant |S|$;
(ii) $\mathcal{D}(\mathcal{T}) \subseteq \mathcal{D}(S)$.

For every $\mathcal{T} \in \mathfrak{T}$, the *size* of $\mathcal{T}$ is denoted:

$$m(\mathcal{T}) = \sum_{A \in \mathcal{T}} |A|.$$

We consider an element $\mathcal{T}_0$ of $\mathfrak{T}$ of minimal size:

$$m(\mathcal{T}_0) = \min\{m(\mathcal{T}) \, ; \, \mathcal{T} \in \mathfrak{T}\}.$$

We next show that $\mathcal{T}_0$ is hereditary. Then, it will ensue that $\mathcal{D}(\mathcal{T}_0) = \mathcal{T}_0$, and hence, using the conditions (i) and (ii):

$$|\mathcal{D}(\mathcal{S})| \geqslant |\mathcal{D}(\mathcal{T}_0)| \geqslant |\mathcal{T}_0| \geqslant |\mathcal{S}|,$$

which is the desired result.

For the proof of the heredity, it suffices to show that if $A \in \mathcal{T}_0$ and $n \in A$, then $A \smallsetminus \{n\} \in \mathcal{T}_0$, and next, to iterate.

To this end, let us define the maps $j_n \colon \mathcal{T}_0 \to \mathcal{P}(\{1, 2, \ldots, N\})$ (for $n = 1, 2, \ldots, N$) by setting, for $A \in \mathcal{T}_0$:

$$\begin{cases} j_n(A) = A \smallsetminus \{n\} & \text{if } n \in A \text{ and } A \smallsetminus \{n\} \notin \mathcal{T}_0 \\ j_n(A) = A & \text{otherwise.} \end{cases}$$

It suffices to show that $j_n(\mathcal{T}_0) \in \mathfrak{T}$, because then $m(\mathcal{T}_0) \leqslant m[j_n(\mathcal{T}_0)]$ (thanks to the minimality of $m(\mathcal{T}_0)$); that is:

$$\sum_{A \in \mathcal{T}_0} |A| \leqslant \sum_{A \in \mathcal{T}_0} |j_n(A)|.$$

However, as $|j_n(A)| \leqslant |A|$, this is only possible if there is equality $j_n(A) = A$. In other words, given the definition of $j_n$, we must have $A \smallsetminus \{n\} \in \mathcal{T}_0$ if $n \in A$, which was our goal.

To show that $j_n(\mathcal{T}_0) \in \mathfrak{T}$, first note that $j_n$ is injective; in fact, note that if $j_n(A) \neq A$, then, in particular, $j_n(A) \notin \mathcal{T}_0$; hence if $j_n(A) = j_n(A')$, with $A, A' \in \mathcal{T}_0$, the only possibilities are $A = A'$ or $A \smallsetminus \{n\} = A' \smallsetminus \{n\}$, with $n \in A \cap A'$; in both cases, we obtain $A = A'$. Thus $|j_n(\mathcal{T}_0)| = |\mathcal{T}_0|$, and hence, since $\mathcal{T}_0 \in \mathfrak{T}$, $|j_n(\mathcal{T}_0)| \geqslant |\mathcal{S}|$, which is the condition (i) for $j_n(\mathcal{T}_0)$ to belong to $\mathfrak{T}$.

For (ii), let $I \in \mathcal{D}[j_n(\mathcal{T}_0)]$; we need to prove that $I \in \mathcal{D}(\mathcal{S})$. For this, it suffices to show $I \in \mathcal{D}(\mathcal{T}_0)$. We have two cases:

*1st case: $n \notin I$*

Then, for every $A \in \mathcal{T}_0$, we have $I \cap A = I \cap (A \smallsetminus \{n\})$; hence, whatever the value of $j_n(A)$, we have $I \cap j_n(A) = I \cap A$. Consequently, as $I$ is pulverized by $j_n(\mathcal{T}_0)$, so it is by $\mathcal{T}_0$, i.e. $I \in \mathcal{D}(\mathcal{T}_0)$.

*2nd case: $n \in I$*

Then let $J \subseteq I$; we must find $A \in \mathcal{T}_0$ such that $J = I \cap A$. Again two subcases are considered:

a) if $n \in J$, then, since $I \in \mathcal{D}[j_n(\mathcal{T}_0)]$, there exists $A \in \mathcal{T}_0$ such that $J = I \cap j_n(A)$; but this requires $n \in j_n(A)$, and hence $j_n(A) = A$, and finally $J = I \cap A$;

b) if $n \in I \setminus J$, let $J' = J \cup \{n\}$; $J' \subseteq I$, and case a) shows that there exists $A' \in \mathcal{T}_0$ such that $J' = I \cap A'$, with, additionally, $j_n(A') = A'$ and $n \in A'$; these latter two conditions require that $A' \setminus \{n\} \in \mathcal{T}_0$; thus indeed $J = I \cap A$, with $A = A' \setminus \{n\} \in \mathcal{T}_0$.

Combining these two cases, we obtain the density of $I$ for $\mathcal{T}_0$.                      □

Here is a consequence, obtained independently (with different terminologies) by Sauer [1972], Shelah [1972] and Vapnik and Červonenkis [1971]:

**Proposition VI.3** (Sauer's Lemma)  *Let $S$ be a subset of $\{-1, 1\}^N$ such that $|S| > \sum_{j<k} \binom{N}{j}$; then there exists a subset $I$ of $\{1, \dots, N\}$, of cardinality $|I| \geqslant k$, such that $P_I(S) = \{-1, 1\}^I$. In other words, if $\Delta$ is the combinatorial density of $S$, then:*

$$|S| \leqslant \sum_{n \leqslant \Delta} \binom{N}{n}.$$

*Proof*  To $S$ we associate the corresponding subset $\mathcal{S}$ in $\mathcal{P}(\{1, \dots, N\})$. By definition of $\Delta$, the number of elements of every subset $I$ dense for $\mathcal{S}$ satisfies $|I| \leqslant \Delta$. Consequently, the set $\mathcal{D}(\mathcal{S})$ of subsets dense for $\mathcal{S}$ has cardinality:

$$|\mathcal{D}(\mathcal{S})| \leqslant \sum_{n \leqslant \Delta} \binom{N}{n},$$

since the latter is the number of subsets of $\{1, \dots, N\}$ having at most $\Delta$ elements.                                                                                □

We also need the following inequalities:

**Proposition VI.4**

(1) *For $1 \leqslant n \leqslant N/2$, we have:*

$$\sum_{j \leqslant n} \binom{N}{j} \leqslant \frac{N^N}{n^n (N-n)^{N-n}} \qquad \text{(Chernov's inequality)}.$$

(2) *For $1 \leqslant n \leqslant N$, we have:*

$$\sum_{0 \leqslant j \leqslant n} \binom{N}{j} \leqslant \left( \frac{Ne}{n} \right)^n.$$

(3) *For $1 \leqslant n \leqslant N$, we have:*

$$\sum_{0 \leqslant j < n} \binom{N}{j} \leqslant N^n.$$

*Proof* Of course we could give an elementary proof of Chernov's inequality; we prefer to consider it within the more general framework of deviation inequalities.

**Lemma VI.5** *Let X be a real random variable such that* $\mathbb{E}(e^{\theta X}) < +\infty$ *for any* $\theta \in \mathbb{R}$. *Set:*

$$H(\theta) = \log \mathbb{E}(e^{\theta X}),$$

*and define, for any* $u \in \mathbb{R}$:

$$H^*(u) = \sup_{\theta \in \mathbb{R}} [\theta u - H(\theta)].$$

*Then:*

$$\begin{cases} \text{if } a \geqslant \mathbb{E}(X) : & \mathbb{P}(X \geqslant a) \leqslant e^{-H^*(a)} \\ \text{if } a \leqslant \mathbb{E}(X) : & \mathbb{P}(X \leqslant a) \leqslant e^{-H^*(a)}. \end{cases}$$

*Proof* The dominated convergence theorem shows that $H$ is indefinitely differentiable on $\mathbb{R}$, and $H'(0) = \mathbb{E}(X)$. Moreover, $H$ is convex because the Cauchy–Schwarz inequality:

$$\mathbb{E}\left[\exp\left(\frac{\theta + \theta'}{2} X\right)\right] \leqslant \left[\mathbb{E}(e^{\theta X}) \mathbb{E}(e^{\theta' X})\right]^{1/2}$$

leads to $H\left(\frac{\theta+\theta'}{2}\right) \leqslant \frac{1}{2}[H(\theta) + H(\theta')]$. Consequently the conjugate function $H^*$ (appearing under various names such as the *Legendre*, *Young* or *Cramér transform*) is positive and vanishes at $u = \mathbb{E}(X)$; moreover:

$$\begin{cases} \text{if } u \geqslant \mathbb{E}(X) : & H^*(u) = \sup_{\theta \geqslant 0} [\theta u - H(\theta)] \\ \text{if } u \leqslant \mathbb{E}(X) : & H^*(u) = \sup_{\theta \leqslant 0} [\theta u - H(\theta)]. \end{cases}$$

If $a \geqslant \mathbb{E}(X)$, then, for any $\theta \geqslant 0$:

$$\mathbb{P}(X \geqslant a) = \mathbb{P}(\theta X \geqslant \theta a) \leqslant e^{-\theta a} \mathbb{E}[e^{\theta X}] = e^{-[\theta a - H(\theta)]},$$

and hence $\mathbb{P}(X \geqslant a) \leqslant e^{-H^*(a)}$. The same reasoning applies to the case $a \leqslant \mathbb{E}(X)$. $\qquad \square$

We now consider a sequence of independent random variables

$$X_1, \ldots, X_N,$$

each having the same distribution as $X$. Then:

$$\mathbb{E}\left[\exp\left(\theta \frac{X_1 + \cdots + X_N}{N}\right)\right] = \exp\left[N H\left(\frac{\theta}{N}\right)\right],$$

and since

$$\sup_{\theta \in \mathbb{R}} \left( \theta u - N H \left( \frac{\theta}{N} \right) \right) = N \sup_{\theta \in \mathbb{R}} \left( \frac{\theta}{N} u - H \left( \frac{\theta}{N} \right) \right) = N H^*(u),$$

we obtain, if $a \leqslant \mathbb{E}(X)$:

$$\mathbb{P} \left( \frac{X_1 + \cdots + X_N}{N} \leqslant a \right) \leqslant e^{-N H^*(a)},$$

and the corresponding inequality if $a \geqslant \mathbb{E}(X)$.

We now apply this to a Bernoulli variable $X$:

$$\mathbb{P}(X = 0) = \mathbb{P}(X = 1) = 1/2.$$

Then:

$$H(\theta) = \log \left( 1 + \frac{1}{2}(e^\theta - 1) \right),$$

and easily, for $u \in \,]0, 1[$:

$$H^*(u) = u \log(2u) + (1 - u) \log 2(1 - u).$$

Consequently, if $n \leqslant N/2$, by taking $a = n/N \leqslant 1/2 = \mathbb{E}(X)$, we obtain:

$$\frac{1}{2^N} \sum_{j \leqslant n} \binom{N}{j} = \mathbb{P} \left( \frac{X_1 + \cdots + X_N}{N} \leqslant a \right)$$

$$\leqslant \left( \frac{1}{(2a)^a} \frac{1}{[2(1-a)]^{1-a}} \right)^N = \frac{1}{2^N} \frac{N^N}{n^n \, (N - n)^{N-n}},$$

which is the inequality (1) of Proposition VI.4.

We prove the other two inequalities by induction.

(2) First note that for $n = N$ the inequality holds, since

$$\sum_{j=0}^{N} \binom{N}{j} = 2^N \leqslant e^N.$$

In particular, (2) is true for $N = 1$. Note that it also holds, for any $N \geqslant 1$, when $n = 1$, as this reduces to $1 + N \leqslant Ne$. Suppose that it holds for $N$, and show that it holds for $N + 1$. By the above, we need only to consider $2 \leqslant n \leqslant N$.

Then:

$$\sum_{j=0}^{n}\binom{N+1}{j} = \sum_{j=0}^{n}\binom{N}{j} + \sum_{j=1}^{n}\binom{N}{j-1} = \sum_{j=0}^{n}\binom{N}{j} + \sum_{j=0}^{n-1}\binom{N}{j}$$

$$\leqslant \left(\frac{Ne}{n}\right)^{n} + \left(\frac{Ne}{n-1}\right)^{n-1}$$

$$= \frac{(Ne)^{n-1}}{n^{n}}\left(Ne + n\left(1 + \frac{1}{n-1}\right)^{n-1}\right)$$

$$\leqslant \frac{(Ne)^{n-1}}{n^{n}}(Ne + ne) = \left(\frac{e}{n}\right)^{n}(N^{n} + nN^{n-1})$$

$$\leqslant \left(\frac{(N+1)\,e}{n}\right)^{n}.$$

(3) For $N = 1$, this is trivial. We pass from $N$ to $N+1$, supposing $n \geqslant 2$, since, for $n = 1$, the inequality is evident:

$$\sum_{j=0}^{n-1}\binom{N+1}{j} = \sum_{j=0}^{n-1}\binom{N}{j} + \sum_{j=1}^{n-1}\binom{N}{j-1} = \sum_{j=0}^{n-1}\binom{N}{j} + \sum_{j=0}^{n-2}\binom{N}{j}$$

$$\leqslant N^{n} + N^{n-1} = N^{n-1}(N+1) \leqslant (N+1)^{n}. \qquad \square$$

## VI.2  Volume Inequalities

The following property, about the concavity behavior of volume, is fundamental:

**Theorem VI.6** (The Brunn–Minkowski Theorem)    *Let A and B be two non-empty compact subsets of $\mathbb{R}^{N}$. Then, for $0 \leqslant \theta \leqslant 1$:*

$$\mathrm{Vol}\,\big(\theta A + (1-\theta)B\big) \geqslant (\mathrm{Vol}\,A)^{\theta}\,(\mathrm{Vol}\,B)^{1-\theta},$$

*and:*

$$\big[\,\mathrm{Vol}(A+B)\big]^{1/N} \geqslant (\mathrm{Vol}\,A)^{1/N} + (\mathrm{Vol}\,B)^{1/N}$$

(Brunn–Minkowski inequality).

We begin with a lemma:

**Lemma VI.7** (The Prékopa–Leindler Inequality)
*Let $f, g, \varphi \colon \mathbb{R}^{N} \to [0, +\infty]$ be measurable functions, and let $0 < \theta < 1$. Assume that, for any $s, t \in \mathbb{R}^{N}$:*

$$\varphi\big(\theta s + (1-\theta)t\big) \geqslant f(s)^{\theta}\,g(t)^{1-\theta}.$$

*Then:*

$$\int_{\mathbb{R}^N} \varphi(t)\, dt \geq \left( \int_{\mathbb{R}^N} f(t)\, dt \right)^\theta \left( \int_{\mathbb{R}^N} g(t)\, dt \right)^{1-\theta}.$$

*Proof* First consider the case $N = 1$. We can, by approximation, assume $f$ and $g$ bounded, and, by homogeneity, that $\|f\|_\infty = \|g\|_\infty = 1$.

First we need to prove the 1-dimensional Brunn–Minkowski inequality:

$$\lambda\big(\theta A + (1-\theta)B\big) \geq \theta\, \lambda(A) + (1-\theta)\, \lambda(B),$$

with $A$, $B$ non-empty compact subsets of $\mathbb{R}$, and $\lambda$ the Lebesgue measure. After translating $A$ and $B$ if necessary, we can assume that $\sup A = \inf B = 0$. Then $\theta A + (1-\theta)B$ contains the two sets $\theta A$ and $(1-\theta)B$, disjoint in measure, which leads to the inequality.

Back to our functions: for $0 \leq a < 1$, we have:

$$\{\varphi \geq a\} \supseteq \theta\{f \geq a\} + (1-\theta)\{g \geq a\};$$

hence, since the sets on both sides are non-empty, the 1-dimensional Brunn–Minkowski inequality gives:

$$\lambda(\{\varphi \geq a\}) \geq \theta\lambda(\{f \geq a\}) + (1-\theta)\lambda(\{g \geq a\}).$$

An integration provides:

$$\int_{\mathbb{R}} \varphi(t)\, dt = \int_0^{+\infty} \lambda(\{\varphi \geq a\})\, da \geq \int_0^1 \lambda(\{\varphi \geq a\})\, da$$

$$\geq \theta \int_{\mathbb{R}} f(t)\, dt + (1-\theta) \int_{\mathbb{R}} g(t)\, dt$$

$$\geq \left( \int_{\mathbb{R}} f(t)\, dt \right)^\theta \left( \int_{\mathbb{R}} g(t)\, dt \right)^{1-\theta}$$

by the arithmetic–geometric mean inequality; thus the case $N = 1$ is proved.

We then reason by induction on $N$. Let $N > 1$ and suppose that the lemma is true for $N - 1$. Given $\varphi, f, g \colon \mathbb{R}^N \to [0, +\infty]$, fix $s \in \mathbb{R}$ and define $\varphi_s, f_s, g_s \colon \mathbb{R}^{N-1} \to [0, +\infty]$ by $\varphi_s(t) = \varphi(s, t)$, $f_s(t) = f(s, t)$, $g_s(t) = g(s, t)$. For $s = \theta s_1 + (1-\theta)s_2$, with $s_1, s_2 \in \mathbb{R}$, we have:

$$\varphi_s\big(\theta t + (1-\theta)u\big) \geq f_{s_1}(t)^\theta\, g_{s_2}(u)^{1-\theta}$$

for any $t, u \in \mathbb{R}^{N-1}$. The induction hypothesis leads to:

$$\int_{\mathbb{R}^{N-1}} \varphi_s(t)\, dt \geq \left( \int_{\mathbb{R}^{N-1}} f_{s_1}(t)\, dt \right)^\theta \left( \int_{\mathbb{R}^{N-1}} g_{s_2}(t)\, dt \right)^{1-\theta}.$$

Applying the case $N = 1$ to these new functions, we obtain:

$$\int_{\mathbb{R}^N} \varphi(t)\, dt = \int_{\mathbb{R}} \left( \int_{\mathbb{R}^{N-1}} \varphi_s(t)\, dt \right) \geqslant \left( \int_{\mathbb{R}^N} f(t)\, dt \right)^\theta \left( \int_{\mathbb{R}^N} g(t)\, dt \right)^{1-\theta},$$

which completes the induction and the proof of the lemma. $\qquad\square$

*Proof of Theorem VI.6* The first inequality follows immediately from the lemma, when we take $\varphi = \mathbb{1}_{\theta A + (1-\theta)B}, f = \mathbb{1}_A$ and $g = \mathbb{1}_B$. Next, with

$$\theta = \frac{(\operatorname{Vol} A)^{1/N}}{(\operatorname{Vol} A)^{1/N} + (\operatorname{Vol} B)^{1/N}},$$

we replace in this inequality $A$ by $A' = \dfrac{1}{(\operatorname{Vol} A)^{1/N}} A$ and $B$ by $B' = \dfrac{1}{(\operatorname{Vol} B)^{1/N}} B$; we obtain:

$$\operatorname{Vol}\left(\theta A' + (1 - \theta)B'\right) \geqslant 1.$$

By homogeneity, this leads to the Brunn–Minkowski inequality, since:

$$\theta A' + (1 - \theta)B' = \frac{1}{(\operatorname{Vol} A)^{1/N} + (\operatorname{Vol} B)^{1/N}} (A + B). \qquad\square$$

**Corollary VI.8** (Urysohn's Inequality) *Denote by $\mathbb{S}^{N-1}$ the unit Euclidean sphere of $\mathbb{R}^N$, and by $\sigma$ its normalized surface measure. Let $K$ be a compact subset of $R^N$, and, for $x \in \mathbb{R}^N$, let:*

$$\|x\|_{K^\circ} = \sup_{y \in K} \langle x, y \rangle.$$

*Then:*

$$\left( \frac{\operatorname{Vol} K}{\operatorname{Vol} B_2^N} \right)^{1/N} \leqslant \int_{\mathbb{S}^{N-1}} \|x\|_{K^\circ}\, d\sigma(x).$$

*Proof* By iterating, we can generalize the Brunn–Minkowski inequality to a finite number of non-empty compact sets $A_1, \ldots, A_k$; for any numbers $m_1, \ldots, m_k \geqslant 0$, we have:

$$\sum_{j=1}^k m_j (\operatorname{Vol} A_j)^{1/N} \leqslant \left[ \operatorname{Vol} \left( \sum_{j=1}^k m_j A_j \right) \right]^{1/N}.$$

Therefore, if $A_t \subseteq \mathbb{R}^N$ depends measurably on a parameter $t$ running through a measure space $(T, \mathcal{T}, \nu)$, then:

$$(1) \qquad \int_T (\operatorname{Vol} A_t)^{1/N}\, d\nu(t) \leqslant \left[ \operatorname{Vol} \left( \int_T A_t\, d\nu(t) \right) \right]^{1/N}.$$

For $(T, \mathcal{T}, \nu)$, we take the orthogonal group $\mathbb{O}(N)$, equipped with its Haar measure, and let $A_t = t(K)$ for every $t \in \mathbb{O}(N)$. Then $\mathrm{Vol}\, A_t = \mathrm{Vol}\, K$ for every $t$. Moreover, as $\int_{\mathbb{O}(N)} t(K)\, d\nu(t)$ is invariant under $\mathbb{O}(N)$, it is a multiple of the Euclidean unit ball $B_2^N$:

(2)
$$\int_{\mathbb{O}(N)} t(K)\, d\nu(t) = \alpha\, B_2^N.$$

From (1), it ensues that:

$$(\mathrm{Vol}\, K)^{1/N} \leqslant \alpha\, (\mathrm{Vol}\, B_2^N)^{1/N}.$$

Now we fix an element $\xi \in \mathbb{S}^{N-1}$; the equality (2) gives:

$$\int_{\mathbb{O}(N)} \xi[t(K)]\, d\nu(t) = \alpha\, [-1, 1].$$

As $\xi[t(K)] = [a_t, b_t]$, with $a_t = \inf_{x \in K} \xi[t(x)]$ and $b_t = \sup_{x \in K} \xi[t(x)]$, this implies:

$$\alpha = \int_{\mathbb{O}(N)} b_t\, d\nu(t) = \int_{\mathbb{O}(N)} \sup_{x \in K} \xi[t(x)]\, d\nu(t) = \int_{\mathbb{O}(N)} \|t^* \xi\|_{K^\circ}\, d\nu(t),$$

which completes the proof, after an integration over $\xi \in \mathbb{S}^{N-1}$:

$$\alpha = \int_{\mathbb{S}^{N-1}} \|x\|_{K^\circ}\, d\sigma(x). \qquad \square$$

**Remark VI.9** Let $\gamma_N$ be the canonical Gaussian probability on $\mathbb{R}^N$, i.e. the tensor product of laws of independent standard Gaussian variables $g_1, \ldots, g_N$ on $\mathbb{R}$. An integration in polar coordinates leads to:

$$\int_{\mathbb{R}^N} \|x\|_{K^\circ}\, d\gamma_N(x) = c_N \int_{\mathbb{S}^{N-1}} \|x\|_{K^\circ}\, d\sigma(x),$$

with:

$$c_N = \int_{\mathbb{R}^N} \left( \sum_{n=1}^{N} x_n^2 \right)^{1/2} d\gamma_N(x) = \mathbb{E}\left[ \left( \sum_{n=1}^{N} g_n^2 \right)^{1/2} \right].$$

To prove Proposition VI.11, we use the minoration:

$$c_N \geqslant \frac{1}{\sqrt{N}}\, \mathbb{E} \sum_{n=1}^{N} |g_n| = \sqrt{\frac{2}{\pi}}\, \sqrt{N}.$$

Later on, we will need the following notation:

**Notation VI.10** Let $T$ be a set equipped with a semi-metric $d$; we denote by $K(\varepsilon) = K(T, d, \varepsilon)$ the largest cardinality of families of points of $T$ whose mutual distances are $> \varepsilon$.

We know (Chapter 1 of this volume, Sub-lemma IV.13) that:

$$K(T, d, \varepsilon) \geqslant N(T, d, \varepsilon')$$

for any $\varepsilon' > \varepsilon$ (the introduction of $\varepsilon' > \varepsilon$ is necessary since the entropy in Chapter 1 was defined with closed balls, whereas open balls are used here).

We have the following majoration:

**Proposition VI.11** *Let $A$ be a compact subset of $\mathbb{R}^N$. For every $\varepsilon > 0$, if $d_2$ is the Euclidean distance, and $g_1, \ldots, g_N$ a standard Gaussian sequence:*

$$K(A, d_2, \varepsilon) \leqslant \left( 1 + \frac{\sqrt{2\pi}}{\varepsilon \sqrt{N}} \, \mathbb{E} \left[ \sup_{x \in A} \sum_{n=1}^N g_n x_n \right] \right)^N.$$

*Proof* It suffices to apply Urysohn's inequality to the compact set $K = A + (\varepsilon/2) B_2^N$. As

$$\|x\|_{K^\circ} = \sup_{y \in K} \langle x, y \rangle = \sup_{a \in A} \langle x, a \rangle + \frac{\varepsilon}{2} \sup_{b \in B_2^N} \langle x, b \rangle = \|x\|_{A^\circ} + \frac{\varepsilon}{2} \|x\|_2,$$

this gives:

$$\frac{\varepsilon}{2} + \int_{\mathbb{S}^{N-1}} \|x\|_{A^\circ} \, d\sigma(x) \geqslant \left( \frac{\mathrm{Vol}(A + (\varepsilon/2) B_2^N)}{\mathrm{Vol}\, B_2^N} \right)^{1/N}.$$

However, noting that

$$\mathrm{Vol}(A + (\varepsilon/2) B_2^N) \geqslant K(A, d_2, \varepsilon) \left( \frac{\varepsilon}{2} \right)^N \mathrm{Vol}(B_2^N),$$

we obtain:

$$1 + \frac{2}{\varepsilon} \int_{\mathbb{S}^{N-1}} \|x\|_{A^\circ} \, d\sigma(x) \geqslant K(A, d_2, \varepsilon)^{1/N}.$$

Remark VI.9 completes the proof. $\qquad\square$

## VI.3 Real Case

First recall (Chapter 8 of Volume 1) that two families $(A_j)_{j \in J}$ and $(B_j)_{j \in J}$ of subsets of a set $T$ are *Boolean independent* if, for any finite and disjoint subsets $K$ and $L$ of $J$, we have:

$$\left(\bigcap_{k\in K}A_k\right)\cap\left(\bigcap_{l\in L}B_l\right)\neq\emptyset.$$

**Theorem VI.12** (The Elton–Pajor Theorem)   *Let $x_1,\ldots,x_N$ be real functions uniformly bounded by 1 on a set $T$. Let $(\varepsilon_1,\ldots,\varepsilon_N)$ be a Bernoulli sequence. Define:*

$$M_N = \mathbb{E}\sup_{t\in T}\left|\sum_{n=1}^{N}\varepsilon_n x_n(t)\right|.$$

*Then there exist a subset $I$ of $\{1,\ldots,N\}$ and a real number $r$ such that, with $c$ and $c'$ two universal constants:*

1)  $|I| \geq [c\,M_N^2/N]$;
2)  *if we set:*

$$A_n = \{t\in T\,;\ x_n(t)\geq r+\beta\}$$
$$B_n = \{t\in T\,;\ x_n(t)\leq r\},$$

   *with $\beta = c'\,(M_N/N)^3$, then the families $(A_n)_{n\in I}$ and $(B_n)_{n\in I}$ are Boolean independent;*
3)  *for every family $(a_n)_{n\in I}$ of real numbers, we have:*

$$\left\|\sum_{n\in I}a_n x_n\right\|_{\ell_\infty(T)} \geq \frac{\beta}{2}\sum_{n\in I}|a_n|.$$

The proof of this theorem requires several lemmas.

**Lemma VI.13**   *Let $S$ be a subset of $[-1,1]^N$. Then, for any $\varepsilon > 0$ (with $\varepsilon < 4$), there exists a subset $I$ of $\{1,\ldots,N\}$, with cardinality*

$$|I| = m \leq [32\log K(S,d_2,\varepsilon\sqrt{N})/3\,\varepsilon^2] + 1,$$

*such that:*

$$K(P_I(S),d_\infty,\varepsilon/2) \geq K(S,d_2,\varepsilon\sqrt{N}),$$

*where $d_2$ and $d_\infty$ are respectively the Euclidean and uniform distances on $\mathbb{R}^N$.*

*Proof*   Denote $K = K(S,d_2,\varepsilon\sqrt{N})$, and let $s_1,\ldots,s_K$ be points of $S$ such that $d_2(s_k,s_l) > \varepsilon\sqrt{N}$ for $k\neq l$. Let us write $s_k = (s_k^n)_{1\leq n\leq N}$ and set, for $1\leq k,l\leq K$:

$$I(k,l) = \{n\leq N\,;\ |s_k^n - s_l^n|\leq\varepsilon/2\}.$$

Then:

$$\varepsilon^2 N \leqslant d_2(s_k, s_l)^2 = \sum_{n \in I(k,l)} |s_k^n - s_l^n|^2 + \sum_{n \notin I(k,l)} |s_k^n - s_l^n|^2$$

$$\leqslant \frac{\varepsilon^2}{4} |I(k,l)| + 4 (N - |I(k,l)|),$$

since $|s_k^n - s_l^n| \leqslant 2$. Thus:

$$|I(k,l)| \leqslant \frac{4 - \varepsilon^2}{4 - (\varepsilon^2/4)} N \leqslant \left(1 - \frac{3}{16} \varepsilon^2\right) N.$$

Now let $X_1, \ldots, X_p$ be $p$ independent random variables uniformly distributed on $\{1, 2, \ldots, N\}$. The preceding estimation of $|I(k,l)|$ leads to:

$$\mathbb{P}\left(X_j \in I(k,l)\right) \leqslant \left(1 - \frac{3}{16} \varepsilon^2\right),$$

and hence:

$$\mathbb{P}\left(\exists\, (k,l)\,;\, X_1, \ldots, X_p \in I(k,l)\right) \leqslant \frac{K(K-1)}{2} \left(1 - \frac{3}{16} \varepsilon^2\right)^p.$$

If the following inequality can be obtained:

(1)  $$\frac{K(K-1)}{2} \left(1 - \frac{3}{16} \varepsilon^2\right)^p < 1,$$

then there exists a subset $I$ of $\{1, \ldots, N\}$, with cardinality $|I| = m \leqslant p$, such that $I \cap I(k,l)^c \neq \emptyset$ for any $k, l$; consequently $\sup_{n \in I} |s_k^n - s_l^n| > \varepsilon/2$, and finally $K(P_I(S), d_\infty, \varepsilon/2) \geqslant K$.

To finish, it only remains to note that the inequality (1) is realised for any integer $p \geqslant 32 \log(K/3\,\varepsilon^2)$, since then:

$$2 \log K + p \log \left(1 - \frac{3}{16} \varepsilon^2\right) < 2 \log K - \frac{3}{16} p \varepsilon^2 \leqslant 0. \qquad \square$$

**Lemma VI.14**  *Let $a \geqslant b \geqslant 1$ and $\varepsilon > 0$, and let $A \subseteq \mathbb{R}^m$ be a finite set of points, of cardinality $|A| \geqslant a^m$, with $d_\infty(s,t) \geqslant \varepsilon$ for every distinct pair of points $s, t \in A$. Then there exist a point $x \in [0, \varepsilon b]^m$ and a subset $B \subseteq x + (\varepsilon b\, \mathbb{Z})^m$, of cardinality $|B| \geqslant (a/b)^m$, such that, for every $s \in B$, there exists $t \in A$ satisfying $d_\infty(s,t) \leqslant \varepsilon/2$.*

*Proof*  Set $D = [0, \varepsilon b]^m$ and:

$$C = \bigcup_{t \in A} \left(t + \frac{\varepsilon}{2} B_\infty^m\right).$$

Since the balls $\left(t + \dfrac{\varepsilon}{2} B_\infty^m\right)$ are disjoint for any $t \in A$, we have:

$$\text{Vol}\, C \geqslant (\varepsilon a)^m.$$

However:

$$\text{Vol}\, C = \sum_{z \in (\varepsilon b\, \mathbb{Z})^m} \int_{(z+D) \cap C} \mathrm{d}x = \int_D \sum_{z \in (\varepsilon b\, \mathbb{Z})^m} \mathbb{1}_{C-z}\, \mathrm{d}x$$

$$= \int_D |(\varepsilon b\, \mathbb{Z})^m \cap (C - x)|\, \mathrm{d}x\,;$$

thus:

$$\int_D |(\varepsilon b\, \mathbb{Z})^m \cap (C - x)|\, \mathrm{d}x \geqslant (\varepsilon a)^m = (a/b)^m \, \text{Vol}\, D\,;$$

hence there exists a point $x \in D$ such that:

$$|(\varepsilon b\, \mathbb{Z})^m \cap (C - x)| \geqslant (a/b)^m,$$

and it suffices to take $B = [x + (\varepsilon b\, \mathbb{Z})^m] \cap C$.                                    $\square$

**Lemma VI.15**  *Let $m, p, k$ be strictly positive integers, with $k \leqslant m$, and let $B$ be a subset of $\{0, 1, 2, \ldots, p\}^m$ such that $|B| > \left(\sum_{j<k} \binom{m}{j}\right)^p$. Then there exist a subset $I$ of $\{1, 2, \ldots, m\}$, with cardinality $|I| \geqslant k$, and an integer $r$ such that, with:*

$$A_n = \{t \in B\,;\ t_n \geqslant r + 1\}$$
$$B_n = \{t \in B\,;\ t_n \leqslant r\},$$

*the families $(A_n)_{n \in I}$ and $(B_n)_{n \in I}$ are Boolean independent.*

*Proof*  For any $t = (t_n)_{n \leqslant m} \in B$, we set, for $0 \leqslant q \leqslant p - 1$:

$$I_q(t) = \{n \leqslant m\,;\ t_n \leqslant q\},$$

and let:

$$C_q = \{I_q(t)\,;\ t \in B\}.$$

Clearly the map:

$$t \in B \longmapsto \left(I_q(t)\right)_{q \leqslant p-1} \in \prod_{q \leqslant p-1} C_q$$

is injective; consequently:

$$\prod_{0 \leqslant q \leqslant p-1} |C_q| \geqslant |B| > \left(\sum_{j<k} \binom{m}{j}\right)^p,$$

and hence there exists an integer $r \leqslant p - 1$ such that:

$$|C_r| > \sum_{j < k} \binom{m}{j}.$$

Then Sauer's lemma ensures the existence of a subset $I$ of $\{1, \ldots, m\}$, with cardinality $|I| \geqslant k$, such that, for every $J \subseteq I$, there exists a $t \in B$ satisfying:

$$I_r(t) \cap I = J.$$

This can also be expressed as $t_n \leqslant r$ for $n \in J$, and $t_n \geqslant r + 1$ for $n \in I \smallsetminus J$; in other words, the families $(A_n)_{n \in I}$ and $(B_n)_{n \in I}$ are Boolean independent. $\square$

**Notation VI.16** Let $x_1, \ldots, x_N$ be functions on a set $T$. For every $\varepsilon > 0$, denote by $\Lambda(\varepsilon) = \Lambda(x_1, \ldots, x_N; \varepsilon)$ the maximum cardinality of subsets $I$ of $\{1, 2, \ldots, N\}$ satisfying the following property:

*There exists a real number $r$ such that, if*

$$A_n = \{t \in T ; x_n(t) \geqslant r + \varepsilon\} \quad and \quad B_n = \{t \in T ; x_n(t) \leqslant r\},$$

*then the families $(A_n)_{n \in I}$ and $(B_n)_{n \in I}$ are Boolean independent.*

Then the following result is the essential combinatorial point in the proof of the Elton–Pajor theorem:

**Proposition VI.17** *Let $x_1, \ldots, x_N$ be functions, uniformly bounded by 1 on a set $T$. Equip $T$ with the semi-metric $d$ defined, for $s, t \in T$, by:*

$$d(s, t) = \left( \sum_{n=1}^{N} |x_n(t) - x_n(s)|^2 \right)^{1/2}.$$

*Then, for $0 < \varepsilon < 2$:*

$$\log K(T, d, \varepsilon \sqrt{N}) \leqslant \frac{96}{\varepsilon} \Lambda(\varepsilon^3 / 128) \log(16/\varepsilon).$$

*Proof* Set $S = \{(x_1(t), \ldots, x_N(t)) ; t \in T\} \subseteq [-1, 1]^N$. By Lemma VI.13, with $K_\varepsilon = K(S, d_2, \varepsilon \sqrt{N})$, there exists a subset $I$ of $\{1, \ldots, N\}$, with cardinality $|I| = m \leqslant [32 \log K_\varepsilon / 3 \varepsilon^2] + 1$, such that:

(1) $$K\big(P_I(S), d_\infty, \varepsilon/2\big) \geqslant K_\varepsilon.$$

We can assume $K_\varepsilon \geqslant 2$; thus, since $\varepsilon < 2$:

(2) $$|I| = m \leqslant 32 \log K_\varepsilon / \varepsilon^2.$$

Using (1), we obtain a subset $A \subseteq P_I(S)$ such that $|A| = K_\varepsilon$ and such that, for every distinct $s, t \in A$, we have:

$$d_\infty(s, t) \geqslant \varepsilon/2.$$

Let us now apply Lemma VI.14 to $A$, with $m = |I|$, $a = |A|^{1/m}$ and $b = \sqrt{a}$ (note that (2) implies $a \geqslant e^{\varepsilon^2/32} > 1$). We thus obtain a point $x_0 = (x_n^0)_{n \in I} \in \mathbb{R}^I$ and a subset $B \subseteq x_0 + \left( (\varepsilon \sqrt{a}/2) \mathbb{Z} \right)^I$ such that:

$$|B| \geqslant |A|^{1/2} = K_\varepsilon^{1/2}$$

and

(3) $\qquad (\forall s \in B), \quad (\exists t \in A) : \quad d_\infty(s,t) \leqslant \varepsilon/4.$

Therefore:

$$B \subseteq \left[ -1 - \frac{\varepsilon}{4}, 1 + \frac{\varepsilon}{4} \right]^I \cap \left( x_0 + \left( \frac{\varepsilon \sqrt{a}}{2} \mathbb{Z} \right)^I \right).$$

However, for any real number $x$, the number of elements in the intersection of the interval $[-1 - \varepsilon/4, 1 + \varepsilon/4]$ with the net $x + (\varepsilon \sqrt{a}/2) \mathbb{Z}$ is at most:

$$\left| \left[ -1 - \frac{\varepsilon}{4}, 1 + \frac{\varepsilon}{4} \right] \cap \left( x + \frac{\varepsilon \sqrt{a}}{2} \mathbb{Z} \right) \right| \leqslant \frac{2 + \varepsilon/2}{\varepsilon \sqrt{a}/2} + 1 \leqslant \frac{4}{\varepsilon} + 2.$$

Moreover, since $\varepsilon < 2$, the length of the interval $[2 + 4/\varepsilon, 1 + 8/\varepsilon]$ is $> 1$, and thus we can find an integer $p \leqslant 8/\varepsilon$ such that this number of elements is $\leqslant p + 1$. Then, we index the points of the sets

$$\left[ -1 - \frac{\varepsilon}{4}, 1 + \frac{\varepsilon}{4} \right] \cap \left( x_n^0 + \frac{\varepsilon \sqrt{a}}{2} \mathbb{Z} \right), \quad n \in I,$$

following the natural order for each coordinate, and thus we can identify $B$ with a subset $B'$ of $\{0, 1, \ldots, p\}^m$.

Now assume that, for an integer $k \leqslant m$, we have:

(4) $\qquad\qquad K_\varepsilon^{1/2} > \left( \sum_{j<k} \binom{m}{j} \right)^p.$

As $|B| \geqslant K_\varepsilon^{1/2}$, we can use Lemma VI.15: there exists a subset $J$ of $I$, of cardinality $|J| \geqslant k$, and a number $r''$ such that:

$$(\forall L \subseteq J) \, (\exists s = (s_n)_{n \in I} \in B') : \quad \begin{cases} s_n \geqslant r'' + 1 & \text{for } n \in L \\ s_n \leqslant r'' & \text{for } n \in J \setminus L. \end{cases}$$

Back to $B$: the way $B$ has been identified with $B'$ provides a real number $r'$ such that:

$$(\forall L \subseteq J) \, (\exists s = (s_n)_{n \in I} \in B) : \quad \begin{cases} s_n \geqslant r' + \varepsilon \sqrt{a}/2 & \text{for } n \in L \\ s_n \leqslant r' & \text{for } n \in J \setminus L. \end{cases}$$

Thanks to (3), this in turn implies:

$$(\forall L \subseteq J) \; (\exists s = (s_n)_{n \in I} \in A):$$

$$\begin{cases} s_n \geqslant r' + \varepsilon \sqrt{a}/2 - \varepsilon/4 & \text{for} \quad n \in L \\ s_n \leqslant r' + \varepsilon/4 & \text{for} \quad n \in J \smallsetminus L. \end{cases}$$

Hence, if we set $\beta = \varepsilon \, (\sqrt{a} - 1)/2, r = r' + \varepsilon/4$,

$$A_n = \{t \in T \, ; \, x_n(t) \geqslant r + \beta\} \quad \text{and} \quad B_n = \{t \in T \, ; \, x_n(t) \leqslant r\},$$

then the families $(A_n)_{n \in I}$ and $(B_n)_{n \in I}$ are Boolean independent. Consequently, the hypothesis (4) implies $\Lambda(\beta) \geqslant k$, and it ensues that:

$$K_\varepsilon^{1/2} \leqslant \left( \sum_{j \leqslant \Lambda(\beta)} \binom{m}{j} \right)^p.$$

Now we use the inequality of Proposition VI.4, valid for $0 \leqslant k \leqslant m$:

$$\sum_{j \leqslant k} \binom{m}{j} \leqslant \left( \frac{me}{k} \right)^k.$$

This leads to:

$$\log K_\varepsilon \leqslant 2\Lambda(\beta) \, p \, \log \left( me/\Lambda(\beta) \right),$$

and, by using the majoration (2) of $m$, and that of $p$, we obtain:

$$\log K_\varepsilon \leqslant 2\Lambda(\beta) \, \frac{8}{\varepsilon} \, \log \left( \frac{32 \, e \, \log K_\varepsilon}{\varepsilon^2 \Lambda(\beta)} \right).$$

Since:

$$x \leqslant \alpha \, \log x \quad \text{and} \quad \alpha \geqslant e \implies x \leqslant 2\alpha \, \log \alpha$$

(the function $u \colon x \mapsto x - \alpha \, \log x$ has a minimum $\alpha - \alpha \, \log \alpha$ at $\alpha$, and $u(2\alpha \, \log \alpha) > 0$), we obtain:

$$\frac{\log K_\varepsilon}{\Lambda(\beta)} \leqslant 2 \, \frac{16}{\varepsilon} \, \log \left( \frac{16 \times 32 \, e}{\varepsilon^3} \right) \leqslant 2 \, \frac{16}{\varepsilon} \, 3 \log \frac{16}{\varepsilon} = \frac{96}{\varepsilon} \, \log \frac{16}{\varepsilon} .$$

Finally, since (2) implies $a \geqslant e^{\varepsilon^2/32}$, then:

$$\beta = \frac{\varepsilon}{2} \, (\sqrt{a} - 1) \geqslant \frac{\varepsilon}{2} \, (e^{\varepsilon^2/64} - 1) \geqslant \frac{\varepsilon^3}{128}.$$

The claimed result is achieved: indeed, by definition of $d$, $(S, d_2)$ and $(T, d)$ are isometric, so $K(T, d, \varepsilon\sqrt{N}) = K(S, d_2, \varepsilon\sqrt{N}) = K_\varepsilon$.                    $\square$

We can now proceed with the proof of the theorem.

*Proof of Theorem VI.12*   We use Dudley's theorem. For this, we set:

$$M' = \mathbb{E}\left\|\sum_{n=1}^{N} g_n x_n\right\|_{\infty} = \mathbb{E}\left(\sup_{(s_n)_n \in S}\left|\sum_{n=1}^{N} g_n s_n\right|\right),$$

where $g_1, \ldots, g_N$ is a standard Gaussian sequence.

Since $\left\|\sum_{n=1}^{N} g_n s_n\right\|_{L^2(\mathbb{P})} \leqslant \sqrt{N}$, then:

$$M' \leqslant b\left(\sqrt{N} + \int_0^{+\infty} \sqrt{\log N(S, d_2, \varepsilon)}\, d\varepsilon\right).$$

Hence, since diam $S \leqslant 2\sqrt{N}$ and $K(S, d_2, \varepsilon) \geqslant N(S, d_2, \varepsilon')$ for any $\varepsilon' > \varepsilon$:

$$M' \leqslant b\left(\sqrt{N} + \int_0^{+\infty} \sqrt{\log K(S, d_2, \varepsilon)}\, d\varepsilon\right)$$

$$= b\left(\sqrt{N} + \int_0^{2\sqrt{N}} \sqrt{\log K(S, d_2, \varepsilon)}\, d\varepsilon\right);$$

hence:

(1) $$\frac{M'}{\sqrt{N}} \leqslant b\left(1 + \int_0^2 \sqrt{\log K_\varepsilon}\, d\varepsilon\right).$$

On the other hand, Proposition VI.11 leads to:

$$K_\varepsilon = K(S, d_2, \varepsilon\sqrt{N}) \leqslant \left(1 + \frac{M'}{N\varepsilon}\sqrt{2\pi}\right)^N;$$

with this majoration, (1) can be transformed. Indeed, we can deduce, for any $x > 0$:

$$\int_0^{xM'/N} \sqrt{\log K_\varepsilon}\, d\varepsilon \leqslant \int_0^{xM'/N} \sqrt{N}\sqrt{\log\left(1 + \sqrt{2\pi}\,\frac{M'}{N\varepsilon}\right)}\, d\varepsilon$$

$$= \frac{M'}{\sqrt{N}}\int_0^x \sqrt{\log\left(1 + \frac{\sqrt{2\pi}}{\varepsilon}\right)}\, d\varepsilon$$

$$\leqslant \frac{M'}{\sqrt{N}}\int_0^x \sqrt{\frac{\sqrt{2\pi}}{\varepsilon}}\, d\varepsilon \leqslant 4\sqrt{x}\,\frac{M'}{\sqrt{N}}.$$

Choosing $x = 1/(8b)^2$, we obtain:

$$\frac{M'}{\sqrt{N}} \leqslant b\left(1 + \int_0^{M'/64b^2N} \sqrt{\log K_\varepsilon}\, d\varepsilon + \int_{M'/64b^2N}^2 \sqrt{\log K_\varepsilon}\, d\varepsilon\right)$$

$$\leqslant b\left(1 + 4\frac{1}{8b}\frac{M'}{\sqrt{N}} + \int_{M'/64b^2N}^2 \sqrt{\log K_\varepsilon}\, d\varepsilon\right),$$

so that:

$$\frac{M'}{\sqrt{N}} \leqslant 2b \left( 1 + \int_{M'/64b^2N}^{2} \sqrt{\log K_\varepsilon} \, d\varepsilon \right).$$

We now use Proposition VI.17:

$$\frac{M'}{\sqrt{N}} \leqslant 2b \left( 1 + \int_{M'/64b^2N}^{2} \sqrt{\frac{96}{\varepsilon} \Lambda \left( \frac{\varepsilon^3}{128} \right) \log \frac{16}{\varepsilon}} \, d\varepsilon \right)$$

$$\leqslant 2b \left( 1 + \sqrt{\Lambda(M'^3/2^{25} b^6 N^3)} \int_{0}^{2} \sqrt{\frac{96}{\varepsilon} \log \frac{16}{\varepsilon}} \, d\varepsilon \right)$$

$$\leqslant C \sqrt{\Lambda(M'^3/2^{25} b^6 N^3)}.$$

As $M_N \leqslant \sqrt{\pi/2} \, M'$, this leads to:

$$\frac{M_N}{\sqrt{N}} \leqslant C' \sqrt{\Lambda(C' M'^3/N^3)}.$$

We can now complete the proof. Indeed, by the definition of $\Lambda$, there exist a subset $I$ of $\{1, \ldots, N\}$, with cardinality

$$|I| \geqslant \Lambda(C' M'^3/N^3) \geqslant \left[ c \frac{M_N^2}{N} \right]$$

and a real number $r$ such that, if $A_n = \{t \in T ; x_n(t) \geqslant r + \beta\}$ and $B_n = \{t \in T ; x_n(t) \leqslant r\}$, with $\beta = C' M'^3/N^3$, and *a fortiori* with $\beta = c' M_N^3/N^3$ (since $M_N \leqslant \sqrt{\pi/2} M'$), the families $(A_n)_{n \in I}$ and $(B_n)_{n \in I}$ are Boolean independent. This implies 1) and 2) of Theorem VI.12.

Item 3) is a consequence of 2), as seen in Chapter 8 (Volume 1). Here again is the argument. Let $(a_n)_{n \in I}$ be a family of real numbers and take $J = \{n \in I ; a_n \geqslant 0\}$. Boolean independence allows us to find two points $s, t \in T$ such that:

$$\begin{cases} x_n(t) \geqslant r + \beta & \text{for } n \in J \\ x_n(t) \leqslant r & \text{for } n \in I \setminus J ; \end{cases}$$

and

$$\begin{cases} x_n(s) \leqslant r & \text{for } n \in J \\ x_n(s) \geqslant r + \beta & \text{for } n \in I \setminus J. \end{cases}$$

For this pair $(s, t)$, we have:

$$\left\| \sum_{n \in I} a_n x_n \right\|_{\ell_\infty(T)} \geqslant \frac{1}{2} \sup_{u,v \in T} \left| \sum_{n \in I} a_n \left( x_n(u) - x_n(v) \right) \right|$$

$$\geqslant \frac{1}{2} \left( \sum_{n \in J} a_n \left( x_n(t) - x_n(s) \right) \right.$$

$$\left. + \sum_{n \in I \setminus J} a_n \left( x_n(t) - x_n(s) \right) \right)$$

$$\geqslant \frac{\beta}{2} \sum_{n \in I} |a_n|. \qquad \qquad \square$$

**Corollary VI.18** (Elton's Theorem)  *For $\delta \in \, ]0, 1]$, constants $c(\delta) > 0$ and $\beta(\delta) > 0$ can be found to ensure the following property: if $x_1, \dots, x_N$ are vectors of the unit ball of a real Banach space satisfying*

$$\mathbb{E} \left\| \sum_{n=1}^{N} \varepsilon_n x_n \right\| \geqslant \delta N,$$

*then there exists a subset $I$ of $\{1, \dots, N\}$, with cardinality $|I| \geqslant c(\delta) N$, such that $(x_n)_{n \in I}$ is $\beta(\delta)$-equivalent to the canonical basis of $\ell_1^{|I|}$:*

$$\left\| \sum_{n \in I} a_n x_n \right\| \geqslant \beta(\delta) \sum_{n \in I} |a_n|$$

*for any real numbers $a_1, \dots, a_N$.*

The proof is an immediate consequence of Theorem VI.12, with $c(\delta) = c \, \delta^2$ and $\beta(\delta) = c' \, \delta^3$. $\qquad \qquad \square$

## VI.4  Complex Case

For the complex case, the same method is used here as in the complex case of Rosenthal's $\ell_1$ theorem (Chapter 8 of Volume 1), with additional combinatorial lemmas to control the dimension.

**Theorem VI.19** (Pajor)  *For $0 < \delta \leqslant 1$, a constant $c(\delta) > 0$ can be found such that, for every finite sequence of functions $z_n = x_n + i y_n$, $1 \leqslant n \leqslant N$, defined on a set $T$, uniformly bounded by 1, and with real parts $x_n$ satisfying:*

$$\left\| \sum_{n=1}^{N} a_n x_n \right\|_{\infty} \geqslant \delta \sum_{n=1}^{N} |a_n|$$

*for any real numbers $a_1, \ldots, a_N \in \mathbb{R}$, then there exists a subset $I$ of $\{1, \ldots, N\}$, of cardinality $|I| \geqslant c(\delta) N$, such that:*

$$\left\| \sum_{n \in I} c_n z_n \right\|_\infty \geqslant \frac{\delta}{2} \sum_{n \in I} |c_n|$$

*for any complex numbers $c_n \in \mathbb{C}$, $n \in I$.*

We immediately obtain the following corollary:

**Corollary VI.20** *For $0 < \delta \leqslant 1$, constants $c(\delta) > 0$ and $\beta(\delta) > 0$ can be found such that, if $z_1, \ldots, z_N$ are vectors of the unit ball of a complex Banach space satisfying:*

$$\mathbb{E} \left\| \sum_{n=1}^{N} \varepsilon_n z_n \right\| \geqslant \delta N,$$

*then there exists a subset $I$ of $\{1, \ldots, N\}$, of cardinality $|I| \geqslant c(\delta) N$, such that $(x_n)_{n \in I}$ is $\beta(\delta)$-equivalent to the canonical basis of $\ell_1^{|I|}$:*

$$\left\| \sum_{n \in I} c_n z_n \right\| \geqslant \beta(\delta) \sum_{n \in I} |c_n|$$

*for any complex numbers $c_n$, $n \in I$.*

*Proof* We consider the vectors as functions on the unit ball of the dual. Let $x_n$ and $y_n$ be respectively the real and imaginary parts of $z_n$, so $z_n = x_n + i y_n$. Since:

$$\mathbb{E} \left\| \sum_{n=1}^{N} \varepsilon_n x_n \right\| + \mathbb{E} \left\| \sum_{n=1}^{N} \varepsilon_n y_n \right\| \geqslant \mathbb{E} \left\| \sum_{n=1}^{N} \varepsilon_n z_n \right\| \geqslant \delta N,$$

by considering the functions $i z_n$ if necessary, we can assume that:

$$\mathbb{E} \left\| \sum_{n=1}^{N} \varepsilon_n x_n \right\| \geqslant \frac{\delta}{2} N.$$

Thanks to Corollary VI.18, there exists a subset $I$ of $\{1, \ldots, N\}$, of cardinality $|I| \geqslant c(\delta) N$, such that:

$$\left\| \sum_{n \in I} a_n x_n \right\| \geqslant \beta(\delta) \sum_{n \in I} |a_n|$$

for any real numbers $a_n$, $n \in I$. It suffices then to use Theorem VI.19 to complete the proof. $\qquad \square$

In particular, we have the following result:

**Corollary VI.21**  *Let $X$ be a complex Banach space. If, when considered as a real Banach space, $X$ contains (isomorphically) $\ell_1^N$ (i.e. there exist $z_1, \ldots, z_N \in X$ such that $\left\| \sum_{n=1}^N a_n z_n \right\| \geqslant \delta \sum_{n=1}^N |a_n|$, for all $a_1, \ldots, a_N \in \mathbb{R}$), then it contains the complex space $\ell_1^{cN}$, where $c$ is a constant depending only on the isomorphism constant $\delta$.*

For the proof of Theorem VI.19, two lemmas are required.

**Lemma VI.22**  *For any integer $p \geqslant 2$, and for any integer $N \geqslant N(p) = 4p^2 \log(2p)$, we have:*

$$p \left[ \sum_{k=0}^{\lceil c(p)N \rceil} \binom{N}{k} \right] \left[ \sum_{k=0}^{\left\lceil \frac{N}{2}\left(1 - \frac{1}{p}\right) \right\rceil} \binom{N}{k} \right] \leqslant 2^{N-1},$$

*where $c(p) = \dfrac{1}{16 p^2 \log(2p)}$.*

*Proof*  The starting point is Chernov's inequality (Proposition VI.4):

$$\sum_{k=0}^{n} \binom{N}{k} \leqslant \frac{N^N}{(N-n)^{N-n} n^n},$$

valid for $0 \leqslant n \leqslant N/2$. If we set:

$$f(\alpha) = -[\alpha \log \alpha + (1-\alpha) \log(1-\alpha)],$$

this becomes:

$$\sum_{k=0}^{\alpha N} \binom{N}{k} \leqslant \exp[N f(\alpha)], \qquad \text{for } 0 \leqslant \alpha \leqslant 1/2.$$

It thus suffices to show that, for $p \geqslant 2$ and $N \geqslant N(p)$, we have:

(1)  $$p \exp\left[ N f(c(p)) + f\left(\frac{p-1}{2p}\right) \right] \leqslant 2^{N-1}.$$

It is elementary to verify that:

$$\begin{cases} f(\alpha) \leqslant \alpha(1 - \log \alpha), & \text{for } 0 < \alpha \leqslant 1 \\ 1 + \log[4 \log(2p)] \leqslant 2 \log(2p), & \text{for } p \geqslant 2. \end{cases}$$

Then we deduce:

$$f[c(p)] \leqslant \frac{1}{16\,p^2\log(2p)} \left[1 + \log\left(16\,p^2\log(2p)\right)\right]$$

$$= \frac{1}{4\,p^2} \, \frac{1 + 2\log(2p) + \log\left(4\log(2p)\right)}{4\log(2p)} \leqslant \frac{1}{4\,p^2}.$$

Now let $g(x) = (1+x)\log(1+x) + (1-x)\log(1-x)$. It is easy to check that:

$$\begin{cases} g(x) \geqslant x^2 & \text{for } 0 \leqslant x \leqslant 1 \\ \log 2 - f\left(\dfrac{p-1}{2p}\right) = \dfrac{1}{2}g\left(\dfrac{1}{p}\right). \end{cases}$$

It ensues that, for $N \geqslant N(p)$:

$$\log 2 - f\left(\frac{p-1}{2p}\right) - \frac{\log(2p)}{N} \geqslant \frac{1}{2p^2} - \frac{1}{4p^2} = \frac{1}{4p^2} \geqslant f[c(p)],$$

which proves both (1) and the lemma. $\qquad\square$

**Lemma VI.23** *Let $E$ be a set and $(E^+, E^-)$ a partition of $E$ such that $|E^+| = p \geqslant 1$ and $|E^-| = q \geqslant 1$. For any integer $N \geqslant 1$, the mapping $Q \colon E^N \to \{-1, +1\}^N$ is defined by:*

$$(Qx)_n = \begin{cases} +1 & \text{if } x_n \in E^+ \\ -1 & \text{if } x_n \in E^-, \end{cases}$$

*where $x = (x_n)_{n\leqslant N} \in E^N$. For $x = (x_n)_{n\leqslant N} \in E^N$ and $e \in E$, define:*

$$\begin{cases} I_e(x) &= \{n \in \{1,\dots,N\}; \; x_n = e\} \\ I^+(x) &= \{n \in \{1,\dots,N\}; \; x_n \in E^+\} \\ I^-(x) &= \{n \in \{1,\dots,N\}; \; x_n \in E^-\}. \end{cases}$$

*Then, for every subset $S$ of $E^N$ with $Q(S) = \{-1,1\}^N$, there exist $e^+ \in E^+$ and $e^- \in E^-$, and a subset $I$ of $\{1,\dots,N\}$, such that:*

(i) $|I| \geqslant c(p)\,c(q)\,N$;

(ii) $\{e^-, e^+\}^I \subseteq P_I(S)$,

*where $c(p) = \dfrac{1}{16\,p^2\log(2p)}$ for $p \geqslant 2$ and $c(1) = 1$, and where $P_I$ is the natural projection from $E^N$ onto $E^I$.*

*Proof* If $Q(S) = \{-1,1\}^N$, there exists a subset $S_1 \subseteq S$ of cardinal $|S_1| = 2^N$ such that $Q(S_1) = \{-1,1\}^N$. Remark that the mapping from $S_1$ into $\mathcal{P}(\{1,\dots,N\})$ that sends $s \in S_1$ to $I^+(s)$ is then bijective. We define:

$$S_2 = \{s \in S_1 \,; \, |I^-(s)| \geqslant N/2\}.$$

Clearly:

$$|S_2| \geqslant 2^{N-1}.$$

For every $s \in S_2$, let $e(s) \in E^+$ be such that:

$$|I_{e(s)}(s)| = \max_{\varepsilon \in E^+} |I_\varepsilon(s)|.$$

Since $|E^+| = p$ and $I^+(s) = \bigcup_{e \in E^+} I_e(s)$, we have:

$$|I_{e(s)}(s)| \geqslant \frac{1}{p} |I^+(s)|.$$

If now, for $e \in E^+$, we set:

$$J_e(s) = I^+(s) \smallsetminus I_e(s),$$

then, for every $s \in S_2$:

$$|J_{e(s)}| \leqslant |I^+(s)| \left(1 - \frac{1}{p}\right) \leqslant \frac{N}{2} \left(1 - \frac{1}{p}\right).$$

Moreover, as $|E^+| = p$, one of the values of $E^+$ must be attained at least $|S_2|/p$ times by the mapping $s \in S_2 \mapsto e(s) \in E^+$; hence there exist a subset $S_3 \subseteq S_2$ and an element $e_0 \in E^+$ such that:

$$\begin{cases} |S_3| \geqslant |S_2|/p \geqslant 2^{N-1}/p \\ |I_{e_0}(s)| = \max_{\varepsilon \in E^+} |I_\varepsilon(s)|, \quad \forall s \in S_3. \end{cases}$$

However, as seen above, for every $s \in S_3$, the cardinality of $J_{e_0}(s)$ is at most $\frac{N}{2}\left(1 - \frac{1}{p}\right)$; thus there are at most

$$\sum_{j=0}^{\frac{N}{2}\left(1-\frac{1}{p}\right)} \binom{N}{j}$$

such distinct subsets $J_{e_0}(s)$ as $s$ runs over $S_3$. Consequently, there exist subsets $S_4 \subseteq S_3$ and $J \subseteq \{1, \ldots, N\}$ such that:

$$\begin{cases} |S_4| \geqslant |S_3| \Big/ \sum_{j=0}^{\frac{N}{2}\left(1-\frac{1}{p}\right)} \binom{N}{j} \\ J_{e_0}(s) = J, \quad \forall s \in S_4. \end{cases}$$

Set

$$R = Q(S_4) \subseteq \{-1, 1\}^N \, ;$$

the above leads to:

$$|R| = |S_4| \geqslant \frac{2^{N-1}}{p \sum_{j=0}^{\frac{N}{2}\left(1-\frac{1}{p}\right)} \binom{N}{j}},$$

and hence, by Lemma VI.22, for $N \geqslant N(p)$:

$$|R| \geqslant \sum_{j=0}^{c(p)N} \binom{N}{j}.$$

We now use Sauer's lemma (Proposition VI.3): there exists a subset $I$ of $\{1,\ldots,N\}$ such that:

$$\begin{cases} |I| \geqslant [c(p)N] + 1 \geqslant c(p)N \\ P_I(R) = \{-1,1\}^I. \end{cases}$$

Making this last equality explicit, we obtain:

$$(\forall K \subseteq I) \quad (\exists s \in S_4) \quad \text{such that} \quad I \cap I^+ = K.$$

In particular, there exists a $s_0 \in S_4$ such that $I \subseteq I^+(s_0)$.

Recall now that, for every $s \in S_4$:

$$J = J_{e_0}(s) = I^+(s) \smallsetminus I_{e_0}(s) \subseteq I^+(s);$$

therefore $I \cap J = \emptyset$, and also, for every $s \in S_4$:

$$I \cap I^+(s) = I \cap I_{e_0}(s).$$

Hence:

$$(*) \qquad (\forall K \subseteq I) \quad (\exists s \in S_4) \quad \text{such that} \quad \begin{cases} s_n = e_0 & \text{for } n \in K \\ s_n \in E^- & \text{for } n \in I \smallsetminus K. \end{cases}$$

Note that if $N \leqslant N(p)$, then $c(p)N \leqslant 1/4$ and, since $Q(S) = \{-1,1\}^N$, there exist $s, s' \in S$ with $s_1 \in E^+$ and $s_1' \in E^-$, so that $(*)$ is satisfied with $I = \{1\}$ and $e_0 = s_1$.

Thus, we have found a subset $I$ of $\{1,\ldots,N\}$ and $e_0 \in E^+$ satisfying:

$$|I| \geqslant c(p)N$$

and:

$$(**) \qquad (\forall K \subseteq I) \quad (\exists s \in S_4) \quad \text{such that} \quad \begin{cases} s_n = e_0 & \text{for } n \in K \\ s_n \in E^- & \text{for } n \in I \smallsetminus K. \end{cases}$$

Now we set $E_1 = E^- \cup \{e_0\}$ and $S' = P_I(S_4) \subseteq E_1^I$. Thanks to $(**)$, we can proceed with the same construction, starting with $S'$ and pulling apart the points of $E^-$. This completes the proof. $\qquad\square$

*Proof of Theorem VI.19*   The proof is identical to the infinite-dimensional case, but we repeat it here for the convenience of the reader.

Let $K$ be the balanced (in the real sense) and $w^*$-closed convex hull of

$$\{(x_n(t))_{n \leqslant N} ; t \in T\} \subseteq \ell_\infty(T, \mathbb{R}).$$

By convexity, for every sequence $(a_n)_{n \leqslant N} \in \ell_1^N(\mathbb{R})$, we have:

$$\sup\left\{ \left|\sum_{n \leqslant N} a_n \kappa_n\right| ; \kappa = (\kappa_n)_{n \leqslant N} \in K \right\} \geqslant \delta \sum_{n=1}^N |a_n|.$$

This means that the map $j: \ell_1^N(\mathbb{R}) \to \mathcal{C}(K)$ sending the $a = (a_n)_{n \leqslant N} \in \ell_1^N(\mathbb{R})$ to the function $\kappa = (\kappa_n)_{n \leqslant N} \mapsto \sum_{n=1}^N a_n \kappa_n$ is a $\delta$-isomorphism. Its adjoint $j^*: \mathcal{M}(T) \to \ell_\infty^N$ is thus surjective, and $j^*(\delta_\kappa) = \kappa$ for every $\kappa \in K$, where $\delta_\kappa$ is the Dirac mass at $\kappa$. The identification of $\delta_\kappa$ and $\kappa$ leads to:

$$\{-\delta, \delta\}^N \subseteq K.$$

Now we consider the balanced (in the real sense) and $w^*$-closed convex hull $H$ of $\{(z_n(t))_{n \leqslant N} ; t \in T\} \subseteq \ell_\infty(T, \mathbb{C})$. Then $\operatorname{Re} H = K$; hence there exists a subset $H'$ of $H$ such that $\operatorname{Re} H' = \{-\delta, \delta\}^N$.

Let $0 < \alpha < \frac{4}{\pi}\delta$; we divide the interval $[-1, 1]$ into $p = \left[\frac{2}{\alpha}\right] + 1$ adjacent intervals $I_1, \ldots, I_p$ of length $\leqslant \alpha$. For any $u \in [-1, 1]$, denote by $k(u)$ the unique integer $k \leqslant p$ such that $u \in I_k$. A map

$$\Psi: H' \longrightarrow \{-p, \ldots, -1, +1, \ldots, p\}^N$$

is defined as follows: if $\kappa = (\zeta_n + i\xi_n)_{n \leqslant N} \in H'$, for any $n \leqslant N$, we set:

$$[\Psi(\kappa)]_n = \operatorname{sign}(\zeta_n)\, k(\xi_n).$$

Let $S = \Psi(H')$ and $E = \{-p, \ldots, -1, +1, \ldots, p\}$, and define the map $Q: E^N \to \{-1, 1\}^N$ by $Q((\theta_n)_n) = ((\operatorname{sign}\theta_n)_n)$.

Since $\operatorname{Re} H' = \{-\delta, \delta\}^N$, we have $Q(S) = \{-1, 1\}^N$, and hence we can apply Lemma VI.23: there exist a subset $I$ of $\{1, \ldots, N\}$ and two distinct integers $k, l \in \{1, 2, \ldots, p\}$ such that:

$$\begin{cases} |I| \geqslant c(p)^2 N \\ P_I(S) \supseteq \{-l, k\}^I. \end{cases}$$

Thus, for each subset $J$ of $I$, there exists $\kappa' \in H'$ such that:

$$\begin{cases} \Psi(\kappa')_n \in (-I_k) & \text{for } n \in J \\ \Psi(\kappa')_n \in I_l & \text{for } n \in I \setminus J. \end{cases}$$

Replacing $J$ by $I \setminus J$, we can also find a $\kappa'' \in H'$ such that:

$$\begin{cases} \Psi(\kappa'')_n \in (-I_k) & \text{for } n \in I \setminus J \\ \Psi(\kappa'')_n \in I_l & \text{for } n \in J. \end{cases}$$

Set $\kappa = (\kappa' + \kappa'')/2$. As $H$ is convex, there exists an element $\kappa = (\zeta_n + i\xi_n)_{n \leqslant N} \in H$ such that, with $a$ denoting the midpoint of the interval $\frac{1}{2}(-I_k + I_l)$, we have:

$$\begin{cases} \zeta_n = \delta & \text{and} \quad |\xi_n - a| \leqslant \alpha/2 & \text{for } n \in J \\ \zeta_n = -\delta & \text{and} \quad |\xi_n + a| \leqslant \alpha/2 & \text{for } n \in I \setminus J. \end{cases}$$

In particular, $|\xi_n - a\,\mathrm{sign}(\zeta_n)| \leqslant \alpha/2$ for any $n \in I$. Now, for every family $(c_n)_{n \in I}$ of complex numbers, we have:

$$\left\| \sum_{n \in I} c_n z_n \right\|_{\ell_\infty(T)} \geqslant \sup \left\{ \left| \sum_{n \in I} c_n \tau_n \right| ; \ (\tau_n)_{n \leqslant N} \in H \right\}$$

$$\geqslant \sup_{\theta_n = \pm 1} \left| \sum_{n \in I} c_n \theta_n \kappa_n \right|$$

$$\geqslant \sup_{\theta_n = \pm 1} \left| \sum_{n \in I} c_n \theta_n \big( \zeta_n + ia\,\mathrm{sign}(\zeta_n) \big) \right|$$

$$- \sum_{n \in I} |\xi_n - a\,\mathrm{sign}(\zeta_n)|\,|c_n|$$

$$\geqslant \sup_{\varepsilon_n = \pm 1} \left| \sum_{n \in I} c_n (\delta + ia)\,\varepsilon_n \right| - \frac{\alpha}{2} \sum_{n \in I} |c_n|$$

$$\geqslant \left( \frac{2}{\pi} |\delta + ia| - \frac{\alpha}{2} \right) \sum_{n \in I} |c_n|.$$

This is because:

$$\max_{\varepsilon_n = \pm 1} \left| \sum_{n \in I} w_n \varepsilon_n \right| \geqslant \frac{2}{\pi} \sum_{n \in I} |w_n|,$$

for every sequence $(w_n)_{n \in I}$ of complex numbers, since the Rademacher functions form a Sidon set of constant $\pi/2$ in the Cantor group (Chapter 6 of Volume 1, Proposition V.2). Thus, *a fortiori*, we obtain:

$$\left\| \sum_{n \in I} c_n z_n \right\|_{\ell_\infty(T)} \geqslant \left( \frac{2}{\pi} \delta - \frac{\alpha}{2} \right) \sum_{n \in I} |c_n| \, ;$$

therefore:

$$\left\| \sum_{n \in I} c_n z_n \right\|_{\ell_\infty(T)} \geqslant \frac{\delta}{2} \sum_{n \in I} |c_n|,$$

after taking $\alpha = \left( \dfrac{4}{\pi} - 1 \right) \delta$. This completes the proof.                    $\square$

Note that the proof provides $c(\delta) = \dfrac{1}{16 \, (15/\delta)^2 \log(15/\delta)}$; indeed, as $0 <$ $\delta \leqslant 1$, then $p = \left[ \dfrac{2}{\alpha} \right] + 1 = \left[ \dfrac{2\pi}{(4 - \pi)\delta} \right] + 1 \leqslant \dfrac{4\pi}{(4 - \pi)\delta} \leqslant 15/\delta.$

## VII Comments

1) A recent reference on Gaussian processes is LIFSHITS. Theorem II.2 is due to Marcus and Shepp [1972].

   The presentation of Brownian motion as the image of the indicator functions of $[0, t]$ by an isometry from $L^2(\mathbb{R}^+)$ to a Gaussian space is found in KAHANE 2, with a very complete study of this Brownian motion, notably of its *slow* points and its *rapid* points. The continuity of its trajectories is obtained thanks to either the trigonometric system or the Haar system.

2) Dudley's majoration theorem was first obtained by Dudley [1967] in a purely Hilbertian context, and for Gaussian variables. The extension (Dudley's abstract theorem) to processes that are not necessarily Gaussian, but satisfy a Lipschitz condition in an Orlicz space, is due to Pisier [1980 b] and provides a very simple proof of the duality theorem $C^{as} - M_{2,\Psi_2}$ (see Chapter 6 of this volume); the initial proof used Preston's theorem (Preston [1971]), no longer necessary.

   The case of an arbitrary Orlicz function $\psi$ leads to an entropy integral $\int_0^{+\infty} \psi^{-1}[N(\varepsilon)] \, d\varepsilon$, at the price of a metric study more subtle than that given here: the nets $R_j$ must be nested, otherwise we are led to $\int_0^{+\infty} \psi^{-1}[N(\varepsilon)^2] \, d\varepsilon$; for $\psi(x) = \Psi_2(x) = e^{x^2} - 1$, the difference is not perceptible as the two integrals are equivalent. For the general case, see Pisier [1980 b] or LEDOUX–TALAGRAND.

3) The use of a continuous partition of unity in the abstract Dudley theorem is in KAHANE 2, and gives a simpler proof of the existence of a continuous version; this use was suggested by A. Fathi.

4) The proof of Fernique's theorem closely follows the one given in KAHANE 2, with notably the use of an auxiliary process on a Cantor tree which greatly clarifies matters. We have proceeded here with some technical simplifications: the auxiliary process is simpler, we need less information about it (that of Proposition V.3 suffices), and the simple Proposition II.2, Chapter 1 of this volume, allows estimations uniform with respect to the parameters. The idea of taking the points four-by-four is in the initial proof of Fernique [1975].

The applications of the Dudley–Fernique theorems to Harmonic Analysis (see Chapter 6, this volume) are essentially due to Pisier (see MARCUS–PISIER).

5) In the non-stationary case, the entropy integral is no longer sufficient to analyze the process. Talagrand [1987] (see also [1992 d]) showed that the adequate replacement is the notion of *majorizing measure* and used this to give a characterization of processes with continuous trajectories. Recently Talagrand [2001], showed that the notion of majorizing measure can in turn be replaced by that of an *adapted partition* (of the set $T$ indexing the process).

6) The Elton–Pajor theorem in the real case, in the form of Corollary VI.18, is due to Elton [1983]. Pajor [1983] (see PAJOR) gave both the proof presented here and that for the complex case. More precise (essentially optimal) estimations of the constants were made by Talagrand [1992 b]; see also Dilworth and Patterson [2003].

For the proof of the Brunn–Minkowski theorem and of Urysohn's inequality, we have followed PISIER 2 and BALL; for the use of the Prékopa–Leindler inequality, see Leindler [1972], Prékopa [1973] and Brascamp and Lieb [1976 a].

PISIER 2 attributes the proof given here of Urysohn's inequality to Milman.

# VIII Exercises

**Exercise VIII.1** In the theorems of Dudley and Fernique, provide the details for the passage to complex Gaussian processes.

**Exercise VIII.2** Let $0 < \beta < 1$, and let $(X_t)_{t\in\mathbb{R}}$ be a Gaussian process such that $\|X_s - X_t\|_2 \leqslant |s - t|^\beta$. Using the Marcus–Shepp theorem, show that $(X_t)_t$ has a continuous version.

**Exercise VIII.3** Let $\alpha \in \, ]0, 2]$. Show that there exists a centered Gaussian process $(X_t)_{t\in[0,1]}$, with covariance $K(s,t) = e^{-|s-t|^\alpha}$. Using Dudley's theorem, show that this process possesses a continuous version.

**Exercise VIII.4**   Set $S_j(f,0) = \sum_{-j}^{j} \widehat{f}(k)$ for $f \in L^1(\mathbb{T})$. Let $(\lambda_n)_{n \geqslant 1}$ be an increasing sequence of integers $> 0$. Suppose that, for a constant $C > 0$:

$$\frac{1}{N} \sum_{n=1}^{N} |S_{\lambda_n}(f,0)| \leqslant C \, \|f\|_\infty \qquad\qquad (*)$$

for any $N \geqslant 1$ and every $f \in \mathcal{C}(\mathbb{T})$.

1) Let $D_n$ be the Dirichlet kernel of order $n$. Show that there exists a constant $K > 0$ such that, for any $N \geqslant 1$:

$$\sup_{\theta_n = \pm 1} \left\| \sum_{n=1}^{N} \theta_n D_{\lambda_n} \right\|_1 \leqslant K N.$$

2) Show that there exists a constant $\mu > 0$ such that $\lambda_N \leqslant e^{\mu \sqrt{N}}$, for any $N \geqslant 1$.

3) Study the converse of 1) and 2). In particular, show that the sequence of all integers $\geqslant 1$ satisfies the condition $(*)$.

**Exercise VIII.5**   Let $(c_n)_{n \geqslant 1}$ be a sequence of complex numbers. We assume that there exists an increasing sequence of real numbers $\lambda_n > 0$ such that $\sum_{n=2}^{+\infty} 1/n(\log n)\lambda_n < +\infty$ and $\sum_{n=1}^{+\infty} |c_n|^2 (\log n)\lambda_n < +\infty$. Set $N_k = 2^{2^k}$, for $k \geqslant 0$.

1) Let $S_k = \left( \sum_{N_k < n \leqslant N_{k+1}} |c_n|^2 \log n \right)^{1/2}$. Show that $\sum_{k=0}^{+\infty} 1/\lambda_{N_k} < +\infty$, then that $\sum_{k=0}^{+\infty} S_k < +\infty$.

2) Let $(Z_n)_{n \geqslant 1}$ be a standard Gaussian sequence or a Bernoulli sequence. Show that there exists a numerical constant $C > 0$ such that, if $e_n(t) = e^{int}$, then:

$$\mathbb{E} \left\| \sum_{N_k < n \leqslant N_{k+1}} c_n Z_n e_n \right\|_\infty \leqslant C \, S_k.$$

Deduce that the series $\sum_{n=1}^{+\infty} c_n Z_n e_n$ is almost surely uniformly convergent.

3) Show that the result of 2) does not hold under the single hypothesis $\sum_{n=1}^{+\infty} |c_n|^2 \log n < +\infty$.

**Remark**   If $\sum_{n=1}^{+\infty} |c_n|^2 \log n < +\infty$, the random power series

$$\sum_{n=1}^{+\infty} c_n Z_n z^n$$

is almost surely in the space *BMOA* of analytic functions in the unit disk such that their boundary values have a bounded mean oscillation (Sledd [1981], Duren [1985]).

**Exercise VIII.6**  With the notation of Exercise VIII.5, for $1/2 < \alpha \leqslant 1$, consider the random series $\sum_{n=1}^{+\infty} n^{-\alpha} Z_n e_n$. Determine the spaces $\mathrm{Lip}_\beta$ to which this series belongs almost surely.

**Exercise VIII.7** (Milman [1982])  Let $x_1, \ldots, x_N$ be functions on a set $T$ taking on only the values $\pm 1$. Define:

$$M_N = \mathbb{E} \left\| \sum_{n=1}^{N} \varepsilon_n x_n \right\|_{\ell_\infty(T)},$$

and $S = \{(x_n(t))_{1 \leqslant n \leqslant N} ; \ t \in T\} \subseteq \{-1, 1\}^N$.

1) Show that, if $\Psi_2(x) = e^{x^2} - 1$:

$$M_N \leqslant \sqrt{\log(|S| + 1)} \sup_{s \in S} \left\| \sum_{n=1}^{N} \varepsilon_n s_n \right\|_{\Psi_2}$$

·  (use the Orlicz–Jensen inequality).
2) Deduce that $M_N \leqslant \sqrt{6N \log(|S| + 1)}$.
3) Show that if $\delta$ is the combinatorial density of $S$, then $|S| \leqslant n^{\delta+1}$ (use Sauer's lemma).
4) Show that there exists a subset $I$ of $\{1, \ldots, N\}$, of cardinality $|I| \geqslant M_N^2/(18N \log N)$, such that $(x_n)_{n \in I}$ is isometrically equivalent to the canonical basis of $\ell_1^I$.

**Exercise VIII.8** (Pisier [1984])  Under the same hypotheses as in Exercise VIII.7, use the Elton–Pajor theorem to obtain $|I| \geqslant c M_N^2/N$.

**Exercise VIII.9**  Let $0 < \xi < 1$, $1 < p \leqslant 2$, and let $z_1, \ldots, z_m$ be vectors of a Banach space $X$ such that $\sum_{j=1}^{m} \|z_j\|^p = m$, and:

$$\int_0^1 \left\| \sum_{j=1}^{m} r_j(t) z_j \right\| dt \geqslant \frac{\xi}{K_{1,2}} m^{1/p^*} \left( \sum_{j=1}^{m} \|z_j\|^p \right)^{1/p},$$

where $K_{1,2}$ is the constant appearing in the Khintchine–Kahane inequality. Set $\xi' = \xi/K_{1,2}$, and:

$$J_1 = \{j \leqslant m; \ \|z_j\| < (2/\xi')^{p^*/p}\}$$
$$J_2 = \{j \leqslant m; \ (2/\xi')^{p^*/p} \leqslant \|z_j\|\}.$$

1) Show that $|J_2| \leqslant (\xi'/2)^{p^*} m$, and then that $\sum_{j \in J_2} \|z_j\| \leqslant \xi' m/2$.
2) Deduce that $|J_1| \geqslant (\xi'/2)^{p^*} m$.

3) Show that:

$$\int_0^1 \left\| \sum_{j \in J_1} r_j(t) \frac{z_j}{\|z_j\|} \right\| dt \geqslant (\xi'/2)^{p^*} |J_1|.$$

4) Use the Elton–Pajor theorem, for $(z_j/\|z_j\|)_{j \in J_1}$, to obtain a constant $c > 1$ and a subset $I \subseteq \{1, \ldots, m\}$ such that:

$$|I| \geqslant m/c \quad \text{and} \quad \text{dist}(\text{span}\,\{z_j\,;\, j \in I\}, \ell_1^{|I|}) \leqslant c.$$

**Remark**   With additional "ingredients", it is possible to deduce from this the following result, due to Milman and Wolfson [1978]: *For any $0 < \delta < 1$, there exist a constant $C \geqslant 1$ and a sequence of integers $k_n \geqslant \beta(\delta) \log n$ such that every space $E$ of dimension $n$ with $\text{dist}(E, \ell_2^n) > \delta \sqrt{n}$ contains a subspace $F \subseteq E$ of dimension $\dim F = k_n$ with $\text{dist}(f, \ell_1^{k_n}) \leqslant C$* (see TOMCZAK-JAEGERMANN, Theorem 30.1).

# 4

## Reflexive Subspaces of $L^1$

### I Introduction

This chapter presents a study of the reflexive subspaces of $L^1$.

Section II characterizes the reflexive subspaces of $L^1$ as those for which the topologies of the norm and of convergence in measure coincide (the Kadeč–Pełczyński theorem); consequently, non-reflexive subspaces are shown to contain complemented subspaces isomorphic to $\ell_1$. Then their local structure is studied: since $L^1$ is weakly sequentially complete (Chapter 7 of Volume 1, Theorem II.6), according to the Rosenthal $\ell_1$ theorem (Chapter 8 of Volume 1), *a priori* its reflexive subspaces are those not containing $\ell_1$. However, much more is obtainable: the reflexive subspaces of $L^1$ are those not containing $\ell_1^n$'s uniformly. Then the Banach spaces not containing $\ell_1^n$'s uniformly are proved to be exactly those with type $p > 1$ (Theorem II.8, due to Pisier); thus each reflexive subspace of $L^1$ has a non-trivial type $p > 1$.

Section III provides several examples of reflexive subspaces. First, for $1 < p \leqslant 2$, the sequences of $p$-stable independent variables are shown to isometrically generate $\ell_p$ in $L^1$. Then comes a study of the $\Lambda(q)$-sets, the reflexive and translation-invariant subspaces of $L^1(\mathbb{T})$.

Section IV is dedicated to a deep theorem of Rosenthal showing that the reflexive subspaces of $L^1$ can be embedded into $L^p$, for some $p > 1$. The proof given here uses Maurey's factorization theorem, which Maurey extracted from the original proof of Rosenthal.

Section V examines the finite-dimensional subspaces of $L^1$, and, more precisely, the dimension $n$ of the $\ell_1^n$ spaces that they can contain (Theorem V.2, of Talagrand). For this, a preliminary study is required, of the $K$-convexity constants of finite-dimensional spaces, and in particular of the finite-dimensional spaces of $L^1$. An auxiliary result is also needed, due to Lewis. The proof then uses the method of selectors.

Throughout this chapter, $L^1$ denotes an infinite-dimensional space $L^1(\Omega, \mathbb{P})$, where $\mathbb{P}$ is an atomless probability measure on a set $\Omega$. Thus it could be the space $L^1(0, 1)$, where $[0, 1]$ is equipped with the Lebesgue measure, or $L^1(G)$, with $G$ an infinite (metrizable) compact Abelian group.

## II Structure of Reflexive Subspaces of $L^1$

### II.1 Reflexive Subspaces and Convergence in Measure

The aim of this subsection is to prove the following:

**Theorem II.1** (The Kadeč–Pełczyński Theorem)    *A subspace $X$ of $L^1$ is reflexive if and only if the topology of the norm and the topology of convergence in measure coincide on $X$.*

**Remark**    The coincidence of these two topologies translates into:

$$(\forall r > 0)\, (\exists \varepsilon > 0) : \left[\forall f \in X : \mathbb{P}(|f| \geqslant \varepsilon) \leqslant \varepsilon \ \Rightarrow \ \|f\|_1 \leqslant r\right].$$

With the choice $r = 1/2$, since $\left\| f/\|f\|_1 \right\|_1 = 1 > 1/2$, we obtain: there exists $\varepsilon_0 > 0$ such that:

$$\mathbb{P}(|f| \geqslant \varepsilon_0 \|f\|_1) > \varepsilon_0, \quad \forall f \in X.$$

Conversely, if this property holds, then, for any $r > 0$, taking $\varepsilon = \inf(\varepsilon_0, r\varepsilon_0)$, we have:

$$\mathbb{P}(|f| \geqslant \varepsilon) \leqslant \varepsilon \ \Rightarrow \ \mathbb{P}(|f| \geqslant \varepsilon) \leqslant \varepsilon_0 \ \Rightarrow \ \varepsilon > \varepsilon_0 \|f\|_1 \ \Rightarrow \ \|f\|_1 \leqslant r.$$

This remark leads to the introduction, for any $\varepsilon > 0$, of the set

$$M_\varepsilon = \left\{ f \in L^1(\Omega); \ \mathbb{P}(|f| \geqslant \varepsilon \|f\|_1) > \varepsilon \right\}.$$

The remark can then be stated in the following form:

**Lemma II.2**    *The topology of the norm and the topology of convergence in measure coincide on $X \subseteq L^1(\Omega)$ if and only if:*

$$(\exists \varepsilon_0 > 0) \qquad X \subseteq M_{\varepsilon_0}.$$

Thus the Kadeč–Pełczyński theorem can be reformulated as follows:

**Theorem II.3** (The Kadeč–Pełczyński Theorem, Reformulation)    *A subspace $X$ of $L^1(\Omega)$ is reflexive if and only if there exists an $\varepsilon_0 > 0$ such that $X \subseteq M_{\varepsilon_0}$.*

Here is a third reformulation:

**Proposition II.4**   *For every subspace $X$ of $L^1(\Omega)$, the following properties are equivalent*:

1) *There exists $\varepsilon_0 > 0$ such that $X \subseteq M_{\varepsilon_0}$.*
2) *For any $\alpha \in \,]0, 1[$, there exists a constant $C_\alpha > 0$ such that*

$$\|f\|_1 \leqslant C_\alpha \|f\|_\alpha$$

   *for every $f \in X$.*
3) *There exists an $\alpha \in \,]0, 1[$ and a constant $C_\alpha > 0$ such that*

$$\|f\|_1 \leqslant C_\alpha \|f\|_\alpha$$

   *for every $f \in X$.*

Note that, for $\alpha < 1$, $\|f\|_\alpha = \left( \int_\Omega |f|^\alpha \right)^{1/\alpha}$ is no longer a norm on $L^1$, but this does not matter.

*Proof*  1) $\Rightarrow$ 2) It suffices to write:

$$\|f\|_\alpha = \left[ \int_\Omega |f|^\alpha \right]^{1/\alpha} \geqslant \left[ \int_{\{|f| \geqslant \varepsilon_0 \|f\|_1\}} |f|^\alpha \right]^{1/\alpha}$$
$$\geqslant \varepsilon_0 \|f\|_1 \big[ \mathbb{P}(\{|f| \geqslant \varepsilon_0 \|f\|_1\}) \big]^{1/\alpha} \geqslant \varepsilon_0^{1+1/\alpha} \|f\|_1 \,.$$

2) $\Rightarrow$ 3) This needs no justification.

3) $\Rightarrow$ 1) Assume that 3) holds. Let $f \in X$ be such that $f \notin M_\varepsilon$, with $\varepsilon > 0$, which we can suppose to be $< 1$. Let $\Omega_\varepsilon = \{|f| \geqslant \varepsilon \|f\|_1\}$; then:

$$\|f\|_\alpha^\alpha = \int_\Omega |f|^\alpha \, d\mathbb{P} = \int_{\Omega_\varepsilon} |f|^\alpha \, d\mathbb{P} + \int_{\Omega_\varepsilon^c} |f|^\alpha \, d\mathbb{P}$$
$$\leqslant \int_{\Omega_\varepsilon} |f|^\alpha \, d\mathbb{P} + \varepsilon^\alpha \|f\|_1^\alpha \leqslant \left[ \int_\Omega |f| \, d\mathbb{P} \right]^\alpha \big[ \mathbb{P}(\Omega_\varepsilon) \big]^{1/\beta} + \varepsilon^\alpha \|f\|_1^\alpha \,,$$

by using Hölder's inequality for the exponent $q = 1/\alpha > 1$, and with $\beta = 1/(1 - \alpha)$ the conjugate exponent of $q$. Since $\mathbb{P}(\Omega_\varepsilon) \leqslant \varepsilon$ when $f \notin M_\varepsilon$, we thus obtain:

$$\|f\|_\alpha^\alpha \leqslant \|f\|_1^\alpha \big( \varepsilon^{1/\beta} + \varepsilon^\alpha \big) = \|f\|_1^\alpha \big( \varepsilon^{1-\alpha} + \varepsilon^\alpha \big),$$

and hence, setting $\delta = \inf \big( 1, (1 - \alpha)/\alpha \big)$:

$$\|f\|_\alpha \leqslant 2^{1/\alpha} \varepsilon^\delta \|f\|_1 \,.$$

Then, since $f \neq 0$ (as $f \notin M_\varepsilon$), condition 3) leads to $1/C_\alpha \leqslant 2^{1/\alpha} \varepsilon^\delta$; that is, $\varepsilon \geqslant 1/(2^{1/\alpha} C_\alpha)^{1/\delta}$. Consequently, if we take $\varepsilon_0$ strictly smaller than this latter value, for example $\varepsilon_0 = (1/2)(2^{1/\alpha} C_\alpha)^{-1/\delta}$, we indeed obtain $f \in M_{\varepsilon_0}$. $\qquad\square$

The proof of the theorem relies on the following essential lemma, to be used twice.

**Lemma II.5** (The Kadeč–Pełczyński Lemma)  *Let $(f_n)_{n \geq 1}$ be a sequence of elements of $L^1(\Omega)$ such that, for every $\varepsilon > 0$, there exists an integer $n(\varepsilon) \geq 1$ for which:*

$$\mathbb{P}\big(\{t \in \Omega \,;\, |f_{n(\varepsilon)}(t)| \geq \varepsilon \, \|f_{n(\varepsilon)}\|_1\}\big) \leq \varepsilon \,.$$

*Then $(f_n)_{n \geq 1}$ contains a subsequence $(g_n)_{n \geq 1}$ such that $\big(g_n / \|g_n\|_1\big)_{n \geq 1}$ is equivalent to the canonical basis of $\ell_1$, and moreover $[g_n, \ n \geq 1] = \overline{\text{span}}\{g_n \,;\, n \geq 1\}$ is complemented in $L^1(\Omega)$.*

**Remark**   The hypothesis means that $\{f_n \,;\, n \geq 1\} \not\subseteq M_\varepsilon$. The lemma hence states that if $X \not\subseteq M_\varepsilon$, for every $\varepsilon > 0$, then $X$ contains a subspace isomorphic to $\ell_1$, and complemented in $L^1(\Omega)$.

Thus the following consequence:

**Corollary II.6** (Kadeč–Pełczyński)   *If $X$ is a non-reflexive subspace of $L^1(\Omega)$, then $X$ contains a subspace isomorphic to $\ell_1$ and complemented in $L^1(\Omega)$.*

Recall that Rosenthal's $\ell_1$ theorem implies that each non-reflexive weakly sequentially complete space contains a subspace isomorphic to $\ell_1$. Here, an additional result is the complementation of this subspace in $L^1(\Omega)$.

Also, since $\ell_1$ is not reflexive, the lemma has an immediate implication: if $X$ is a reflexive subspace of $L^1(\Omega)$, then there exists an $\varepsilon_0 > 0$ such that $X \subseteq M_{\varepsilon_0}$, i.e. the topologies of the norm and of convergence in measure coincide on $X$, which is the necessary condition of Theorem II.1.

*Proof of Lemma II.5*   Denote:

$$\Omega(\varepsilon) = \big\{t \in \Omega \,;\, |f_{n(\varepsilon)}(t)| \geq \varepsilon \, \|f_{n(\varepsilon)}\|_1\big\} \,;$$

by hypothesis, $\mathbb{P}[\Omega(\varepsilon)] \leq \varepsilon$. Moreover, as $|f_{n(\varepsilon)}| < \varepsilon \, \|f_{n(\varepsilon)}\|_1$ on $\Omega(\varepsilon)^c$, then:

$$\int_{\Omega(\varepsilon)} \frac{|f_{n(\varepsilon)}|}{\|f_{n(\varepsilon)}\|_1} \, d\mathbb{P} = \int_\Omega \frac{|f_{n(\varepsilon)}|}{\|f_{n(\varepsilon)}\|_1} \, d\mathbb{P} - \int_{\Omega(\varepsilon)^c} \frac{|f_{n(\varepsilon)}|}{\|f_{n(\varepsilon)}\|_1} \, d\mathbb{P} \geq 1 - \varepsilon \,.$$

Setting:

$$\Omega_1 = \Omega(1/4^2) \quad \text{and} \quad f_{n_1} = f_{n(1/4^2)} \,,$$

we thus obtain a subset $\Omega_1 \subseteq \Omega$ and an integer $n_1 \geq 1$ such that:

$$\mathbb{P}(\Omega_1) \leq \frac{1}{4^2} \quad \text{and} \quad \int_{\Omega_1} \frac{|f_{n_1}|}{\|f_{n_1}\|_1} \, d\mathbb{P} \geq 1 - \frac{1}{4^2} \,.$$

Moreover, the absolute continuity of the integral leads to the existence of an $\varepsilon_1 < 1/4^3$ such that:

$$\mathbb{P}(A) \leqslant \varepsilon_1 \implies \int_A \frac{|f_{n_1}|}{\|f_{n_1}\|_1}\, d\mathbb{P} \leqslant \frac{1}{4^3} \, ;$$

hence $\Omega_2 = \Omega(\varepsilon_1)$ and $f_{n_2} = f_{n(\varepsilon_1)}$ satisfy:

$$\mathbb{P}(\Omega_2) \leqslant \frac{1}{4^3}, \quad \int_{\Omega_2} \frac{|f_{n_2}|}{\|f_{n_2}\|_1}\, d\mathbb{P} \geqslant 1 - \frac{1}{4^3} \quad \text{and} \quad \int_{\Omega_2} \frac{|f_{n_1}|}{\|f_{n_1}\|_1}\, d\mathbb{P} \leqslant \frac{1}{4^3}\, .$$

With this method, we construct a subsequence $(g_n)_{n \geqslant 1}$ of $(f_n)_{n \geqslant 1}$, and sets $\Omega_n$, $n \geqslant 1$, such that $\mathbb{P}(\Omega_n) \leqslant 1/4^{n+1}$ and:

$$\int_{\Omega_n} \frac{|g_n|}{\|g_n\|_1}\, d\mathbb{P} \geqslant 1 - \frac{1}{4^{n+1}} \quad \text{and} \quad \int_{\Omega_n} \frac{|g_j|}{\|g_j\|_1}\, d\mathbb{P} \leqslant \frac{1}{4^{n+1}} \quad \text{for } j \leqslant n-1.$$

We separate these sets $\Omega_n$ by setting:

$$E_n = \Omega_n \smallsetminus \bigcup_{k=n+1}^{+\infty} \Omega_k \quad \text{and} \quad h_n = \frac{g_n}{\|g_n\|_1} \mathbb{1}_{E_n}\, .$$

Then:

$$\left\| \frac{g_n}{\|g_n\|_1} - h_n \right\|_1 = \int_{E_n^c} \frac{|g_n|}{\|g_n\|_1}\, d\mathbb{P} = \int_{\Omega_n^c} \frac{|g_n|}{\|g_n\|_1}\, d\mathbb{P} + \int_{\Omega_n \smallsetminus E_n} \frac{|g_n|}{\|g_n\|_1}\, d\mathbb{P}$$

$$\leqslant \frac{1}{4^{n+1}} + \sum_{k=n+1}^{+\infty} \int_{\Omega_k} \frac{|g_n|}{\|g_n\|_1}\, d\mathbb{P}$$

$$\leqslant \frac{1}{4^{n+1}} + \sum_{k=n+1}^{+\infty} \frac{1}{4^{k+1}} \leqslant \frac{1}{4^n}\, .$$

However, since:

$$1 \geqslant \|h_n\|_1 = \int_{E_n} \frac{|g_n|}{\|g_n\|_1}\, d\mathbb{P} \geqslant \int_{\Omega_n} \frac{|g_n|}{\|g_n\|_1}\, d\mathbb{P} - \sum_{k=n+1}^{+\infty} \int_{\Omega_k} \frac{|g_n|}{\|g_n\|_1}\, d\mathbb{P}$$

$$\geqslant \left( 1 - \frac{1}{4^{n+1}} \right) - \sum_{k=n+1}^{+\infty} \frac{1}{4^{k+1}} \geqslant 1 - \frac{1}{4^n}\, ,$$

we obtain:

$$\left\| \frac{g_n}{\|g_n\|_1} - \frac{h_n}{\|h_n\|_1} \right\|_1 \leqslant \left\| \frac{g_n}{\|g_n\|_1} - h_n \right\|_1 + \left\| h_n - \frac{h_n}{\|h_n\|_1} \right\|_1$$

$$\leqslant \frac{1}{4^n} + (1 - \|h_n\|_1) \leqslant \frac{2}{4^n}\, .$$

Now, the functions $h_n/\|h_n\|_1$ have *disjoint supports* $S_n$, so they generate a subspace of $L^1(\Omega)$ isometric to $\ell_1$, and complemented by a projection $Q$ of norm 1:

$$Q(f) = \sum_{n=1}^{+\infty} \left[ \int_{S_n} \frac{\operatorname{sgn} h_n(\omega)}{\|h_n\|_1} f(\omega) \, d\mathbb{P}(\omega) \right] h_n \, .$$

Now let $\varphi_n \in \big[L^1(\Omega)\big]^*$ be the linear functionals of norm 1, extending the coordinate linear functionals of the basic sequence $(h_n/\|h_n\|_1)_{n \geqslant 1}$. Since:

$$\sum_{n=1}^{+\infty} \|\varphi_n\|_\infty \left\| \frac{g_n}{\|g_n\|_1} - \frac{h_n}{\|h_n\|_1} \right\|_1 \leqslant \sum_{n=1}^{+\infty} \frac{2}{4^n} < 1 \, ,$$

by the Bessaga–Pełczyński equivalence theorem (Chapter 2 of Volume 1), the sequence $(g_n/\|g_n\|_1)_{n \geqslant 1}$ is equivalent to the canonical basis of $\ell_1$. More precisely, if $A \colon L^1(\Omega) \to L^1(\Omega)$ is defined by:

$$Af = (f - Qf) + \sum_{n=1}^{+\infty} \varphi_n(Qf) \frac{g_n}{\|g_n\|_1} \, ,$$

then $\|A - \operatorname{Id}\| < 1$, and thus $A$ is an isomorphism from $L^1(\Omega)$ onto itself, which, for every $n \geqslant 1$, maps $h_n/\|h_n\|_1$ to $g_n/\|g_n\|_1$. Moreover, $R = AQA^{-1}$ is a projection from $L^1(\Omega)$ onto $[g_n \, ; \; n \geqslant 1]$. $\qquad\square$

*Proof of the sufficient condition of Theorem II.1*  Let $X$ be a non-reflexive subspace of $L^1(\Omega)$. Its unit ball $B_X$ is thus not uniformly integrable:

$$\limsup_{a \to +\infty} \left( \sup_{f \in B_X} \int_{\{|f| \geqslant a\}} |f| \, d\mathbb{P} \right) = \delta > 0 \, .$$

Hence there exists a sequence of real positive numbers $(a_n)_{n \geqslant 1}$, increasing to infinity, such that $a_n \geqslant 2^n$ and

$$\delta - \frac{\delta}{2^n} < \sup_{f \in B_X} \int_{\{|f| \geqslant a_n\}} |f| \, d\mathbb{P} \leqslant \delta + \frac{\delta}{2^n} \, , \qquad \forall n \geqslant 1 \, .$$

Thus there exists a sequence $(f_n)_{n \geqslant 1}$ in $B_X$ such that:

$$\delta - \frac{\delta}{2^n} < \int_{\{|f_n| \geqslant a_n\}} |f_n| \, d\mathbb{P} \leqslant \delta + \frac{\delta}{2^n} \, , \qquad \forall n \geqslant 1 \, .$$

Set:

$$\begin{cases} g_n = f_n . \mathbb{I}_{\{|f_n| \geqslant a_n\}} \\ h_n = f_n . \mathbb{I}_{\{|f_n| < a_n\}} = f_n - g_n \, . \end{cases}$$

Then:

a) For every $\varepsilon > 0$, $\|f_n\|_1 \leqslant 1$ leads to:

$$\mathbb{P}(|g_n| \geqslant \varepsilon) \leqslant \mathbb{P}(|g_n| \neq 0) = \mathbb{P}(|f_n| \geqslant a_n) \leqslant \frac{1}{a_n} \leqslant \frac{1}{2^n},$$

and hence $(g_n)_{n \geqslant 1}$ converges in measure to $0$.

Moreover, as $\delta - \dfrac{\delta}{2^n} \leqslant \|g_n\|_1 \leqslant \delta + \dfrac{\delta}{2^n}$, then:

$$\frac{\delta}{2} \leqslant \|g_n\|_1 \leqslant \frac{3\delta}{2}, \qquad \forall n \geqslant 1.$$

b) Furthermore, $(h_n)_{n \geqslant 1}$ is uniformly integrable: indeed, if $a \geqslant a_j$,

$$\sup_{n \geqslant 1} \int_{\{|h_n| \geqslant a\}} |h_n| \, d\mathbb{P} \leqslant \sup_{n \geqslant 1} \int_{\{|h_n| \geqslant a_j\}} |h_n| \, d\mathbb{P}$$

$$= \sup_{n > j} \int_{\{|h_n| \geqslant a_j\}} |h_n| \, d\mathbb{P}$$

$$\text{since for } n \leqslant j: |h_n| < a_n \leqslant a_j$$

$$= \sup_{n > j} \int_{\{a_n > |f_n| \geqslant a_j\}} |f_n| \, d\mathbb{P}$$

$$= \sup_{n > j} \left( \int_{\{|f_n| \geqslant a_j\}} |f_n| \, d\mathbb{P} - \int_{\{|f_n| \geqslant a_n\}} |f_n| \, d\mathbb{P} \right)$$

$$\leqslant \sup_{n > j} \left[ \left( \delta + \frac{\delta}{2^j} \right) - \left( \delta - \frac{\delta}{2^j} \right) \right]$$

$$\leqslant \frac{\delta}{2^{j-1}} \xrightarrow[j \to +\infty]{} 0.$$

The set $\{h_n \, ; \, n \geqslant 1\}$ is hence relatively weakly compact, and, by the Eberlein–Šmulian theorem, a weakly convergent subsequence $(h_{n_k})_{k \geqslant 1}$ can be extracted.

Thus a subsequence $(f_{n_k})_{k \geqslant 1}$ of $(f_n)_{n \geqslant 1}$ has been obtained, that can be written:

$$f_{n_k} = g_{n_k} + h_{n_k},$$

where $(g_{n_k})_{k \geqslant 1}$ converges to $0$ in measure and $(h_{n_k})_{k \geqslant 1}$ weakly converges. This is a version of the *subsequence splitting lemma* (see Exercise VII.2). However, there is no reason for these two subsequences to be in $X$.

To resolve this problem, Lemma II.5 is used once again. Indeed, the sequence $(g_{n_k})_{k \geqslant 1}$ satisfies the hypotheses: for any $\varepsilon > 0$, we have:

$$\mathbb{P}\big(|g_{n_k}| \geqslant \varepsilon \|g_{n_k}\|_1\big) \leqslant \mathbb{P}(g_{n_k} \neq 0) = \mathbb{P}(|f_{n_k}| \geqslant a_{n_k}) \leqslant \frac{1}{a_{n_k}} \leqslant \varepsilon$$

for $k$ large enough. Hence a subsequence $(g'_l)_{l \geqslant 1}$ can be extracted such that $(g'_l/\|g'_l\|_1)_{l \geqslant 1}$ is $C$-equivalent to the canonical basis of $\ell_1$. The corresponding subsequence $(h'_l)_{l \geqslant 1}$ still weakly converges. As the weak closure of a convex set is equal to its closure in norm, there exist convex combinations:

$$w_m = \sum_{l=l_m+1}^{l_{m+1}} \lambda_l^{(m)} h'_l,$$

with $l_0 = 0 < l_1 < l_2 < \ldots$, $\lambda_l^{(m)} \geqslant 0$ and $\sum_{l=l_m+1}^{l_{m+1}} \lambda_l^{(m)} = 1$, which converge in norm. We denote their limit by $h$. Now we consider the corresponding convex combinations of the sequences $(f'_l)_{l \geqslant 1}$ and $(g'_l)_{l \geqslant 1}$:

$$\begin{cases} u_m = \displaystyle\sum_{l=l_m+1}^{l_{m+1}} \lambda_l^{(m)} f'_l \\[2mm] v_m = \displaystyle\sum_{l=l_m+1}^{l_{m+1}} \lambda_l^{(m)} g'_l \ . \end{cases}$$

Observe that the sequence $(v_m)_{m \geqslant 1}$ still converges in measure to $0$; indeed, even though in general the convex combinations of a sequence that converges in measure to $0$ does not converge in measure, here we have almost sure convergence, since, for every $\varepsilon > 0$:

$$\mathbb{P}(|v_m| \geqslant \varepsilon) \leqslant \sum_{l=l_m+1}^{l_{m+1}} \mathbb{P}(|g'_l| \geqslant \varepsilon) \leqslant \sum_{l=l_m+1}^{l_{m+1}} \frac{1}{2^l} \leqslant \frac{1}{2^{l_m}},$$

and hence $v_m \xrightarrow[m \to +\infty]{a.s.} 0$. Consequently $u_m = v_m + w_m \xrightarrow[m \to +\infty]{\mathbb{P}} h$. However:

$$\|u_m - h\|_1 \geqslant \|v_m\|_1 - \|w_m - h\|_1$$

$$= \left\| \sum_{l=l_m+1}^{l_{m+1}} \lambda_l^{(m)} \|g'_l\|_1 \frac{g'_l}{\|g'_l\|_1} \right\|_1 - \|w_m - h\|_1$$

$$\geqslant \frac{1}{C} \sum_{l=l_m+1}^{l_{m+1}} \lambda_l^{(m)} \|g'_l\|_1 - \|w_m - h\|_1 \geqslant \frac{1}{C} \frac{\delta}{2} - \|w_m - h\|_1 ;$$

therefore:

$$\liminf_{m \to +\infty} \|u_m - h\|_1 \geqslant \frac{1}{C} \frac{\delta}{2},$$

and hence $(u_m)_{m \geq 1}$ does not converge in norm (otherwise, it would have to converge to $h$, since it converges in probability to $h$).

It is now possible to draw a conclusion, even though *a priori* $h \notin X$; indeed, finally, the topologies of convergence in measure and of convergence in norm do not coincide on $X$, since $u_m \in X$, and the sequence $(u_m)_{m \geq 1}$ is Cauchy in measure, but not in norm. $\qquad \square$

## II.2 Local Structure of Reflexive Subspaces of $L^1$

From Corollary II.6, it ensues that every non-reflexive subspace of $L^1(\Omega)$ contains $\ell_1$. *A fortiori*, it contains $\ell_1^n$'s uniformly. A Banach space $X$ is said to *contain $\ell_1^n$'s uniformly* if a constant $C > 0$ can be found such that, for every $n \geq 1$, there exists an isomorphism $T_n \colon \ell_1^n \to X_n \subseteq X$ with $\|T_n\| . \|T_n^{-1}\| \leq C$. In general, this property is far from sufficient for $X$ to contain $\ell_1$, but for the subspaces of $L^1(\Omega)$, this in fact holds.

**Theorem II.7** *If a subspace $X$ of $L^1(\Omega)$ contains $\ell_1^n$'s uniformly, then there exists $\varepsilon_0 > 0$ such that, for every $N \geq 1$, there exist disjoint subsets $A_1, \ldots, A_N$ of $\Omega$ such that*:

$$\sup_{f \in B_X} \left( \int_{A_n} |f| \, d\mathbb{P} \right) \geq \varepsilon_0, \quad n = 1, 2, \ldots, N.$$

*In particular, $X$ is not reflexive.*

*Proof* Let us now prove the first assertion.

As $X$ contains $\ell_1^n$'s uniformly, a constant $C > 0$ can be found such that, for every $n \geq 1$, there are $f_1, \ldots, f_n \in B_X$ with:

$$\frac{1}{C} \sum_{k=1}^n |\alpha_k| \leq \int_\Omega \left| \sum_{k=1}^n \alpha_k f_k \right| d\mathbb{P}$$

for every choice of scalars $\alpha_1, \ldots, \alpha_n$. We now show that the statement of the theorem is obtained with $\varepsilon_0 = 1/(2C^2)$. Let $N \geq 1$, and $n \geq (2C^2 - 1)N$. Let $(r_k)_{k \geq 1}$ be the sequence of Rademacher functions. Then:

$$\frac{1}{C} \sum_{k=1}^n |\alpha_k| \leq \int_0^1 \left( \int_\Omega \left| \sum_{k=1}^n \alpha_k r_k(t) f_k \right| d\mathbb{P} \right) dt$$

$$\leq \int_\Omega \left( \int_0^1 \left| \sum_{k=1}^n \alpha_k r_k(t) f_k \right|^2 dt \right)^{1/2} d\mathbb{P}$$

$$= \int_\Omega \left( \sum_{k=1}^n |\alpha_k f_k|^2 \right)^{1/2} d\mathbb{P}$$

$$\leqslant \int_\Omega \left( \max_k |\alpha_k f_k| \right)^{1/2} \left( \sum_{k=1}^n |\alpha_k f_k| \right)^{1/2} d\mathbb{P}$$

$$\leqslant \left( \int_\Omega \max_k |\alpha_k f_k| \, d\mathbb{P} \right)^{1/2} \left( \int_\Omega \sum_{k=1}^n |\alpha_k f_k| \, d\mathbb{P} \right)^{1/2}$$

$$\leqslant \left( \int_\Omega \max_k |\alpha_k f_k| \, d\mathbb{P} \right)^{1/2} \left( \sum_{k=1}^n |\alpha_k| \right)^{1/2} ;$$

therefore:

$$\frac{1}{C^2} \sum_{k=1}^n |\alpha_k| \leqslant \int_\Omega \left( \max_k |\alpha_k f_k| \right) d\mathbb{P} .$$

In particular:

$$\frac{n}{C^2} \leqslant \int_\Omega \left( \max_k |f_k| \right) d\mathbb{P} .$$

Then let $B_1, \ldots, B_n$ be disjoint sets (possibly empty), whose union is $\Omega$, and such that:

$$\left( \max_k |f_k| \right) \mathbb{I}_{B_j} = |f_j| \, \mathbb{I}_{B_j} , \qquad \text{for } j = 1, 2, \ldots, n .$$

We have:

$$\frac{n}{C^2} \leqslant \sum_{j=1}^n \int_{B_j} |f_j| \, d\mathbb{P} .$$

Let $N_1$ be the number of indices $j$ such that:

$$(*) \qquad \int_{B_j} |f_j| \, d\mathbb{P} \geqslant \frac{1}{2C^2} = \varepsilon_0 .$$

Since $\|f_j\|_1 \leqslant 1$, we have:

$$\frac{n}{C^2} \leqslant \frac{n - N_1}{2C^2} + N_1 ,$$

or: $N_1 \geqslant \frac{n}{2C^2 - 1} \geqslant N$.

We take for $A_1, \ldots, A_N$ $N$ of the $B_j$'s for which $(*)$ holds.

Then $X$ is not reflexive because its unit ball $B_X$ cannot be uniformly absolutely continuous; indeed, for any $\delta > 0$, with an $N > 1/\delta$, we can find some $A_n$, $1 \leqslant n \leqslant N$, such that $\mathbb{P}(A_n) \leqslant \delta$, because the subsets $A_1, \ldots, A_N$ are pairwise disjoint. $\qquad \square$

It is known that a Banach space containing $\ell_1^n$'s uniformly cannot have any type $> 1$. Remarkably, the converse is true:

**Theorem II.8** (Pisier)  *If a Banach space $X$ does not contain $\ell_1^n$'s uniformly, then it has some type $p > 1$.*

An immediate consequence is the following:

**Corollary II.9**  *Every reflexive subspace of $L^1(\Omega)$ possesses a non-trivial type $p > 1$.*

The proof of the theorem is essentially based on the following lemma (where the $r_j$'s are the Rademacher functions):

**Lemma II.10**  *Let $X$ be a Banach space. Denote by:*

$$\gamma_n(X) = \inf\left\{ \gamma > 0 \,;\; \left\| \sum_{j=1}^n r_j x_j \right\|_{L^2(X)} \leqslant \gamma \left( \sum_{j=1}^n \|x_j\|^2 \right)^{1/2}, \right.$$
$$\left. \forall\, x_1, \ldots, x_n \in X \right\}$$

*the norm of the mapping*

$$(x_1, \ldots, x_n) \in \ell_2(X) \mapsto \sum_{j=1}^n r_j x_j \in L^2([0,1]\,;\, X)\,.$$

*Then:*

a) *the sequence $(\gamma_n(X))_{n \geqslant 1}$ is sub-multiplicative:*

$$\gamma_{nk}(X) \leqslant \gamma_n(X)\, \gamma_k(X)\,;$$

b) *if there exists an $n_0 \geqslant 1$ such that $\gamma_{n_0}(X) < \sqrt{n_0}$, then $X$ has a type $p > 1$.*

Note that we always have:

$$\gamma_n(X) \leqslant \sqrt{n}$$

by the Cauchy–Schwarz inequality: $\sum_{j=1}^n \|x_j\| \leqslant \sqrt{n}\left( \sum_{j=1}^n \|x_j\|^2 \right)^{1/2}$.

*Proof*

a) Fix $n$ and $k$. Given an $\varepsilon > 0$, we choose $x_1, \ldots, x_{nk} \in X$ such that:

$$\left( \int_0^1 \left\| \sum_{j=1}^{nk} r_j(t) x_j \right\|^2 dt \right)^{1/2} > \left( \gamma_{nk}(X) - \varepsilon \right) \left( \sum_{j=1}^{nk} \|x_j\|^2 \right)^{1/2}.$$

For $s = 0, 1, \ldots, (k-1)$, we define:

$$\phi_s(\theta) = \sum_{j=sn+1}^{(s+1)n} r_j(\theta) x_j.$$

Then, for any $\theta$:

$$\int_0^1 \left\| \sum_{s=0}^{k-1} r_s(t) \phi_s(\theta) \right\|^2 dt \leqslant \gamma_k^2(X) \sum_{s=0}^{k-1} \|\phi_s(\theta)\|^2.$$

Since the Rademacher functions are symmetric, an integration with respect to $\theta$ leads to:

$$\int_0^1 \left\| \sum_{j=1}^{nk} r_j(t) x_j \right\|^2 dt = \int_0^1 \int_0^1 \left\| \sum_{s=0}^{k-1} r_s(t) \phi_s(\theta) \right\|^2 dt\, d\theta$$

$$\leqslant \gamma_k^2(X) \sum_{s=0}^{k-1} \int_0^1 \|\phi_s(\theta)\|^2 \, dt\, d\theta$$

$$\leqslant \gamma_k^2(X) \sum_{s=0}^{k-1} \gamma_n^2(X) \sum_{j=sn+1}^{(s+1)n} \|x_j\|^2$$

$$= \gamma_k^2(X)\, \gamma_n^2(X) \sum_{j=1}^{nk} \|x_j\|^2,$$

which proves the sub-multiplicativity.

b) Assume that $\gamma_{n_0}(X) < \sqrt{n_0}$. First note that this is possible only if $n_0 \geqslant 2$. Then the number $p_0 > 1$ is defined by:

$$\gamma_{n_0}(X) = n_0^{\frac{1}{p_0} - \frac{1}{2}}.$$

We have $1 < p_0 \leqslant 2$, and we now show that $X$ has type $p$ for any $p$ satisfying $1 < p < p_0$. For this, we select an arbitrary finite sequence $x_1, \ldots, x_l$ of elements of $X$ for which $\sum_{j=1}^{l} \|x_j\|^p = 1$. Next, we define the subsets of indices:

$$A_s = \left\{ j \geqslant 1 \,;\, n_0^{-\frac{s+1}{p}} \leqslant \|x_j\| \leqslant n_0^{-\frac{s}{p}} \right\}.$$

Since $\sum_{j \in A_s} \|x_j\|^p \leqslant 1$, then $\mathrm{card}(A_s) = |A_s| \leqslant n_0^{s+1}$; hence:

$$\left\| \sum_{j=1}^{l} r_j x_j \right\|_{L^2(X)} = \left( \int_0^1 \left\| \sum_{j=1}^{l} r_j(t) x_j \right\|^2 dt \right)^{1/2}$$

$$\leqslant \sum_{s=0}^{+\infty} \left( \int_0^1 \left\| \sum_{j \in A_s} r_j(t) x_j \right\|^2 dt \right)^{1/2}$$

$$\leqslant \sum_{s=0}^{+\infty} \gamma_{|A_s|}(X) \left( \sum_{j \in A_s} \|x_j\|^2 \right)^{1/2}$$

$$\leqslant \sum_{s=0}^{+\infty} \gamma_{|A_s|}(X) \sqrt{|A_s|}\, n_0^{-\frac{s}{p}}.$$

However the sub-multiplicativity of $(\gamma_n)_{n \geqslant 1}$ implies

$$\gamma_{n_0^{s+1}} \leqslant (\gamma_{n_0})^{s+1} = n_0^{(s+1)\left(\frac{1}{p_0} - \frac{1}{2}\right)},$$

and consequently, since the sequence $(\gamma_n)_{n \geqslant 1}$ is also non-decreasing, we obtain:

$$\left\| \sum_{j=1}^{l} r_j x_j \right\|_{L^2(X)} \leqslant \sum_{s=0}^{+\infty} n_0^{(s+1)\left(\frac{1}{p_0} - \frac{1}{2}\right)} \sqrt{n_0^{s+1}}\, n_0^{-\frac{s}{p}}$$

$$= n_0^{\frac{1}{p_0}} \sum_{s=0}^{+\infty} n_0^{-s\left(\frac{1}{p} - \frac{1}{p_0}\right)} = C < +\infty,$$

because $p < p_0$. The proof of the lemma is thus complete.    □

We now proceed to prove the theorem.

*Proof of Theorem II.8*

1) According to the lemma, if there does not exist any $p > 1$ for which $X$ has type $p$, then necessarily:

$$\gamma_n(X) = \sqrt{n} \quad \text{for all } n \geqslant 1.$$

This can also be written:

$$(*) \quad \begin{cases} (\forall\, n \geqslant 1)\ (\forall\, \varepsilon > 0)\ (\exists\, x_1, \dots, x_n \in X): \\[2mm] (1 - \varepsilon)\sqrt{n} \left( \sum_{j=1}^{n} \|x_j\|^2 \right)^{1/2} < \left\| \sum_{j=1}^{n} r_j x_j \right\|_{L^2(X)}. \end{cases}$$

As $\left\| \sum_{j=1}^n r_j x_j \right\|_{L^2(X)} \leqslant \sum_{j=1}^n \|x_j\|$, we thus obtain:

$$(1-\varepsilon)\sqrt{n} \left( \sum_{j=1}^n \|x_j\|^2 \right)^{1/2} \leqslant \sum_{j=1}^n \|x_j\|,$$

which implies, for $\varepsilon$ "sufficiently" small (depending on $n$), that the $\|x_j\|$ are, for $1 \leqslant j \leqslant n$, "almost constant", More precisely:

**Lemma II.11**  *If $a_1, \ldots, a_n \geqslant 0$, and:*

$$(1-\varepsilon)\sqrt{n} \left( \sum_{j=1}^n a_j^2 \right)^{1/2} \leqslant \sum_{j=1}^n a_j,$$

*then, denoting $\alpha = \left( \sum_{j=1}^n a_j^2 \right)^{1/2}$, we have, for $j = 1, 2, \ldots, n$:*

$$\alpha \left( \frac{1}{\sqrt{n}} - \sqrt{2\varepsilon} \right) \leqslant a_j \leqslant \alpha \left( \frac{1}{\sqrt{n}} + \sqrt{2\varepsilon} \right).$$

*Proof*   Let $a = (a_1, \ldots, a_n)$ and $\mathbb{I} = (1, \ldots, 1) \in \mathbb{R}^n$. Then:

$$\left\| a - \frac{\|a\|_2}{\sqrt{n}} \mathbb{I} \right\|_2^2 = \|a\|_2^2 - \frac{2\|a\|_2}{\sqrt{n}} \langle a \mid \mathbb{I} \rangle + \|a\|_2^2 = 2\|a\|_2^2 - \frac{2\|a\|_2}{\sqrt{n}} \|a\|_1$$

$$\leqslant 2\|a\|_2^2 - 2(1-\varepsilon)\|a\|_2^2 = 2\varepsilon \|a\|_2^2. \qquad \square$$

Consequently:

$$(**)\quad \begin{cases} (\forall n \geqslant 1) \quad (\forall \varepsilon > 0) \quad (\exists x_1, \ldots, x_n \in X) \text{ such that} \\[4pt] \qquad \|x_j\| = 1, \, j = 1, 2, \ldots, n, \text{ and} \\[4pt] \qquad (1-\varepsilon)n \leqslant \left\| \sum_{j=1}^n r_j x_j \right\|_{L^2(X)}. \end{cases}$$

Indeed, thanks to the strict inequality ensuring that the $x_n$'s are not all null, we can assume in $(*)$ that $\sum_{j=1}^n \|x_j\|^2 = n$. Thus, by Lemma II.11:

$$1 - \sqrt{2n\varepsilon} \leqslant \|x_j\| \leqslant 1 + \sqrt{2n\varepsilon},$$

and, setting $x_j' = x_j / \|x_j\|$, thanks to $(*)$ and the contraction principle, we obtain:

$$(1-\varepsilon)\, n \leqslant \left\| \sum_{j=1}^n r_j \|x_j\| \, x_j' \right\|_{L^2(X)} \leqslant (1 + \sqrt{2n\varepsilon}) \left\| \sum_{j=1}^n r_j x_j' \right\|_{L^2(X)},$$

hence (∗∗) since $\dfrac{1-\varepsilon}{1+\sqrt{2n\varepsilon}} \underset{\varepsilon\to 0}{\to} 1$. Now note that

$$\left\| \sum_{j=1}^{n} r_j x_j \right\|_{L^2(X)}^2 \leqslant \frac{1}{2^n} \min_{\theta_j=\pm 1} \left\| \sum_{j=1}^{n} \theta_j x_j \right\|^2 + \left(1-\frac{1}{2^n}\right) \left(\sum_{j=1}^{n} \|x_j\|\right)^{1/2},$$

because

$$\left\| \sum_{j=1}^{n} r_j x_j \right\|_{L^2(X)}^2 = \int_0^1 \left\| \sum_{j=1}^{n} r_j(t) x_j \right\|^2 dt = \frac{1}{2^n} \sum_{\theta_j=\pm 1} \left\| \sum_{j=1}^{n} \theta_j x_j \right\|^2,$$

and because if the sequence $(\theta_j^0)_{1\leqslant j\leqslant n}$ takes on the minimum,

$$\sum_{(\theta_j)_j \neq (\theta_j^0)_j} \left\| \sum_{j=1}^{n} \theta_j x_j \right\|^2 \leqslant \sum_{(\theta_j)_j \neq (\theta_j^0)_j} \left(\sum_{j=1}^{n} \|x_j\|\right)^2$$

$$= (2^n - 1) \left(\sum_{j=1}^{n} \|x_j\|\right)^2.$$

Thus, (∗∗) implies:

$$(1-\varepsilon)^2 n^2 \leqslant \frac{1}{2^n} \min_{\theta_j=\pm 1} \left\| \sum_{j=1}^{n} \theta_j x_j \right\|^2 + \left(1-\frac{1}{2^n}\right) n^2 ;$$

therefore:

$$\min_{\theta_j=\pm 1} \left\| \sum_{j=1}^{n} \theta_j x_j \right\|^2 \geqslant n^2 \big[1 - \varepsilon(2-\varepsilon)2^n\big] \geqslant (n-\delta)^2 ,$$

for an arbitrary $\delta > 0$, once $\varepsilon > 0$ is selected small enough.

2) Now select $\omega = \big(\varepsilon_1(\omega), \dots, \varepsilon_n(\omega)\big) = (\theta_1, \dots, \theta_n) \in \{-1, +1\}^n$, and a linear functional $x^* = x_\omega^* \in X^*$ with $\|x^*\| = 1$ such that:

$$x^* \left( \sum_{j=1}^{n} \theta_j x_j \right) = \left\| \sum_{j=1}^{n} \theta_j x_j \right\| \geqslant n - \delta .$$

This can also be written:

$$\delta \geqslant \sum_{j=1}^{n} \left[ 1 - x^*(\theta_j x_j) \right] = \mathrm{Re} \left( \sum_{j=1}^{n} \left[ 1 - x^*(\theta_j x_j) \right] \right)$$

$$= \sum_{j=1}^{n} \left[ 1 - \mathrm{Re}\, x^*(\theta_j x_j) \right] .$$

For any $j$, $1 - \mathrm{Re}\, x^*(\theta_j x_j) \geqslant 0$, thus:

$$\delta \geqslant 1 - \mathrm{Re}\, x^*(\theta_j x_j) ;$$

and consequently:

$$\left| 1 - x^*(\theta_j x_j) \right|^2 = 1 + |x^*(\theta_j x_j)|^2 - 2\, \mathrm{Re}\, x^*(\theta_j x_j)$$

$$\leqslant 2 - 2\, \mathrm{Re}\, x^*(\theta_j x_j) \leqslant 2\delta .$$

Hence, for any scalars $a_1, \ldots, a_n \in \mathbb{R}$ or $\mathbb{C}$, we obtain:

$$\left\| \sum_{j=1}^{n} a_j x_j \right\| \geqslant \max_{\omega} \left| \sum_{j=1}^{n} a_j x_\omega^*(x_j) \right|$$

$$\geqslant \max_{\omega} \left( \left| \sum_{j=1}^{n} a_j \varepsilon_j(\omega) \right| - \sum_{j=1}^{n} |a_j| \left| x_\omega^*(x_j) - \varepsilon_j(\omega) \right| \right)$$

$$\geqslant \max_{\omega} \left( \left| \sum_{j=1}^{n} a_j \varepsilon_j(\omega) \right| - \sqrt{2\delta} \sum_{j=1}^{n} |a_j| \right)$$

$$\geqslant \frac{2}{\pi} \sum_{j=1}^{n} |a_j| - \sqrt{2\delta} \sum_{j=1}^{n} |a_j| ,$$

since $\{\varepsilon_1, \ldots, \varepsilon_n\}$ is a Sidon set with constant $\leqslant \pi/2$ in the dual group of $\{-1, +1\}^n$

$$\geqslant \left( \frac{2}{\pi} - \sqrt{2\delta} \right) \sum_{j=1}^{n} |a_j| ,$$

and hence $X$ indeed contains $\ell_1^n$'s uniformly.                                        □

## III  Examples of Reflexive Subspaces of $L^1$

Two types of examples of reflexive subspaces of $L^1$ are presented here. The first are constructed via Probability, and the second via methods of Harmonic Analysis.

## III.1 Stable Variables

This section deals with *real* spaces. As we have already seen (see, for example, Chapter 1 of Volume 1, Corollary III.6), if $(G_n)_{n \geqslant 1}$ is a sequence of independent Gaussian random variables $\mathcal{N}(0, 1)$, then the subspace it generates in $L^1(\Omega)$ is isometric to $\ell_2$. More generally, we will show the following:

**Theorem III.1**    *For any $q$ with $1 \leqslant q \leqslant 2$, $L^1(\Omega)$ contains a subspace isometric to $\ell_q$.*

The case $q = 1$ is easy (Chapter 2 of Volume 1, Exercise V.2). For $q > 1$, we use the notion of $q$-stable random variables, seen in Chapter 5 (Volume 1), Subsection II.2. Recall the following result (see Chapter 5, Theorem II.10):

**Theorem III.2**    *For any $q \in \,]0, 2]$, there exists a probability distribution $\sigma_q$ on $\mathbb{R}$ such that:*

$$\int_{\mathbb{R}} e^{itu} \, d\sigma_q(u) = e^{-|t|^q}, \qquad \forall t \in \mathbb{R}.$$

**Definition III.3**    A random variable $Z$ is said to be *$q$-stable* if its distribution is $\sigma_q$, i.e. if its characteristic function is $\Phi_Z(t) = e^{-|t|^q}$.

Also recall that the 2-stable variables are Gaussian variables. The following result, already seen in Chapter 5 (Volume 1), Theorem II.10, is reproduced here for convenience:

**Lemma III.4**    *If $Z \colon \Omega \to \mathbb{R}$ is $q$-stable, then $Z \in L^r(\Omega, \mathbb{P})$ for any $r < q$.*

Note that the case $p = 2$ is different in that $G \in L^r(\Omega, \mathbb{P})$ for any $r < +\infty$ when $G$ is Gaussian.

*Proof*    First, note that:

$$|u|^r = C_r \int_0^{+\infty} \frac{(1 - \cos ut)}{t^{1+r}} \, dt \,,$$

where

$$C_r^{-1} = \int_0^{+\infty} \frac{1 - \cos v}{v^{1+r}} \, dv \quad (< +\infty \text{ for } r < 2)\,.$$

Then:

$$\int_\Omega |Z|^r \, d\mathbb{P} = \int_{\mathbb{R}} |u|^r \, d\sigma_q(u) = \int_{\mathbb{R}} \left( C_r \int_0^{+\infty} \frac{(1 - \cos ut)}{t^{1+r}} \, dt \right) d\sigma_q(u)$$

$$= C_r \int_0^{+\infty} \left( \int_{\mathbb{R}} (1 - \cos ut) \, d\sigma_q(u) \right) \frac{dt}{t^{1+r}}$$

$$= C_r \int_0^{+\infty} \left( \int_{\mathbb{R}} 1 - \operatorname{Re}(e^{iut}) \, d\sigma_q(u) \right) \frac{dt}{t^{1+r}}$$

$$= C_r \int_0^{+\infty} \left( 1 - e^{-|t|^q} \right) \frac{dt}{t^{1+r}} = S_{q,r}^r < +\infty. \qquad \square$$

**Proposition III.5** *If $Z_1, \ldots, Z_n$ are independent $q$-stable variables (with $0 < q \leqslant 2$), then, for any $a_1, \ldots, a_n \in \mathbb{R}$ and for $1 \leqslant r < q$:*

$$\left\| \sum_{k=1}^n a_k Z_k \right\|_{L^r(\Omega)} = S_{q,r} \left( \sum_{k=1}^n |a_k|^q \right)^{1/q},$$

*with $S_{q,r} = \|Z_1\|_{L^r(\Omega)}$.*

*Proof* We can assume $\sum_{k=1}^n |a_k|^q = 1$. Set $Z = \sum_{k=1}^n a_k Z_k$. By independence, we obtain:

$$\Phi_Z(t) = \mathbb{E}(e^{itZ}) = \mathbb{E}\left( e^{it \sum_{k=1}^n a_k Z_k} \right) = \prod_{k=1}^n \mathbb{E}(e^{ita_k Z_k})$$

$$= \prod_{k=1}^n e^{-|a_k t|^q} = \exp\left( -|t|^q \sum_{k=1}^n |a_k|^q \right) = e^{-|t|^q},$$

so that $Z$ is $q$-stable, and hence $\|Z\|_{L^r(\Omega)} = S_{q,r}$. $\qquad \square$

The selection of an infinite sequence of independent $q$-stable random variables leads to the result stated in Theorem III.1, and even a bit more:

**Corollary III.6** *For $1 \leqslant r < q \leqslant 2$, $L^r(\Omega)$ contains a subspace isometric to $\ell_q$.*

## III.2 $\Lambda(q)$ Sets

Here we consider complex Banach spaces. Recall that if $\Omega = \{-1, +1\}^{\mathbb{N}^*}$ is the Cantor group, and $(\varepsilon_n)_{n \geqslant 1}$ the sequence of Rademacher functions (or equivalently, the sequence of projections $\varepsilon_n(\omega) = \omega_n \in \{-1, +1\}$), then $\mathfrak{R} = \{\varepsilon_n; \; n \geqslant 1\}$ is a Sidon set in $\Gamma = \widehat{\Omega}$. Another property of $\mathfrak{R}$ is given by the *Khintchine inequalities*: for any trigonometric polynomial $P \in \mathcal{P}_{\mathfrak{R}}$ with spectrum in $\mathfrak{R}$:

$$\|f\|_q \leqslant \sqrt{q} \, \|f\|_2$$

for any $q \geqslant 2$ (and $q < +\infty$). Therefore the norms $\|\cdot\|_q$ and $\|\cdot\|_2$ are equivalent, and $L_{\mathfrak{R}}^q = L_{\mathfrak{R}}^2$. Another easy consequence, to be seen later, is $L_{\mathfrak{R}}^q = L_{\mathfrak{R}}^1$; thus this space appears as a reflexive subspace of $L^1(\Omega)$.

This result can also be "transferred" to all Sidon sets, as we will show. We begin with a definition.

**Definition III.7**  Let $G$ be a (metrizable) compact Abelian group, and $\Lambda \subseteq \Gamma = \widehat{G}$ a subset of its dual group. The set $\Lambda$ is said to be a $\Lambda(q)$-*set* (with $0 < q < +\infty$) if there exist $r < q$ ($r > 0$) and a constant $C > 0$ such that:

$$\|f\|_q \leqslant C \|f\|_r, \qquad \forall f \in \mathcal{P}_\Lambda .$$

We will only use this notion for $q \geqslant 1$. As already mentioned at the beginning of this chapter, for $0 < r < 1$, the quantity:

$$\|f\|_r = \left( \int_G |f(t)|^r \, dm(t) \right)^{1/r}$$

is not a norm, but this does not matter. Also the constant $C = C(q,r)$ depends on $q$ and on $r$, but there is no mention of $r < q$ in the definition, for the following reason:

**Proposition III.8**  *If there exists $r < q$ such that $\|f\|_q \leqslant C_r \|f\|_r$, then, for any $s < q$, we have $\|f\|_q \leqslant C_s \|f\|_s$.*

*Proof*  This is clear if $s \geqslant r$. For $s < r$, this ensues from Hölder's inequality, written as follows: if

$$\frac{1}{r} = \frac{\theta}{s} + \frac{1-\theta}{q}, \qquad \text{with } 0 < \theta < 1 ,$$

then: $\|f\|_r \leqslant \|f\|_s^\theta \|f\|_q^{1-\theta}$.  $\square$

**Corollary III.9**  *If $\Lambda$ is $\Lambda(q)$, then it is $\Lambda(r)$ for any $r \leqslant q$.*

**Remark**  As seen in Section II (Proposition II.4 and Theorem II.3), a set $\Lambda$ is $\Lambda(1)$ if and only if $L^1_\Lambda$ is reflexive. That such a set is in fact $\Lambda(q)$ for some $q > 1$ will be seen in the next section, as a consequence of Rosenthal's theorem.

**Theorem III.10**  (The Rudin Transfer Theorem)  *Let $\Omega = \{-1,+1\}^{\mathbb{N}^*}$ be the Cantor group, and $(\varepsilon_n)_{n \geqslant 1}$ the Rademacher functions. Then, for every (metrizable) compact Abelian group $G$, for every Sidon set $\Lambda = \{\gamma_k ; \ k \geqslant 1\} \subseteq \Gamma = \widehat{G}$, for any $a_1, \ldots, a_n \in \mathbb{C}$ and for $1 \leqslant q < +\infty$, we have:*

$$\frac{1}{2S(\Lambda)} \left\| \sum_{k=1}^n \varepsilon_k a_k \right\|_{L^q(\Omega)} \leqslant \left\| \sum_{k=1}^n a_k \gamma_k \right\|_{L^q(G)} \leqslant 2S(\Lambda) \left\| \sum_{k=1}^n \varepsilon_k a_k \right\|_{L^q(\Omega)} ,$$

*where $S(\Lambda)$ is the Sidon constant of $\Lambda$.*

Thanks to the Khintchine inequalities, this implies:

**Theorem III.11** (Rudin)  *Every Sidon set* $\Lambda$ *is a* $\Lambda(q)$*-set for any* $q < +\infty$. *More precisely, there exists a constant* $C > 0$ *such that, for* $2 \leqslant q < +\infty$, *and for every* $f \in \mathcal{P}_\Lambda$, *we have:*

$$\|f\|_q \leqslant C\,S(\Lambda)\sqrt{q}\,\|f\|_2.$$

**Remark**  The converse question – whether a set $\Lambda$ which is $\Lambda(q)$ for any $q < +\infty$, and satisfies $\|f\|_q \leqslant K\sqrt{q}\,\|f\|_2$ for any $q \geqslant 2$, and every $f \in \mathcal{P}_\Lambda$, is a Sidon set – remained open for a long time. It was settled by Pisier in 1978 (Pisier [1978 b]). His proof introduced the use of Gaussian processes in Harmonic Analysis and will be presented in Chapter 6 of this volume. In 1983, he gave a simpler proof, still using Gaussian processes, in terms of extractions of quasi-independent sets (Pisier [1983 a], [1983 b] and [1983 c]). In Chapter 5 of this volume, another proof of the extraction of these quasi-independent sets will be given: it is due to Bourgain [1985 a] and [1985 b], and uses selectors instead of Gaussian processes.

*Proof*  For every $\omega \in \Omega$, the sequence $\big(\varepsilon_k(\omega)\big)_{k \geqslant 1}$ is bounded, and thus, since $\Lambda$ is a Sidon set, there exists a measure $\mu_\omega \in \mathcal{M}(G)$ such that $\|\mu_\omega\| \leqslant S(\Lambda)$ and:

$$\widehat{\mu_\omega}(\gamma_k) = \varepsilon_k(\omega), \qquad 1 \leqslant k \leqslant n.$$

Set:

$$f(t) = \sum_{k=1}^{n} a_k \gamma_k(t), \qquad t \in G,$$

and, for $\omega \in \Omega$:

$$f_\omega(t) = \sum_{k=1}^{n} \varepsilon_k(\omega)\, a_k \gamma_k(t), \qquad t \in G.$$

The *essential fact* is:

$$f_\omega = f * \mu_\omega \quad \text{and} \quad f = f_\omega * \mu_\omega.$$

Then, the second equality leads to:

$$\|f\|_{L^q(G)} \leqslant \|f_\omega\|_{L^q(G)} \|\mu_\omega\| \leqslant S(\Lambda) \|f_\omega\|_{L^q(G)};$$

hence, by an integration with respect to $\omega$:

$$\|f\|_{L^q(G)}^q \leqslant S(\Lambda)^q \int_\Omega \|f_\omega\|_{L^q(G)}^q \, d\mathbb{P}(\omega).$$

However:

$$
\begin{aligned}
\int_{\Omega} \| f_{\omega} \|_{L^q(G)}^q \, d\mathbb{P}(\omega) &= \int_{\Omega} \left( \int_G \left| \sum_{k=1}^n \varepsilon_k(\omega) a_k \gamma_k(t) \right|^q \, dm(t) \right) d\mathbb{P}(\omega) \\
&= \int_G \left( \int_{\Omega} \left| \sum_{k=1}^n \varepsilon_k(\omega) a_k \gamma_k(t) \right|^q \, d\mathbb{P}(\omega) \right) dm(t) \\
&\leqslant \int_G \left( \int_{\Omega} 2^q \left| \sum_{k=1}^n a_k \varepsilon_k(\omega) \right|^q \, d\mathbb{P}(\omega) \right) dm(t) \\
&= 2^q \left\| \sum_{k=1}^n a_k \, \varepsilon_k \right\|_{L^q(\Omega)}^q ,
\end{aligned}
$$

since $|\gamma_k(t)| = 1$, and since the random variables $X_k = a_k \varepsilon_k$, $1 \leqslant k \leqslant n$, are symmetric, and hence form a 2-unconditional basic sequence in the complex space $L^q(\Omega)$. Consequently:

$$
\| f \|_{L^q(G)} \leqslant 2S(\Lambda) \left\| \sum_{k=1}^n \varepsilon_k a_k \right\|_{L^q(\Omega)} .
$$

Similarly, the other inequality gives:

$$
\| f_{\omega} \|_{L^q(G)} \leqslant S(\Lambda) \| f \|_{L^q(G)} ,
$$

so that:

$$
\int_{\Omega} \| f_{\omega} \|_{L^q(G)}^q \, d\mathbb{P}(\omega) \leqslant S(\Lambda)^q \| f \|_{L^q(G)}^q ,
$$

and we write this time:

$$
\begin{aligned}
\int_{\Omega} \| f_{\omega} \|_{L^q(G)}^q \, d\mathbb{P}(\omega) &= \int_{\Omega} \left( \int_G \left| \sum_{k=1}^n \varepsilon_k(\omega) a_k \gamma_k(t) \right|^q \, dm(t) \right) d\mathbb{P}(\omega) \\
&= \int_G \left( \int_{\Omega} \left| \sum_{k=1}^n \varepsilon_k(\omega) a_k \gamma_k(t) \right|^q \, d\mathbb{P}(\omega) \right) dm(t) \\
&= \int_G \left\| \sum_{k=1}^n \gamma_k(t) a_k \varepsilon_k \right\|_{L^q(\Omega)}^q \, dm(t) .
\end{aligned}
$$

Then the 2-unconditionality of the symmetric independent random variables $Y_k = \overline{\gamma_k(t)} \, a_k \varepsilon_k$, $1 \leqslant k \leqslant n$, and the equality $\gamma_k(t) Y_k = a_k \varepsilon_k$, lead to:

$$\left\|\sum_{k=1}^{n} a_k \varepsilon_k\right\|_{L^q(\Omega)} = \left\|\sum_{k=1}^{n} \gamma_k(t) Y_k\right\|_{L^q(\Omega)} \leqslant 2 \left\|\sum_{k=1}^{n} Y_k\right\|_{L^q(\Omega)}$$

$$= 2 \left\|\sum_{k=1}^{n} \gamma_k(t) a_k \varepsilon_k\right\|_{L^q(\Omega)},$$

and finally:

$$\frac{1}{2} \left\|\sum_{k=1}^{n} a_k \varepsilon_k\right\|_{L^q(\Omega)} \leqslant S(\Lambda) \|f\|_{L^q(G)}. \qquad \square$$

**Remark** By using the Khintchine–Kahane inequalities instead of those of Khintchine, we can replace the $a_k \in \mathbb{C}$ by $x_k \in X$, for any Banach space $X$. The vectorial transfer theorem thus obtained allows, for example, the definition of the type and cotype of Banach spaces by replacing the sequence of Rademacher random variables $(\varepsilon_k)_{k \geqslant 1}$ by an arbitrary Sidon set.

In conclusion, a property of "smallness" for the $\Lambda(q)$-sets (mesh condition) is presented, analogous to that given in Chapter 6 (Volume 1), Theorem V.11, for Sidon sets.

**Theorem III.12** *If $\Lambda$ is a $\Lambda(q)$-set contained in $\mathbb{Z}$, with $q > 2$, and if*

$$\|f\|_q \leqslant C_q \|f\|_2, \qquad \forall f \in \mathcal{P}_\Lambda,$$

*then:*

$$\alpha_\Lambda(N) \leqslant 9 \, C_q^2 \, N^{2/q},$$

*where $\alpha_\Lambda(N)$ is the maximum number of elements of $\Lambda$ that can be contained in an arithmetic progression of length $N$.*

Note that the two preceding theorems allow us to recuperate the mesh condition for the Sidon sets:

**Corollary III.13** *If $\Lambda$ is a Sidon set contained in $\mathbb{Z}$, then:*

$$\alpha_\Lambda(N) \leqslant CS(\Lambda)^2 \log N.$$

*Proof* Indeed, as a Sidon set is $\Lambda(q)$ for any $q > 2$, with a constant $C_q = C_0 S(\Lambda) \sqrt{q}$, it suffices to choose $q = 2 \log N$ ($> 2$ for $N \geqslant 3$), which leads to $N^{2/q} = e$, and

$$\alpha_\Lambda(N) \leqslant 18 \, e \, C_0^2 S(\Lambda)^2 \log N. \qquad \square$$

*Proof of the theorem*   Consider the Fejér kernel:

$$K_N(t) = \sum_{n=-N}^{N} \left(1 - \frac{|n|}{N}\right) e^{int}.$$

Since:

$$\|K_N\|_\infty = \sum_{n=-N}^{N} \left(1 - \frac{|n|}{N}\right) = N \quad \text{and} \quad \|K_N\|_1 = 1,$$

we have:

$$\int_{\mathbb{T}} K_n^{q^*}(t)\, dm(t) \leqslant \int_{\mathbb{T}} N^{q^*-1} K_N(t)\, dm(t) = N^{q^*-1},$$

hence $\|K_N\|_{q^*} \leqslant N^{1-1/q^*} = N^{1/q}$.

Now let $A = \{a+b, a+2b, \ldots, a+Nb\}$ be an arithmetic progression of length $N$. Since $\Lambda - a$ has the same properties as $\Lambda$, we can assume $a = 0$.

If $N$ is even, $N = 2s$, and we define:

$$Q(t) = e^{isbt} K_N(bt).$$

Then

$$Q(t) = \sum_{n=-N}^{N} \left(1 - \frac{|n|}{N}\right) e^{i(n+s)bt} = \sum_{k=-s}^{3s} \left(1 - \frac{|k-s|}{2s}\right) e^{ikbt};$$

therefore:

$$0 \leqslant k \leqslant N = 2s \implies |k-s| \leqslant s \implies \widehat{Q}(kb) \geqslant \frac{1}{2}.$$

If $N$ is odd, $N = 2s+1$, we define:

$$Q(t) = e^{isbt} K_N(bt) = \sum_{k=-s-1}^{3s+1} \left(1 - \frac{|k-s|}{2s+1}\right) e^{ikbt}.$$

Then:

$$0 \leqslant k \leqslant N = 2s+1 \implies -s \leqslant k-s \leqslant s+1$$
$$\implies \widehat{Q}(kb) \geqslant 1 - \frac{s+1}{2s+1} = \frac{s}{2s+1} \geqslant \frac{1}{3}.$$

In both cases, we thus obtain:

$$\widehat{Q}(kb) \geqslant \frac{1}{3} \quad \text{for } 0 \leqslant k \leqslant N.$$

Now we write:

$$\Lambda \cap A = \{\lambda_1, \ldots, \lambda_r\},$$

and set:

$$f(t) = \sum_{j=1}^{r} e^{i\lambda_j t} \, ;$$

then:

$$\frac{r}{3} \leqslant \sum_{j=1}^{r} \widehat{Q}(\lambda_j) = \int_{\mathbb{T}} f(-t) Q(t) \, dm(t) \leqslant \|f\|_q \|Q\|_{q^*}$$

$$\leqslant C_q \|f\|_2 \|K_N\|_{q^*} \leqslant C_q \sqrt{r} N^{1/q} \, ,$$

i.e. $r \leqslant 9 \, C_q^2 N^{2/q}$, and the proof is complete.                  $\square$

**Remark**   It can be shown (Meyer [1968 b], Theorem 3, page 558; see also Bonami [1970], Corollary 4, page 361) that, for example:

$$\Lambda = \{3^k + 3^l \, ; \; 1 \leqslant k \leqslant l\}$$

is a $\Lambda(q)$-set for any $q < +\infty$; however, as

$$|\Lambda \cap \{1, 3, 2.3, \ldots 3^k.3\}| \approx 1 + 2 + \cdots + k \approx k^2 \approx \left(\log(3^k)\right)^2,$$

it is not a Sidon set.


## IV   Maurey's Factorization Theorem and Rosenthal's Theorem

The following result is proved in this section:

**Theorem IV.1** (Rosenthal's Theorem)   *Let $X$ be a reflexive subspace of $L^1(\Omega, \mathbb{P})$. Then, there exist $q \in \, ]1, 2]$, a probability density $\Delta$ and a constant $K > 0$ such that:*

$$(*)  \qquad \left( \int_{\Omega} \left| \frac{f}{\Delta} \right|^q \Delta \, d\mathbb{P} \right)^{1/q} \leqslant K \|f\|_1 , \qquad \forall f \in X.$$

*In particular $X$ is isomorphic to a subspace of $L^q(\Omega, \Delta.\mathbb{P})$.*

A probability density is a positive function $\Delta \geqslant 0$ such that $\int_{\Omega} \Delta \, d\mathbb{P} = 1$.

The proof is a result of the following factorization theorem (a version of it, useful for Theorem IV.1, can already be found in Rosenthal [1973], Theorem 1, as was pointed out to us by one of the referees):

**Theorem IV.2** (The Maurey Factorization Theorem) *Let X be a Banach space having a type $p > 1$. Then, for any pair $(q, r)$ with $0 < r < q < p$, and every operator $T: X \to L^r(\Omega, \mathbb{P})$, there exist a probability density $\Delta$ and a constant $K > 0$ such that:*

$$\left( \int_\Omega \left| \frac{Tx}{\Delta^{1/r}} \right|^q \Delta \, d\mathbb{P} \right)^{1/q} \leqslant K \|x\|, \quad \forall x \in X.$$

In other words, this theorem means that the operator $T$ can be factorized as follows:

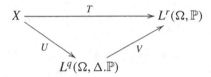

where:

$$U(x) = \frac{T(x)}{\Delta^{1/r}} \quad \text{and} \quad V(f) = f.\Delta^{1/r},$$

with $\|U\| \leqslant K$ and $\|V\| \leqslant 1$. The operator $T$ is said to *factorize strongly through $L^q(\Omega, \Delta.\mathbb{P})$*.

First, let us see how Rosenthal's theorem can easily be deduced.

*Proof of Rosenthal's Theorem*   It has been proved that any reflexive subspace of $L^1(\Omega, \mathbb{P})$ has a type $p > 1$. Hence, it suffices to apply the Maurey factorization theorem to the canonical injection

$$T = j: X \longrightarrow L^1(\Omega, \mathbb{P}),$$

and then, since $j$ is an isomorphism between $X$ and $j(X)$, the relation $j = VU$ implies that $U$ is an isomorphism from $X$ onto $U(X)$.   $\square$

Note that in these two theorems we have to add a density to $\mathbb{P}$. This is no longer the case for $\Lambda(1)$-sets :

**Corollary IV.3** (Bachelis–Ebenstein)   *Let $G$ be a (metrizable) compact Abelian group and $\Lambda$ a $\Lambda(1)$-set in $\Gamma = \widehat{G}$. Then there exists $q > 1$ such that $\Lambda$ is a $\Lambda(q)$-set.*

Note that $L^1_\Lambda(G) = L^q_\Lambda(G)$ ensues.

*Proof*   We have already seen that if $\Lambda$ is $\Lambda(1)$, then $L^1(G)$ is reflexive (Remark following Corollary III.9). In (∗) of Rosenthal's theorem, for the function $f \in X = L^1_\Lambda(G)$ we select a character $\gamma \in \Gamma$; then we get:

$$\Delta^{1-q} \in L^1(G).$$

Setting $h = \Delta^{1/q-1}$, we thus have $h \in L^q(G)$, and:

$$\widehat{h}(0) = \int_G \Delta^{1/q-1}(t)\, dm(t) > 0.$$

Next, for every $f \in L^1_\Lambda(G)$, we define:

$$(Mf)(t) = h(t)f(t).$$

Then the inequality $(*)$ becomes:

$$\|Mf\|_q = \left(\int_G |h(t)f(t)|^q\, dm(t)\right)^{1/q} \leqslant K\|f\|_1,$$

so that $M$ defines a continuous operator $M\colon L^1_\Lambda(G) \to L^q(G)$. For each $f \in L^1_\Lambda(G)$, the mapping:

$$\tau \in G \longmapsto \big(M(f_{-\tau})\big)_\tau \in L^q(G)$$

(where $g_\tau(t) = g(t - \tau)$) is continuous; thus the following vector-valued integral can be defined:

$$\widetilde{M}(f) = \int_G \big(M(f_{-\tau})\big)_\tau\, dm(\tau).$$

Then:
– on one hand:

$$\|\widetilde{M}(f)\|_q \leqslant \|M(f)\|_q \leqslant K\|f\|_1,$$

– on the other, for every $\phi \in L^{q^*}(G)$:

$$\langle \widetilde{M}(f), \phi \rangle = \left\langle \int_G \big(M(f_{-\tau})\big)_\tau\, dm(\tau), \phi \right\rangle = \left\langle \int_G (hf_{-\tau})_\tau\, dm(\tau), \phi \right\rangle$$

$$= \int_G \langle h_\tau f, \phi \rangle\, dm(\tau)$$

$$= \int_G \left(\int_G h(t - \tau)f(t)\phi(t)\, dm(t)\right)\, dm(\tau)$$

$$= \int_G h(u)\langle f, \phi \rangle\, dm(u) = \widehat{h}(0)\langle f, \phi \rangle.$$

Then, as $\widehat{h}(0) > 0$, we obtain:

$$\|f\|_q \leqslant \frac{K}{\widehat{h}(0)}\|f\|_1,$$

and hence $\Lambda$ is a $\Lambda(q)$-set.                                    $\square$

To show the factorization theorem, we need the following criterion:

**Proposition IV.4**   *Let $T: X \to L^r(\Omega, \mathbb{P})$ $(0 < r < +\infty)$, and $q > r$. If there exists $K > 0$ such that:*

$$\left( \int_\Omega \left( \sum_{j=1}^n |Tx_j|^q \right)^{r/q} d\mathbb{P} \right)^{1/r} \leqslant K \left( \sum_{j=1}^n \|x_j\|^q \right)^{1/q}$$

*for all $x_1, \ldots, x_n \in X$, then $T$ can be strongly factorized through the space $L^q(\Omega, \Delta.\mathbb{P})$ for some probability density $\Delta$.*

The converse is true and trivial.

**Remark**   For $n = 1$, this condition is none other than the continuity of $T$:

$$\|T\|_r \leqslant K\|x\| \,.$$

*Proof* (Pisier)   Let $r < q < +\infty$ (the case $q = +\infty$ is easy: it suffices to take $\Delta = \sup_{x \in B_X} |Tx|^r$). We define:

$$K_n = \sup \left\{ \left( \int_\Omega \left( \sum_{j=1}^n |Tx_j|^q \right)^{r/q} d\mathbb{P} \right)^{1/r} \;;\; \sum_{j=1}^n \|x_j\|^q \leqslant 1 \right\},$$

and

$$K = \lim_{n \to +\infty} \uparrow K_n \,.$$

We select $x_1^n, \ldots, x_n^n$ such that:

$$\sum_{j=1}^n \|x_j^n\|^q \leqslant \frac{1}{K_n^q}\left( 1 + \frac{1}{n} \right) \quad \text{and} \quad \int_\Omega \left( \sum_{j=1}^n |Tx_j^n|^q \right)^{r/q} d\mathbb{P} = 1 \,,$$

and define:

$$f_n = \left( \sum_{j=1}^n |Tx_j^n|^q \right)^{r/q} \,.$$

Then clearly $f_n \geqslant 0$ and $\int_\Omega f_n \, d\mathbb{P} = 1$. Moreover:

**Lemma IV.5**   *The sequence $(f_n)_{n \geqslant 1}$ is uniformly integrable.*

*Proof*   If not, there would exist $\varepsilon_0 > 0$ and $A_n \subseteq \Omega$, $n \geqslant 1$, such that:

$$\mathbb{P}(A_n) \xrightarrow[n \to +\infty]{} 0 \quad \text{and} \quad \int_{A_n} f_n \, d\mathbb{P} \geqslant \varepsilon_0 \,.$$

As every finite subset of $L^1(\Omega, \mathbb{P})$ is uniformly absolutely continuous, we could construct a subsequence $(A_{n_k})_{k \geqslant 1}$ such that:

$$\int_{A_{n_l}} f_{n_k} \, d\mathbb{P} \leqslant \frac{\varepsilon_0}{4^l} \qquad \text{for } l \geqslant k + 1$$

and:

$$\int_{A_{n_k}} f_{n_k} \, d\mathbb{P} \geqslant \varepsilon_0 \, .$$

Setting $B_k = A_{n_k} \setminus \bigcup_{l \geqslant k+1} A_{n_l}$, we would obtain a sequence of pairwise disjoint sets such that:

$$\int_{B_k} f_{n_k} \, d\mathbb{P} \geqslant \frac{\varepsilon_0}{2} \, .$$

Then, for any $N \geqslant 1$:

$$\frac{N\varepsilon_0}{2} \leqslant \sum_{k=1}^{N} \int_{B_k} f_{n_k} \, d\mathbb{P} = \int_{\Omega} \left( \sum_{k=1}^{N} \left( f_{n_k} \mathbb{I}_{B_k} \right)^{q/r} \right)^{r/q} d\mathbb{P}$$

$$\text{(thanks to the disjunction)}$$

$$\leqslant \int_{\Omega} \left( \sum_{k=1}^{N} f_{n_k}^{q/r} \right)^{r/q} d\mathbb{P} = \int_{\Omega} \left( \sum_{k=1}^{N} \sum_{j=1}^{n_k} |T x_j^{n_k}|^q \right)^{r/q} d\mathbb{P}$$

$$\leqslant K \left( \sum_{k=1}^{N} \sum_{j=1}^{n_k} \|x_j^{n_k}\|^q \right)^{r/q} \leqslant K \left( \sum_{k=1}^{N} \frac{1}{K_{n_k}^q} \left( 1 + \frac{1}{n_k} \right) \right)^{r/q}$$

$$\leqslant \left( \frac{K}{K_1} \right)^r (2N)^{r/q} \, ,$$

which is not possible for $N$ large enough, since $r/q < 1$. $\qquad\square$

Extracting a subsequence if necessary, we can thus assume that $(f_n)_{n \geqslant 1}$ converges weakly, to some function $f \in L^1(\Omega, \mathbb{P})$. Again, $f \geqslant 0$ and $\int_{\Omega} f \, d\mathbb{P} = 1$.

Now, for every $x \in X$ such that $\|x\| = 1$, any $t \geqslant 0$ and any $n \geqslant 1$, we have:

$$\int_{\Omega} \left( f_n^{q/r} + t^q |Tx|^q \right)^{r/q} d\mathbb{P} = \int_{\Omega} \left( \sum_{j=1}^{n} |T x_j^n|^q + |T(tx)|^q \right)^{r/q} d\mathbb{P}$$

$$\leqslant K_{n+1}^r \left( \sum_{j=1}^{n} \|x_j^n\|^q + t^q \right)^{r/q}$$

$$\leqslant K_{n+1}^r \left[ \frac{1}{K_n^q} \left( 1 + \frac{1}{n} \right) + t^q \right]^{r/q} \, .$$

Let $n$ tend to infinity and observe what happens to this inequality. Set $v(\omega) = t^r|(Tx)(\omega)|^r$, and $s = q/r > 1$. The pair $(f, v)$ can be considered as an element of $L^1(\Omega, \mathbb{P}; \ell_s^2)$, and thus we can choose $(a, b) \in L^\infty(\Omega, \mathbb{P}; \ell_{s^*}^2)$ such that $|a|^{s^*} + |b|^{s^*} \leqslant 1$ almost everywhere and $\int_\Omega (f^s + v^s)^{1/s} \, d\mathbb{P} = \int_\Omega (af + bv) \, d\mathbb{P}$. Then:

$$\int_\Omega \left(f^{q/r} + t^q|Tx|^q\right)^{r/q} d\mathbb{P} = \int_\Omega (af + bv) \, d\mathbb{P} = \lim_{n \to +\infty} \int_\Omega (af_n + bv) \, d\mathbb{P}$$

$$\leqslant \varlimsup_{n \to +\infty} \int_\Omega \left(f_n^{q/r} + t^q|Tx|^q\right)^{r/q} d\mathbb{P}.$$

Hence:

$$\int_\Omega \left(f^{q/r} + t^q|Tx|^q\right)^{r/q} d\mathbb{P} \leqslant (1 + K^q t^q)^{r/q}.$$

However, for $t = 0$, this inequality is an equality; thus:

$$\frac{d}{dt} \left[\int_\Omega \left(f^{q/r} + t^q|Tx|^q\right)^{r/q} d\mathbb{P} - (1 + K^q t^q)^{r/q}\right]_{t=0} \leqslant 0;$$

that is:

$$\int_\Omega f^{1-\frac{q}{r}}|Tx|^q \, d\mathbb{P} \leqslant K^q,$$

and finally, for every $x \in X$:

$$\left[\int_\Omega \left|\frac{Tx}{f^{1/r}}\right|^q f \, d\mathbb{P}\right]^{1/q} \leqslant K\|x\|. \qquad \square$$

*Proof of the factorization theorem*  Let $T: X \to L^r(\Omega, \mathbb{P})$. We check that $T$ satisfies the conditions of the criterion. Let $x_1, \ldots, x_n \in X$. Let $Z_1, \ldots, Z_n$ be $q$-stable independent random variables. Then:

$$\int_\Omega \left(\sum_{k=1}^n |(Tx_k)(\omega)|^q\right)^{r/q} d\mathbb{P}(\omega)$$

$$= \frac{1}{S_{q,r}^r} \int_\Omega \left(\int_\Omega \left|\sum_{k=1}^n (Tx_k)(\omega)Z_k(\theta)\right|^r d\mathbb{P}(\theta)\right) d\mathbb{P}(\omega)$$

$$= \frac{1}{S_{q,r}^r} \int_\Omega \left(\int_\Omega \left|T\left(\sum_{k=1}^n Z_k(\theta)x_k\right)(\omega)\right|^r d\mathbb{P}(\omega)\right) d\mathbb{P}(\theta)$$

$$\leqslant \frac{\|T\|^r}{S_{q,r}^r} \int_\Omega \left\|\sum_{k=1}^n Z_k(\theta)x_k\right\|^r d\mathbb{P}(\theta).$$

However, the variables $Z_1, \ldots, Z_n$ are symmetric; thus:

$$\int_\Omega \left\| \sum_{k=1}^n Z_k(\theta) x_k \right\|^r \, d\mathbb{P}(\theta) = \int_\Omega \left( \int_0^1 \left\| \sum_{k=1}^n r_k(t) Z_k(\theta) x_k \right\|^r \, dt \right) d\mathbb{P}(\theta)$$

$$\leqslant K_r^r \tau_p(X)^r \int_\Omega \left( \sum_{k=1}^n \|Z_k(\theta) x_k\|^p \right)^{r/p} d\mathbb{P}(\theta),$$

where $K_r$ is the constant of the Khintchine–Kahane inequality, and $\tau_p(X)$ the type-$p$ constant of $X$.

Now let us consider $p$-stable independent variables $W_1, \ldots, W_n$; we have:

$$\int_\Omega \left( \sum_{k=1}^n \|Z_k(\theta) x_k\|^p \right)^{r/p} d\mathbb{P}(\theta)$$

$$= \frac{1}{S_{p,r}^r} \int_\Omega \left( \int_\Omega \left| \sum_{k=1}^n (Z_k(\theta) \|x_k\|) W_k(\xi) \right|^r d\mathbb{P}(\xi) \right) d\mathbb{P}(\theta)$$

$$= \frac{1}{S_{p,r}^r} \int_\Omega \left( \int_\Omega \left| \sum_{k=1}^n (W_k(\xi) \|x_k\|) Z_k(\theta) \right|^r d\mathbb{P}(\theta) \right) d\mathbb{P}(\xi)$$

$$= \frac{S_{q,r}^r}{S_{p,r}^r} \int_\Omega \left( \sum_{k=1}^n |W_k(\xi)|^q \|x_k\|^q \right)^{r/q} d\mathbb{P}(\xi)$$

$$\leqslant \left( \frac{S_{q,r}}{S_{p,r}} \right)^r \left( \int_\Omega \sum_{k=1}^n |W_k(\xi)|^q \|x_k\|^q \, d\mathbb{P}(\xi) \right)^{r/q}$$

(by Hölder's inequality, given that $q/r > 1$); hence:

$$\int_\Omega \left( \sum_{k=1}^n \|Z_k(\theta) x_k\|^p \right)^{r/p} d\mathbb{P}(\theta) = \left( \frac{S_{q,r}}{S_{p,r}} \right)^r S_{p,q}^r \left( \sum_{k=1}^n \|x_k\|^q \right)^{r/q}.$$

Thus the criterion is satisfied and the proof of the factorization theorem is complete.                                                                          □

Note that the end of the proof is superfluous if the stable type $p$ of $X$ is used instead of the Rademacher type $p$ (see Chapter 5 of Volume 1).

**Remark**   The proof shows that the constant $K$ in the Maurey factorization theorem can be taken $\leqslant C(p, q, r) \|T\| \tau_p(X)$, with $C(p, q, r) = K_r S_{p,q}/S_{p,r}$.

On the other hand, if $p = 2$ ($X$ of type 2), we can take $q = p = 2$ instead of $q < p$ since the Gaussian variables are square integrable. In fact, Rademacher variables can be used instead of Gaussian variables. In particular, for $r = 1$:

**Corollary IV.6** *Let $X$ be a Banach space of type 2. Then, for every operator $T: X \to L^1(\Omega, \mathbb{P})$, there exists a probability density $\Delta$ such that:*

$$\left( \int_\Omega \left| \frac{Tx}{\Delta} \right|^2 \Delta \, d\mathbb{P} \right)^{1/2} \leqslant \sqrt{2} \, \|T\| \, \tau_2(X) \, \|x\|, \qquad \forall x \in X.$$

*Proof*  We have:

$$\int_\Omega \left( \sum_{k=1}^n |(Tx_k)(\omega)|^2 \right)^{1/2} d\mathbb{P}(\omega)$$

$$\leqslant \sqrt{2} \int_\Omega \left( \int_0^1 \left| \sum_{k=1}^n r_k(t)(Tx_k)(\omega) \right| dt \right) d\mathbb{P}(\omega)$$

$$= \sqrt{2} \int_0^1 \left( \int_\Omega \left| T\left( \sum_{k=1}^n r_k(t)x_k \right)(\omega) \right| d\mathbb{P}(\omega) \right) dt$$

$$= \sqrt{2} \int_0^1 \left\| T\left( \sum_{k=1}^n r_k(t)x_k \right) \right\|_1 dt$$

$$\leqslant \sqrt{2} \, \|T\| \int_0^1 \left\| \sum_{k=1}^n r_k(t)x_k \right\| dt$$

$$\leqslant \sqrt{2} \, \|T\| \, \tau_2(X) \left( \sum_{k=1}^n \|x_k\|^2 \right)^{1/2}. \qquad \square$$

# V  Finite-Dimensional Subspaces of $L^1$

## V.1  Statement of the Result

First, we recall the following simple fact (already seen in Chapter 6 of Volume 1, Proposition III.6):

**Proposition V.1**  *If $X$ is a finite-dimensional subspace of $L^1$, then, for any $\varepsilon > 0$, there exist $M \geqslant 1$ and a subspace $Y$ of $\ell_1^M$ such that $\mathrm{dist}(X, Y) \leqslant 1 + \varepsilon$.*

It suffices to approximate the functions of $X$ by simple functions, uniformly on $B_X$. In fact, the functions which are constant on the elements of a finite partition of $[0, 1]$ span a space isometric to some space $\ell_1^M$. In other more "sophisticated" terms, the conditional expectations with respect to the $\sigma$-algebras generated by the dyadic intervals converge pointwise to the identity.

This proposition does not give a clue about the value of $M$ as a function of $\varepsilon$ and $\dim X$. The best result currently known (2004) seems to be:

**Theorem V.2** (Talagrand)    *If* $X$ *is a finite-dimensional subspace of* $L^1$ *with* dim $X = n$, *then, for any* $\varepsilon > 0$, *there exist* $N \geqslant 1$ *and a subspace* $Y$ *of* $\ell_1^N$ *such that* dist$(X, Y) \leqslant 1 + \varepsilon$ *and*:

$$N \leqslant C \frac{K(X)^2}{\varepsilon^2} n \,,$$

*where* $C$ *is a numerical constant and* $K(X)$ *the K-convexity constant of* $X$.

Note that this only gives a good estimation of $N$ when $\varepsilon$ is small. For large values of $\varepsilon$, the correct order of magnitude of $N$ does not seem to be known.

The end of this chapter is devoted to the proof of this theorem. Prior to this, estimations are provided for the $K$-convexity constant that appears in the statement of the theorem.

## V.2 *K*-Convexity

Recall (Chapter 5 of Volume 1, Definition IV.6) that a Banach space $X$ is said to be *K-convex* if the natural projection of $L^2(X) = L^2([0, 1] \,; X)$ onto Rad$(X)$ is continuous, where Rad$(X)$ is the (closed) subspace generated in $L^2(X)$ by all functions of the form $\sum_{k=1}^n r_k x_k$, for $x_1, \ldots, x_n \in X$, $n \geqslant 1$, and $(r_k)_{k \geqslant 1}$ the sequence of Rademacher functions. The *K-convexity constant* of $X$ is the norm of this projection. In other words, if $f \in L^2(X)$, and if:

$$f_k = \int_0^1 f(t)\, r_k(t)\, \mathrm{d}t \,,$$

then, for any $n \geqslant 1$:

$$\left\| \sum_{k=1}^n r_k f_k \right\|_{L^2(X)} \leqslant K(X) \|f\|_{L^2(X)} \,.$$

Also recall that Pisier [1982] showed the following deep result: $X$ is $K$-convex if and only it does not contain $\ell_1^n$'s uniformly (and hence if and only if it has a type $p > 1$).

Every finite-dimensional space is of course $K$-convex, but we can be more precise:

**Theorem V.3** (Pisier)    *For every Banach space* $X$ *of dimension* $n \geqslant 2$, *the K-convexity constant of* $X$ *satisfies*:

$$K(X) \leqslant C \log\left[1 + \mathrm{dist}\,(X, \ell_2^n)\right] \leqslant C' \log n \,.$$

The right-hand inequality results from dist$(X, \ell_2^n) \leqslant \sqrt{n}$ (see the F. John theorem, Chapter 6 of Volume 1, Theorem IV.4); however dist$(X, \ell_2^n) \leqslant n$ suffices here. We set $d_X = \mathrm{dist}\,(X, \ell_2^n)$.

*Proof* (Bourgain and Milman) Every $f \in L^2([0, 1]; X)$ can be written as the sum of its Fourier–Walsh series:

$$f(t) = \sum_{A \subseteq \mathbb{N}^*} w_A(t) \widehat{f}(A),$$

where:

$$\widehat{f}(A) = \int_0^1 w_A(t) f(t) \, dt,$$

and where $w_A, A \subseteq \mathbb{N}^*, A$ finite, are the Walsh functions:

$$w_A(t) = \prod_{j \in A} r_j(t).$$

For $j \geqslant 0$, we denote:

$$R_j(f) = \sum_{|A| = j} w_A \widehat{f}(A),$$

so that $K(X) = \|R_1\|$. For $|s| \leqslant 1/2$, we consider the Riesz product:

$$\mu_s = \prod_{j \geqslant 1} (1 + s r_j).$$

Then:

$$\|f\|_{L^2(X)} \geqslant \|f * \mu_s\|_{L^2(X)} = \left\| \sum_{j \geqslant 0} s^j R_j(f) \right\|_{L^2(X)}.$$

Let $J = \left[ \log_2 d_X \right] + 2$; then

$$\|f\|_{L^2(X)} \geqslant \left\| \sum_{j \geqslant 0}^{J} s^j R_j(f) \right\|_{L^2(X)} - \sum_{j = J+1}^{+\infty} |s|^j \|R_j(f)\|_{L^2(X)}.$$

Now, when $X$ is a Hilbert space $H$, $R_j$ is an orthogonal projection of $L^2([0, 1]; H)$; thus:

$$\|R_j(f)\|_{L^2(X)} \leqslant d_X \|f\|_{L^2(X)},$$

and hence:

$$\sum_{j = J+1}^{+\infty} |s|^j \|R_j(f)\|_{L^2(X)} \leqslant \frac{1}{2^J} d_X \|f\|_{L^2(X)} \leqslant \frac{1}{2} \|f\|_{L^2(X)}.$$

To minorize the other term, we use the following lemma:

**Lemma V.4**   *For any $J \geqslant 2$, there exists a measure $v$ on $[-1/2, +1/2]$ such that $\|v\| \leqslant 2J$, and*

$$\int_{-1/2}^{1/2} s \, dv(s) = 1 \quad and \quad \int_{-1/2}^{1/2} s^j \, dv(s) = 0 \quad for \ 0 \leqslant j \leqslant J, j \neq 1.$$

*Proof*  Let $\mathcal{P}_J$ be the space of polynomials of degree $\leqslant J$, defined on $[-1/2, 1/2]$. The linear functional $\varphi \colon \mathcal{P}_J \to \mathbb{C}$ defined by $\varphi(P) = P'(0)$ is continuous and of norm $\leqslant 2J$, by Bernstein's inequality (Chapter 6 of Volume 1, Lemma V.12), applied to the trigonometric polynomial $Q(t) = P(\frac{1}{2} \sin t)$: $|P'(0)| = 2 |Q'(0)| \leqslant 2J \|Q\|_\infty = 2J \|P\|_\infty$. We extend $\varphi$ to $\mathcal{C}([-1/2, 1/2])$ by the Hahn–Banach theorem, which provides $v$.  □

*End of the Proof of Theorem V.3*   Thus, integrating with respect to $s$, we obtain:

$$4J \|f\|_{L^2(X)} \geqslant \int_{-1/2}^{1/2} \left\| \sum_{j=0}^{J} s^j R_j(f) \right\|_{L^2(X)} d|v|(s)$$

$$\geqslant \left\| \sum_{j=0}^{J} \left( \int_{-1/2}^{1/2} s^j \, dv(s) \right) R_j(f) \right\|_{L^2(X)} = \|R_1(f)\|_{L^2(X)}.$$   □

Bourgain [1984 b] showed that, in general, this cannot be improved (see Chapter 5 of this Volume, Section IV). Nevertheless, regarding the subspaces of $L^1$, there is a better estimate:

**Theorem V.5** (Pisier)   *For every subspace $X$ of $L^1$ of finite dimension $n \geqslant 2$, we have:*

$$K(X) \leqslant C \sqrt{\log n},$$

*where $C$ is a numerical constant.*

Here is an immediate consequence:

**Corollary V.6**   *In Talagrand's theorem, we have:*

$$N \leqslant C \frac{n \log n}{\varepsilon^2}.$$

To show this theorem, we first need the $K$-convexity constant of the $L^p$ spaces.

**Proposition V.7**   *If* $1 < p < +\infty$*, then:*

$$K\big(L^p(\Omega)\big) \leqslant \max\{\sqrt{p}, \sqrt{p^*}\},$$

*where $p^*$ is the conjugate exponent of $p$.*

*Proof*   First recall the following inequality, valid for every positive measurable function $g$ and for $\alpha \geqslant 1$:

$$(*) \quad \left[\int_\Omega \left(\int_0^1 g(\omega, t)\, dt\right)^\alpha d\mathbb{P}(\omega)\right]^{1/\alpha} \leqslant \int_0^1 \left(\int_\Omega [g(\omega, t)]^\alpha\, d\mathbb{P}(\omega)\right)^{1/\alpha} dt.$$

As the notion of $K$-convexity is self-dual: $K(X^*) = K(X)$, it suffices to give the proof for $p \geqslant 2$. With the preceding notation, for $f \in L^2([0, 1]; L^p(\Omega))$, and with $f_j = f_{\{j\}}$, we have:

$$\|R_1(f)\|^2_{L^2(L^p)} = \int_0^1 \|R_1(f)(t)\|^2_{L^p(\Omega)}\, dt$$

$$= \int_0^1 \left(\int_\Omega \left|\sum_{k=1}^{+\infty} r_k(t) f_k(\omega)\right|^p d\mathbb{P}(\omega)\right)^{2/p} dt$$

$$\leqslant \left[\int_0^1 \left(\int_\Omega \left|\sum_{k=1}^{+\infty} r_k(t) f_k(\omega)\right|^p d\mathbb{P}(\omega)\right) dt\right]^{2/p}$$

$$\text{by Hölder's inequality because } p/2 \geqslant 1$$

$$= \left[\int_\Omega \left(\int_0^1 \left|\sum_{k=1}^{+\infty} r_k(t) f_k(\omega)\right|^p dt\right) d\mathbb{P}(\omega)\right]^{2/p}$$

$$\leqslant p \left[\int_\Omega \left(\sum_{k=1}^{+\infty} |f_k(\omega)|^2\right)^{p/2} d\mathbb{P}(\omega)\right]^{2/p}$$

$$\text{by Khintchine's inequalities}$$

$$\leqslant p \left[\int_\Omega \left(\sum_{A \subseteq \mathbb{N}^*} |f_A(\omega)|^2\right)^{p/2} d\mathbb{P}(\omega)\right]^{2/p}$$

$$= p \left[\int_\Omega \left(\int_0^1 |[f(\omega)](t)|^2\, dt\right)^{p/2} d\mathbb{P}(\omega)\right]^{2/p}$$

$$\text{by Parseval's identity}$$

$$\leqslant p \int_0^1 \left( \int_\Omega |[f(\omega)](t)|^p \, d\mathbb{P}(\omega) \right)^{2/p} dt$$

$$\text{by } (*) \text{ with } \alpha = p/2 \geqslant 1$$

$$= p \, \|f\|_{L^2(L^p)}^2 \, . \qquad\qquad \square$$

*Proof of Theorem V.5*   We are going to embed $X$ in an $L^p(\Omega)$-space, with an appropriate $p < 2$, in order to exploit our estimation of the $K$-convexity constant of $L^p$. To this end, we use the Maurey factorization theorem. This requires information about the type-$p$ constant $\tau_p(X)$ of $X$. We reason by interpolation, starting with $p = 2$. Indeed, thanks to the F. John theorem (Chapter 6 of Volume 1, Theorem IV.4), we know that $\mathrm{dist}\,(X, \ell_2^n) \leqslant \sqrt{n}$, hence $\tau_2(X) \leqslant \sqrt{n}$. Then, from Corollary IV.6, for every operator $T \colon X \to L^1(\Omega, \mathbb{P})$, there exists a probability density satisfying:

$$\left( \int_\Omega \left| \frac{Tf}{\Delta} \right|^2 \Delta \, d\mathbb{P} \right)^{1/2} \leqslant \sqrt{2} \, \|T\| \, \tau_2(X) \, \|f\|_1 \,, \quad \forall f \in X.$$

For $1 < p < 2$, write $\dfrac{1}{p} = \dfrac{1-\theta}{1} + \dfrac{\theta}{2}$, with $0 < \theta < 1$; then:

$$\left\| \frac{Tf}{\Delta} \right\|_{L^p(\Delta.\mathbb{P})} \leqslant \left\| \frac{Tf}{\Delta} \right\|_{L^1(\Delta.\mathbb{P})}^{1-\theta} \left\| \frac{Tf}{\Delta} \right\|_{L^2(\Delta.\mathbb{P})}^{\theta}$$

$$= \left( \int_\Omega \left| \frac{Tf}{\Delta} \right| \Delta\mathbb{P} \right)^{1-\theta} \left( \int_\Omega \left| \frac{Tf}{\Delta} \right|^2 \Delta\mathbb{P} \right)^{\theta/2}$$

$$\leqslant \left( \|T\| \, \|f\|_1 \right)^{1-\theta} . \left( \sqrt{2} \, \|T\| \, \tau_2(X) \, \|f\|_1 \right)^{\theta}$$

$$= 2^{\theta/2} \, \|T\| \, \tau_2(X)^{\theta} \, \|f\|_1 \,.$$

We thus obtain a factorization:

with:

$$U(f) = \frac{T(f)}{\Delta} \quad \text{and} \quad V(g) = g.\Delta$$

and:

$$\|V\| \, \|U\| \leqslant 2^{\theta/2} \, \tau_2(X)^{\theta} \leqslant (2n)^{\theta/2}.$$

Applying this to $T = j$, the canonical injection from $X$ into $L^1(\Omega, \mathbb{P})$, we obtain a subspace $Y = U(X)$ of $L^p(\Omega, \Delta.\mathbb{P})$, isomorphic to $X$, with:

$$\text{dist}(X, Y) \leqslant (2n)^{\theta/2}.$$

Hence:

$$K(X) \leqslant \text{dist}(X, Y) K(Y) \leqslant \text{dist}(X, Y) K[L^p(\Delta.\mathbb{P})] \leqslant (2n)^{\theta/2} \sqrt{p^*},$$

where $p^*$ is the conjugate exponent of $p$. It only remains to note that $\theta/2 = 1/p^*$, and to optimize by taking $p^* = \log n \, (> 2 \text{ for } n \geqslant 8)$, to obtain:

$$K(X) \leqslant \exp\left(1 + \frac{\log 2}{\log n}\right) \sqrt{\log n} \leqslant C \sqrt{\log n}$$

for a numerical constant $C \leqslant 4$ ( for $n \geqslant 8$). $\qquad\square$

Instead of the $K$-convexity constant relative to the Rademacher functions, it is often more convenient to use $K_g(X)$ relative to the Gaussian variables: this is the smallest constant $K$ satisfying:

$$\left\| \sum_{k=1}^n g_k f_k \right\|_{L^2(X)} \leqslant K \, \|f\|_{L^2(X)},$$

for any $n \geqslant 1$ and every $f \in L^2(X)$, and with $f_k = \mathbb{E}(g_k f)$, $(g_k)_{k \geqslant 1}$ a sequence of independent standard Gaussian variables. These two constants are actually equivalent:

**Theorem V.8** (Figiel)    *For every K-convex X Banach space*:

$$\frac{2}{\pi} K(X) \leqslant K_g(X) \leqslant K(X).$$

*Proof*

1) We first show the inequality $K(X) \leqslant (\pi/2) K_g(X)$.

The proof is based on the comparison principle (Chapter 4 of Volume 1, Theorem IV.4), stating that, for $x_1, \ldots, x_n \in X$:

$$\left( \int_0^1 \left\| \sum_{k=1}^n r_k(t) x_k \right\|^2 dt \right)^{1/2} \leqslant \sqrt{\frac{\pi}{2}} \left( \int_\Omega \left\| \sum_{k=1}^n g_k x_k \right\|^2 d\mathbb{P} \right)^{1/2}.$$

Let $x_1, \ldots, x_n \in X$ and define the operators $\Gamma_n$ on $L^2(\Omega, \mathbb{P}; X)$ and $R_n$ on $L^2([0, 1]; X)$ by:

$$\Gamma_n f = \sum_{k=1}^{n} g_k \left( \int_{\Omega} g_k f \, d\mathbb{P} \right) \quad \text{and} \quad R_n f = \sum_{k=1}^{n} r_k \left( \int_0^1 r_k(t) f(t) \, dt \right),$$

for $f$ in $L^2(\Omega, \mathbb{P}; X)$ and $L^2([0,1]; X)$ respectively.

Fix $f \in L^2([0,1]; X)$, and set:

$$\varphi = \sum_{k=1}^{n} g_k \left( \int_0^1 r_k(t) f(t) \, dt \right) \in L^2(\Omega, \mathbb{P}; X).$$

Next, we select $\varphi^* \in L^2(\Omega, \mathbb{P}; X^*)$ such that $\|\varphi^*\|_{L^2(X^*)} = 1$ and $\int_{\Omega} \langle \varphi^*, \varphi \rangle \, d\mathbb{P} = \|\varphi\|_{L^2(X)}$. Then:

$$\|R_n f\|_{L^2(X)} = \left( \int_0^1 \left\| \sum_{k=1}^{n} r_k(t) \left( \int_0^1 r_k(\tau) f(\tau) \, d\tau \right) \right\|^2 dt \right)^{1/2}$$

$$\leqslant \sqrt{\frac{\pi}{2}} \left( \int_{\Omega} \left\| \sum_{k=1}^{n} g_k \left( \int_0^1 r_k(\tau) f(\tau) \, d\tau \right) \right\|^2 d\mathbb{P} \right)^{1/2}$$

$$= \sqrt{\frac{\pi}{2}} \, \|\varphi\|_{L^2(X)} = \sqrt{\frac{\pi}{2}} \int_{\Omega} \langle \varphi^*(\omega), \varphi(\omega) \rangle \, d\mathbb{P}(\omega)$$

$$= \sqrt{\frac{\pi}{2}} \int_0^1 \left\langle \sum_{k=1}^{n} r_k(t) \left( \int_{\Omega} g_k \, \varphi^* \, d\mathbb{P} \right), f(t) \right\rangle dt$$

$$\leqslant \sqrt{\frac{\pi}{2}} \|f\|_{L^2(X)} \left( \int_0^1 \left\| \sum_{k=1}^{n} r_k(t) \left( \int_{\Omega} g_k \, \varphi^* \, d\mathbb{P} \right) \right\|^2 dt \right)^{1/2}$$

$$\leqslant \left( \sqrt{\frac{\pi}{2}} \right)^2 \|f\|_{L^2(X)} \left( \int_{\Omega} \left\| \sum_{k=1}^{n} g_k \left( \int_{\Omega} g_k \, \varphi^* \, d\mathbb{P} \right) \right\|^2 d\mathbb{P} \right)^{1/2}$$

$$= \frac{\pi}{2} \|f\|_{L^2(X)} \|\Gamma_n^* \varphi^*\|_{L^2(X^*)} \leqslant \frac{\pi}{2} \|f\|_{L^2(X)} \|\Gamma_n\|,$$

hence the result.

2) For the second inequality, the central limit theorem (Chapter 5 of Volume 1, Theorem II.5) is used in order to approximate the Rademacher functions by Gaussian variables. Indeed, with a fixed integer $n \geqslant 1$, for $k = 1, \ldots, n$ and $m = 1, 2, \ldots$, we define:

$$\rho_{k,m} = \frac{1}{\sqrt{m}} \sum_{l=1}^{m} r_{(k-1)m+l}.$$

Since the variables $\rho_{1,m}, \ldots, \rho_{n,m}$ are centered and independent, the vector $(\rho_{1,m}, \ldots, \rho_{n,m})$ converges in distribution to $(g_1, \ldots, g_n)$, where $g_1, \ldots, g_n$ are independent standard Gaussian variables. Furthermore, we have:

$$\lim_{m \to +\infty} \int_0^1 \psi(\rho_{1,m}, \ldots, \rho_{n,m}) \, dt = \int_\Omega \psi(g_1, \ldots, g_n) \, d\mathbb{P}$$

for every function $\psi : \mathbb{R}^n \to \mathbb{R}$ such that:

$$\psi(s_1, \ldots, s_n) \exp\left(-\sum_{k=1}^n |s_k|\right) \to 0 \quad \text{when} \quad \sum_{k=1}^n |s_k| \longrightarrow +\infty,$$

and this *uniformly* when the following norm is bounded:

$$\|\psi\|_B = \sup_{(s_1, \ldots, s_n) \in \mathbb{R}^n} |\psi(s_1, \ldots, s_n)| \exp\left(-\sum_{k=1}^n |s_k|\right)$$

(see Chapter 5 of Volume 1, Theorem II.5).

For $f \in L^2(\Omega, \mathbb{P}; X)$, we want to estimate:

$$\|\Gamma_n f\| = \left( \int_\Omega \left\| \sum_{k=1}^n g_k \left( \int_\Omega g_k f \, d\mathbb{P} \right) \right\|^2 d\mathbb{P} \right)^{1/2}.$$

Let $x_k = \int_\Omega g_k f \, d\mathbb{P} \in X$, $1 \leqslant k \leqslant n$, and apply the central limit theorem with:

$$\psi(s_1, \ldots, s_n) = \left\| \sum_{k=1}^n s_k x_k \right\|^2,$$

for which $\|\psi\|_B \leqslant (4/e^2) \max_{1 \leqslant k \leqslant n} \|x_k\|^2$. Then:

$$\|\Gamma_n f\| = \left( \int_\Omega \psi(g_1, \ldots, g_n) \, d\mathbb{P} \right)^{1/2}$$

$$= \lim_{m \to +\infty} \left( \int_0^1 \psi(\rho_{1,m}, \ldots, \rho_{n,m}) \, dt \right)^{1/2}$$

$$= \lim_{m \to +\infty} \left( \int_0^1 \left\| \sum_{k=1}^n \rho_{k,m}(t) \left( \int_\Omega g_k f \, d\mathbb{P} \right) \right\|^2 dt \right)^{1/2}.$$

Let $\varphi_m^* \in L^2(\Omega, \mathbb{P}; X^*)$ of norm $\|\varphi_m^*\|_{L^2(X^*)} = 1$ be such that:

$$\left( \int_0^1 \left\| \sum_{k=1}^n \rho_{k,m}(t) \left( \int_\Omega g_k f \, d\mathbb{P} \right) \right\|^2 dt \right)^{1/2}$$

$$= \int_0^1 \left\langle \varphi_m^*(t), \sum_{k=1}^n \rho_{k,m}(t) \left( \int_\Omega g_k f \, d\mathbb{P} \right) \right\rangle dt.$$

Then:

$$\|\Gamma_n f\| \leqslant \|f\|_{L^2(X)} \lim_{m \to +\infty} \left[ \int_\Omega \left\| \sum_{k=1}^n g_k \left( \int_0^1 \rho_{k,m}(t) \, \varphi_m^*(t) \, dt \right) \right\|^2 d\mathbb{P} \right]^{1/2}.$$

Again using the central limit theorem, this time with

$$\psi^*(s_1, \ldots, s_n) = \left\| \sum_{k=1}^n s_k \left( \int_0^1 \rho_{k,m}(t) \, \varphi_m^*(t) \, dt \right) \right\|^2,$$

we obtain:

$$\|\Gamma_n f\| \leqslant \|f\|_{L^2(X)} \times \lim_{m \to +\infty} \lim_{m' \to +\infty}$$

$$\left( \int_0^1 \left\| \sum_{k=1}^n \rho_{k,m'}(t') \left( \int_0^1 \rho_{k,m}(t) \, \varphi_m^*(t) \, dt \right) \right\|^2 dt' \right)^{1/2}.$$

Now the limit in $m'$ is uniform with respect to $m$, since:

$$\left\| \int_0^1 \rho_{k,m}(t) \, \varphi_m^*(t) \, dt \right\|_{X^*} \leqslant \|\varphi_m^*\|_{L^2(X^*)} \|\rho_{k,m}\|_2$$

$$= \|\varphi_m^*\|_{L^2(X^*)} \frac{1}{\sqrt{m}} \left( \sum_{l=1}^m \|r_{(k-1)m+l}\|_2^2 \right)^{1/2}$$

$$= 1.$$

Consequently:

$$\|\Gamma_n f\| \leqslant \|f\|_{L^2(X)} \times$$

$$\lim_{m \to +\infty} \left( \int_0^1 \left\| \sum_{k=1}^n \rho_{k,m}(t') \left( \int_0^1 \rho_{k,m}(t) \, \varphi_m^*(t) \, dt \right) \right\|^2 dt' \right)^{1/2}.$$

Next, for each $m$, with $\varphi_m^* = \varphi^*$, the triangle inequality in $L^2(X^*)$ leads to:

$$\left( \int_0^1 \left\| \sum_{k=1}^n \rho_{k,m}(t') \left( \int_0^1 \rho_{k,m}(t)\, \varphi^*(t)\, \mathrm{d}t \right) \right\|^2 \mathrm{d}t' \right)^{1/2}$$

$$= \left( \int_0^1 \left\| \sum_{k=1}^n \sum_{l'=1}^m \sum_{l=1}^m \frac{1}{m} r_{(k-1)m+l'}(t') \left( \int_0^1 r_{(k-1)m+l}(t)\, \varphi^*(t)\, \mathrm{d}t \right) \right\|^2 \mathrm{d}t' \right)^{1/2}$$

$$\leqslant \frac{1}{m} \sum_{l'=1}^m \left( \int_0^1 \left\| \sum_{k=1}^n \sum_{l=1}^m \left( \int_0^1 r_{(k-1)m+l'}(t')\, r_{(k-1)m+l}(t)\, \varphi^*(t)\, \mathrm{d}t \right) \right\|^2 \mathrm{d}t' \right)^{1/2}.$$

However, for each $t'$, the variable $r_{(k-1)m+l'}(t')\, r_{(k-1)m+l}\, \varphi^*$ has the same distribution as $r_{(k-1)m+l}(t')\, r_{(k-1)m+l}\, \varphi^*$; hence:

$$\int_0^1 \left\| \sum_{k=1}^n \sum_{l=1}^m \left( \int_0^1 r_{(k-1)m+l'}(t')\, r_{(k-1)m+l}(t)\, \varphi^*(t)\, \mathrm{d}t \right) \right\|^2 \mathrm{d}t'$$

$$= \int_0^1 \left\| \sum_{k=1}^n \sum_{l=1}^m \left( \int_0^1 r_{(k-1)m+l}(t')\, r_{(k-1)m+l}(t)\, \varphi^*(t)\, \mathrm{d}t \right) \right\|^2 \mathrm{d}t' \, ;$$

consequently:

$$\left( \int_0^1 \left\| \sum_{k=1}^n \rho_{k,m}(t') \left( \int_0^1 \rho_{k,m}(t)\, \varphi^*(t)\, \mathrm{d}t \right) \right\|^2 \mathrm{d}t' \right)^{1/2}$$

$$\leqslant \left( \int_0^1 \left\| \sum_{k=1}^n \sum_{l=1}^m r_{(k-1)m+l}(t') \left( \int_0^1 \left\| r_{(k-1)m+l}(t)\, \varphi^*(t)\, \mathrm{d}t \right) \right) \right\|^2 \mathrm{d}t' \right)^{1/2}$$

$$= \left( \int_0^1 \left\| \sum_{j=1}^{nm} r_j(t') \left( \int_0^1 r_j(t)\, \varphi^*(t)\, \mathrm{d}t \right) \right\|^2 \mathrm{d}t' \right)^{1/2}$$

$$\leqslant K(X^*) \|\varphi^*\|_{L^2(X^*)} = K(X).$$

Finally, letting $m$ tend to infinity, we obtain:

$$\|\Gamma_n f\| \leqslant \|f\|_{L^2(X)} \leqslant K(X) \|f\|_{L^2(X)} \, ;$$

hence:

$$K_g(X) \leqslant K(X),$$

which completes the proof. $\qquad\qquad\qquad\qquad\qquad\qquad\qquad\qquad\quad$ $\square$

## V.3 An Auxiliary Result

The following result is required:

**Theorem V.9** (Lewis)   *Let $X$ be an $n$-dimensional subspace of a real space $L^1(S, \mu)$ ($\mu$ $\sigma$-finite). Then there exist a probability $\nu$ on $S$ and a subspace $Y$ of $L^1(\nu)$, isometric to $X$, with a basis $(\psi_j)_{1 \leqslant j \leqslant n}$ orthogonal in $L^2(\nu)$, satisfying:*

$$\sum_{j=1}^{n} \psi_j^2 = 1 \quad and \quad \|\psi_j\|_{L^2(\nu)} = \frac{1}{\sqrt{n}}, \qquad 1 \leqslant j \leqslant n.$$

*Proof*   First, we equip the product $[L^1(\mu)]^n$ with a norm $\| \cdot \|_1$ by setting, for $\vec{f} = (f_1, \ldots, f_n)$:

$$\|\vec{f}\|_1 = \|F\|_1,$$

where $F$ is the square function:

$$F(t) = \left( \sum_{k=1}^{n} f_k(t)^2 \right)^{1/2}.$$

The dual of $[L^1(\mu)]^n$ can be identified with $[L^\infty(\mu)]^n$, equipped with the norm $\|(g_1, \ldots, g_n)\|_\infty = \|G\|_\infty$, with $G = (g_1^2 + \cdots + g_n^2)^{1/2}$. Indeed, for every continuous linear functional $\Psi$ on $[L^1(\mu)]^n$, there exist $g_1, \ldots, g_n \in L^\infty(\mu)$ such that:

$$\Psi(\vec{f}) = \Psi_{\vec{g}}(\vec{f}) = \sum_{k=1}^{n} \int_S f_k \, g_k \, \mathrm{d}\mu,$$

and the Cauchy–Schwarz inequality leads to:

$$(*) \qquad \begin{aligned} |\Psi_{\vec{g}}(\vec{f})| &\leqslant \int_S \sum_{k=1}^{n} |f_k(t)| \, |g_k(t)| \, \mathrm{d}\mu(t) \\ &\leqslant \int_S F G \, \mathrm{d}\mu \leqslant \|F\|_1 \|G\|_\infty = \|\vec{f}\|_1 \|\vec{g}\|_\infty; \end{aligned}$$

hence $\|\Psi_{\vec{g}}\| \leqslant \|\vec{g}\|_\infty$. To show that, in fact, it is an equality, we select, for a given $\vec{g}$ and for any $\varepsilon > 0$, a function $h \geqslant 0$ such that $\|h\|_1 = 1$, $h(t) = 0$ when $G(t) = 0$, and $\int_S h \, G \, \mathrm{d}\mu \geqslant \|G\|_\infty (1 - \varepsilon)$.

Set $f_k(t) = \dfrac{g_k(t)}{G(t)} h(t)$ if $G(t) \neq 0$, and $f_k(t) = 0$ otherwise. Then the square function $F$ of $\vec{f}$ is equal to $h$, and hence $\|\vec{f}\|_1 = 1$. As $\Psi_{\vec{g}}(\vec{f}) = \int_S h \, G \, \mathrm{d}\mu \geqslant \|G\|_\infty (1 - \varepsilon)$, we indeed have $\|\Psi_{\vec{g}}\| = \|\vec{g}\|_\infty$.

The probability $v$ and the functions $\psi_j$ are now obtained by maximizing a determinant, as was done to show the existence of the Lewis ellipsoid (Chapter 1 of this volume, Lemma IV.3).

Given a basis $\mathcal{B}$ of the space $X$ of dimension $n$, the determinant $\det_\mathcal{B}$ $(h_1,\ldots,h_n)$ of the $n$ vectors $h_1,\ldots,h_n \in X$ is a continuous function, and, as the unit ball of $X^n \subseteq [L^1(\mu)]^n$ is compact, $f_1,\ldots,f_n \in X$ can be found such that:

$$\Delta = \max\{\det_\mathcal{B}(h_1,\ldots,h_n)\,;\ \|\vec{h}\|_1 \leqslant n\} = \det_\mathcal{B}(f_1,\ldots,f_n)\,.$$

Note that necessarily $\Delta > 0$, so that $f_1,\ldots,f_n$ are linearly independent, and hence form a basis of $X$. Moreover $\|\vec{f}\|_1 = n$, since otherwise we could multiply $\vec{f}$ by a number $> 1$ to obtain a larger determinant, all while keeping the norm $\|.\|_1 \leqslant n$. Also note (even though this is not used here) that, with a change of the basis $\mathcal{B}$ to another basis $\mathcal{B}'$ having the same orientation, the functions $f_1,\ldots,f_n$ remain maximal for the new determinants, since they are all multiplied by the same constant $> 0$ (the determinant of the matrix of the change of basis).

Let us fix maximizing functions $f_1,\ldots,f_n$. With the choice of $\mathcal{B} = \{f_1,\ldots,f_n\}$ as the basis, we have $\Delta = 1$. Now let $\varphi_1,\ldots,\varphi_n \in X^*$ be the dual basis of $f_1,\ldots,f_n$; and we extend $\varphi_1,\ldots,\varphi_n$ to $g_1,\ldots,g_n \in L^\infty(\mu)$ with the same norm. Let $\Psi = \Psi_{\vec{g}}$ be the associated linear functional on $[L^1(\mu)]^n$. We prove that $\|\Psi\|_{(X^n)^*} = 1$.

Indeed, since $\Psi(\vec{f}) = \sum_{k=1}^n \langle \varphi_k, f_k \rangle = n$, and $\|\vec{f}\|_1 = n$, we have $\|\Psi\| \geqslant 1$. To reach the equality, note that, for every non-null $\vec{h} \in X^n$ and any $t > 0$ small enough, we have:

$$\det_\mathcal{B}\left(n\frac{\vec{f}+t\vec{h}}{\|\vec{f}+t\vec{h}\|_1}\right) \leqslant \det_\mathcal{B}(\vec{f}) = 1\,,$$

so that:

$$n^n \det_\mathcal{B}(\vec{f}+t\vec{h}) \leqslant \|\vec{f}+t\vec{h}\|_1^n \leqslant \left(\|\vec{f}\|_1 + t\|\vec{h}\|_1\right)^n = \left(n+t\|\vec{h}\|_1\right)^n\,.$$

Now, if we denote by $A$ the linear mapping sending each $f_k$ to $h_k$, then:

$$\frac{\det_\mathcal{B}(\vec{f}+t\vec{h})-1}{t} = \frac{\det(I+tA)-1}{t}$$

$$\underset{t\to 0}{\to} \operatorname{tr}(A) = \sum_{k=1}^n \langle \varphi_k, h_k \rangle = \Psi(\vec{h})\,;$$

we thus obtain:

$$\Psi(\vec{h}) \leqslant \|\vec{h}\|_1\,,$$

and hence $\|\Psi\| \leqslant 1$.

To complete the proof, note that $\{f_1, \ldots, f_n\}$ is a basis of $X$; thus the functions of $X$ vanish where the square function $F$ vanishes. We can thus assume that $F(t) > 0$ for every $t \in S$. Now the preceding results lead to:

$$n = \Psi_{\vec{g}}(\vec{f}) \leqslant \|\Psi_{\vec{g}}\| \, \|\vec{f}\|_1 = 1 \times n;$$

we thus obtain equality, and consequently all the inequalities in $(*)$ are equalities.

Equality in the last inequality of $(*)$ implies that $G(t) = 1$ for (almost) all $t \in S$ (since $F(t) \neq 0$). On the other hand, equality in the Cauchy–Schwarz inequality implies:

$$g_k(t) = u_t f_k(t), \quad \forall k = 1, \ldots, n,$$

with $u_t > 0$. As $G(t) = 1$, we obtain:

$$1 = G(t)^2 = \sum_{k=1}^{n} g_k(t)^2 = u_t^2 \sum_{k=1}^{n} f_k(t)^2 = u_t^2 \, F(t)^2,$$

and hence $u_t = 1/F(t)$.

Therefore, if $\Phi = \dfrac{1}{n} F$, the measure $\nu = \Phi.\mu$ is a probability on $S$, and the mapping $T \colon h \mapsto \Phi^{-1} h$ is an isometry from $X$ onto a subspace $Y$ of $L^1(\nu)$. Then $g_k = u f_k = (1/F) f_k = \frac{1}{n} T(f_k) \in Y$, and:

$$\int_S g_k \, g_l \, d\nu = \int_S \frac{f_k}{F} \, g_l \, \frac{F}{n} \, d\mu = \frac{1}{n} \langle f_k, \varphi_l \rangle = \begin{cases} 0 & \text{if } k \neq l \\ \dfrac{1}{n} & \text{if } k = l. \end{cases}$$

Thus the functions $g_1, \ldots, g_n$ are orthogonal in $L^2(\nu)$ and have norm $\|g_k\|_{L^2(\nu)} = 1/\sqrt{n}$; moreover:

$$\sum_{k=1}^{n} g_k^2 = G^2 = 1.$$

Hence, the functions $\psi_k = g_k$, $1 \leqslant k \leqslant n$, satisfy the conditions required in the theorem. $\qquad\square$

## V.4 Proof of Talagrand's Theorem

According to the Elton–Pajor theorem (formulated as in Corollary VI.21 of Chapter 3, this volume), it suffices to prove the result when the scalar field is $\mathbb{R}$. Moreover, by Proposition V.1, we can assume that $X \subseteq \ell_1^M$ for a certain $M < +\infty$. Our aim is to reduce this dimension $M$. The construction is probabilistic.

Choose selectors, i.e. independent random variables $\xi_k$, $1 \leqslant k \leqslant M$, such that:

$$\mathbb{P}(\xi_k = 1) = \mathbb{P}(\xi_k = 0) = 1/2,$$

and define a random operator $U_\omega \colon \ell_1^M \to \ell_1^M$ by:

$$U_\omega\big((x_k)_{1 \leqslant k \leqslant M}\big) = \big(2\xi_k(\omega) x_k\big)_{1 \leqslant k \leqslant M}.$$

This operator replaces certain coordinates by zero and doubles the others. We seek an $\omega$ for which $\big\| U_{\omega|X} \big\| \, \big\| (U_{\omega|X})^{-1} \big\|$ is close to 1. First, $U_\omega(\ell_1^M) \cong \ell_1^{M_\omega}$, where:

$$M_\omega = \mathrm{card}\{k \leqslant M \, ; \, \xi_k(\omega) = 1\} = \sum_{k=1}^{M} \xi_k(\omega),$$

and next, since $\mathbb{E}\big(\sum_{k=1}^{M} \xi_k\big) = M/2$, we have $M_\omega \leqslant 3M/5$ with probability $\geqslant 1/6$. In what follows, instead of the $\xi_k$'s, we use the Rademacher variables $\varepsilon_k = 2\xi_k - 1$, for which:

$$\mathbb{P}(\varepsilon_k = -1) = \mathbb{P}(\varepsilon_k = +1) = 1/2.$$

**Lemma V.10** *For $X$ a subspace of $\ell_1^M$, denote:*

$$A_X(\omega) = \sup_{x \in B_X} \left| \sum_{k=1}^{M} \varepsilon_k(\omega) \, |x_k| \right|.$$

*Then, when $\mathbb{E}(A_X) \leqslant 1/16$, there exist $M_1 \leqslant 3M/5$ and a subspace $E$ of $\ell_1^{M_1}$ such that:*

$$\mathrm{dist}\,(X, E) \leqslant 1 + 32\mathbb{E}(A_X).$$

*Proof* For $x = (x_k)_{1 \leqslant k \leqslant M}$, let:

$$Z_x(\omega) = \| U_\omega(x) \| - \|x\|.$$

Then:

$$Z_x(\omega) = \sum_{k=1}^{M} \big(1 + \varepsilon_k(\omega)\big) |x_k| - \sum_{k=1}^{M} |x_k| = \sum_{k=1}^{M} \varepsilon_k |x_k| \, ;$$

hence:

$$A_X(\omega) = \sup_{x \in B_X} |Z_x(\omega)|$$

and the restriction $R_\omega$ of $U_\omega$ to $X$ satisfies:

$$\|R_\omega\| \leqslant 1 + A_X(\omega) \quad \text{and} \quad \|R_\omega^{-1}\| \leqslant \frac{1}{1 - A_X(\omega)} \, ;$$

thus, when $A_X(\omega) \leqslant 1/2$, we have:

$$\text{dist}\left(X, U_\omega(X)\right) \leqslant 1 + 4A_X(\omega).$$

However:

$$\mathbb{P}\left(A_X \leqslant 8\,\mathbb{E}(A_X)\right) = 1 - \mathbb{P}\left(A_X > 8\,\mathbb{E}(A_X)\right) \geqslant 1 - \frac{1}{8} = \frac{7}{8};$$

thus, when $\mathbb{E}(A_X) \leqslant 1/16$, we have $A_X(\omega) \leqslant 1/2$ with probability $\geqslant 7/8$. As $M_\omega \leqslant 3M/5$ with probability $\geqslant 1/6$, and as $7/8 + 1/6 > 1$, by taking $E = R_\omega(X) = U_\omega(X)$ we reach the stated result.                    $\square$

The idea is now to iterate the method to keep diminishing the dimension. For this, we use Lewis' theorem, shown in Subsection V.3. This provides us with a probability measure $\nu$ on $\{1, \ldots, M\}$ and a subspace $Y$ of $L^1(\nu)$ isometric to $X$ with a basis $(\psi_j)_{1 \leqslant j \leqslant n}$ orthogonal in $L^2(\nu)$ such that $\sum_{j=1}^n \psi_j^2 = 1$ and $\|\psi_j\|_{L^2(\nu)} = 1/\sqrt{n}$. Note that replacing $\{1, \ldots, M\}$ by $\{1, \ldots, M'\}$, where $M' \leqslant 3M/2$, allows us to assume that:

$$\nu_k = \nu(\{k\}) \leqslant \frac{2}{M} \qquad \text{for any } k.$$

Indeed, when $\nu_k \geqslant 2/M$ for $k = k_1, \ldots, k_r$, we can split each of these points into $\left[\nu_k \frac{M}{2}\right] + 1$ points of equal mass, producing $M'$ points, with:

$$M' \leqslant \sum_{s=1}^r \left(\nu_{k_s}\frac{M}{2} + 1\right) + (M - r) \leqslant \frac{3M}{2}$$

(since $\nu_{k_1} + \cdots + \nu_{k_r} \leqslant 1$). Diminishing $M'$ if necessary, we can also assume that $\nu_k > 0$ for $1 \leqslant k \leqslant M'$. Now denote:

$$H_X = \mathbb{E} \sup_{y \in B_Y} \left|\sum_{k=1}^{M'} \nu_k \varepsilon_k |y(k)|\right|.$$

The essential point is to find a good estimate for $H_X$. Indeed, if $Z$ is the image of $Y$ by the map:

$$\begin{array}{ccc} L^1(\nu) & \longrightarrow & \ell_1^{M'} \\ y = \big(y(k)\big)_{1 \leqslant k \leqslant M'} & \longmapsto & \big(\nu_k\, y(k)\big)_{1 \leqslant k \leqslant M'}, \end{array}$$

$Z$ is isometric to $Y$, and $\mathbb{E}(A_Z) = H_X$. Lemma V.10 applied to the subspace $Z$ of $\ell_1^{M'}$ hence leads to:

**Lemma V.11** *If* $H_X \leqslant 1/16$, *there exists a subspace* $W$ *of* $\ell_1^{M_1}$, *with* $M_1 \leqslant 3M'/5 \leqslant 9M/10$, *such that:*

$$\text{dist}\,(X, W) = \text{dist}\,(Y, W) = \text{dist}\,(Z, W) \leqslant 1 + 32H_X.$$

The crucial point is thus the following lemma:

**Lemma V.12** *There exists a constant* $C_0 \leqslant 2\sqrt{\pi}$ *such that:*

$$H_X \leqslant C_0 K(X) \sqrt{\frac{n}{M}} .$$

*Proof* We replace the Rademacher variables by Gaussian variables, in order to exploit their rotation-invariance. Indeed, if we set $\varphi_k(y) = y(k)$ for $y \in B_Y$, then $H_X = \mathbb{E} \big\| \sum_{k=1}^{M'} \varepsilon_k v_k |\varphi_k| \big\|_\infty$, and the contraction principle (Chapter 4 of Volume 1, Theorem IV.4) gives:

$$H_X \leqslant \sqrt{\frac{\pi}{2}} \, \mathbb{E} \left\| \sum_{k=1}^{M'} v_k \, g_k |\varphi_k| \right\|_\infty = \sqrt{\frac{\pi}{2}} \, \mathbb{E} \sup_{y \in B_Y} \left| \sum_{k=1}^{M'} v_k \, g_k \, |y(k)| \right| .$$

Now, if we set:

$$X_y = \sum_{k=1}^{M'} v_k |y(k)| \, g_k \quad \text{and} \quad Y_y = \sum_{k=1}^{M'} v_k y(k) \, g_k ,$$

since $\big| \, |y(k)| - |z(k)| \, \big| \leqslant |y(k) - z(k)|$, we obtain:

$$\|X_y - X_z\|_2 \leqslant \|Y_y - Y_z\|_2 ,$$

and hence:

$$\mathbb{E} \sup_{y \in B_Y} |X_y| = \mathbb{E} \sup_{y \in B_Y} |X_y - X_0| \leqslant \mathbb{E} \sup_{y,z \in B_Y} (X_y - X_z)$$

$$\leqslant \mathbb{E} \sup_{y,z \in B_Y} (Y_y - Y_z) \leqslant 2 \, \mathbb{E} \sup_{y \in B_Y} |Y_y| .$$

The lemma of Slepian–Sudakov (Chapter 1 of this volume, Theorem III.4) thus provides:

$$H_X \leqslant 2 \sqrt{\frac{\pi}{2}} \, \mathbb{E} \sup_{y \in B_Y} \left| \sum_{k=1}^{M'} v_k g_k \, y(k) \right| ,$$

and we seek a majoration of this latter expectation.

We use the functions $\psi_j$, $1 \leqslant j \leqslant n$, orthogonal in $L^2(v)$, obtained in Lewis' theorem. Denote by $\langle . \, , . \rangle$ the scalar product in $L^2(v)$ and let $f \in L^\infty(\Omega, \mathbb{P}; Y) = L^\infty(Y)$, of norm 1, with:

$$\sup_{y \in B_Y} \left\langle \sum_{j=1}^{n} g_j(\omega)\psi_j, y \right\rangle = \left\langle \sum_{j=1}^{n} g_j(\omega)\psi_j, f(\omega) \right\rangle.$$

Then:

$$\mathbb{E}\left( \sup_{y \in B_Y} \left\langle \sum_{j=1}^{n} g_j\psi_j, y \right\rangle \right) = \sum_{j=1}^{n} \langle \psi_j, \mathbb{E}(g_j f) \rangle.$$

Setting $f_j = \mathbb{E}(g_j f)$, the inequalities $K_g(Y) \leqslant K(Y) = K(X)$ and $\|f\|_{L^2(Y)} \leqslant 1$ lead to:

$$\left\| \sum_{j=1}^{n} g_j f_j \right\|_{L^2(Y)} \leqslant K(X).$$

We thus obtain:

$$
\begin{aligned}
\sum_{j=1}^{n} \langle \psi_j, f_j \rangle &= \int_{\{1,\dots,M'\}} \left( \sum_{j=1}^{n} \psi_j(t) f_j(t) \right) d\nu(t) \\
&\leqslant \int_{\{1,\dots,M'\}} \left( \sum_{j=1}^{n} [f_j(t)]^2 \right)^{1/2} d\nu(t)
\end{aligned}
$$

by the Cauchy–Schwarz inequality, because $\displaystyle\sum_{j=1}^{n} \psi_j^2 = 1$

$$
\begin{aligned}
&= \left( \mathbb{E}\int_{\{1,\dots,M'\}} \left| \sum_{j=1}^{n} g_j f_j(t) \right|^2 d\nu(t) \right)^{1/2} \\
&= \left\| \sum_{j=1}^{n} g_j f_j \right\|_{L^2(Y)} \leqslant K(X).
\end{aligned}
$$

Then, a normalization of the $\psi_j$'s gives:

$$\mathbb{E}\left( \sup_{y \in B_Y} \left\langle \sum_{j=1}^{n} g_j \sqrt{n}\psi_j, y \right\rangle \right) \leqslant \sqrt{n}\, K(X).$$

Now, if we set $v_k = \mathbb{1}_{\{k\}}$, the sequence $\left( \frac{1}{\sqrt{\nu_k}} v_k \right)_{1 \leqslant k \leqslant M'}$ is an orthonormal basis of $L^2(\nu)$ (recall that $\nu_k = \nu(\{k\}) > 0$). As $(\sqrt{n}\psi_j)_{1 \leqslant j \leqslant n}$ is an orthonormal basis of $Y \subseteq L^2(\nu)$, the rotation-invariance of Gaussian vectors gives:

$$\mathbb{E}\left(\sup_{y\in B_Y}\left\langle\sum_{j=1}^{n}g_j\sqrt{n}\psi_j,y\right\rangle\right) = \mathbb{E}\left(\sup_{y\in B_Y}\left\langle\sum_{k=1}^{M'}\frac{1}{\sqrt{v_k}}g_k v_k,y\right\rangle\right).$$

It ensues that the preceding inequality can be written:

$$\mathbb{E}\left(\sup_{y\in B_Y}\left\langle\sum_{k=1}^{M'}\frac{1}{\sqrt{v_k}}g_k v_k,y\right\rangle\right) \leqslant \sqrt{n}\,K(X).$$

It only remains to use the contraction principle, since $v_k \leqslant 2/M$:

$$\mathbb{E}\left(\sup_{y\in B_Y}\left|\sum_{k=1}^{M'}v_k g_k y(k)\right|\right) = \mathbb{E}\left(\sup_{y\in B_Y}\left\langle\sum_{k=1}^{M'}g_k v_k,y\right\rangle\right)$$

$$\leqslant \max_{1\leqslant k\leqslant M'}\sqrt{v_k}\,\mathbb{E}\left(\sup_{y\in B_Y}\left\langle\sum_{k=1}^{M'}\frac{1}{\sqrt{v_k}}g_k v_k,y\right\rangle\right)$$

$$\leqslant \sqrt{\frac{2}{M}}\sqrt{n}\,K(X),$$

and hence:

$$H_X \leqslant 2\sqrt{\pi}\,K(X)\sqrt{\frac{n}{M}}. \qquad \square$$

*End of the proof of Theorem V.2*   We can now complete the proof of the theorem. As mentioned above, we apply Lemma V.11 repeatedly. For $X \subseteq L^1$ of dimension $n$, and for any $\alpha > 0$, there exist an integer $M \geqslant n$ and $E_0 \subseteq \ell_1^M$ such that dist $(X, E_0) \leqslant 1+\alpha$. Then $K(E_0) \leqslant (1+\alpha)\,K(X)$. Let $C_0$ $(\leqslant 2\sqrt{\pi})$ be the constant of Lemma V.12, and set $C_1 = 96\,C_0$. We can assume that $0 < \alpha \leqslant 1/30$. We also assume that $M \geqslant (C_1/\alpha)^2 K(X)^2\, n$ (indeed, otherwise, game over!).

Now use Lemma V.11 to construct a sequence of integers $(M_j)_{j\geqslant 0}$ such that $M_0 = M$ and $M_{j+1} \leqslant 9M_j/10$, and also subspaces $E_j \subseteq \ell_1^{M_j}$ $(j \geqslant 0)$ such that, for any $j \geqslant 0$, we have:

$$\text{dist}\,(E_j, E_{j+1}) \leqslant 1 + C_1 K(X)\sqrt{\frac{n}{M_j}}\,;$$

we stop at $j_0$, the first index for which $M_{j_0} \leqslant (C_1/\alpha)^2 K(X)^2\, n$. Indeed, suppose that $j < j_0$ and that $M_0,\ldots,M_j$ and $E_0,\ldots,E_j$ have already been constructed. As $j < j_0$, we have $C_1 K(X)\sqrt{\dfrac{n}{M_j}} \leqslant \alpha$, and hence:

$$\prod_{l=0}^{j}\left(1+C_1 K(X)\sqrt{\frac{n}{M_l}}\right) \leqslant \prod_{l=0}^{j}\left(1+C_1 K(X)\left(\frac{9}{10}\right)^{\frac{j-l}{2}}\sqrt{\frac{n}{M_j}}\right)$$

$$\leqslant \prod_{k=0}^{+\infty}\left[1+\alpha\left(\frac{3}{\sqrt{10}}\right)^k\right]$$

$$\leqslant e^{20\alpha} \leqslant e^{2/3} \leqslant 2 \quad \text{(because } \alpha \leqslant 1/30\text{)}.$$

In particular, it ensues that:

$$K(E_j) \leqslant (1+\alpha)\,e^{20\alpha} K(X) \leqslant 3\,K(X),$$

and hence:

$$H_{E_j} \leqslant C_0\,K(E_j)\sqrt{\frac{n}{M_j}} \leqslant 3\,C_0 K(X)\sqrt{\frac{n}{M_j}} \leqslant 3\frac{C_1}{96}K(X)\sqrt{\frac{n}{M_j}} \leqslant \frac{\alpha}{32}$$

(since $j < j_0$). In particular, $H_{E_j} \leqslant 1/16$, and thus there exist a number $M_{j+1} \leqslant 9M_j/10$ and a subspace $E_{j+1} \subseteq \ell_1^{M_{j+1}}$ such that:

$$\text{dist}\,(E_j, E_{j+1}) \leqslant 1 + 32\,C_0\,K(E_j)\sqrt{\frac{n}{M_j}}$$

$$\leqslant 1 + 32\,C_0.3K(X)\sqrt{\frac{n}{M_j}} = 1 + C_1\,K(X)\sqrt{\frac{n}{M_j}},$$

which completes the induction. It only remains to see that:

$$\text{dist}\,(E_0, E_{j_0}) \leqslant \prod_{j=0}^{j_0-1}\left(1+C_1 K(X)\sqrt{\frac{n}{M_j}}\right) \leqslant e^{20\alpha},$$

and hence:

$$\text{dist}\,(X, E_{j_0}) \leqslant (1+\alpha)\,e^{20\alpha}$$

to complete the proof of the theorem.                                □

# VI  Comments

1) The space $L^1$ has a very rich structure, and its study would require a dedicated book (or possibly many books!). It is not entirely elucidated, as the open problem of complemented subspaces attests: *is it true that every complemented subspace of $L^1$ is isomorphic either to $\ell_1$ or to $L^1$?* (Bourgain [1981] nonetheless showed that *there exist subspaces of $L^1$ isomorphic to $\ell_1$, but not complemented*). For more information on this problem, refer to Talagrand [1990 b].

For a description of the properties of the $L^p$ spaces, see Alspach and Odell [2001], as well as Johnson and Schechtman [2001].

2) Aldous [1981] showed this: *every subspace of $L^1$ contains an $\ell_p$ space for some $p \geqslant 1$* (see also Maurey [1980 b]). The ideas developed in Aldous' article led Krivine and Maurey [1981] to introduce the notion of *stable Banach spaces*. A presentation of Aldous' theorem using this notion can be found in GUERRE-DELABRIÈRE, Chapter III. Every stable Banach space is weakly sequentially complete (Guerre and Lapresté [1981]). Guerre and Lévy [1983] showed that *every infinite-dimensional subspace $X$ of $L^1$ contains, for any $\varepsilon > 0$, $\ell_{p(X)}$ $(1 + \varepsilon)$-isomorphically, where $p(X) = \sup\{p \leqslant 2 \,;\, X$ is of type $p\}$* (see GUERRE-DELABRIÈRE, Chapter IV). For other properties of stable spaces, see Raynaud [1981 a], [1981 b] and [1983].

3) The results of Subsection II.1 are due to Kadeč and Pełczyński [1962] (see also Pisier [1973 b]). A version of these results for integrable vector-valued functions was given by Bourgain [1979 b] (see also Pisier [1978 f] and Talagrand [1984 a]).

Since the topology of convergence in measure coincides with that of the norm on the reflexive subspaces of $L^1$, these are hence closed in measure. A more general class of subspaces of $L^1$ is obtained with those whose *unit ball* is closed in measure; these were called *nicely placed* by G. Godefroy [1984 b]; we have already mentioned these in Section VI of Chapter 7 (Volume 1). There we saw that the space $H^1(\mathbb{D})$ of holomorphic functions in the unit disk $\mathbb{D}$ with integrable radial limits (which is isometric to the space $L^1_{\mathbb{N}}(\mathbb{T}) = \{f \in L^1(\mathbb{T}) \,;\, \widehat{f}(n) = 0, \, \forall n < 0\}$) is an example of a nicely placed subspace (Godefroy [1984 b]).

The result at the base of their study is due to Bukhvalov and Lozanovski [1978]:

**Theorem** (The Bukhvalov–Lozanovski Theorem) *Let $C$ be a bounded closed convex subset of $L^1$. Then $C$ is closed in measure if and only if $P(\overline{C}^{w^*}) = C$.*

In this statement, $P$ denotes the Hewitt–Yosida projection. This projection is defined as follows: the bidual $L^{1**}$ of $L^1 = L^1(0,1)$ is identified with the finitely additive measures absolutely continuous with respect to the Lebesgue measure; if $L^1_s$ denotes the subspace of purely finitely additive measures, there is a decomposition $L^{1**} = L^1 \oplus_1 L^1_s$ inducing the projection $P \colon L^{1**} \to L^1$ (Hewitt and Yosida [1952]).

This theorem implies that a subspace $E$ of $L^1$ is nicely placed if and only if $P(E^{\perp\perp}) = E$. An interesting property of nicely placed subspaces of

$L^1$ is that the quotients of $L^1$ by such subspaces are weakly sequentially complete (Godefroy [1984 b] and [1988]); this leads to a new proof (Godefroy [1984 b]) of the weakly sequentially complete nature of $L^1/H^1$, due, independently, to Havin [1973] and Mooney [1972].

For a general presentation, refer to HARMAND–WERNER–WERNER, Chapter IV.

Other properties of these spaces, as well as some examples, can be found in Godefroy [1984 b] and [1988], Li [1987] and [1988], Pfitzner [1993] and Godefroy, Kalton and Li [1996] and [2000].

4) Theorem II.8 is due to Pisier [1973 b] and [1974 b].

5) Theorem III.1 seems to have been pointed out by Kadeč [1958].

The $\Lambda(p)$-sets were introduced by Rudin [1960], who also proved the elementary properties given in Section IV. While it is easy to see that the set of squares $\{n^2\,;\ n \in \mathbb{N}^*\}$ is not $\Lambda(4)$ (Exercise VII.12), it is not known if it is $\Lambda(2)$, or even $\Lambda(1)$.

A stronger notion of *completely bounded* $\Lambda(p)$-*sets* was introduced by Harcharras [1999].

In Li [1998], Neuwirth [1999] and in Li, Queffélec and Rodríguez-Piazza [2002], constructions of sets $\Lambda \subseteq \mathbb{N}$ can be found that are $\Lambda(p)$ for any $p < +\infty$, but "large" in the sense that the space $L_\Lambda^\infty(\mathbb{T})$ is not separable.

The Rudin transfer theorem, exhibiting the Rademacher functions as the prototype of Sidon sets, was further studied by Pisier [1978 e], and then by Asmar and Montgomery-Smith [1993]: any series $\sum_{n=1}^{+\infty} \gamma_n x_n$, where the $x_n$'s are elements of a Banach space and $\{\gamma_n\,;\ n \geqslant 1\}$ a Sidon set in a discrete Abelian group, has the same distribution (in the probabilistic sense) as the series where the Rademacher functions replace the $\gamma_n$'s (see also Pełczyński [1988]). See also Nazarov [1996] (unpublished work).

6) In the original proof by Rosenthal [1973] of Theorem IV.1, the notion of type, even though subjacent, is not yet clearly enunciated, and reading this article is quite difficult. It was Maurey [1972 a], [1972 b], [1972 c], [1973 b] and [1973 c] (see also MAUREY) who simplified matters, introducing the notions of type and cotype of Banach spaces (notions introduced independently by Hoffmann-Jørgensen [1973], in another context, for the study of Probability in Banach spaces), notably proving the factorization theorem (Theorem IV.2), generalizing on one hand Rosenthal [1973], Theorem 1, and on the other Nikishin [1970] (see MAUREY, and Maurey [2003 a]).

Corollary IV.3 can be found in Bachelis and Ebenstein [1974].

Actually Rosenthal [1973] showed that if $X$ is a subspace of $L^p(\Omega, \mathbb{P})$, with $1 \leqslant p < 2$, not containing any subspace isomorphic to $\ell_p$, then there exists $p'$ such that $p < p' < 2$ and $X$ is isomorphic to a subspace of $L^{p'}(\Omega, \Delta.\mathbb{P})$; this requires a bit more work than the case $p = 1$ presented here.

It follows that if $\Lambda$ is a $\Lambda(p)$-set with $1 \leqslant p < 2$, it is in fact $\Lambda(p')$ for some $p' \in ]p, 2[$.

For any even integer $p = 2n \geqslant 4$, Rudin [1960] constructed $\Lambda(p)$ subsets of $\mathbb{Z}$ that are not $\Lambda(p + \varepsilon)$, for any $\varepsilon > 0$. No essential progress was made until Bourgain [1989] used an extremely difficult probabilistic construction to obtain, for every discrete Abelian group $\Gamma$, and for any $p > 2$, some $\Lambda(p)$ subsets of $\Gamma$ that are not $\Lambda(p + \varepsilon)$ for any $\varepsilon > 0$.

Somewhat later, Talagrand [1995] gave another proof, highlighting the crucial nature of a convexity property of the $L^p$-spaces for $p > 2$.

It is still not known if every $\Lambda(p)$-set with $p < 2$ is in fact $\Lambda(2)$, nor if every $\Lambda(2)$-set is in fact $\Lambda(p)$ for some $p > 2$.

The proof given here of Proposition IV.4 is due to Pisier [1986 a] (we have followed WOJTASZCZYK, § III.H), but, as Pisier has told us, it was already known by P. Ørno.

7) Theorem V.2 is due to Talagrand [1990 a]. There is also a version for finite-dimensional subspaces of $L^p$ with $p > 1$; however, the estimations of the dimension are different, and are not the same for $1 < p < 2$ and for $p > 2$. See, for example, LEDOUX–TALAGRAND, § 15.5.

This result improves that of Schechtman [1987] and of Bourgain, Lindenstrauss and Milman [1989].

Theorem V.3 is due to Pisier [1980 a]; he gave another proof in Pisier [1981 a]. The proof presented here is due to Bourgain and Milman [1987]; we have followed TOMCZAK-JAEGERMANN.

Pisier [1980 a] (see also Pisier [1980 b], last remark) provides an estimation $K(X) \leqslant C\sqrt{\log \dim X}$ (if $\dim X \geqslant 2$) valid for every finite-dimensional space $X$ with an unconditional basis. Theorem V.5 is stated there, but without proof; the proof given here was communicated to us by Pisier.

The proof of Theorem V.8 is due to Figiel, in 1979 (see TOMCZAK–JAEGERMANN, page 112). The left-hand inequality can be found in Figiel and Tomczak-Jaegermann [1979]. The proof of Lewis' theorem given here (Theorem V.9) is due to Lorentz and Tomczak-Jaegermann [1984] (see also WOJTASZCZYK, § III.B). The statement by Lewis [1978] was slightly different; the form presented here is taken from Bourgain, Lindenstrauss and Milman [1989].

For a survey of the properties of finite-dimensional subspaces of $L^p$-spaces, see Johnson and Schechtman [2001].

## VII Exercises

**Exercise VII.1**   Show that if a reflexive subspace of $L^1$ is complemented, then it is finite-dimensional (use Exercise VII.2 of Chapter 7, Volume 1).

**Exercise VII.2**   For every bounded subset $F \subseteq L^1$, set, for $\varepsilon > 0$:

$$\eta(F, \varepsilon) = \sup\left\{ \int_A |f|\, d\mathbb{P} \,;\, f \in F,\, A \in \mathcal{A},\, \mathbb{P}(A) \leqslant \varepsilon \right\},$$

and $\eta(F) = \lim_{\varepsilon \to 0} \eta(F, \varepsilon)$.

1) Show that:

$$\eta(F) = \limsup_{M \to +\infty} \left[ \sup_{f \in F} \int_{\{|f| \geqslant M\}} |f|\, d\mathbb{P} \right],$$

    and that $\eta(F) = 0$ if and only if $F$ is uniformly integrable.

2) Show that if $(g_j)_{j \geqslant 1}$ is a bounded sequence in $L^1$, and $(E_j)_{j \geqslant 1}$ a sequence of measurable subsets such that $\mathbb{P}(E_j) \xrightarrow[j \to +\infty]{} 0$, then:

$$\eta\big((g_j)_{j \geqslant 1}\big) \geqslant \varlimsup_{n \to +\infty} \int_{E_n} |g_n|\, d\mathbb{P}.$$

    Now let $(f_j)_{j \geqslant 1}$ be a bounded sequence in $L^1$.
    Set $\eta(\varepsilon) = \eta\big((f_j)_{j \geqslant 1}, \varepsilon\big)$ and $\eta = \eta\big((f_j)_{j \geqslant 1}\big)$.

3) Show that there exist integers $n_j \geqslant 1$ and measurable subsets $E_j \subseteq \Omega$ such that $\mathbb{P}(E_j) \leqslant 1/2^j$ and:

$$\int_{E_j} |f_{n_j}|\, d\mathbb{P} > \eta\left(\frac{1}{2^j}\right) - \frac{1}{2^j}.$$

    Then deduce the existence of a subsequence $(\tilde{f}_j)_{j \geqslant 1}$ of $(f_j)_{j \geqslant 1}$ such that:

$$\lim_{j \to +\infty} \int_{E_j} |\tilde{f}_j|\, d\mathbb{P} = \eta.$$

4) Show that the sequence $\big(\tilde{f}_j\, \mathbb{1}_{E_j^c}\big)_{j \geqslant 1}$ is uniformly integrable (assume the opposite, and use the preceding question, replacing $(f_j)_{j \geqslant 1}$ by $\big(\tilde{f}_j\, \mathbb{1}_{E_j^c}\big)_{j \geqslant 1}$).

5) Show that there exist integers $1 = n_1 < n_2 < \cdots$ such that, for $1 \leqslant j \leqslant k$:

$$\int_{\bigcup_{l \geqslant n_{k+1}} E_l} |\tilde{f}_{n_j}|\, d\mathbb{P} \leqslant \frac{1}{2^{k+1}}.$$

6) Set $G_j = E_{n_j} \setminus \bigcup_{k > j} E_{n_k}$.

    a) Show that $\big(\tilde{f}_{n_j}\, \mathbb{1}_{G_j^c}\big)_{j \geqslant 1}$ is uniformly integrable.
    b) Show that $\big\|\tilde{f}_{n_j}\, \mathbb{1}_{G_j}\big\|_1 \xrightarrow[j \to +\infty]{} \eta$.

7) Deduce that, for every bounded sequence $(f_j)_{j \geqslant 1}$ of $L^1$, there exists a subsequence $(f'_j)_{j \geqslant 1}$ that can be written $f'_j = u_j + d_j$, where $(u_j)_{j \geqslant 1}$ is uniformly integrable and $(d_j)_{j \geqslant 1}$ is a sequence of functions with disjoint supports satisfying $\|d_j\|_1 \xrightarrow[j \to +\infty]{} \eta = \eta((f_j)_{j \geqslant 1})$, with these two sequences disjoint (*subsequence splitting lemma*).

For a vector-valued version of this decomposition, see Talagrand [1984 a].

**Exercise VII.3**   Let $(f_j)_{j \geqslant 1}$ be a sequence in the unit ball of $L^1$ such that $\eta = \eta((f_j)_{j \geqslant 1}) > 0$.

Prove the existence of a subsequence $(f'_j)_{j \geqslant 1}$ such that, for every $\varepsilon > 0$, there exists $N \geqslant 1$ such that $(f'_j)_{j \geqslant N}$ is $(\frac{1}{\eta} + \varepsilon)$-equivalent to the canonical basis of $\ell_1$. (*Hint*: use the *subsequence splitting lemma*: choose, for $\varepsilon' > 0$ small enough, $\delta > 0$ so that $\int_A |u_j| \, d\mathbb{P} \leqslant \varepsilon'$ for any $j \geqslant 1$ when $\mathbb{P}(A) \leqslant \delta$, then $N \geqslant 1$ large enough so that $\mathbb{P}\left(\bigcup_{j \geqslant N} \operatorname{supp} d_j\right) \leqslant \delta$ and $\|d_j\|_1 \geqslant \eta - \varepsilon'$ when $j \geqslant N$.)

**Exercise VII.4**   A Banach space $X$ is said to have the *strong Schur property*, or, more precisely, the $K$-*strong Schur property*, with $K \geqslant 1$, if, for any $\delta > 0$, every sequence $(x_n)_{n \geqslant 1}$ in the unit ball $B_X$ of $X$ which is $\delta$-separated (i.e. $\|x_n - x_k\| \geqslant \delta$ for $n \neq k$) contains, for any $\varepsilon > 0$, a subsequence $(x'_n)_{n \geqslant 1}$ that is $(\frac{2K}{\delta} + \varepsilon)$-equivalent to the canonical basis of $\ell_1$:

$$\left\| \sum_{k=1}^{n} a_k x'_k \right\| \geqslant \frac{1}{(2K/\delta) + \varepsilon} \sum_{k=1}^{n} |a_k|$$

for any scalars $a_1, \ldots, a_n$.

1) Show that $\ell_1$ has the 1-strong Schur property.
2) Show that every space with the strong Schur property has the Schur property.

**Exercise VII.5** (Rosenthal [1979])   Let $X$ be a subspace of $L^1$.

1) Show that if there exists a constant $K \geqslant 1$ such that, for every $\delta$-separated sequence $(f_n)_{n \geqslant 1}$ in $B_X$ we have $\eta((f_n)_{n \geqslant 1}) \geqslant \delta/K$, then $X$ has the $K$-strong Schur property (use Exercise VII.3).
2) Show that if $X$ has the Schur property and is complemented in $L^1$, with a projection $P$, then, for every $\delta$-separated sequence $(f_n)_{n \geqslant 1}$ in $B_X$, we have $\eta((f_n)_{n \geqslant 1}) \geqslant \delta/(2\|P\|)$ (use the *subsequence splitting lemma*). Deduce that $X$ has the $(2\|P\|)$-strong Schur property.
3) Assume that $X$ has the $K$-strong Schur property. Let $(f_n)_{n \geqslant 1}$ be a $\delta$-separated sequence in $B_X$, and $\eta = \eta((f_n)_{n \geqslant 1})$. Let $(f'_n)_{n \geqslant 1}$ be a

subsequence of $(f_n)_{n\geqslant 1}$, with $f'_n = u_n + d_n$, as in the *subsequence splitting lemma* (note that $u_n$ and $d_n$ have no reason to be in $X$).

Show that, for any $\varepsilon > 0$, we can assume that $\|d_n\|_1 \leqslant \eta + \varepsilon$ for any $n \geqslant 1$, and that $u_n \xrightarrow[n\to+\infty]{w} u$.

Set $g_n = (f'_n - u)/2$. Show that we can assume that $(g_n)_{n\geqslant 1}$ is $\frac{2K+\varepsilon}{(\delta/2)}$-equivalent to the canonical basis of $\ell_1$, and that there exist numbers $\lambda_1, \ldots, \lambda_N \geqslant 0$ such that $\sum_{n=1}^N \lambda_n = 1$ and

$$\left\| \sum_{n=1}^N \lambda_n (u_n - u) \right\|_1 \leqslant \varepsilon.$$

Deduce that $\eta \geqslant \delta/(2K)$.

4) a) Let $(f_n)_{n\geqslant 1}$ be a sequence in $B_X$, and let $\eta = \eta\big((f_n)_{n\geqslant 1}\big)$. Suppose that $f_n = u_n - d_n$, where $(u_n)_{n\geqslant 1}$ and $(d_n)_{n\geqslant 1}$ are as in the *subsequence splitting lemma*. Moreover, suppose that $(u_n)_{n\geqslant 1}$ is $\alpha$-separated. Show that, for every $\varepsilon > 0$, there exists $N \geqslant 1$ such that $(f_n)_{n\geqslant N}$ is $(2\eta + \alpha - \varepsilon)$-separated (proceed as in Exercise VII.3).

   b) Deduce that if $X$ has the 1-strong Schur property, then the unit ball $B_X$ of $X$ is relatively compact for the topology of convergence in measure.

5) Show that if the unit ball $B_X$ of $X$ is relatively compact in measure, then, for every $\delta$-separated sequence $(f_n)_{n\geqslant 1}$ in $B_X$, we have $\delta \leqslant 2\,\eta\big((f_n)_{n\geqslant 1}\big)$ (use the *subsequence splitting lemma*). Deduce that $X$ has the 1-strong Schur property (use Exercise VII.3).

**Exercise VII.6** (Johnson and Odell [1974])   Recall (Chapter 2 of Volume 1, Subsection IV.7)) that a Banach space $X$ has an *FDD* if it contains a sequence $(X_n)_{n\geqslant 1}$ of finite-dimensional subspaces such that every $x \in X$ can be uniquely written as $x = \sum_{n=1}^{+\infty} x_n$, with $x_n \in X_n$ for any $n \geqslant 1$. Then there exists (Chapter 2, Exercise V.3) a smallest constant $C \geqslant 1$ such that $\left\| \sum_{k=1}^n x_k \right\| \leqslant C \left\| \sum_{k=1}^m x_k \right\|$ for any $n \leqslant m$, and every $x_1 \in X_1, \ldots, x_m \in X_m$, called the *constant of the FDD*. The *FDD* is said to be unconditional when the convergence of $x = \sum_{n=1}^{+\infty} x_n$, with $x_n \in X_n$, is unconditional.

Let $X$ be a subspace of $L^1 = L^1(\Omega, \mathbb{P})$ with an *FDD* $(X_n)_{n\geqslant 1}$, and having the $K$-strong Schur property. By Exercise VII.5, we thus have $\eta\big((f_n)_{n\geqslant 1}\big) \geqslant \delta/(2K)$ for every $\delta$-separated sequence $(f_n)_{n\geqslant 1}$ in the unit ball of $X$.

1) Show that, for any $M > 0$, there exists $n \geqslant 1$ such that

$$\int_{\{|y|\geqslant M\,\|y\|_1\}} |y|\,d\mathbb{P} \geqslant \frac{1}{4\,CK}\,\|y\|_1$$

for every $y \in \bigoplus_{k=n+1}^{+\infty} X_k$. (*Hint*: assume the contrary; a $(1/C)$-separated sequence $(y_k)_{k \geqslant 1}$ of elements of norm 1 could then be constructed, while $\eta((y_k)_{k \geqslant 1}) \leqslant 1/(4\,CK)$, which is not possible.)

2) Set $a = 1/(8\,CK)$. Show that we can construct a strictly increasing sequence of integers $(n_j)_{j \geqslant 1}$, with $n_1 = 1$, and a decreasing sequence of numbers $\delta_j > 0$ such that:

   (a) for every $y \in X_1 \oplus \cdots \oplus X_{n_j}$, we have $\int_A |y| \leqslant (a/2^j)\,\|y\|_1$ for every measurable subset $A$ such that $\mathbb{P}(A) \leqslant \delta_j$ (use the absolute uniform continuity of the unit ball of $X_1 \oplus \cdots \oplus X_{n_j}$);

   (b) for every $y \in \bigoplus_{k=n_{j+2}}^{+\infty} X_k$, there exists a measurable subset $A$ such that $\mathbb{P}(A) \leqslant \delta_j$ and $\int_A |y|\,d\mathbb{P} \geqslant 2a\,\|y\|_1$.

3) Deduce that, with $Y_1 = X_1 \oplus \cdots \oplus X_{n_2}$ and $Y_j = X_{n_j+1} \oplus \cdots \oplus X_{n_{j+1}}$ for $j \geqslant 2$, then for every $y_j \in Y_j$ there exists a measurable subset $A_j$ such that $\mathbb{P}(A_j) \leqslant \delta_j$, $\|y_j\,\mathbb{1}_{A_j}\|_1 \geqslant 2a\,\|y_j\|_1$ and $\|y_j\,\mathbb{1}_{A_k}\|_1 \leqslant (a/2^k)\,\|y_j\|_1$ for any $k \geqslant j+2$.

4) Suppose additionally that the *FDD* of $X$ is unconditional, of unconditional constant $\beta$. Let $(y_j)_{j \geqslant 1}$ be as in 3).

   a) Set $D_j = A_{2j} \smallsetminus \bigcup_{k \geqslant j+1} A_{2k}$.
      Show that $\|y_{2j}\,\mathbb{1}_{D_j}\|_1 \geqslant a\,\|y_{2j}\|_1$.

   b) By using Khintchine's inequalities, show that:
   $$\left\|\sum_{k=1}^{n} y_{2k}\right\|_1 \geqslant \frac{1}{\beta\sqrt{2}} \int \left(\sum_{k=1}^{n} |y_{2k}|^2\right)^{1/2} d\mathbb{P},$$

   and deduce that
   $$\left\|\sum_{k=1}^{n} y_{2k}\right\|_1 \geqslant \frac{a}{\beta\sqrt{2}} \sum_{k=1}^{n} \|y_{2k}\|_1.$$

   (*Hint*: note that the $D_k$'s are disjoint.

   c) Similarly, show:
   $$\left\|\sum_{k=1}^{n} y_{2k-1}\right\|_1 \geqslant \frac{a}{\beta\sqrt{2}} \sum_{k=1}^{n} \|y_{2k-1}\|_1.$$

   d) Deduce that
   $$\left\|\sum_{k=1}^{n} y_k\right\|_1 \geqslant \frac{a}{2\,\beta^2\sqrt{2}} \sum_{k=1}^{n} \|y_k\|_1.$$

   e) Conclude that $X$ is isomorphic to a subspace of $\ell_1$.

**Exercise VII.7** (Rosenthal [1979], Bourgain [1979 a], Bourgain and Rosenthal [1980])

1) Let $(Y_n)_{n \geqslant 1}$ be a sequence of symmetric independent random variables in $L^1$, such that:

   (i) $\mathbb{E}|Y_n| = 1$ for any $n \geqslant 1$;

   (ii) $\mathbb{E}\big| |Y_n| - 1\!\!1 \big| \geqslant \delta > 0$ for any $n \geqslant 1$.

   a) Show that:

$$\left\| \sum_{k=1}^n c_k \left(|Y_k| - 1\!\!1\right) \right\|_1 \leqslant 2\,(1 + \sqrt{2}) \left\| \sum_{k=1}^n c_k\, Y_k \right\|_1$$

   for every $c_1, \ldots, c_n \in \mathbb{R}$. (*Hint*: use the fact that the variables $|Y_k| - 1\!\!1$, being independent and centered, form a 2-unconditional basic sequence, and hence have the same distribution as $(|Y_k| - 1\!\!1) \otimes r_k$, with $(r_k)_{k \geqslant 1}$ the sequence of Rademacher functions; next note that $\big\| \sum_{k=1}^n c_k\, r_k \big\|_1 \leqslant \big( \sum_{k=1}^n c_k^2 \big)^{1/2} \leqslant \sqrt{2}\, \big\| \sum_{k=1}^n c_k\, Y_k \big\|_1$, since $(Y_k)_{k \geqslant 1}$ is 1-unconditional.)

   b) Similarly, show that:

$$\left\| \sum_{k=1}^n c_k \left(|Y_k| - 1\!\!1\right) \otimes r_k \right\|_{L^1(\Omega \times [0,1])} \geqslant \frac{\delta}{\sqrt{2}} \left\| \sum_{k=1}^n c_k\, r_k \right\|_{L^1([0,1])}$$

   (use the fact that $\big((|Y_k| - 1\!\!1) \otimes r_k\big)_{k \geqslant 1}$ is 1-unconditional).

   Deduce that:

$$\left\| \sum_{k=1}^n c_k\, Y_k \right\|_1 \leqslant 2 \left(1 + \frac{\sqrt{2}}{\delta}\right) \left\| \sum_{k=1}^n c_k \left(|Y_k| - 1\!\!1\right) \right\|_1.$$

   c) Deduce that the sequences $(Y_k)_{k \geqslant 1}$ and $(|Y_k| - 1\!\!1)_{k \geqslant 1}$ are $K$-equivalent, with $K \leqslant 4 + 2\sqrt{2}$.

2) For any random variable $X$, denote its characteristic function by $\Phi_X$. Recall that if $X^s$ is the symmetrization of $X$, then $\Phi_{X^s} = |\Phi_X|^2$ (Chapter 1 of Volume 1, Lemma III.10).

   Show that, for any $\varepsilon > 0$, there exists $\delta > 0$ such that, for every random variable $X$ satisfying $|\Phi_X(t)| \geqslant 1 - \delta$ when $|t| \leqslant 1$, we have $\mathbb{P}(|X - c| > \varepsilon) \leqslant \varepsilon$, for a suitable $c \in \mathbb{R}$. (*Hint*: if not, a sequence of random variables $X_n$ could be constructed such that $|\Phi_{X_n}(t)| \xrightarrow[n \to +\infty]{} 1$, uniformly for $t \in [-1, 1]$, and such that $\mathbb{P}(|X_n - c| > \varepsilon) > \varepsilon$ for any $n \geqslant 1$ and any $c \in \mathbb{R}$; then show that $(\Phi_{X_n^s})_{n \geqslant 1}$ would converge uniformly to 1, and that $(X_n^s)_{n \geqslant 1}$ would converge in probability to 0 (use Exercise VII.11 of Chapter 1, Volume 1), and obtain a contradiction by using the Fubinization principle, Proposition III.7 of Chapter 1, Volume 1.)

3) Show that, for any symmetric real random variable $Y$, we have
$|\Phi_Y| = |\operatorname{Re} \Phi_{|Y|}| \leqslant |\Phi_{|Y|}|$.

4) Let $1 < p < 2$, and let $Y_p$ be a $p$-stable random variable such that $\|Y_p\|_1 = 1$ and $\Phi_{Y_p}(t) = e^{-c_p |t|^p}$ for any $t \in \mathbb{R}$.

 a) Show that $c_p \underset{p \to 1}{\longrightarrow} 0$ (if not, there would be a sequence $c_{p_n} \underset{n \to +\infty}{\longrightarrow} c > 0$; then use the Paul Lévy continuity theorem (Chapter 1 of Volume 1, Theorem II.6) to obtain a contradiction).

 b) Deduce that $Y_p \underset{p \to 1}{\overset{\mathbb{P}}{\longrightarrow}} 0$, and that $\| \, |Y_p| - \mathbb{1} \|_1 \underset{p \to 1}{\longrightarrow} 2$.

5) Let $Z_1, \ldots, Z_k$ be $p$-stable independent variables, of norm $\|Z_j\|_1 = 1$. For $a_1, \ldots, a_k \in \mathbb{R}$, set:

$$V = \sum_{j=1}^{k} a_j \left( |Z_j| - \mathbb{1} \right).$$

 Suppose that $a_1, \ldots, a_k \in \mathbb{R}$ are small enough so that $\|V\|_1 \leqslant 1$.

 a) Show that $|\Phi_V(t)| \geqslant e^{-c_p K^p}$, where $K$ is the constant of Question 1) c), for any $t \in [-1, 1]$.

 b) Deduce that, for every $\delta > 0$, there exists $\alpha > 0$ such that $|\Phi_V(t)| \geqslant 1 - \delta$ for any $t \in [-1, 1]$ and any $p \in \,]1, 1 + \alpha]$.

 c) Show that, for every $\varepsilon > 0$, there exists $\alpha > 0$ such that $d_{\mathbb{P}}(V, \mathbb{R} \, \mathbb{1}) \leqslant \varepsilon$ for any $p \in \,]1, 1 + \alpha]$ and every $V = \sum_{j=1}^{k} a_j \left( |Z_j| - \mathbb{1} \right)$ of norm $\|V\|_1 \leqslant 1$, with $k \geqslant 1$ arbitrary, and $d_{\mathbb{P}}$ a distance defining the topology of convergence in probability (use 2)b)).

**Exercise VII.8** (Godefroy, Kalton and Li [1996])

1) Use Exercise VII.7 to construct a sequence $(p_n)_{n \geqslant 1}$ decreasing to 1, and intervals $I_n \subseteq \mathbb{N}^*$, of length $|I_n| \underset{n \to +\infty}{\longrightarrow} +\infty$, with $\max I_n < \min I_{n+1}$, such that, if $Z_j$ are independent $p_n$-stable variables for $j \in I_n$, and if all the variables $Z_j$ for $j \in I = \bigcup_{n \geqslant 1} I_n$ are independent, we have:

 (i) $d_{\mathbb{P}}(V, \mathbb{R} \, \mathbb{1}) \leqslant 1/2^n$ for every $V = \sum_{j \in I_n} a_j Z_j$ such that $\|V\|_1 \leqslant 1$;

 (ii) $\left\| \dfrac{1}{|I_n|} \sum_{j \in I_n} |Z_j| - \mathbb{1} \right\|_1 \leqslant 1/n$.

 (*Hint*: use the strong law of large numbers; see Chapter 1 of Volume 1, Exercise VII.4, or use Question 1) of Exercise VII.7 in a direct calculation.)

2) Let $E$ be the subspace of $L^1$ generated by $\mathbb{1}$ and by the $U_j = |Z_j| - \mathbb{1}$, for $j \in I$.

 a) Show that $\mathcal{B} = \{\mathbb{1}\} \cup \{U_j \, ; \, j \in I\}$ is a $C$-unconditional basis of $E$, with $C \leqslant 3$, and is boundedly complete.

b) Let $V_l = \alpha_0(V_l)\, \mathbb{1} + \sum_{j\in I}\alpha_j(V_l)\, U_j$ be a sequence of elements of the unit ball $B_E$ of $E$ such that $\alpha_j(V_l) \xrightarrow[l\to+\infty]{} \alpha_j \in \mathbb{R}$ for any $j \in \{0\} \cup I$. Show that we can define $W = \alpha_0\, \mathbb{1} + \sum_{j\in I}\alpha_j\, U_j \in E$ (use the fact that $\mathcal{B}$ is boundedly complete), and that $\|W\|_1 \leqslant C$.

c) Show that $d_{\mathbb{P}}(V_l - W, \mathbb{R}\,\mathbb{1}) \xrightarrow[l\to+\infty]{} 0$, and deduce that $V_l$ can be written
$$V_l = W + \gamma_l\, \mathbb{1} + H_l, \text{ with } \gamma_l \in \mathbb{R} \text{ such that } \overline{\lim}_{l\to+\infty} |\gamma_l| \leqslant 1 + C, \text{ and}$$
with $H_l \xrightarrow[l\to+\infty]{\mathbb{P}} 0$.

d) Deduce that there exist $\gamma \in \mathbb{R}$ and a subsequence $(V'_l)_{l\geqslant 1}$ that converges in probability to $W + \gamma\, \mathbb{1}$.

3) Deduce that the unit ball $B_E$ of $E$ is compact in measure.

4) Show that $\varphi(\mathbb{1}) = 0$ for every linear functional $\varphi \in E^*$ that is continuous in measure on $B_E$.

In Godefroy, Kalton and Li [1996], it is shown that, even though this space $E$ can be embedded in $\ell_1$ (by Exercise VII.6), the topology of convergence in probability is not locally convex on $B_E$, and there exists $\varepsilon_0 > 0$ such that, for every quotient $Q$ of $c_0$, we have $d(E, Q^*) \geqslant 1 + \varepsilon_0$. This construction is based upon the more complicated one by Bourgain and Rosenthal [1980] of a subspace of $L^1$ whose ball is relatively compact in measure but does not have the Radon–Nikodým property. It is not known if every subspace of $L^1$ whose ball is compact in measure has the Radon–Nikodým property. In Godefroy, Kalton and Li [2000], an analogous construction, using the $p$-stable variables themselves, rather than their absolute values, provides a subspace $X$ of $L^1$, with a ball compact in measure, for which the topology of convergence in measure is locally convex, and, for any $\varepsilon > 0$, is $(1 + \varepsilon)$-isomorphic to a weak*-closed subspace of $\ell_1$; nevertheless it is far from the "trivial" $\ell_1$-subspaces, i.e. those generated by a countable measurable partition $\mathcal{S}$: we have $\sup_{f\in B_X} \|\mathbb{E}^{\mathcal{S}} Jf - Jf\|_1 \geqslant 1$ for every isometry $J\colon X \to L^1$. Also, this space $X$ satisfies the following property: for any $\varepsilon > 0$, there exists an isomorphism $T_\varepsilon\colon X \to L^1$, with $\|T_\varepsilon\|\,\|T_\varepsilon^{-1}\| \leqslant 1 + \varepsilon$, such that $\|T_\varepsilon - J\| \geqslant 1/2$ for every isometry $J\colon X \to L^1$. Consequently, thanks to a theorem of Alspach [1983], $T_\varepsilon$ has no extension to $L^1$ with good control of the norm, even though every isometry $T\colon X \to L^1$ can be extended to an isometry $\widetilde{T}\colon L^1 \to L^1$ (Hardin [1981]; see also Rudin [1976], Lusky [1978] and Plotkin [1974]).

**Exercise VII.9** Let $G$ be a compact metrizable Abelian group and $\Lambda$ a subset of the dual group $\Gamma = \widehat{G}$, and let $1 \leqslant p < +\infty$. Show that if $L^p_\Lambda$ is complemented in $L^p(G)$, then it is so with a translation-invariant projection $P\colon L^p(G) \to L^p_\Lambda$, i.e. such that $P(f_x) = [P(f)]_x$, where $f_x = \tau_x f$ is the translation of $f$ by $x \in G$. (*Hint:* if $Q: L^p(G) \rightarrow L^p_\Lambda$ is a projection, show that $Pf = \int_G [Q(f_x)]_{-x}\, dm(x)$ fits the bill; this is called the *Rudin averaging method*.)

**Exercise VII.10** Let $G$ be a compact metrizable Abelian group and $\Lambda$ a subset of the dual group $\Gamma = \widehat{G}$, and let $1 < p < +\infty$.

1) Suppose $L^p_\Lambda$ isomorphic to $\ell_2$.

    a) Using the Rudin averaging method, show that the subspaces $L^p_E$, for $E \subseteq \Lambda$, are complemented in $L^p_\Lambda$, with projections commuting with the translations, and with uniformly bounded norms.

    b) Deduce the existence of a constant $C > 0$ such that:

$$\left\| \sum_{\gamma \in \Lambda} \varepsilon_\gamma\, a_\gamma\, \gamma \right\|_p \leqslant C \left\| \sum_{\gamma \in \Lambda} a_\gamma\, \gamma \right\|_p$$

    for every almost null sequence $(a_\gamma)_{\gamma \in \Lambda}$ of complex numbers, and every choice of signs $\varepsilon_\gamma = \pm 1$.

    c) Deduce that $\Lambda$ is a $\Lambda(r)$-set, with $r = \max(p, 2)$.

    d) Show the converse.

2) Show similarly that $L^p_\Lambda$ is isomorphic to $\ell_2$ and is complemented in $L^p(G)$ if and only if $\Lambda$ is a $\Lambda(s)$-set, with $s = \max(p, p^*)$.

3) Show that if $\Lambda_1$ and $\Lambda_2$ are two $\Lambda(p)$-sets, with $p > 2$, then $\Lambda_1 \cup \Lambda_2$ is also $\Lambda(p)$.

4) Show that, for $1 < p < 2$, $L^p_\Lambda$ can be isomorphic to $\ell_2$ without being complemented in $L^p(G)$. (*Hint:* for $2 < s < p^*$, use the existence of sets that are $\Lambda(s)$ but not $\Lambda(t)$ for any $t > s$; such sets was constructed by Bourgain [1989], see also Talagrand [1995].)

5) Let $\Lambda \subseteq \mathbb{Z}$ be such that $\Lambda^+ = \Lambda \cap \mathbb{N}$ and $\Lambda^- = \Lambda \cap (\mathbb{Z} \smallsetminus \mathbb{N})$ are $\Lambda(p)$-sets, with $p \geqslant 1$. Show that $\Lambda$ is $\Lambda(p)$. (*Hint:* for $p > 1$, use the Marcel Riesz theorem: see Chapter 7, Volume 1; for $p = 1$, use Kolmogorov's theorem: Theorem III.6 of Chapter 7, Volume 1 or Corollary IV.3 of this chapter.)

**Exercise VII.11** Let $1 < p < +\infty$, let $G_1, G_2$ be two compact metrizable Abelian groups, and let $\Lambda_1 \subseteq \Gamma_1 = \widehat{G}_1$, $\Lambda_2 \subseteq \Gamma_2 = \widehat{G}_2$. Show that if $\Lambda_1$ and $\Lambda_2$ are two $\Lambda(p)$-sets, then $\Lambda_1 \times \Lambda_2$ is also a $\Lambda(p)$-set, in the product $\Gamma_1 \times \Gamma_2$ (use Minkowski's inequality for the integrals). When $G_2 = G_1 = G$, is it true that $\Lambda_1 + \Lambda_2$ is $\Lambda(p)$ in $\widehat{G}$?

**Exercise VII.12** By expanding the fourth power of $\left\| \sum_{n=1}^{N} e^{in^2 t} \right\|_4$, show that this expression is not $O(\sqrt{N})$, and deduce that the set of perfect squares is not a $\Lambda(4)$-set.

**Exercise VII.13**   For $\Lambda \subseteq \mathbb{Z}$ and $n \in \mathbb{Z}$, denote by $r_s(\Lambda, n)$ the number of $s$-tuples $(s \geqslant 1)$ $(n_1, \ldots, n_s) \in \Lambda^s$ such that $n_1 + \cdots + n_s = n$; in other words, $r_s(\Lambda, n)$ is the number of decompositions of $n$ as the sum of $s$ elements of $\Lambda$, distinct or not, each decomposition counted as many times as there are permutations, with repetition, of the $s$ elements of $\Lambda$.

1) Let $\Lambda \subseteq \mathbb{Z}$. Suppose that, for an $s \geqslant 2$, there exists $M = M_s > 0$ such that $r_s(\Lambda, n) \leqslant M$ for any $n \in \mathbb{Z}$. Let $P \in \mathcal{P}_\Lambda(\mathbb{T})$, and $Q = P^s$. Show that, for any $n \in \mathbb{Z}$, we have:

$$|\widehat{Q}(n)|^2 = \left| \sum_{n_1 + \cdots + n_s = n} \widehat{P}(n_1) \ldots \widehat{P}(n_s) \right|^2$$

$$\leqslant M \sum_{n_1 + \cdots + n_s = n} |\widehat{P}(n_1)|^2 \ldots |\widehat{P}(n_s)|^2 .$$

Deduce that $\Lambda$ is a $\Lambda(2s)$ set, with constant $\leqslant M^{1/2s}$.

2) Let $(n_j)_{j \geqslant 1}$ be a sequence of integers $\geqslant 1$ such that $n_{j+1}/n_j \geqslant 2s + 1$. Show that:

$$S_2 = \{\pm n_j \pm n_k \, ; \, k > j \geqslant 1\}$$

is a $\Lambda(2s)$ set.

3) Construct a subset of $\mathbb{N}$ that is $\Lambda(p)$ for every $p < +\infty$, but is not a Sidon set.

4) Let $\lambda_j \geqslant 1$ be integers such that $\lambda_{j+1}/\lambda_j \geqslant 3$ for any $j \geqslant 1$. For $N \geqslant 1$, set:

$$P_N(t) = \prod_{j=1}^{N} (1 + \cos \lambda_j t) .$$

Show that $\|P_N\|_2 = (3/2)^{N/2}$, and deduce that the set:

$$\Lambda = \bigcup_{k \geqslant 1} \{\pm \lambda_{j_1} \pm \lambda_{j_2} \pm \cdots \pm \lambda_{j_k} \, ; \, j_1 > j_2 > \cdots > j_k \geqslant 1\}$$

is not a $\Lambda(p)$-set for $p \geqslant 2$.

**Exercise VII.14** (Proof of the Bachelis–Ebenstein theorem without using Rosenthal's theorem, K. Hare [1988])

Let $G$ be a compact metrizable Abelian group and let $\Lambda \subseteq \Gamma = \widehat{G}$, and let $1 \leqslant p < q$.

*Part I* – The aim here is to prove this:

*If $\Lambda$ is not $\Lambda(q)$, then, for any integer $n \geqslant 1$ and any real number $a \in \, ]0, 1[$, there exist trigonometric polynomials $f_1, \ldots, f_n \in \mathcal{P}_\Lambda$ such that:*

$$\left\| \sum_{k=1}^{n} \alpha_k f_k \right\|_p \geqslant a \left( \sum_{k=1}^{n} |\alpha_k|^q \right)^{1/q}$$

*for any $\alpha_1, \ldots, \alpha_n \in \mathbb{C}$.*

To simplify, $L_\Lambda^p$ is said to *a-contain* $\ell_q^n$.

1) Fix $\delta < 1$ such that $\delta - n^{1/q^*}(1 - \delta^q)^{1/q} \geqslant a$. Show that there exist a finite subset $F$ of $\Lambda$ and $f \in \mathcal{P}_F$ such that $\|f\|_p = 1$, $\|f\|_q = M > \left( \dfrac{n+1}{1 - \delta^q} \right)^{1/p}$, and $\|g\|_q \leqslant M \|g\|_p$ for every $g \in \mathcal{P}_F$.

2) For $l \leqslant n$, suppose that we have constructed translated functions $f_1, \ldots, f_l$ of $f_1 = f$ and disjoint subsets $A_1, \ldots, A_l \subseteq G$ such that $m(A_j) \leqslant 1/M^p$ and $\left\| f_j \mathbb{1}_{A_j} \right\|_q > \delta M$, for $1 \leqslant j \leqslant l$ (observe that, for $l = 1$, the set $A_1 = \{|f| > M\}$ answers the question). Set $B = \bigcup_{j=1}^{l} A_j$; show that:

$$\int_G \left( \int_{B^c} |f(x-y)|^q \, dm(x) \right) dm(y) \geqslant M^q (1 - nM^{-p}).$$

Deduce the existence of a $y \in G$ such that, when setting $f_{l+1}(x) = f_y(x) = f(x-y)$, we have:

$$\int_{B^c} |f_{l+1}(x)|^q \, dm(x) \geqslant M^q (1 - nM^{-p}).$$

Now let $A_{l+1} = \{|f_{l+1}| > M\} \cap B^c$; show that $m(A_{l+1}) \leqslant 1/M^p$ and $\|f_{l+1} \mathbb{1}_{A_{l+1}}\|_q > \delta M$.

3) With the translates $f_1, \ldots, f_n$ and the sets $A_1, \ldots, A_n$ constructed as indicated in 2), set $\varphi_k = f_k \mathbb{1}_{A_k}$. Show that:

$$\left\| \sum_{j=1}^{n} \alpha_j \varphi_j \right\|_q \geqslant \delta M \left( \sum_{j=1}^{n} |\alpha_j|^q \right)^{1/q},$$

and next that:

$$\left\| \sum_{j=1}^{n} \alpha_j f_j \right\|_q \geqslant a M \left( \sum_{j=1}^{n} |\alpha_j|^q \right)^{1/q}.$$

Conclude that $\left\| \sum_{j=1}^{n} \alpha_j f_j \right\|_p \geqslant a M \left( \sum_{j=1}^{n} |\alpha_j|^q \right)^{1/q}$.

*Part II* – The aim here is to show this:

*For $a, \delta > 0$, $1 \leqslant p < 2$ and $n \in \mathbb{N}^*$, there exists $N = N(a, \delta, p, n)$ such that if $L_\Lambda^p$ $a$-contains $\ell_p^N$, then we can find non-zero $f_1, \ldots, f_n \in \mathcal{P}_\Lambda$ and measurable disjoint subsets $A_1, \ldots, A_n$ such that:*

$$\int_{A_j} |f_j|^p \, dm \geqslant \frac{a^{\frac{2p}{2-p}}}{1 + \delta} \|f_j\|_p^p, \qquad 1 \leqslant j \leqslant n.$$

1) Let $f_1, \ldots, f_N \in \mathcal{P}_\Lambda$ be such that $\left\| \sum_{j=1}^N \alpha_j f_j \right\|_p \geqslant a \left( \sum_{j=1}^N |\alpha_j|^p \right)^{1/p}$ and $\|f_j\|_p = 1$, $1 \leqslant j \leqslant N$.

Set $f^* = \max(|f_1|, \ldots, |f_N|)$ and $S = \left( \sum_{j=1}^N |f_j|^2 \right)^{1/2}$. Show that:

$$a^p N \leqslant \int_G S^p \, dm \leqslant \left( \int_G f^{*p} \, dm \right)^{\frac{2-p}{p}} \left( \int_G \sum_{j=1}^N |f_j|^p \right)^{p/2}.$$

2) Show that $a^p N \leqslant N^{p/2} \left( \int_G f^{*p} \, dm \right)^{(2-p)/p}$, and then

$$\int_G f^{*p} \, dm \geqslant a^{2p/(2-p)} N.$$

3) For $1 \leqslant j \leqslant N$, let $A_j$ be the set of $x \in G$ such that $j$ is the first index for which $|f_j(x)| = f^*(x)$. Show that

$$\sum_{j=1}^N \int_{A_j} |f_j|^p \, dm \geqslant a^{2p/(2-p)} N.$$

4) Let

$$J = \left\{ j \leqslant N ; \int_{A_j} |f_j|^p \, dm \geqslant \frac{a^{2p/(2-p)}}{1 + \delta} \right\};$$

show that $|J| \geqslant N a^{2p/(2-p)} \delta / (1 + \delta)$, and conclude.

*Part III* – Here we show a property of weak uniform integrability of $\Lambda(p)$-sets. Let $C > 0$ be a constant for which $\|f\|_p \leqslant C \|f\|_{p/2}$ for every $f \in \mathcal{P}_\Lambda$.

1) Let $A$ be a measurable subset of measure $m(A) \leqslant \delta < 1$, and let $f \in \mathcal{P}_\Lambda$ be such that $\|f\|_p = 1$. Show that

$$\frac{1}{C^{p/2}} \leqslant \int_G |f|^{p/2} \, dm \leqslant \delta^{1/2} + \left( 1 - \int_A |f|^p \, dm \right)^{1/2}.$$

2) Let $n$ be an integer $> 4\,C^p$. Show that if $m(A) \leqslant 1/n$, then

$$\int_A |f|^p \, dm \leqslant \left(1 - \frac{1}{n}\right) \|f\|_p^p$$

for every $f \in \mathcal{P}_\Lambda$.

*Part IV* – In this step, we assemble the preceding parts to conclude.

Let $\Lambda$ be a $\Lambda(p)$-set, with $1 \leqslant p < 2$. With the number $C > 0$ as given in Part III, let $n$ be an integer $> 4\,C^p$, $a = (1 - 1/n^2)^{(2-p)/2p} < 1$ and $N = N(a, 1/n, p, n)$ as in Part II.

1) Suppose that $L_\Lambda^p$ $a$-contains $\ell_p^N$. Show the existence of non-null functions $f_1, \ldots, f_n \in \mathcal{P}_\Lambda$, and measurable disjoint subsets $A_1, \ldots, A_n$ such that $\int_{A_j} |f_j|^p \, dm \geqslant (1 - 1/n) \|f_j\|_p^p$ for $1 \leqslant j \leqslant n$. Show that the measure of one of the $A_j$'s must be $\leqslant 1/n$, and reach a contradiction. Hence $L_\Lambda^p$ does not $a$-contain $\ell_p^N$.

2) Let $q > p$ be such that $N^{\frac{1}{q} - \frac{1}{p}} \geqslant \sqrt{a}$. Show that if $L_\Lambda^p$ $\sqrt{a}$-contains $\ell_q^N$, then $L_\Lambda^p$ $a$-contains $\ell_p^N$. Thus, it is false that $L_\Lambda^p$ $\sqrt{a}$-contains $\ell_q^N$.

3) With $q$ adjusted as in 2), use Part I to conclude that $\Lambda$ is a $\Lambda(q)$-set.

**Exercise VII.15** We wish to show directly that for every space $X$ of dimension $n \geqslant 2$, we have $K_g(X) \leqslant C \log n$, by using the Hermite functions instead of the Walsh functions.

The *Hermite polynomials* $(h_n)_{n \geqslant 0}$ of one variable are defined by the formula:

$$\exp\left(\lambda x - \frac{\lambda^2}{2}\right) = \sum_{n=0}^{+\infty} h_n(x) \frac{\lambda^n}{n!},$$

so that $h_0(x) = 1, h_1(x) = x$, etc.; $(h_n)_{n \geqslant 0}$ is an orthogonal basis of $L^2(\mathbb{R}, \gamma_1)$, where $\gamma_1$ is the Gaussian measure $(2\pi)^{-1/2} \exp(-x^2/2) \, dx$. Now, Hermite polynomials of several variables are defined as follows: if $A$ is the set of eventually null sequences of integers and if $\alpha = (\alpha_1, \ldots, \alpha_n, \ldots) \in A$ and $x = (x_1, \ldots, x_n, \ldots) \in \mathbb{R}^{\mathbb{N}^*}$, set:

$$H_\alpha(x) = h_{\alpha_1}(x_1) \cdots h_{\alpha_n}(x_n) \cdots .$$

The $H_\alpha$'s form an orthogonal basis of $\mathcal{H} = L^2(\mathbb{R}^{\mathbb{N}^*}, \gamma)$, where $\gamma = \gamma_1^{\otimes \mathbb{N}^*}$. Denote by $\mathcal{H}_k$ the closed subspace of $\mathcal{H} = L^2(\mathbb{R}^{\mathbb{N}^*}, \gamma)$ generated by the $H_\alpha$ for $|\alpha| = \sum_{n=1}^{+\infty} \alpha_n = k$; $\mathcal{H}_k$ is called the *k-th Wiener chaos*. Denote by $Q_k$ the orthogonal projection from $\mathcal{H}$ onto $\mathcal{H}_k$.

1) Show that $T(\varepsilon) = \sum_{k=0}^{+\infty} \varepsilon^k Q_k$, $-1 \leqslant \varepsilon \leqslant 1$, is a semi-group of positive contractions on $\mathcal{H}$, by showing that, for $f \in \mathcal{H}$:

$$[T(\varepsilon)f](x) = \int_{\mathbb{R}^{N^*}} f\left(\varepsilon x + \sqrt{1 - \varepsilon^2}\, y\right) d\gamma(y).$$

In what follows, $X$ is a Banach space isomorphic to a Hilbert space. Denote by $d_X$ the Banach–Mazur distance from $X$ to a Hilbert space, and:

$$Q'_k = Q_k \otimes \text{Id}_X, \qquad T'(\varepsilon) = \sum_{k=0}^{+\infty} \varepsilon^k Q'_k = T(\varepsilon) \otimes \text{Id}_X.$$

2) Show that $\|T'(\varepsilon)\| \leqslant 1$ for $-1 \leqslant \varepsilon \leqslant 1$.
3) Show that $\|Q'_k\| \leqslant d_X$ for any $k \geqslant 0$.
4) Show that if $|\varepsilon| \leqslant 1/2$, then:

$$\left\| \sum_{k>n} \varepsilon^k Q'_k \right\| \leqslant d_X/2^n \quad \text{and} \quad \left\| \sum_{k=0}^{n} \varepsilon^k Q'_k \right\| \leqslant 1 + (d_X/2^n).$$

5) Let $Y$ be a Banach space, and let $P(\varepsilon) = \sum_{k=0}^{n} \varepsilon^k x_k$ be a polynomial with coefficients $x_k \in Y$. Draw inspiration from the proof in the text (use of Bernstein's inequality) to show that:

$$\|x_1\| \leqslant 2n \sup_{|\varepsilon| \leqslant 1/2} \|P(\varepsilon)\|.$$

6) Show that $\|Q'_1\| \leqslant C \log(1 + d_X)$, where $C$ is a numerical constant, and hence that the Gaussian $K$-convexity constant of $X$ satisfies:

$$K_g(X) \leqslant C \log(1 + d_X).$$

# 5

# The Method of Selectors.
# Examples of Its Use

## I Introduction

The method of *selectors* is based on the following idea: given a sequence of independent Bernoulli variables $(\varepsilon_n)_{n \geqslant 1}$ (taking the values 0 and 1), select the random set $I_\omega = \{n \geqslant 1 \, ; \, \varepsilon_n(\omega) = 1\}$. Apparently, this method was introduced by Cramér [1935] and [1937] and later on by Erdös [1955] (see Erdös and Rényi [1960], pages 83–84), and has often been used ever since (see notably Katznelson and Malliavin [1966] and Katznelson [1973]). The term "selectors" seems to have been introduced by Bourgain. This method has already been applied, notably in Chapters 2 and 4 of this volume (see also Chapter 7 of Volume 1, Exercise VII.13). In this chapter, various new examples illustrate this method:

- a characterization of Sidon sets based on the extraction in every finite subset of particular subsets, namely quasi-independent sets, of proportional cardinality (Section II)
- a result on the majoration of a sum of sines, in relation to the vector-valued Hilbert transform (Section III)
- a minoration of the $K$-convexity constant for finite-dimensional spaces (Section IV).

## II Extraction of Quasi-Independent Sets

This section shows how the functional notion of Sidon sets can be characterized by arithmetical properties, namely the presence, in every finite subset of the set in question, of proportional-sized quasi-independent subsets. This easily leads to both the theorem of Drury on the stability of Sidon sets under union, and that of Pisier characterizing the Sidon sets as the $\Lambda(p)$-sets with a constant on the order of $\sqrt{p}$. Pisier's original proof of this latter result was much more sophisticated, using the continuity of Gaussian processes; this proof will be given in Chapter 6 of this volume. In this section, we show that, from every finite subset of $\mathbb{Z}$, a quasi-independent subset can be extracted, with a size controlled from below. To present this result due to Pisier ([1983 a], [1983 b] and implicitly in [1981 b]), the proof of Bourgain [1985 a] is used, which combines probabilistic methods with selectors and combinatorial arguments. As a consequence, Rider's theorem is quite easily derived.

To clarify, the framework presented here is that of the circle group $G = \mathbb{T}$; however, everything translates word for word to the general case of a (metrizable) compact Abelian group.

## II.1 Quasi-Independent sets

**Definition II.1**    A subset $B$ of $\mathbb{Z}$ is said to be *quasi-independent* if, for every finite subset $B' \subseteq B$, the relation

$$\sum_{n \in B'} \theta_n n = 0$$

with $\theta_n \in \{-1, 0, 1\}$ is only possible when $\theta_n = 0$ for every $n \in B'$.

In other words – *this is the important point*:

*Every integer $N \in \mathbb{Z}$ that can be written as the sum (of a finite number) of elements of $B$ can be written like this in* one single *way only.*

For example, every Hadamard set, with a constant $\geqslant 3$, is quasi-independent. The quasi-independent sets are typical examples of Sidon sets:

**Proposition II.2**    *Every quasi-independent set $B$ is a Sidon set, and the Sidon constants of quasi-independent sets are uniformly bounded: $S(B) \leqslant 8$.*

Note that the main open question concerning Sidon sets is:

**Question:** Is every Sidon set in $\mathbb{Z}$ a finite union of quasi-independent sets?

Partial answers to this question were given by Malliavin and Malliavin-Brameret [1967], Pisier [1981 a] (Corollary 3.3) and Bourgain [1983 b].

*Proof*    Set $B = \{\lambda_n, n \geqslant 1\}$. For $N$ fixed, we consider the Riesz product:

$$\mu = \prod_{j=1}^{N}(1 + \cos \lambda_j t) = \prod_{j=1}^{N}\left(1 + \frac{1}{2}\,\mathrm{e}^{i\lambda_j t} + \frac{1}{2}\,\mathrm{e}^{-i\lambda_j t}\right) \in \mathcal{M}(\mathbb{T}).$$

Since $\mu \geqslant 0$ and $B$ is quasi-independent, then $\|\mu\| = \widehat{\mu}(0) = 1$. Moreover, for $s \geqslant 1$ and $n \in \mathbb{Z}$, we denote by $R_s(n)$ the number of subsets $\{j_1, \ldots, j_s\}$, with $1 \leqslant j_1 < j_2 < \ldots < j_s \leqslant N$, such that:

$$n = \pm \lambda_{j_1} \pm \lambda_{j_2} \pm \cdots \pm \lambda_{j_s};$$

then: $\mu = \sum_{n \in \mathbb{Z}} \left[ \sum_{s \geqslant 1} 2^{-s} R_s(n) \right] e_n$; hence:

$$\sum_{s=1}^{+\infty} 2^{-s} R_s(n) = \widehat{\mu}(n) \leqslant \|\mu\| \leqslant 1.$$

Now the Riesz product is used "à la Rider": for $0 < a < 1$ and $\varphi_1, \ldots, \varphi_N \in \mathbb{R}$, we set:

$$\mu_a = \prod_{j=1}^{N} \left[ 1 + a \cos(\lambda_j t - \varphi_j) \right] - \prod_{j=1}^{N} \left[ 1 - a \cos(\lambda_j t - \varphi_j) \right]$$

$$= \prod_{j=1}^{N} \left( 1 + \frac{a}{2} e^{-i\varphi_j} e^{i\lambda_j t} + \frac{a}{2} e^{i\varphi_j} e^{-i\lambda_j t} \right)$$

$$- \prod_{j=1}^{N} \left( 1 - \frac{a}{2} e^{-i\varphi_j} e^{i\lambda_j t} - \frac{a}{2} e^{i\varphi_j} e^{-i\lambda_j t} \right).$$

Then $\|\mu_a\| \leqslant 2$. Moreover, an expansion for $1 \leqslant j \leqslant N$ and the fact that $\sum_{s=1}^{+\infty} 2^{-s} R_s(n) \leqslant 1$ lead to:

$$\widehat{\mu}_a(\lambda_j) = a e^{-i\varphi_j} + r_j,$$

where

$$|r_j| \leqslant 2 \sum_{s \geqslant 3} \left( \frac{a}{2} \right)^s R_s(\lambda_j) \leqslant 2 a^3 \sum_{s \geqslant 3} 2^{-s} R_s(\lambda_j) \leqslant 2 a^3.$$

Then let $f = \sum_{j=1}^{N} a_j e_{\lambda_j}$ be a trigonometric polynomial with spectrum in $B$. It can be written:

$$f(t) = \sum_{j=1}^{N} |a_j| e^{i\varphi_j} e^{i\lambda_j t}.$$

Let us consider the measures $\mu_a$ constructed above, with the coefficients $\varphi_1, \ldots, \varphi_N$. Then:

$$\sum_{j=1}^{N} |a_j| = \sum_{j=1}^{N} e^{-i\varphi_j} \widehat{f}(\lambda_j)$$

$$= \frac{1}{a} \sum_{j=1}^{N} \widehat{\mu}_a(\lambda_j)\widehat{f}(\lambda_j) + \sum_{j=1}^{N} \left[ e^{-i\varphi_j} - \frac{1}{a}\widehat{\mu}_a(\lambda_j) \right]\widehat{f}(\lambda_j);$$

thus:

$$\sum_{j=1}^{N} |a_j| \leqslant \frac{1}{a} \|f\|_\infty \|\mu_a\| + 2\,a^2 \sum_{j=1}^{N} |a_j|,$$

so finally:

$$\sum_{j=1}^{N} |a_j| \leqslant \frac{2}{a(1-2a^2)} \|f\|_\infty.$$

The best value possible is $a = 1/\sqrt{6}$, hence $S(B) \leqslant 3\sqrt{6} \leqslant 8$. $\qquad\square$

Note that the best estimation of the Sidon constant of quasi-independent sets does not seem to be known.

## II.2  Characterization of Sidon Sets

The main theorem of this section is:

**Theorem II.3** (Pisier)  *A subset $\Lambda$ of $\mathbb{Z}$ is a Sidon set if and only if there exists a constant $\delta > 0$ such that, from every finite subset $A$ of $\Lambda$ (with $|A| \geqslant 2$), a quasi-independent subset $B$ can be extracted with $|B| \geqslant \delta|A|$.*

The role of the condition $|A| \geqslant 2$ is to prevent $A$ from being reduced to $\{0\}$; for in this case, it could not contain any quasi-independent subset. Before tackling the proof, let us see how Drury's theorem follows immediately.

**Corollary II.4** (Drury's Theorem)  *The union of two Sidon sets is also a Sidon set.*

*Proof*  If $\Lambda_1$ and $\Lambda_2$ are Sidon sets, let $A$ be a finite subset of $\Lambda_1 \cup \Lambda_2$. It can be assumed that:

$$|A \cap \Lambda_1| \geqslant |A \cap \Lambda_2| \geqslant 2.$$

As $\Lambda_1$ is Sidon, there exists a quasi-independent subset $B \subseteq A \cap \Lambda_1$ such that:

$$|B| \geqslant \delta|A \cap \Lambda_1| \geqslant \frac{\delta}{2}|A|.$$

The set $\Lambda_1 \cup \Lambda_2$ is thus indeed Sidon. $\qquad\square$

In order to prove the necessary condition of Theorem II.3, first a theorem of extraction of quasi-independent subsets (Theorem II.6) is shown. For this, the following notation is introduced:

**Notation II.5**   For every finite subset $A$ of $\mathbb{Z}$, denote:

$$\psi_A = \sup_{q \geqslant 2} \frac{\left\| \sum_{n \in A} e_n \right\|_q}{\sqrt{q}} .$$

With $e_A = \sum_{n \in A} e_n$, then $\psi_A$ is equivalent to the norm of $e_A$ in the Orlicz space $L^{\Psi_2}(\mathbb{T})$ associated to the function:

$$\Psi_2(x) = e^{x^2} - 1,$$

as can be seen by expanding this function as a power series, and using Stirling's formula.

**Theorem II.6**   *There exists a numerical constant $K > 0$ such that, from every finite subset $A \subseteq \mathbb{Z}$, with $|A| \geqslant 2$, a quasi-independent subset $B \subseteq A$ can be extracted satisfying*:

$$|B| \geqslant K \left( \frac{|A|}{\psi_A} \right)^2 .$$

This theorem is implicit in Pisier [1981 a] (via Pisier [1983 a] and [1983 b]), and was explicitly stated by Rodríguez-Piazza [1991]. Its proof is essentially that of Bourgain [1985 a] for the necessary condition of Pisier's theorem.

Let us see how the necessary condition of Theorem II.3 can be deduced:

*Proof of the necessary condition of Theorem II.3*   We know that every Sidon set $\Lambda$ is $\Lambda(q)$ for any $q < +\infty$, and, more precisely, that:

$$\|f\|_q \leqslant C S(\Lambda) \sqrt{q} \, \|f\|_2$$

for every $f \in \mathcal{P}_\Lambda$, and $2 \leqslant q < +\infty$ (Chapter 4 of this volume, Theorem III.11). Consequently, for every finite subset $A \subseteq \Lambda$:

$$\psi_A \leqslant C S(\Lambda) |A|^{1/2},$$

and hence, by Theorem II.6, when $|A| \geqslant 2$, a quasi-independent subset $B$ can be extracted, with cardinality:

$$|B| \geqslant \frac{K}{C^2 S(\Lambda)^2} |A|. \qquad \square$$

*Proof of Theorem II.6*   First note that we can assume that $|A|/\psi_A \geqslant 128$ (otherwise, it suffices to choose for $B$ a set containing a single non-null element of $A$); the conclusion of Theorem II.6 is then reached, with $K \leqslant 1/(128)^2 = 1/2^{14}$.

The proof of Theorem II.6 is then done in two steps: first, a probabilistic method is used to extract a set that is "almost" quasi-independent; then, from the set thus obtained, a deterministic method is used to extract a "truly" quasi-independent subset. For this, *selectors* $\xi_n$, for $n \in A$, are brought into play; these are independent Bernoulli random variables satisfying:

$$\mathbb{P}(\xi_n = 1) = \delta \quad \text{and} \quad \mathbb{P}(\xi_n = 0) = 1 - \delta.$$

These are selectors in the sense that from $A$ a random subset is to be extracted, given by:

$$D = D(\omega) = \{n \in A \,;\, \xi_n(\omega) = 1\}.$$

The following notions are required:

**Definition II.7**   Let $D$ be a subset of $\mathbb{Z}$ not containing 0. A sequence $(\theta_k)_{k \in D}$, with $\theta_k \in \{-1, 0, 1\}$ is said to be a *relation* in $D$ if only a finite number of $\theta_k$'s are non-null: $\sum_{k \in D} |\theta_k| < +\infty$ and

$$\sum_{k \in D} \theta_k \, k = 0.$$

The number $\sum_{k \in D} |\theta_k|$ is called the *length of the relation*. The equality $\sum_{k \in D} \theta_k \, k = 0$ is also called *a relation in $D$*.

A set $B \subseteq \mathbb{Z}^*$ is hence quasi-independent if and only if it does not possess any relation of length $> 0$.

The crucial point is the following lemma, where $[\,.\,]$ denotes the integer part.

**Lemma II.8**   *Let $A$ be a finite subset of $\mathbb{Z}$ with $|A|/\psi_A \geqslant 128$, and let*

$$\delta = \frac{1}{2^{10}} \frac{|A|}{\psi_A^2} \quad \text{and} \quad l = \left[ \frac{1}{4} \delta |A| \right] = \left[ \frac{1}{2^{12}} \left( \frac{|A|}{\psi_A} \right)^2 \right].$$

*Then, there exists a subset $D \subseteq A$, with $|D| \geqslant \delta |A|/2$, having only relations of length $\leqslant l$.*

Indeed, once we have such a subset $D$, we select a relation $(\theta_k)_{k \in D}$ of *maximal length* in $D$, and set:

$$D' = \{k \in D \,;\, \theta_k \neq 0\}.$$

Then, $B = D \smallsetminus D'$ is quasi-independent. Indeed, if $(\theta'_k)_{k \in B}$ is a relation in $B$, then:

$$\sum_{k \in B} \theta'_k \, k = 0 \, ;$$

hence

$$\sum_{k \in B} \theta'_k \, k + \sum_{k \in D'} \theta_k \, k = 0 + 0 = 0;$$

thus the sequence formed by the $\theta'_k$ for $k \in B$ and the $\theta_k$ for $k \in D'$ is a relation in $D$, of length

$$\sum_{k \in B} |\theta'_k| + \sum_{k \in D'} |\theta_k|.$$

As the length $\sum_{k \in D'} |\theta_k|$ is maximal, this is only possible when $\theta'_k = 0$ for every $k \in B$. This quasi-independent set $B$ is suitable because:

$$|B| = |D| - |D'| \geqslant \frac{1}{2} \delta |A| - l$$

$$\geqslant \frac{1}{2} \delta |A| - \frac{1}{4} \delta |A| = \frac{1}{4} \delta |A| = \frac{1}{2^{12}} \left( \frac{|A|}{\psi_A} \right)^2. \qquad \square$$

*Proof of Lemma II.8* Let $\xi_n$, $n \in A$, be independent real random variables with:

$$\mathbb{P}(\xi_n = 1) = \delta \quad \text{and} \quad \mathbb{P}(\xi_n = 0) = 1 - \delta.$$

Note that, since:

$$\psi_A \geqslant \frac{1}{\sqrt{2}} \left\| \sum_{n \in A} e_n \right\|_2 = \frac{1}{\sqrt{2}} |A|^{1/2},$$

then $|A|/\psi_A^2 \leqslant 2$, and hence $\delta \leqslant 1/2^9 < 1$. For each $\omega \in \Omega$, we define the random polynomials:

$$F_\omega(t) = \sum_{n=l+1}^{|A|} \sum_{\substack{S \subseteq A \\ |S|=n}} \prod_{k \in S} \xi_k(\omega) \left( e^{ikt} + e^{-ikt} \right).$$

The independence leads to:

$$\iint_{\Omega \times \mathbb{T}} F_\omega(t) \, dm(t) \, d\mathbb{P}(\omega) = \sum_{n=l+1}^{|A|} \delta^n \left[ \sum_{\substack{S \subseteq A \\ |S|=n}} \int_{\mathbb{T}} \prod_{k \in S} \left( e^{ikt} + e^{-ikt} \right) dm(t) \right].$$

a) The important point is that:

$$N(\omega) = \int_{\mathbb{T}} F_\omega(t)\, dm(t)$$

is an integer: it is the *number of relations of length* $\geq l+1$ existing in $D(\omega) = \{k \in A \,;\, \xi_k(\omega) = 1\}$. Indeed, on one hand:

$$\int_{\mathbb{T}} e^{i(\sum_{k \in S} \theta_k\, k)t}\, dm(t) = 1 \text{ or } 0,$$

depending on the existence – or non-existence – of $\theta_k = \pm 1$ such that $\sum_{k \in S} \theta_k\, k = 0$; on the other hand, each term of the form

$$\prod_{k \in S} \left(e^{ikt} + e^{-ikt}\right)$$

can be written:

$$\sum_{\theta_k = \pm 1} e^{i(\sum_{k \in S} \theta_k\, k)t}.$$

b) Now, when $|S| = n$, each term of type:

$$\int_{\mathbb{T}} \prod_{k \in S} \left(e^{ikt} + e^{-ikt}\right) dm(t)$$

appears $n!$ times in the expansion of:

$$\int_{\mathbb{T}} \left[\sum_{k \in A} \left(e^{ikt} + e^{-ikt}\right)\right]^n dm(t).$$

We thus have:

$$0 \leq \int_{\Omega} N(\omega)\, d\mathbb{P}(\omega) = \iint_{\Omega \times \mathbb{T}} F_\omega(t)\, dm(t)\, d\mathbb{P}(\omega)$$

$$\leq \sum_{n=l+1}^{|A|} \frac{\delta^n}{n!} \int_{\mathbb{T}} \left[\sum_{k \in A} \left(e^{ikt} + e^{-ikt}\right)\right]^n dm(t)$$

$$\leq \sum_{n=l+1}^{|A|} \frac{2^n \delta^n}{n!} \left\| \sum_{k \in A} e_k \right\|_n^n \leq \sum_{n=l+1}^{|A|} \frac{2^n \delta^n}{n!} \left(\sqrt{n}\, \psi_A\right)^n$$

$$\leq \sum_{n=l+1}^{|A|} \left(\frac{6\delta\, \psi_A}{\sqrt{n}}\right)^n \quad \text{since } n! \geq \left(\frac{n}{3}\right)^n$$

$$\leq \sum_{n=l+1}^{+\infty} \left(\frac{6\delta\, \psi_A}{\sqrt{l+1}}\right)^n \leq \sum_{n=l+1}^{+\infty} \frac{1}{2^n} = \frac{1}{2^l},$$

since

$$\frac{6\,\delta\,\psi_A}{\sqrt{l+1}} \leqslant \frac{6\,\delta\,\psi_A}{\sqrt{\delta|A|/2}} = 12\left(\frac{\delta}{|A|}\right)^{1/2}\psi_A = \frac{12}{2^5} = \frac{3}{8} \leqslant \frac{1}{2}\,.$$

Consequently:

$$\mathbb{P}\big(N(\omega) \geqslant 1\big) \leqslant \int_\Omega N(\omega)\,d\mathbb{P}(\omega) \leqslant \frac{1}{2^l}\,.$$

c) It remains to apply the following lemma, with $X_k = \xi_k - \delta$,

$$\sigma^2 = |A|\,\delta \geqslant |A|\,\delta\,(1-\delta) = |A|\mathbb{E}(X_1^2) = \sum_{k\in A}\mathbb{E}(X_k^2),$$

and $a = |A|\,\delta/2$:

**Lemma II.9** *Let $X_1,\ldots,X_N$ be real independent centered random variables with $|X_k| \leqslant 1$. Then, if $\sigma^2 \geqslant \mathbb{E}(X_1^2) + \cdots + \mathbb{E}(X_N^2)$, we have, for any $a > 0$:*

$$\mathbb{P}\big(|X_1 + \cdots + X_N| \geqslant a\big) \leqslant 2\exp\left(-\frac{a^2}{2(a+\sigma^2)}\right).$$

Indeed, since

$$|D(\omega)| = \sum_{k\in A}\xi_k(\omega),$$

this gives:

$$\mathbb{P}\big(|D(\omega)| \leqslant |A|\,\delta/2\big) \leqslant \mathbb{P}\big[\,\big||D(\omega)| - |A|\,\delta\,\big| \geqslant |A|\,\delta/2\,\big]$$
$$\leqslant 2\exp\left(-|A|\delta/12\right),$$

and hence, as: $|A|\,\delta = \dfrac{1}{2^{10}}\left(\dfrac{|A|}{\psi_A}\right)^2 \geqslant \dfrac{2^{14}}{2^{10}} = 16$:

$$\mathbb{P}\big(|D(\omega)| \leqslant |A|\,\delta/2\big) \leqslant 2e^{-4/3} < 0,53 < 9/16.$$

Moreover, $l+1 \geqslant \delta|A|/4 \geqslant 4$ implies $\mathbb{P}\big(N(\omega) \geqslant 1\big) \leqslant 1/8$. Hence, we have simultaneously $|D(\omega)| \geqslant |A|\,\delta/2$ and $N(\omega) = 0$, with probability $\geqslant 1 - (9/16 + 1/8) = 5/16 > 0$.

This completes the proof of Lemma II.8. $\qquad\square$

*Proof of Lemma II.9* The proof could be deduced from Lemma VI.5 of Chapter 3 (this volume), but instead we present a simple direct proof. For $\lambda > 0$, we have:

$$e^{\lambda X_k} = 1 + \lambda X_k + \sum_{j\geqslant 2}\frac{\lambda^j}{j!}X_k^j \leqslant 1 + \lambda X_k + \sum_{j\geqslant 2}\frac{\lambda^j}{j!}X_k^2.$$

Hence:

$$\mathbb{E}(e^{\lambda X_k}) \leqslant 1 + \sum_{j \geqslant 2} \frac{\lambda^j}{j!} \mathbb{E} X_k^2 = 1 + \mathbb{E} X_k^2 (e^\lambda - \lambda - 1) \leqslant \exp\left[\mathbb{E} X_k^2 (e^\lambda - \lambda - 1)\right].$$

Markov's inequality then provides:

$$\mathbb{P}(X_1 + \cdots + X_N \geqslant a) \leqslant e^{-\lambda a} \prod_{k=1}^{N} \mathbb{E}(e^{\lambda X_k}) \leqslant e^{-\lambda a} \exp[\sigma^2 (e^\lambda - \lambda - 1)].$$

An optimization with respect to $\lambda$ leads to $\lambda = \log(1 + a/\sigma^2)$, and:

$$\mathbb{P}(X_1 + \cdots + X_N \geqslant a) \leqslant \exp\left[a - (a + \sigma^2) \log\left(1 + \frac{a}{\sigma^2}\right)\right]$$
$$= \exp\left[a + (a + \sigma^2) \log(1 - u)\right],$$

where $u = a/(a + \sigma^2)$. It ensues that:

$$\mathbb{P}(X_1 + \cdots + X_N \geqslant a) \leqslant \exp\left[a - (a + \sigma^2)\left(u + \frac{u^2}{2}\right)\right]$$

$$= \exp\left(-\frac{a^2}{2(a + \sigma^2)}\right),$$

and hence:

$$\mathbb{P}(|X_1 + \cdots + X_N| \geqslant a) \leqslant 2 \exp\left(-\frac{a^2}{2(a + \sigma^2)}\right). \qquad \square$$

## II.3 Proof of the Sufficient Condition

The proof ensues from the following characterization:

**Theorem II.10** (Bourgain)    *Let $\Lambda$ be a subset of $\mathbb{Z}$. Suppose there exists $\delta > 0$ such that, from every finite subset $A$ of $\Lambda$, a quasi-independent subset $B \subseteq A$ can be extracted satisfying $|B| \geqslant \delta|A|$. Then, there exists a constant $\delta_1 > 0$ ($\delta_1 = C\delta$) such that, for every finite subset $A$ of $\Lambda$, and for every sequence $(a_n)_{n \in A}$ of complex numbers, a quasi-independent subset $B \subseteq A$ can be extracted satisfying:*

$$\sum_{n \in B} |a_n| \geqslant \delta_1 \sum_{n \in A} |a_n|.$$

This immediately implies the sufficient condition thanks to Rider's theorem (Chapter 6 of Volume 1, Theorem V.18). Indeed, let $f \in \mathcal{P}_\Lambda$; for $A$, we take the spectrum of $f$ and set $a_n = \widehat{f}(n)$, $n \in A$. There exists a quasi-independent subset $B \subseteq A$ as indicated in the theorem. Then, as the Sidon constant of $B$ is bounded by a universal constant, we have:

$$\sum_{n \in A} |a_n| \leqslant \frac{1}{\delta_1} \sum_{n \in B} |a_n| \leqslant C \left[\!\!\left[ \sum_{n \in B} a_n e_n \right]\!\!\right] \leqslant C \left[\!\!\left[ \sum_{n \in A} a_n e_n \right]\!\!\right] = C[\![f]\!],$$

and Rider's theorem states that $\Lambda$ is a Sidon set.

In fact, it is more interesting to give a proof that does not use Rider's theorem; this theorem will then follow as a corollary of Pisier's characterization.

*Proof of the sufficient condition of Theorem II.3 based on Theorem II.10* Let us go back to the proof of Proposition II.2. First note that:

$$\sum_{n \in A} (\operatorname{Re} a_n) \, \mathrm{e}^{int} = \frac{f(t) + \overline{f(-t)}}{2} = g(t),$$

and hence, as $\|g\|_\infty \leqslant \|f\|_\infty$, the $a_n$'s can be assumed real, up to the loss of a factor 2. We denote $\theta_n = \operatorname{sgn} a_n$, and consider the following measures (where $0 < a < 1$):

$$\mu_a = \prod_{k \in B} \left[ 1 + a \, \theta_k \cos(kt) \right] - \prod_{k \in B} \left[ 1 - a \, \theta_k \cos(kt) \right]$$

$$= \prod_{k \in B} \left( 1 + \frac{a}{2} \theta_k \mathrm{e}^{ikt} + \frac{a}{2} \mathrm{e}^{-ikt} \right) - \prod_{k \in B} \left( 1 - \frac{a}{2} \theta_k \mathrm{e}^{ikt} - \frac{a}{2} \mathrm{e}^{-ikt} \right).$$

We have seen that $\|\mu_a\| \leqslant 2$, and that, for $n \in B$:

$$|\widehat{\mu_a}(n) - a \theta_n| \leqslant 2 a^3.$$

Here $\widehat{\mu_a}(-n) = \widehat{\mu_a}(n)$, so this inequality is in fact valid when $n \in B \cup (-B)$.

Now consider the $n \notin B \cup (-B)$ for which $\widehat{\mu_a}(n) \neq 0$. Such a $\widehat{\mu_a}(n)$ comes necessarily from the product of at least two terms of the type $a \, \theta_k \, \mathrm{e}^{\pm ikt}$ (note that $\widehat{\mu_a}(0) = 0$, and also, in passing, that a quasi-independent set cannot contain 0). For these $n$, we thus have:

$$|\widehat{\mu_a}(n)| \leqslant 2 \left[ \left( \frac{a}{2} \right)^2 + \sum_{\substack{s \geqslant 3 \\ s \text{ odd}}} \left( \frac{a}{2} \right)^s R_s(n) \right] \leqslant 2 \left[ \frac{a^2}{4} + a^3 \sum_{s \geqslant 3} 2^{-s} R_s(n) \right]$$

$$\leqslant 2 \left( \frac{a^2}{4} + a^3 \right) \leqslant 3 \, a^2.$$

Denoting $B' = B \cup (-B)$, we thus obtain:

$$\delta_1 \sum_{n \in A} |a_n| \leqslant \sum_{n \in B} |a_n| \leqslant \sum_{n \in B' \cap A} |a_n| = \sum_{n \in B' \cap A} \theta_n \widehat{f}(n)$$

$$= \frac{1}{a} \sum_{B' \cap A} \widehat{\mu_a}(n) \widehat{f}(n) + \sum_{n \in B' \cap A} \left( \theta_n - \frac{1}{a} \widehat{\mu_a}(n) \right) \widehat{f}(n)$$

$$= \frac{1}{a} \sum_{n \in A} \widehat{\mu_a}(n) \widehat{f}(n) - \frac{1}{a} \sum_{n \in A \setminus B'} \widehat{\mu_a}(n) \widehat{f}(n)$$

$$+ \sum_{n \in B' \cap A} \left( \theta_n - \frac{1}{a} \widehat{\mu_a}(n) \right) \widehat{f}(n)$$

$$\leqslant \frac{1}{a} \|\mu_a\| \, \|f\|_\infty + \frac{1}{a} 3 \, a^2 \sum_{n \in A \setminus B'} |\widehat{f}(n)| + \frac{1}{a} 2 \, a^3 \sum_{n \in B' \cap A} |\widehat{f}(n)|$$

$$\leqslant \frac{2}{a} \|f\|_\infty + 3 \, a \sum_{n \in A} |\widehat{f}(n)|, \qquad \text{as } a < 1.$$

Hence, with $a = \delta_1/6$:

$$\sum_{n \in A} |\widehat{f}(n)| \leqslant (24/\delta_1^2) \|f\|_\infty. \qquad \square$$

We now proceed to the proof of Theorem II.10.

The crucial point is the following lemma:

**Lemma II.11** (Bourgain)   *There exists a numerical constant $R > 10$ such that if $\Lambda_1, \ldots, \Lambda_J$ are finite quasi-independent subsets of $\mathbb{Z}$, pairwise disjoint, and satisfying:*

$$\frac{|\Lambda_{j+1}|}{|\Lambda_j|} \geqslant R, \qquad j = 1, 2, \ldots, J - 1,$$

*then subsets $\Lambda_j' \subseteq \Lambda_j$ ($1 \leqslant j \leqslant J$) can be extracted so that:*

a) $|\Lambda_j'| \geqslant (1/10)|\Lambda_j|$, $1 \leqslant j \leqslant J$;

b) *the union $\bigcup_{1 \leqslant j \leqslant J} \Lambda_j'$ is quasi-independent.*

The proof revisits the method used for the necessary condition, but is "*a bit more complicated*" (as Bourgain himself said in Bourgain [1985 b]). Note that in fact it leads to $R > 4 \times 10^5$. First let us see how this provides a proof of Theorem II.10.

*Proof of Theorem II.10*   Fix a finite subset $A$ and complex numbers $a_n$, $n \in A$ such that $\sum_{n \in A} |a_n| = 1$. With $R$ the constant appearing in Lemma II.11, define, for $k = 0, 1, \ldots$:

$$A_k = \left\{ n \in A \,;\, \frac{1}{R^{k+1}} < |a_n| \leqslant \frac{1}{R^k} \right\}.$$

As $A$ is finite, there exists $K \geqslant 1$ such that $A_k$ is empty for $k > K$. By hypothesis, for each $k \geqslant 0$, a quasi-independent subset $A'_k$ of $A_k$ can be extracted satisfying:

$$|A'_k| \geqslant \delta |A_k|.$$

Since:

$$1 = \sum_{k \geqslant 0} \sum_{n \in A_k} |a_n| \leqslant \sum_{k \geqslant 0} \frac{1}{R^k} |A_k| \leqslant \frac{1}{\delta} \sum_{k \geqslant 0} \frac{1}{R^k} |A'_k| \leqslant \frac{R}{\delta} \sum_{k \geqslant 0} \sum_{n \in A'_k} |a_n|,$$

we obtain:

$$\sum_{n \in P} |a_n| + \sum_{n \in I} |a_n| \geqslant \frac{\delta}{R},$$

after setting:

$$P = \bigcup_{k \text{ even}} A'_k \quad \text{and} \quad I = \bigcup_{k \text{ odd}} A'_k,$$

and we can, for example, assume that:

$$\sum_{n \in P} |a_n| \geqslant \frac{\delta}{2R}.$$

Next we define $k_1, k_2, \ldots, k_J$ by setting $k_1 = 0$, and, for $1 \leqslant j \leqslant J - 1$:

$$k_{j+1} = \min\{ k > k_j \,;\, |A'_{2k}| \geqslant R \,|A'_{2k_j}| \}.$$

Then, since:

$$\frac{1}{R^{2k+1}} < |a_n| \leqslant \frac{1}{R^{2k}} \quad \text{for } n \in A'_{2k},$$

we obtain:

$$\sum_{j \geqslant 1} \sum_{k_j < k < k_{j+1}} \sum_{n \in A'_{2k}} |a_n| \leqslant \sum_{j \geqslant 1} \sum_{k_j < k < k_{j+1}} \frac{1}{R^{2k}} |A'_{2k}|$$

$$\leqslant \sum_{j \geqslant 1} \sum_{k > k_j} \frac{1}{R^{2k}} \cdot R \,|A'_{2k_j}|$$

$$= \frac{R}{R^2 - 1} \sum_{j \geqslant 1} \frac{1}{R^{2k_j}} |A'_{2k_j}|$$

$$\leqslant \frac{R^2}{R^2 - 1} \sum_{j \geqslant 1} \sum_{n \in A'_{2k_j}} |a_n|$$

$$\leqslant 2 \sum_{j \geqslant 1} \sum_{n \in A'_{2k_j}} |a_n|;$$

hence:

$$\frac{\delta}{2R} \leqslant \sum_{n \in P} |a_n| = \sum_{j \geqslant 1} \sum_{k_j < k < k_{j+1}} \sum_{n \in A'_{2k}} |a_n| + \sum_{j \geqslant 1} \sum_{n \in A'_{2k_j}} |a_n|$$

$$\leqslant 3 \sum_{j \geqslant 1} \sum_{n \in A'_{2k_j}} |a_n|.$$

Now, since

$$\frac{|A'_{2k_{j+1}}|}{|A'_{2k_j}|} \geqslant R$$

for any $j \geqslant 1$, Lemma II.11 provides sets $B_j \subseteq A'_{2k_j}$ with

$$|B_j| \geqslant \frac{1}{10} |A'_{2k_j}|$$

and such that

$$B = \bigcup_{j \geqslant 1} B_j$$

is quasi-independent. This completes the proof of the theorem, since:

$$\sum_{n \in B} |a_n| \geqslant \sum_{j \geqslant 1} \frac{1}{R^{2k_j+1}} |B_j| \geqslant \frac{1}{10R} \sum_{j \geqslant 1} \frac{1}{R^{2k_j}} |A'_{2k_j}|$$

$$\geqslant \frac{1}{10R} \sum_{j \geqslant 1} \sum_{n \in A'_{2k_j}} |a_n| \geqslant \frac{1}{10R} \cdot \frac{\delta}{6R} = \frac{\delta}{60R^2}. \qquad \square$$

We now prove Lemma II.11:

*Proof of Lemma II.11*   The proof presented by Bourgain was rather "condensed"; we are grateful to Myriam Déchamps for explaining the details to us.

For every finite subset $A$ of $\mathbb{Z}$ and any integer $d \geqslant 1$, we denote:

$$W_d(A) = \left\{ \sum_{n \in A} s_n n \, ; \, s_n \in \mathbb{Z} \text{ and } \sum_{n \in A} |s_n| \leqslant d \right\}.$$

*Step 1.* It suffices to show that, for each fixed $j$ ($1 \leqslant j \leqslant J$), a subset $\widetilde{\Lambda}_j \subseteq \Lambda_j$ can be extracted such that:

(i) $|\widetilde{\Lambda}_j| \geqslant \frac{1}{5}|\Lambda_j|$;

(ii) *if* $r_1, \ldots, r_J$ *are numbers satisfying*:

    (a) $r_j = \sum_{n \in \widetilde{\Lambda}_j} \theta_n n$, *with* $\theta_n \in \{-1, 0, 1\}$ *and* $\sum_{n \in \widetilde{\Lambda}_j} |\theta_n| \geqslant \frac{1}{2}|\widetilde{\Lambda}_j|$;

    (b) *for* $k \neq j$: $r_k \in W_{d_{j,k}}(\Lambda_k)$, *with* $d_{j,k} = \frac{|\Lambda_j|}{|\Lambda_k|} \sum_{n \in \widetilde{\Lambda}_j} |\theta_n|$,

    *then* $r_1 + \cdots + r_J \neq 0$.

Indeed, let us see how this leads to the proof of the lemma. For $1 \leqslant j \leqslant J$, we select $\Lambda'_j$ as follows:

*First case.* Suppose that there exist $\widetilde{r}_1, \ldots, \widetilde{r}_J$ such that:

(a') $\widetilde{r}_j = \sum_{n \in \widetilde{\Lambda}_j} \theta_n n \neq 0$, with $\theta_n \in \{-1, 0, 1\}$;

(b') for $k \neq j$: $\widetilde{r}_k \in W_{\widetilde{d}_{j,k}}(\Lambda_k)$ with $\widetilde{d}_{j,k} = \frac{|\Lambda_j|}{|\Lambda_k|} \sum_{n \in \widetilde{\Lambda}_j} |\theta_n|$;

(c') $\widetilde{r}_1 + \cdots + \widetilde{r}_J = 0$.

Then, condition (ii) implies: $\sum_{n \in \widetilde{\Lambda}_j} |\theta_n| < \frac{1}{2}|\widetilde{\Lambda}_j|$. Next, we select $\widetilde{r}_1^0, \ldots, \widetilde{r}_j^0, \ldots, \widetilde{r}_J^0$ satisfying the conditions (a'), (b') and (c'), and such that $\widetilde{r}_j^0$ have support

$$\overset{\approx}{\Lambda}_j = \{n \in \widetilde{\Lambda}_j \, ; \, \theta_n^0 \neq 0\}$$

of maximum length $|\overset{\approx}{\Lambda}_j|$.

We have $|\overset{\approx}{\Lambda}_j| < |\widetilde{\Lambda}_j|/2$, so if we take $\Lambda'_j = \widetilde{\Lambda}_j \smallsetminus \overset{\approx}{\Lambda}_j$, we obtain:

$$|\Lambda'_j| = |\widetilde{\Lambda}_j| - |\overset{\approx}{\Lambda}_j| \geqslant \frac{1}{2}|\widetilde{\Lambda}_j| \geqslant \frac{1}{10}|\Lambda_j|.$$

*Second case.* If no $\widetilde{r}_1, \ldots, \widetilde{r}_J$ can be found as in the first case, we choose to take $\Lambda'_j = \widetilde{\Lambda}_j$.

We now show that the union:

$$\Lambda' = \bigcup_{1 \leqslant j \leqslant J} \Lambda'_j$$

is indeed quasi-independent.

If not, then there would exist integers $\overset{\approx}{r}_1, \ldots, \overset{\approx}{r}_J$ such that:

$$\overset{\approx}{r}_1 + \cdots + \overset{\approx}{r}_J = 0,$$

with, for $1 \leqslant j \leqslant J$:

$$\widetilde{\widetilde{r}}_j = \sum_{n \in \Lambda'_j} \theta_n n, \qquad \theta_n \in \{-1, 0, 1\},$$

and with the numbers $\theta_n$ not all null for $n \in \Lambda' = \bigcup_{1 \leqslant j \leqslant J} \Lambda'_j$. This last condition can be written: $\max_{1 \leqslant j \leqslant J} \sum_{n \in \Lambda'_j} |\theta_n| > 0$. We select $j_0$ such that:

$$|\Lambda_{j_0}| \sum_{n \in \Lambda'_{j_0}} |\theta_n| = \max_{1 \leqslant j \leqslant J} |\Lambda_j| \sum_{n \in \Lambda'_j} |\theta_n|.$$

Note that, if $|\Lambda_1| < R$, we necessarily have $j_0 \geqslant 2$ (since $|\Lambda_j| \geqslant R|\Lambda_1|$ if $j \geqslant 2$). In any case, for $k \neq j_0$, the maximality leads to:

$$\widetilde{\widetilde{r}}_k \in W_{\widetilde{\widetilde{d}}_{j_0,k}}(\Lambda_k),$$

with

$$\widetilde{\widetilde{d}}_{j_0,k} = \frac{|\Lambda_{j_0}|}{|\Lambda_k|} \sum_{n \in \Lambda'_{j_0}} |\theta_n|.$$

Since $\widetilde{\widetilde{r}}_{j_0} \neq 0$, it ensues that $\widetilde{\widetilde{\Lambda}}_{j_0}$ is not empty. Then we consider the numbers $\widetilde{r}^0_1, \ldots, \widetilde{r}^0_{j_0}, \ldots, \widetilde{r}^0_J$ satisfying (a′), (b′) and (c′) and defining $\widetilde{\widetilde{\Lambda}}_{j_0}$. As the support of $\widetilde{r}^0_{j_0}$ is $\widetilde{\widetilde{\Lambda}}_{j_0}$ and that of $\widetilde{\widetilde{r}}_{j_0}$ is $\Lambda'_{j_0} = \widetilde{\Lambda}_{j_0} \smallsetminus \widetilde{\widetilde{\Lambda}}_{j_0}$, these supports are disjoint.

Now, as $\widetilde{\Lambda}_{j_0}$ is quasi-independent, the condition $\widetilde{\widetilde{r}}_{j_0} \neq 0$ implies that:

$$\widetilde{\widetilde{r}}_{j_0} + \widetilde{r}^0_{j_0} = \sum_{n \in \Lambda'_{j_0}} \theta_n n + \sum_{n \in \widetilde{\widetilde{\Lambda}}_{j_0}} \theta^0_n n \neq 0.$$

Next we consider the relation:

$$\widetilde{\widetilde{r}}_1 + \widetilde{r}^0_1 + \cdots + \widetilde{\widetilde{r}}_J + \widetilde{r}^0_J = \left(\widetilde{\widetilde{r}}_1 + \cdots + \widetilde{\widetilde{r}}_J\right) + \left(\widetilde{r}^0_1 + \cdots + \widetilde{r}^0_J\right) = 0.$$

For $k \neq j_0$, we have:

$$\widetilde{\widetilde{r}}_k + \widetilde{r}^0_k \in W_{\widetilde{\widetilde{d}}_{j_0,k} + \widetilde{d}^0_{j_0,k}}(\Lambda_k),$$

with

$$\widetilde{\widetilde{d}}_{j_0,k} + \widetilde{d}^0_{j_0,k} = \frac{|\Lambda_{j_0}|}{|\Lambda_k|} \sum_{n \in \Lambda'_{j_0}} |\theta_n| + \frac{|\Lambda_{j_0}|}{|\Lambda_k|} \sum_{n \in \widetilde{\widetilde{\Lambda}}_{j_0}} |\theta^0_n| = \frac{|\Lambda_{j_0}|}{|\Lambda_k|} \sum_{n \in \widetilde{\Lambda}_{j_0}} |\theta_n|$$

(we have set $\theta_n = \theta_n^0$ for $n \in \overset{\approx}{\Lambda}_{j_0}$). This means that the sequence

$$\overset{\approx}{r}_1 + \tilde{r}_1^0, \ldots, \overset{\approx}{r}_J + \tilde{r}_J^0$$

satisfies the conditions (a′), (b′) and (c′) (corresponding to the index $j_0$), even though the length of $\overset{\approx}{r}_{j_0} + \tilde{r}_{j_0}^0$ is strictly greater than that of $\tilde{r}_{j_0}^0$. This contradiction shows that $\Lambda'$ is indeed quasi-independent.

*Step 2.* Construction of $\widetilde{\Lambda}_j$, for $1 \leqslant j \leqslant J$.

First note that, for $j = 1$, if $|\Lambda_1| < R$, it suffices to take $\widetilde{\Lambda}_1 = \Lambda_1$. In fact, in this case, for each $r_1 = \sum_{n \in \Lambda_1} \theta_n n$ with $\theta_n \in \{-1, 0, 1\}$ and $\sum_{n \in \Lambda_1} |\theta_n| \geqslant \frac{1}{2}|\Lambda_1|$, for any $k \geqslant 2$, we have:

$$d_{1,k} = \frac{|\Lambda_1|}{|\Lambda_k|} \sum_{n \in \Lambda_1} |\theta_n| < \frac{1}{R} \cdot R = 1;$$

hence $d_{1,k} = 0$ and thus $W_{d_{1,k}}(\Lambda_k) = \{0\}$; therefore, if $r_k \in W_{d_{1,k}}(\Lambda_k)$ for $k \geqslant 2$, we have:

$$r_1 + r_2 + \cdots + r_J = r_1 \neq 0.$$

From now on, we thus assume $|\Lambda_1| \geqslant R$. In this case, the construction of $\widetilde{\Lambda}_j$, for $1 \leqslant j \leqslant J$ fixed, is based on a random method. We use independent Bernoulli random variables $\xi_n, n \in \Lambda_j$ satisfying

$$\mathbb{P}(\xi_n = 1) = 1/4 \quad \text{and} \quad \mathbb{P}(\xi_n = 0) = 3/4,$$

and set:

$$\widetilde{\Lambda}_j(\omega) = \{n \in \Lambda_j \, ; \, \xi_n(\omega) = 1\}.$$

As in the proof of Lemma I.7, we introduce a random polynomial that here must take into account the $\Lambda_k$'s for $k \neq j$. Hence we set:

$$F_\omega(t) = \sum_{\substack{\frac{|\Lambda_j|}{10} \leqslant l \leqslant |\Lambda_j|}} \sum_{\substack{S \subseteq \Lambda_j \\ |S| = l}} \prod_{n \in S} \xi_n(\omega) \, (e^{int} + e^{-int}) \prod_{\substack{1 \leqslant k \leqslant J \\ k \neq j}} \sum_{r \in W_{d_k(l)}(\Lambda_k)} e^{irt},$$

where $d_k(l) = l \frac{|\Lambda_j|}{|\Lambda_k|}$. Then $\int_{\mathbb{T}} F_\omega(t) \, dm(t)$ is the number of relations $r_1 + \cdots + r_j + \cdots + r_J = 0$ with:

$$r_j = \sum_{n \in \widetilde{\Lambda}_j(\omega)} \theta_n n, \quad \theta_n \in \{-1, 0, 1\}, \quad \text{and} \quad \sum_{n \in \widetilde{\Lambda}_j(\omega)} |\theta_n| = l \geqslant \frac{|\Lambda_j|}{10}.$$

For $k \neq j$:

$$r_k \in W_{d_k(l)}(\Lambda_k),$$

with

$$d_{k(l)} = \frac{|\Lambda_j|}{|\Lambda_k|} \sum_{n \in \widetilde{\Lambda}_j(\omega)} |\theta_n|.$$

It hence suffices to show that there exists $\omega \in \Omega$ such that:

$$\int_{\mathbb{T}} F_\omega(t)\, dm(t) = 0 \quad \text{and} \quad |\widetilde{\Lambda}_j(\omega)| \geqslant \frac{1}{5} |\Lambda_j|.$$

For this, in turn, it suffices to show that if we set:

$$\Omega_1 = \left\{ \omega \in \Omega \,;\, \int_{\mathbb{T}} F_\omega(t)\, dm(t) = 0 \right\}$$

and:

$$\Omega_2 = \left\{ \omega \in \Omega \,;\, |\widetilde{\Lambda}_j(\omega)| \geqslant \frac{1}{5} |\Lambda_j| \right\},$$

then $\mathbb{P}(\Omega_1 \cap \Omega_2) > 0$.

*Step 3.* We proceed by contradiction.

Suppose that $\mathbb{P}(\Omega_1 \cap \Omega_2) = 0$. Then $\Omega_2$ is almost surely contained in $\Omega_1^c$, and hence $\mathbb{P}(\Omega_2) \leqslant \mathbb{P}(\Omega_1^c)$. Let us evaluate $\mathbb{P}(\Omega_2)$. We have:

$$\frac{|\Lambda_j|}{4} = \mathbb{E}\left( \sum_{n \in \Lambda_j} \xi_n \right) = \int_\Omega |\Lambda_j(\omega)|\, d\mathbb{P}(\omega) \leqslant \frac{|\Lambda_j|}{5} + |\Lambda_j|\, \mathbb{P}(\Omega_2);$$

thus $\mathbb{P}(\Omega_2) \geqslant 1/20$. Then, as $\int_{\mathbb{T}} F_\omega(t)\, dm(t)$ is non-negative (a non-negative integer), we have:

$$\int_\Omega \left( \int_{\mathbb{T}} F_\omega(t)\, dm(t) \right) d\mathbb{P}(\omega) = \int_{\Omega_1^c} \left( \int_{\mathbb{T}} F_\omega(t)\, dm(t) \right) d\mathbb{P}(\omega)$$

$$\geqslant \mathbb{P}(\Omega_1^c) \geqslant \mathbb{P}(\Omega_2) \geqslant \frac{1}{20}.$$

Next, we seek a majoration for:

$$I = \int_\Omega \left( \int_{\mathbb{T}} F_\omega(t)\, dm(t) \right) d\mathbb{P}(\omega)$$

to reach the desired contradiction.

*Step 4.* Majoration of $I$.

We interchange the order of integration. Thanks to the independence of the $\xi_n$, we obtain:

$$I = \int_{\mathbb{T}} \left( \int_{\Omega} F_\omega(t)\, d\mathbb{P}(\omega) \right) dm(t)$$

$$= \sum_{\frac{|\Lambda_j|}{10} \leqslant l \leqslant |\Lambda_j|} \frac{1}{4^l} \int_{\mathbb{T}} \left( \sum_{\substack{S \subseteq \Lambda_j \\ |S|=l}} \prod_{n \in S} (e^{int} + e^{-int}) \prod_{k \neq j} \sum_{r \in W_{d_{k(l)}}(\Lambda_k)} e^{irt} \right) dm(t)$$

$$\leqslant \sum_{\frac{|\Lambda_j|}{10} \leqslant l \leqslant |\Lambda_j|} \frac{1}{2^l} \int_{\mathbb{T}} \left( \prod_{n \in \Lambda_j} \left( 1 + \frac{e^{int} + e^{-int}}{2} \right) \prod_{k \neq j} \sum_{r \in W_{d_{k(l)}}(\Lambda_k)} e^{irt} \right) dm(t)$$

because the expansion of the product:

$$\prod_{n \in \Lambda_j} \left( 1 + \frac{e^{int} + e^{-int}}{2} \right)$$

reveals the sum $\dfrac{1}{2^l} \displaystyle\sum_{\substack{S \subseteq \Lambda_j \\ |S|=l}} \prod_{n \in S} (e^{int} + e^{-int})$, plus other terms, and each term

contributes a non-negative quantity to the integral. However as $\Lambda_j$ is quasi-independent, the $L^1$-norm of the Riesz product

$$\prod_{n \in \Lambda_j} \left( 1 + \frac{e^{int} + e^{-int}}{2} \right)$$

is equal to 1; hence:

$$\int_{\mathbb{T}} \left[ \prod_{n \in \Lambda_j} \left( 1 + \frac{e^{int} + e^{-int}}{2} \right) \prod_{k \neq j} \sum_{r \in W_{d_{k(l)}}(\Lambda_k)} e^{irt} \right] dm(t)$$

$$\leqslant \left\| \prod_{k \neq j} \sum_{r \in W_{d_{k(l)}}(\Lambda_k)} e^{irt} \right\|_\infty = \prod_{k \neq j} |W_{d_{k(l)}}(\Lambda_k)|$$

$$\leqslant \prod_{k \neq j} |W_{d_k}(\Lambda_k)|$$

after setting $d_k = d_k(|\Lambda_j|) = \dfrac{|\Lambda_j|^2}{|\Lambda_k|}$ .

Now we use the following result:

**Sub-Lemma II.12**   *For every non-empty finite subset F of $\mathbb{Z}$, and any integer $d \geqslant 1$, we have*:

$$|W_d(F)| \leqslant \begin{cases} \left(4\mathrm{e}\,\dfrac{|F|}{d}\right)^d & \text{if } d \leqslant |F| \\[3mm] \left(4\mathrm{e}\,\dfrac{d}{|F|}\right)^{|F|} & \text{if } |F| \leqslant d. \end{cases}$$

Then, as $d_k \leqslant |\Lambda_k|$ if and only if $|\Lambda_j| \leqslant |\Lambda_k|$, or in other words, given the conditions $|\Lambda_{k+1}| \geqslant R\,|\Lambda_k|$, if and only if $j \leqslant k$, we obtain:

$$|W_{d_k}(\Lambda_k)| \leqslant \begin{cases} \left(4\mathrm{e}\,\dfrac{|\Lambda_k|^2}{|\Lambda_j|^2}\right)^{|\Lambda_j|^2/|\Lambda_k|} & \text{if } j < k \\[3mm] \left(4\mathrm{e}\,\dfrac{|\Lambda_j|^2}{|\Lambda_k|^2}\right)^{|\Lambda_k|} & \text{if } k < j. \end{cases}$$

The inequalities $\log x \leqslant 2\sqrt{x}$ (valid for $x \geqslant 1$) and $4\mathrm{e} < 16$ lead to:

$$\log |W_{d_k}(\Lambda_k)| \leqslant \begin{cases} 2\,\dfrac{|\Lambda_j|^2}{|\Lambda_k|} \times 2\sqrt{4\dfrac{|\Lambda_k|}{|\Lambda_j|}} = 8\,\dfrac{|\Lambda_j|^{3/2}}{|\Lambda_k|^{1/2}} & \text{if } j < k \\[3mm] 2\,|\Lambda_k| \times 2\sqrt{4\dfrac{|\Lambda_j|}{|\Lambda_k|}} = 8\,|\Lambda_k|^{1/2}\,|\Lambda_j|^{1/2} & \text{if } k < j. \end{cases}$$

Hence:

$$\prod_{k \neq j} |W_{d_k}(\Lambda_k)| \leqslant \exp\left(8\,|\Lambda_j|\left[\sum_{k<j}\left(\frac{|\Lambda_k|}{|\Lambda_j|}\right)^{1/2} + \sum_{j<k}\left(\frac{|\Lambda_j|}{|\Lambda_k|}\right)^{1/2}\right]\right)$$

$$\leqslant \exp\left(8\,|\Lambda_j|\left[\sum_{k<j}\left(\frac{1}{R^{j-k}}\right)^{1/2} + \sum_{k<j}\left(\frac{1}{R^{k-j}}\right)^{1/2}\right]\right)$$

$$\leqslant \exp\left(8\,|\Lambda_j| \times \frac{2}{\sqrt{R}} \times \frac{1}{1-1/\sqrt{R}}\right)$$

$$\leqslant \exp\left(8\,|\Lambda_j| \times \frac{3}{\sqrt{R}}\right) \quad (\text{as } R > 10).$$

Thus, we obtain:

$$I \leqslant \exp\left(24\,\frac{|\Lambda_j|}{\sqrt{R}}\right)\sum_{l \geqslant |\Lambda_j|/10} \frac{1}{2^l} \leqslant \exp\left(24\,\frac{|\Lambda_j|}{\sqrt{R}}\right) \times 2 \times 2^{-|\Lambda_j|/10}$$

$$= 2\exp\left[|\Lambda_j|\left(-\frac{\log 2}{10} + \frac{24}{\sqrt{R}}\right)\right].$$

Then we select:

$$R = 4 \left( \frac{240}{\log 2} \right)^2 ;$$

consequently:

$$I \leqslant 2 \exp \left( -\frac{\log 2}{20} |\Lambda_j| \right) \leqslant 2 \exp \left( -\frac{\log 2}{20} |\Lambda_1| \right).$$

It has been assumed that $|\Lambda_1| \geqslant R$, so this gives:

$$I \leqslant 2 \exp \left( -\frac{\log 2}{20} R \right).$$

This contradicts:

$$I \geqslant \frac{1}{20} ;$$

therefore the proof of Lemma II.11 is complete. $\qquad\square$

To finish, only the sub-lemma remains to be proved.

*Proof of sub-lemma II.12* Set $N = |F|$. We have:

$$
\begin{aligned}
|W_d(F)| \leqslant & \left| \{(k_1, \ldots, k_N) \in \mathbb{Z}^N ; |k_1| + \cdots + |k_N| \leqslant d \} \right| \\
\leqslant & \inf(2^d, 2^N). \left| \{(k_1, \ldots, k_N) \in \mathbb{N}^N ; k_1 + \cdots + k_N \leqslant d \} \right| \\
= & \inf(2^d, 2^N). \left| \{(k_1, \ldots, k_N, k_{N+1}) \in \mathbb{N}^{N+1} ; \right. \\
& \qquad\qquad\qquad\qquad \left. k_1 + \cdots + k_N + k_{N+1} = d \} \right|.
\end{aligned}
$$

However, if we denote:

$$a_d = \left| \{(k_1, \ldots, k_N, k_{N+1}) \in \mathbb{N}^{N+1} ; k_1 + \cdots + k_N + k_{N+1} = d \} \right|,$$

then, for $|x| < 1$:

$$
\sum_{d=0}^{+\infty} a_d x^d = \left( \sum_{k_1=0}^{+\infty} x^{k_1} \right) \cdots \left( \sum_{k_N=0}^{+\infty} x^{k_N} \right) \left( \sum_{k_{N+1}=0}^{+\infty} x^{k_{N+1}} \right)
$$

$$
= \left( \sum_{k=0}^{+\infty} x^k \right)^{N+1} = \frac{1}{(1-x)^{N+1}} ;
$$

therefore:

$$a_d = \binom{N+d}{d}.$$

We now use Robbins' version of Stirling's formula (Robbins [1955]):

$$\sqrt{2\pi} \, n^{n+\frac{1}{2}} e^{-n} e^{\frac{1}{12n+1}} < n! < \sqrt{2\pi} \, n^{n+\frac{1}{2}} e^{-n} e^{\frac{1}{12n}}.$$

For $N \geqslant 1$, and $d \geqslant 1$ (note that $a_0 = 1$), we obtain:

$$a_d \leqslant \frac{e^{\frac{1}{12(N+d)}}}{\sqrt{2\pi}} \left(\frac{1}{N} + \frac{1}{d}\right)^{1/2} \left(\frac{N+d}{d}\right)^d \left(\frac{N+d}{N}\right)^N$$

$$\leqslant \frac{e^{1/24}}{\sqrt{2\pi}} \sqrt{2} \left(\frac{N+d}{d}\right)^d \left(\frac{N+d}{N}\right)^N$$

$$\leqslant \left(\frac{N+d}{d}\right)^d \left(\frac{N+d}{N}\right)^N ;$$

however, for $N \leqslant d$:

$$\left(\frac{N+d}{d}\right)^d = \left(1 + \frac{N}{d}\right)^d \leqslant e^N \quad \text{and} \quad \left(\frac{N+d}{N}\right)^N \leqslant 2^N \left(\frac{d}{N}\right)^N,$$

and then symmetrically for $d \leqslant N$; hence the result. □

## II.4 Applications

The following theorem is a consequence of Theorem II.3:

**Theorem II.13** (Pisier)  *If $\Lambda \subseteq \mathbb{Z}$ is a $\Lambda(q)$-set for every $q < +\infty$, with a constant $O(\sqrt{q})$:*

$$\|f\|_q \leqslant C\sqrt{q}\,\|f\|_2, \quad \forall f \in \mathcal{P}_\Lambda,$$

*then $\Lambda$ is a Sidon set.*

**Remark**  The stated condition can also be expressed as $L_\Lambda^{\Psi_2} = L_\Lambda^2$. This theorem will also be obtained in a different manner in Chapter 6 (Theorem IV.1).

*Proof*  Let $A$ be a finite subset of $\Lambda$, containing at least two elements. By hypothesis:

$$\psi_A \leqslant C \left\|\sum_{n \in A} e_n\right\|_2 = C |A|^{1/2}.$$

By Theorem II.6, there exists a quasi-independent set $B \subseteq A$ such that:

$$|B| \geqslant K \left(\frac{|A|}{\psi_A}\right)^2 \geqslant \frac{K}{C^2} |A|.$$

Then, thanks to Theorem II.3, $\Lambda$ is a Sidon set. □

Also, Rider's theorem (Chapter 6 of Volume 1, Theorem V.18) can be derived here. Recall that $[\![P]\!]_R = \mathbb{E}\|\sum_{n \in \mathbb{Z}} \varepsilon_n \widehat{P}(n)\, e_n\|_\infty$, where $(\varepsilon_n)_{n \in \mathbb{Z}}$ is an independent sequence of Rademacher variables.

**Theorem II.14** (Rider's Theorem)  *If*

$$\delta \sum_{n \in \Lambda} |\widehat{P}(n)| \leqslant [\![P]\!]_R, \quad \forall P \in \mathcal{P}_\Lambda,$$

*then $\Lambda$ is a Sidon set.*

*Proof*  In fact we only use this inequality when $\widehat{P}(n) = 0$ or $1$; in other words:

$$(*) \qquad\qquad \delta|A| \leqslant \left[\!\!\left[\sum_{n \in A} e_n\right]\!\!\right]_R$$

for every finite set $A$. As in the proof given in Chapter 6 (Volume 1), we use the dual:

$$\left[L^1\big(\Omega_{|A|}; \mathcal{C}(\mathbb{T})\big)\right]^* = L^\infty\big(\Omega_{|A|}; \mathcal{M}(\mathbb{T})\big).$$

For every $\omega \in \Omega_{|A|}$, by $(*)$ there exists a measure $\mu_\omega \in \mathcal{M}(\mathbb{T})$ such that:

$$\begin{cases} \|\mu_\omega\| \leqslant 1 \\ \langle \mu_\omega, \varepsilon_n(\omega)e_n \rangle \geqslant 0 \\ \displaystyle\sum_{n \in A} \langle \mu_\omega, \varepsilon_n(\omega)e_n \rangle \geqslant \delta|A|. \end{cases}$$

Next, we will need the following two lemmas:

**Lemma II.15**  *Let $A$ be a finite set, and let $\alpha_n$ be numbers satisfying $0 \leqslant \alpha_n \leqslant 1$ for $n \in A$. If there exists $\delta > 0$ such that:*

$$\sum_{n \in A} \alpha_n \geqslant \delta|A|,$$

*then there exists $B \subseteq A$ such that $|B| \geqslant (\delta/2)|A|$ and $\alpha_n \geqslant \delta/2$ for any $n \in B$.*

**Lemma II.16**  *If $p$ is an even integer, $p = 2l$, then:*

$$|\widehat{f}(n)| \leqslant \widehat{g}(n), \ \forall n \in \mathbb{Z} \implies \|f\|_{2l} \leqslant \|g\|_{2l}.$$

To prove Lemma II.15, it suffices to take $B = \{n \in A \,;\, \alpha_n \geqslant \delta/2\}$. Then:

$$\delta|A| \leqslant \sum_{n \in A} \alpha_n \leqslant \sum_{n \in B} \alpha_n + (\delta/2)\,|A \smallsetminus B| \leqslant |B| + (\delta/2)|A|.$$

For Lemma II.16, it suffices to expand $\|f\|_{2l}^{2l}$ and $\|g\|_{2l}^{2l}$.

Then, Lemma II.15, with $\alpha_n = \langle \mu_\omega, \varepsilon_n(\omega)e_n \rangle$, and Lemma II.16, with $p = 2l$, imply:

$$\left\| \sum_{n \in B} \varepsilon_n(\omega)\langle \mu_\omega, e_n \rangle e_n \right\|_{L^p(\mathbb{T})} \geqslant \frac{\delta}{2} \left\| \sum_{n \in B} e_n \right\|_{L^p(\mathbb{T})}.$$

Moreover, for any $p \geqslant 1$:

$$
\left\| \sum_{n \in B} \varepsilon_n(\omega) \langle \mu_\omega, e_n \rangle e_n \right\|_{L^p(\mathbb{T})}^p = \int_{\mathbb{T}} \left| \sum_{n \in B} \varepsilon_n(\omega) \langle \mu_\omega, e_n \rangle e^{int} \right|^p \, dm(t)
$$

$$
\leqslant \int_{\mathbb{T}} \left( \int_{\mathbb{T}} \left| \sum_{n \in B} \varepsilon_n(\omega) e^{inx} e^{int} \right|^p \, d|\mu_\omega|(x) \right) dm(t)
$$

$$
= \int_{\mathbb{T}} \left( \int_{\mathbb{T}} \left| \sum_{n \in B} \varepsilon_n(\omega) e^{inx} e^{int} \right|^p \, dm(t) \right) d|\mu_\omega|(x)
$$

$$
= \int_{\mathbb{T}} \left( \int_{\mathbb{T}} \left| \sum_{n \in B} \varepsilon_n(\omega) e^{inu} \right|^p \, dm(u) \right) d|\mu_\omega|(x)
$$

$$
\leqslant \int_{\mathbb{T}} \left| \sum_{n \in B} \varepsilon_n(\omega) e^{inu} \right|^p \, dm(u).
$$

An integration with respect to $\omega$ provides:

$$
\int_\Omega \left\| \sum_{n \in B} \varepsilon_n(\omega) \langle \mu_\omega, e_n \rangle e_n \right\|_{L^p(\mathbb{T})} d\mathbb{P}(\omega)
$$

$$
\leqslant \left[ \int_\Omega \left\| \sum_{n \in B} \varepsilon_n(\omega) \langle \mu_\omega, e_n \rangle e_n \right\|_{L^p(\mathbb{T})}^p d\mathbb{P}(\omega) \right]^{1/p}
$$

$$
\leqslant \left[ \int_\Omega \left( \int_{\mathbb{T}} \left| \sum_{n \in B} \varepsilon_n(\omega) e^{inu} \right|^p \, dm(u) \right) d\mathbb{P}(\omega) \right]^{1/p}
$$

$$
= \left[ \int_{\mathbb{T}} \left( \int_\Omega \left| \sum_{n \in B} \varepsilon_n(\omega) e^{inu} \right|^p \, d\mathbb{P}(\omega) \right) dm(u) \right]^{1/p}
$$

$$
\leqslant C\sqrt{p} \, |B|^{1/2}
$$

for $p \geqslant 2$, by the Khintchine inequalities.

Finally, for any even integer $p = 2l$, we obtain:

$$
\frac{\delta}{2} \left\| \sum_{n \in B} e_n \right\|_{L^p(\mathbb{T})} \leqslant C\sqrt{p} \, |B|^{1/2}.
$$

If the constant $C$ is increased by a factor $\sqrt{2}$, this inequality remains valid for any $p \geqslant 2$. This can also be written:

$$
\frac{\delta}{2} \psi_B \leqslant C|B|^{1/2}.
$$

Finally, it only remains to see that, according to Theorem II.6, a quasi-independent set $B_1$ can be extracted from $B$ such that:

$$|B_1| \geqslant K \left( \frac{|B|}{\psi_B} \right)^2 \geqslant K \left( \frac{\delta |B|^{1/2}}{2C} \right)^2 = \frac{K}{4C^2} \delta^2 |B| \geqslant \frac{K}{8C^2} \delta^3 |A|.$$

By Theorem II.3, $\Lambda$ is hence a Sidon set. $\qquad\square$

# III Sums of Sines and Vectorial Hilbert Transforms

## III.1 Introduction

Recall that to every trigonometric polynomial $f = \sum_{n \in \mathbb{Z}} a_n e_n$ (with $e_n(t) = e^{int}$) corresponds its *conjugate function*

$$\mathcal{H}f = \tilde{f} = \sum_{n \in \mathbb{Z}} (-i)\,(\mathrm{sign}\,n)\,a_n\,e_n$$

(Chapter 7 of Volume 1, Definition III.1). This can also be written as:

$$f(t) = \frac{\alpha_0}{2} + \sum_{n=1}^{+\infty} (\alpha_n \cos nt + \beta_n \sin nt)$$

and:

$$\tilde{f}(t) = \sum_{n=1}^{+\infty} (\alpha_n \sin nt - \beta_n \cos nt);$$

then

$$f(t) + i\tilde{f}(t) = \frac{\alpha_0}{2} + \sum_{n=1}^{+\infty} (\alpha_n - i\beta_n)\,e^{int}$$

is of analytic type. The norm of the mapping $\mathcal{H} \colon L^2(\mathbb{T}) \to L^2(\mathbb{T})$ is 1. For $f \in L^1(\mathbb{T}), \tilde{f}$ is given by a singular integral (see KATZNELSON).

Given a Banach space $E, x_1, \ldots, x_k \in E$ and functions $f_1, \ldots, f_k \in L^2(\mathbb{T})$, let us consider $f = \sum_{j=1}^{k} f_j \otimes x_j \in L^2(\mathbb{T}, E)$. The conjugate function can be defined as $\tilde{f} = \sum_{j=1}^{n} \tilde{f_j} \otimes x_j$. This vector-valued Hilbert transform remains bounded on $L^2(\mathbb{T}, E)$ if and only if $E$ is *UMD* (Burkholder [1983] and Bourgain [1983 a]).

In the case of $E = \ell_1^N$, by analyzing the above singular integral with the aid of a Calderón–Zygmund decomposition, B. Virot [1981] showed that:

$$\|\mathcal{H}\|_{L^2(\ell_1^N) \to L^2(\ell_1^N)} \ll \log N$$

(where $A \ll B$ denotes that $A \leqslant c B$, for a numerical constant $c > 0$). Of course, $\|\mathcal{H}\|_{L^2(\ell_2^N) \to L^2(\ell_2^N)} = 1$, and since the distance (in the sense of Banach–Mazur) between $\ell_1^N$ and $\ell_2^N$ is maximal, namely $d(\ell_1^N, \ell_2^N) = \sqrt{N}$, it would be heuristically plausible to suspect that:

$$\|\mathcal{H}\|_{L^2(E) \to L^2(E)} \ll \log N$$

for every space $E$ of dimension $N$. In fact, this is not the case at all: even though clearly:

$$\|\mathcal{H}\|_{L^2(E) \to L^2(E)} \leqslant d(E, \ell_2^N) \leqslant \sqrt{N}$$

always holds, nevertheless it is possible to have:

$$\|\mathcal{H}\|_{L^2(E) \to L^2(E)} \gg N^{1/3},$$

as an example will now illustrate.

### III.2  Sums of Sines

The example can easily be deduced from the following result:

**Theorem III.1** (Bourgain)  *For any integer $N \geqslant 1$, there exists a subset $\Lambda \subseteq \mathbb{N}^*$, of cardinality $|\Lambda| = N$, such that*:

$$\left\| \sum_{k \in \Lambda} \sin kx \right\|_\infty \leqslant C_0 \, N^{2/3},$$

*where $C_0$ is a numerical constant.*

### Remarks

1) Exercise VI.2 shows that explicit examples of such $\Lambda$ are hard to find.
2) Obvious minorations are:

$$\left\| \sum_{k \in \Lambda} \sin kx \right\|_\infty \geqslant \left\| \sum_{k \in \Lambda} \sin kx \right\|_2 = \sqrt{N}/\sqrt{2}$$

for every set $\Lambda$ of cardinality $N$. S. Konyagin [1997] obtained a slightly better result:

$$\left\| \sum_{k \in \Lambda} \sin kx \right\|_\infty \geqslant \delta \sqrt{N} \sqrt{\frac{\log N}{\log \log N}},$$

for a numerical constant $\delta > 0$.

**Corollary III.2** *There exists a Banach space E, of dimension 2N, for which*:

$$\|\mathcal{H}\|_{L^2(E)\to L^2(E)} \geq C_0^{-1} N^{1/3}.$$

*Proof* Let $\Lambda = \{\lambda_1 < \cdots < \lambda_N\}$ be the set obtained in the theorem, and set:

$$\varphi(t) = \sum_{j=1}^{N} \sin \lambda_j t.$$

Let $E = \mathcal{P}_{(-\Lambda)\cup\Lambda}$ be the space of dimension $2N$ generated by the translates $\varphi_\theta(t) = \varphi(t - \theta)$ of $\varphi$. Define $\Phi \in L^2(E)$ by:

$$\Phi(\theta) = \varphi_\theta.$$

Set $c_j(t) = \cos \lambda_j t$ and $s_j(t) = \sin \lambda_j t$, so that $c_j, s_j \in E$. Then:

$$\varphi_\theta(t) = \sum_{j=1}^{N} \sin \lambda_j t \cos \lambda_j \theta - \sin \lambda_j \theta \cos \lambda_j t,$$

and hence:

$$\Phi(\theta) = \sum_{j=1}^{N} (\cos \lambda_j \theta) s_j - (\sin \lambda_j \theta) c_j.$$

Consequently:

$$(\mathcal{H}\Phi)(\theta) = \sum_{j=1}^{N} (\sin \lambda_j \theta) s_j + (\cos \lambda_j \theta) c_j.$$

Hence, $(\mathcal{H}\Phi)(\theta) = \psi_\theta$, with $\psi = \sum_{j=1}^{N} c_j$.

Since:

$$\|\Phi\|_{L^2(E)} = \left( \int_0^{2\pi} \|\varphi_\theta\|_\infty^2 \frac{d\theta}{2\pi} \right)^{1/2} = \|\varphi\|_\infty \leq C_0 N^{2/3}$$

and since:

$$\|\mathcal{H}\Phi\|_{L^2(E)} = \left( \int_0^{2\pi} \|\psi_\theta\|_\infty^2 \frac{d\theta}{2\pi} \right)^{1/2} = \|\psi\|_\infty = N,$$

we hence obtain:

$$\|\mathcal{H}\|_{L^2(E)\to L^2(E)} \geq N/(C_0 N^{2/3}) = C_0^{-1} N^{1/3},$$

as stated. $\qquad\square$

*Proof of Theorem III.1*   The proof combines the method of selectors with the well-known deterministic inequality:

$$(*)\qquad\left|\sum_{k=1}^{m}\frac{\sin kx}{k}\right|\leqslant 2,\qquad \forall x\in\mathbb{R},\ \forall m\geqslant 1.$$

Select an integer $n>N$, to be adjusted later. To profit from $(*)$, we introduce selectors $\xi_1,\ldots,\xi_n$ of mean $\mathbb{E}(\xi_k)=C_n/k$, where $C_n$ is a constant depending on the number of selectors, to be selected further on, and consider the random set:

$$\Lambda=\Lambda(\omega)=\{k\leqslant n\,;\,\xi_k(\omega)=1\},$$

of cardinality $|\Lambda|=\sum_{k=1}^{n}\xi_k$. Should we prove that:

$$\mathbb{E}\left(\left\|\sum_{k=1}^{n}\xi_k\,\sin kx\right\|_{\infty}\right)\ll C_n+\sqrt{C_n}\,\log n$$

(where the notation $A\ll B$ again means, as in Subsection II.1, that $A\leqslant cB$ for some numerical constant $c>0$), since:

$$\mathbb{E}(|\Lambda|)=\sum_{k=1}^{n}\mathbb{E}(\xi_k)\gg C_n\,\log n\,,$$

then we will have, *on average*:

$$\frac{\left\|\sum_{k\in\Lambda}\sin kx\right\|_{\infty}}{|\Lambda|}\ll\frac{C_n+\sqrt{C_n}\,\log n}{C_n\,\log n}=\frac{1}{\log n}+\frac{1}{\sqrt{C_n}}\cdot$$

This last quantity, always larger than $1/\log n$, will be $\leqslant 2/\log n$ if, for example, $C_n$ is chosen so that $\sqrt{C_n}=\log n$, i.e. $C_n=(\log n)^2$. This choice, however, presents a slight technical inconvenience: $\mathbb{P}(\xi_k=1)=C_n/k=(\log n)^2/k$ would be $>1$ for small values of $k\ldots$! There are two ways to fix this problem. One way is to replace $\mathbb{P}(\xi_k=1)=(\log n)^2/k$ by $\mathbb{P}(\xi_k=1)=(\log k)^2/k$, and then verifying, thanks to an Abel transformation, that $\left\|\sum_{k=1}^{n}\frac{(\log k)^2}{k}\,\sin kx\right\|_{\infty}\ll(\log n)^2$, but this hides the basic idea of the proof. The second way is to take the indices $k$ for which $(\log n)^2<k$. The second option is adopted here, but for technical reasons, we in fact take $k\geqslant\sqrt{n}$ (which implies $(\log n)^2<k$ if $n$ is large enough; for example $n\geqslant 3^8$).

The following lemma is admitted temporarily:

**Lemma III.3**  *Given selectors $\xi_k$ of mean $\mathbb{E}(\xi_k) = C_n/k$ ($\sqrt{n} \leqslant k \leqslant n$), then:*

$$\mathbb{E}\left(\left\|\sum_{\sqrt{n}\leqslant k\leqslant n} \xi_k \sin kx\right\|_\infty\right) \ll C_n + \sqrt{C_n}\, \log n.$$

In this case:

$$\mathbb{E}\left(\left\|\sum_{k\in\Lambda} \sin kx\right\|_\infty\right) \leqslant a\,(\log n)^2,$$

for a numerical constant $a > 0$, and Markov's inequality gives:

$$\mathbb{P}\left(\left\|\sum_{k\in\Lambda} \sin kx\right\|_\infty \leqslant 3a\,(\log n)^2\right) \geqslant 1 - \frac{1}{3} = \frac{2}{3}.$$

Moreover, Tchebychev's inequality and the inequalities:

$$V(|\Lambda|) \leqslant \sum_{\sqrt{n}\leqslant k\leqslant n} \mathbb{E}(\xi_k) \leqslant C_n \int_1^n \frac{dx}{x} \leqslant C_n \log n,$$

if $n \geqslant 4$, lead to:

$$\mathbb{P}\big(\big|\,|\Lambda| - \mathbb{E}(|\Lambda|)\big| > t\big) \leqslant \frac{C_n \log n}{t^2},$$

for any $t > 0$; hence, since $C_n = (\log n)^2$:

$$\mathbb{P}\big(\big|\,|\Lambda| - \mathbb{E}(|\Lambda|)\big| \leqslant (\log n)^2\big) \geqslant 1 - \frac{1}{\log 5} > \frac{1}{3},$$

for $n \geqslant 5$. Thus there exists an $\omega_0 \in \Omega$ such that:

$$\begin{cases} \left\|\displaystyle\sum_{k\in\Lambda(\omega_0)} \sin kx\right\|_\infty \leqslant 3a\,(\log n)^2 \\[2mm] |\Lambda(\omega_0)| = \mathbb{E}(|\Lambda|) + \mathrm{O}\big((\log n)^2\big) = \dfrac{1}{2}\,(\log n)^3 + \mathrm{O}\big((\log n)^2\big), \end{cases}$$

since:

$$\mathbb{E}(|\Lambda|) = \sum_{\sqrt{n}\leqslant k\leqslant n} \mathbb{E}(\xi_k) = \sum_{\sqrt{n}\leqslant k\leqslant n} \frac{C_n}{k} = C_n\left(\log n - \log\sqrt{n} + \mathrm{O}(1)\right)$$

$$= \frac{1}{2}\,(\log n)^3 + \mathrm{O}\big((\log n)^2\big).$$

Now we choose $n = \left[\exp\big((2N)^{1/3}\big)\right]$ ($\geqslant 3^8$ if $N \geqslant 340$); thus the preceding inequalities become:

$$\begin{cases} \left\| \sum_{k \in \Lambda(\omega_0)} \sin kx \right\|_{\infty} \ll N^{2/3} \\ |\Lambda(\omega_0)| = N + O(N^{2/3}). \end{cases}$$

It only remains to correct $\Lambda(\omega_0)$, by adding or removing $O(N^{2/3})$ elements to obtain a set whose cardinality is exactly $N$. This does not modify the first estimation (or at least only modifies the implicit constant), so the proof of Theorem III.1 is now complete.                      $\square$

*Proof of Lemma III.3*   We center the variables, by writing:

$$\sum_{\sqrt{n} \leqslant k \leqslant n} \xi_k \sin kx = \sum_{\sqrt{n} \leqslant k \leqslant n} (\xi_k - \mathbb{E}\xi_k) \sin kx + C_n \sum_{\sqrt{n} \leqslant k \leqslant n} \frac{\sin kx}{k}.$$

Then the inequality $\left| \sum_{k=1}^{m} \frac{\sin kx}{k} \right| \leqslant 2$ leads to:

$$\left\| \sum_{\sqrt{n} \leqslant k \leqslant n} \xi_k \sin kx \right\|_{\infty} \leqslant \left\| \sum_{\sqrt{n} \leqslant k \leqslant n} (\xi_k - \mathbb{E}\xi_k) \sin kx \right\|_{\infty} + 4\, C_n.$$

Let $X_k = \xi_k - \mathbb{E}\xi_k$, $X_k^s$ be the symmetrization of $X_k$, and let $(\varepsilon_k)_{k \geqslant 1}$ be a Rademacher sequence; then:

$$\mathbb{E}\left( \left\| \sum_k X_k \sin kx \right\|_{\infty} \right) \leqslant \mathbb{E}\left( \left\| \sum_k X_k^s \sin kx \right\|_{\infty} \right)$$

$$= \mathbb{E}_{\omega,\omega'}\left( \left\| \sum_k \varepsilon_k(\omega') X_k^s(\omega) \sin kx \right\|_{\infty} \right)$$

by the symmetry of the $X_k^s$

$$\ll \sqrt{\log n}\, \mathbb{E}\left[ \left( \sum_k |X_k^s|^2 \right)^{1/2} \right]$$

by the majoration theorem for Bernoulli processes and by Bernstein's inequality; see the proof of Theorem V.11 of Chapter 6 (Volume 1). Hence:

$$\mathbb{E}\left( \left\| \sum_k X_k \sin kx \right\|_{\infty} \right) \ll \sqrt{\log n} \left[ \mathbb{E}\left( \sum_k |X_k^s|^2 \right) \right]^{1/2}$$

$$= \sqrt{\log n}\, \sqrt{2 \sum_k V(X_k)}$$

$$= \sqrt{\log n} \, \sqrt{2} \sqrt{\sum_k V(\xi_k)}$$

$$\ll \sqrt{\log n} \sqrt{\sum_k \mathbb{E}(\xi_k)}$$

$$\text{because } V(\xi_k) \leqslant \mathbb{E}(\xi_k^2) = \mathbb{E}(\xi_k)$$

$$\ll \sqrt{\log n} \sqrt{\sum_{k \leqslant n} \frac{C_n}{k}}$$

$$\ll \sqrt{C_n} \, \log n;$$

thus the lemma is proved. $\qquad\qquad\qquad\qquad\qquad\qquad\qquad\qquad\square$

## IV Minoration of the *K*-Convexity Constant

In Chapter 4 of this volume (Theorem V.3), the majoration $K(X) \leqslant C \log(1+d_X)$ was seen to always hold, for the *K*-convexity constant of a space $X$ of dimension $n$ (where $d_X$ is the distance of $X$ to $\ell_2^n$)). Thus $K(X) \leqslant C' \log n$ for $n \geqslant 2$.

A question arises naturally: is this the best possible? In the case of subspaces of $L^1$, we already saw (Theorem V.5) that we can improve the estimation: $\log n$ is replaced by $\sqrt{\log n}$. We also pointed out that Pisier ([1980 a] and [1981 c]) showed that the same can be done when $X$ possesses a 1-unconditional basis.

The aim of this section is to show that in fact the estimation $K(X) \leqslant C \log n$ is optimal. More precisely:

**Theorem IV.1** (Bourgain)   *There exists a constant $\delta > 0$ such that, for any integer $n \geqslant 1$, it is possible to find a normed space X, of dimension n, for which $K(X) \geqslant \delta \log n$.*

*Proof*   We seek a space $X$ of dimension $n$, an integer $N$ and a family $(x_A)_{A \subseteq \{1,\ldots,N\}}$ of vectors of $X$ such that, with $G = \{-1,1\}^N$ the Cantor group of order $N$:

1) $$\left( \int_G \left\| \sum_A w_A(t) \, x_A \right\|^2 dm(t) \right)^{1/2} \sim 1,$$

2) $$\left( \int_G \left\| \sum_{j=1}^N \varepsilon_j(t) \, x_j \right\|^2 dm(t) \right)^{1/2} \sim \delta \log n,$$

where the $w_A$ are the Walsh functions and the $\varepsilon_j = w_{\{j\}}$ the Rademacher functions (we denote $x_{\{j\}} = x_j$). The idea is to find a space in which $\left\| \sum_A w_A(t) x_A \right\|$ does not depend on $t \in G$, for a good choice of $x_A$. For this, we seek a space of the form $\mathcal{C}_\Lambda(G)$, where $\Lambda$ is a subset of the dual $\Gamma = \widehat{G} = \{ w_A \, ; \, A \subseteq \{1, \dots, N\} \}$ of $G$; indeed, if $x_A = c_A \, w_A$, with $c_A \in \mathbb{C}$ for $A \in \Lambda$, and $x_A = 0$ for $A \notin \Lambda$, then, given the translation invariance of the norm $\| \cdot \|_\infty$, we would have:

$$\left\| \sum_{A \in \Lambda} w_A(t) \, x_A \right\|_\infty = \left\| \sum_{A \in \Lambda} c_A \, w_A \right\|_\infty$$

and

$$\left\| \sum_{\{j\} \in \Lambda} \varepsilon_j(t) \, x_j \right\|_\infty = \left\| \sum_{\{j\} \in \Lambda} c_j \, \varepsilon_j \right\|_\infty .$$

Hence we would be brought to seek a set $\Lambda$ such that:

1') $\left\| \sum_{A \in \Lambda} c_A \, w_A \right\|_\infty \sim 1,$

2') $\left\| \sum_{\{j\} \in \Lambda} c_j \, \varepsilon_j \right\|_\infty \sim \delta \, \log |\Lambda|.$

Let us begin the construction. We take an integer $n$, which can be assumed $\geqslant 81\,938$; let

$$N = \left[ \frac{(\log n)^2}{16} \right]$$

be the integer part of $(\log n)^2 / 16$, so that we have:

$$e^{4\sqrt{N}} \leqslant n \leqslant e^{8\sqrt{N}}.$$

Let $G = \{-1, 1\}^N$ be the Cantor group of order $N$. We are going to construct $\Lambda \subseteq \widehat{G}$ such that $|\Lambda| \leqslant e^{4\sqrt{N}}$ and $K(\mathcal{C}_\Lambda) \geqslant \alpha \sqrt{N}$ for a numerical constant $\alpha > 0$. Then, with $X$ a space of dimension $n$ containing $\mathcal{C}_\Lambda$, we will have:

$$K(X) \geqslant K(\mathcal{C}_\Lambda) \geqslant \alpha \sqrt{N} \geqslant \frac{\alpha}{8} \log n,$$

which is the desired result.

We start with $Y = \mathcal{C}(G)$, and introduce the imaginary Riesz product:

$$\mathcal{R} = \prod_{j=1}^{N} \left( 1 + i \, \frac{\varepsilon_j}{\sqrt{N}} \right),$$

for which:

$$|\mathcal{R}(t)| = \prod_{j=1}^{N} \left(1 + \frac{\varepsilon_j^2(t)}{N}\right)^{1/2} = \left(1 + \frac{1}{N}\right)^{N/2} \leqslant \sqrt{e} \leqslant 2,$$

and hence $\|\mathcal{R}\|_\infty \leqslant 2$. An expansion of this product leads to:

$$\mathcal{R} = \sum_{A \subseteq \{1,\ldots,N\}} i^{|A|} N^{-|A|/2} w_A,$$

so that:

$$\operatorname{Im} \mathcal{R} = \sum_{A \subseteq \{1,\ldots,N\}} c_A w_A,$$

with $c_\emptyset = 0$ and:

$$(*) \qquad \begin{cases} c_j = c_{\{j\}} = 1/\sqrt{N} & \text{for } 1 \leqslant j \leqslant N \\ |c_A| \leqslant N^{-|A|/2} & \text{for any } A. \end{cases}$$

Note that $c_A$ depends only on $|A|$. Then $\left\|\sum_A c_A w_A\right\|_\infty \leqslant \|\mathcal{R}\|_\infty \leqslant 2$, whereas $\left\|\sum_{j=1}^{N} c_j \varepsilon_j\right\|_\infty = \frac{1}{\sqrt{N}} \left\|\sum_{j=1}^{N} \varepsilon_j\right\|_\infty = \sqrt{N}$.

However, this $\sqrt{N}$ is of the order of $\sqrt{\log(\dim Y)}$, and not of the order of $\log(\dim Y)$. The idea is thus to rarefy the spectrum of $\operatorname{Im} \mathcal{R}$ to obtain a set $\Lambda$ (of cardinality $\leqslant e^{4\sqrt{N}}$) containing all the singletons $\{j\}$, and for which $\left\|\sum_{A \in \Lambda} c_A w_A\right\|_\infty \leqslant \alpha$ (i.e. we do not modify $\|\operatorname{Im} \mathcal{R}\|_\infty$ too much). Then:

$$K(\mathcal{C}_\Lambda) \geqslant \frac{\left\|\sum_{j=1}^{N} c_j \varepsilon_j\right\|_\infty}{\left\|\sum_{A \in \Lambda} c_A w_A\right\|_\infty} \geqslant \alpha \sqrt{N},$$

as stated above.

We achieve this rarefaction using selectors $\xi_A$, $A \subseteq \{1,\ldots,N\}$, with expectation $\sigma_A$:

$$\mathbb{P}(\xi_A = 1) = \sigma_A, \qquad \mathbb{P}(\xi_A = 0) = 1 - \sigma_A.$$

Let:

$$\Lambda_\omega = \{A \subseteq \{1,\ldots,N\} \,;\, \xi_A(\omega) = 1\}$$

be the corresponding random set. We consider the random polynomial:

$$\varphi_\omega = \sum_{A \subseteq \{1,\ldots,N\}} \xi_A(\omega) \frac{c_A}{\sigma_A} w_A \in \mathcal{C}_{\Lambda_\omega}.$$

To avoid changing $\| \operatorname{Im} \mathcal{R} \|_\infty$ too much, as desired, we arrange to modify nothing for small values of $|A|$. Moreover, as $c_A$ depends only on $|A|$, it is judicious to chose $\sigma_A$ depending only on $|A|$. To be precise, we take:

$$\mathbb{P}(\xi_A = 1) = \begin{cases} 1 & \text{for} \quad |A| < m \\ \delta_k = \dfrac{2^k}{k!} & \text{for} \quad |A| = k \geqslant m, \end{cases}$$

with $m \geqslant 2$ to be determined later. Note that the sets $\Lambda_\omega$ contain all the singletons $\{j\}$, since $\sigma_{\{j\}} = 1$.

We then proceed in two steps.

*Step 1. Majoration of* $\mathbb{E}_\omega(|\Lambda_\omega|)$.

This is based on the following lemma:

**Lemma IV.2** *For any $a > 0$ and any $N \geqslant 1$, then:*

$$\sum_{k=0}^{N} \binom{N}{k} \frac{a^k}{k!} \leqslant \exp\left(2\sqrt{Na}\right).$$

*Proof* Note that:

$$\binom{N}{k} \frac{a^k}{k!} = \frac{N(N-1)\cdots(N-k+1)}{(k!)^2} a^k \leqslant \frac{N^k}{(k!)^2} a^k = \left( \frac{(\sqrt{Na})^k}{k!} \right)^2;$$

hence:

$$\sum_{k=0}^{N} \binom{N}{k} \frac{a^k}{k!} \leqslant \sum_{k=0}^{+\infty} \left( \frac{(\sqrt{Na})^k}{k!} \right)^2 \leqslant \left( \sum_{k=0}^{+\infty} \frac{(\sqrt{Na})^k}{k!} \right)^2$$
$$= \left( \exp\sqrt{Na} \right)^2 = \exp(2\sqrt{Na}). \qquad \square$$

Therefore:

$$\mathbb{E}(|\Lambda_\omega|) = \sum_{A \subseteq \{1,\dots,N\}} \mathbb{E}(\xi_A) = \sum_{|A| < m} 1 + \sum_{m \leqslant k \leqslant N} \sum_{|A| = k} \frac{2^k}{k!}$$
$$\leqslant \sum_{k < m} \binom{N}{k} + \sum_{k=0}^{N} \binom{N}{k} \frac{2^k}{k!} \leqslant N^m + \exp\left(2\sqrt{2N}\right),$$

after the use of Lemma IV.2 to bound the second sum, and of Proposition VI.4 (3) of Chapter 3 (this volume) to bound the first. We now adjust $m$, by taking:

$$m = \left\lceil \log_2 N \right\rceil.$$

Note that $m \geqslant 3$, since $n \geqslant 81\,938$ implies $N = \left[ (\log n)^2 / 16 \right] \geqslant 8$. We obtain $N^m \leqslant \mathrm{e}^{(\log N)^2} \leqslant \mathrm{e}^{3\sqrt{N}}$, so that:

$$\mathbb{E}(|\Lambda_\omega|) \leqslant 2\,\mathrm{e}^{3\sqrt{N}}.$$

*Step 2. Majoration in mean of* $\|\varphi_\omega\|_\infty$.

We have:

$$\varphi_\omega - \mathrm{Im}\,\mathcal{R} = \varphi_\omega - \mathbb{E}(\varphi_\omega) = \sum_{A \subseteq \{1,\dots,N\}} \frac{c_A}{\sigma_A}\,(\xi_A - \sigma_A)\,w_A;$$

hence, by $(*)$:

$$\|\varphi_\omega - \mathrm{Im}\,\mathcal{R}\|_\infty \leqslant \sum_{m \leqslant k \leqslant N} \left\| \sum_{|A|=k} \frac{c_A}{\sigma_A}\,(\xi_A - \sigma_A)\,w_A \right\|_\infty$$

$$\leqslant \sum_{m \leqslant k \leqslant N} \frac{N^{-k/2}}{\delta_k} \left\| \sum_{|A|=k} (\xi_A - \sigma_A)\,w_A \right\|_\infty.$$

It is convenient to isolate a lemma.

**Lemma IV.3** *For any* $k \geqslant 1$, *we have*:

$$\mathbb{E} \left\| \sum_{|A|=k} (\xi_A - \sigma_A)\,w_A \right\|_\infty \ll \sqrt{N} \sqrt{\delta_k \binom{N}{k}}.$$

*Proof* The proof is very similar to that of Lemma III.3. As the variables $(\xi_A - \sigma_A)$ are independent and centered, we inject Rademacher variables $r_A$ according to the contraction principle, and then use the majoration principle of Rademacher processes, with the understanding that the norm $\| \, . \, \|_\infty$ refers to the least upper bound over the $2^N$ elements of $G$. This gives:

$$\mathbb{E} \left\| \sum_{|A|=k} (\xi_A - \sigma_A)\,w_A \right\|_\infty$$

$$\leqslant \iint \left\| \sum_{|A|=k} (\xi_A(\omega) - \sigma_A)\,r_A(\omega')\,w_A \right\|_\infty \mathrm{d}\mathbb{P}(\omega)\,\mathrm{d}\mathbb{P}(\omega')$$

$$\leqslant 2\,C \sqrt{\log 2^N} \int_\Omega \left( \sum_{|A|=k} (\xi_A(\omega) - \sigma_A)^2 \right)^{1/2} \mathrm{d}\mathbb{P}(\omega)$$

$$\ll \sqrt{N} \left( \int_\Omega \sum_{|A|=k} (\xi_A(\omega) - \sigma_A)^2\,\mathrm{d}\mathbb{P}(\omega) \right)^{1/2}.$$

Since:

$$\int_\Omega \sum_{|A|=k} (\xi_A(\omega) - \sigma_A)^2 \, d\mathbb{P}(\omega) = \sum_{|A|=k} \int_\Omega (\xi_A(\omega) - \sigma_A)^2 \, d\mathbb{P}(\omega)$$

$$= \sum_{|A|=k} V(\xi_A) = \sum_{|A|=k} (\delta_k - \delta_k^2)$$

$$\leqslant \sum_{|A|=k} \delta_k = \binom{N}{k} \delta_k,$$

this proves the lemma.                                                                 □

It ensues that:

$$\mathbb{E}\|\varphi_\omega - \operatorname{Im}\mathcal{R}\|_\infty \leqslant C \sum_{m \leqslant k \leqslant N} \frac{N^{-k/2}}{\delta_k} \sqrt{N} \sqrt{\delta_k} \sqrt{\binom{N}{k}}$$

$$\ll \sqrt{N} \sum_{m \leqslant k \leqslant N} \sqrt{\delta_k^{-1} N^{-k} \binom{N}{k}}.$$

Then note that:

$$\delta_k^{-1} N^{-k} \binom{N}{k} \leqslant \frac{k!}{2^k} N^{-k} \frac{N^k}{k!} = 2^{-k};$$

hence:

$$\mathbb{E}\|\varphi_\omega - \operatorname{Im}\mathcal{R}\|_\infty \ll \sqrt{N} \sum_{k \geqslant m} 2^{-k/2} \ll \sqrt{N} \, 2^{-m/2} = \sqrt{N} \, 2^{-m} \ll 1,$$

because of the value of $m = [\log_2 N]$, whose choice is thus justified. Therefore:

$$\mathbb{E}\|\varphi_\omega\|_\infty \leqslant \|\operatorname{Im}\mathcal{R}\|_\infty + \mathbb{E}\|\varphi_\omega - \operatorname{Im}\mathcal{R}\|_\infty \leqslant 2 + O(1) = C.$$

By Markov's inequality, Steps 1 and 2 thus give a probability of at least $1/2$ to have:

$$|\Lambda_\omega| \leqslant 8 \, e^{3\sqrt{N}} \quad \text{and} \quad \|\varphi_\omega\|_\infty \leqslant 4 \, C.$$

It only remains to select such an $\omega = \omega_0$ and take $\Lambda = \Lambda_{\omega_0}$ to complete the proof of the theorem, as indicated at the beginning; indeed note that $n \geqslant 81\,938$ implies $N \geqslant 8$, and hence $8 \, e^{3\sqrt{N}} \leqslant e^{4\sqrt{N}}$.                                 □

# V  Comments

Theorem II.3 is due to Pisier ([1983 a]; see [1983 b] and [1983 c]). The proof uses Gaussian processes and Theorem II.13 (and hence Rider's theorem). The proof given here is due to Bourgain [1985 a] (see also Bourgain [1985 b]).

The proof of Theorem II.14 presented here can be found in Bourgain [1985 b].

Theorem III.1 is due to Bourgain [1983 c] (see also KAHANE 2, Chapter 6, §6) and provides an answer to a question of Virot [1981], mentioned in Subsection III.1. This remains the best majoration known to this day. The only subsequent progress has been the minoration of Konyagin [1997], already mentioned in the text.

Even given the somewhat arbitrary nature of the selectors used, it is not possible to do better with this method; indeed, Bourgain [1983 c] showed that:

*If $\Lambda$ is an arbitrary finite subset of $\mathbb{Z}^+$, and $\xi_k$, $k \in \Lambda$, are arbitrary selectors such that $1 - \mathbb{E}(\xi_k) \geqslant c > 0$ (to prevent them from being completely deterministic), then:*

$$\left( \mathbb{E} \left\| \sum_{k \in \Lambda} \xi_k \sin kx \right\|_\infty^2 \right)^{1/2} \geqslant \delta(c) \left( \mathbb{E}_\omega(|\Lambda(\omega)|) \right)^{2/3},$$

*where $\delta(c) > 0$ depends only on c.*

In other words, even with a random choice within a particular set $\Lambda$, and even with optimized expectations of the selectors, this method will never achieve better than the exponent $2/3$.

In fact we have (see POLYA–SZEGÖ)

$$\sup_{x,n} \left| \sum_{k=1}^n \frac{\sin kx}{k} \right| = \int_0^\pi \frac{\sin t}{t} \, dt = 1,85 \ldots$$

There is a related inequality, much more difficult to prove:

*For any odd integer d, there exists a constant $C_d$ such that:*

$$\left| \sum_{k=1}^n \frac{\sin k^d x}{k} \right| \leqslant C_d$$

*for any $x \in \mathbb{R}$ and $n \geqslant 1$*

(Arkhipov and Oskolkov [1987]). This enables us to show that in Theorem III.1, the set $\Lambda$ can be chosen among the $d$-th powers of integers (see Exercise VI.2).

J. Bourgain also became interested in a closely related problem, the *problem of cosines*: if

$$f(x) + i\widetilde{f}(x) = \sum_{k=1}^N e^{in_k x},$$

we wish to estimate:

$$M = - \min_{x \in [0, 2\pi]} f(x).$$

Bourgain [1986] showed that, if $a_1 = \cdots = a_N = 1$:

$$M \geqslant 2^{(\log N)^{\varepsilon}}$$

for a suitable $\varepsilon > 0$; to this day, this is the best estimation known. Papadopoulos [1998] related this problem to that of the sum of sines: *when* $|a_k| \geqslant 1$ *for* $1 \leqslant k \leqslant N$, *then* $M \|\widetilde{f}\|_{\infty} \geqslant c \dfrac{N}{\log N}$.

Theorem IV.1 was proved in Bourgain [1984 b]. Lemma IV.2 appears, for other reasons, in PAJOR.

# VI Exercises

**Exercise VI.1** (Rodríguez-Piazza [1991])   The aim is to prove the following theorem:

*There exists a constant $K_0 > 0$ such that, for every finite subset $A$ of $\mathbb{Z}$ (with $A \neq \{0\}$), we have:*

$$\frac{1}{K_0} q(A) \leqslant \|i_A\|^2 \leqslant K_0 \, q(A),$$

*where*

$$q(A) = \max \big\{ |B| \; ; \; B \subseteq A \text{ and } B \text{ quasi-independent} \big\},$$

*and where $i_A \colon L_A^2(\mathbb{T}) \to C_A^{as}(\mathbb{T})$ is the identity.*

Hence $\|i_A\|$ is the smallest constant $C > 0$ for which:

$$[\![f]\!]_R \leqslant C \|f\|_2 \quad \forall f \in \mathcal{P}_A = L_A^2.$$

We will assume that $|A| \geqslant 2$.

*Part 1.*

1) Let $B \subseteq A$ be a quasi-independent set such that $|B| = q(A)$. Show that $|B| \leqslant 8 [\![f]\!]_R$ for $f(t) = \sum_{k \in A} e^{ikt}$, and deduce that $\dfrac{1}{64} q(A) \leqslant \|i_A\|^2$.

2) Let $P$ be a trigonometric polynomial such that:

    (i) $\|P\|_1 = 1$;

    (ii) $\dfrac{1}{4e} \leqslant \widehat{P}(k) \leqslant 1$, $\forall k \in A$;

    (iii) $\|P\|_{\infty} \leqslant 2^{5eq(A)}$

(such a polynomial will be constructed in the third part).

a) Show that, for every $f \in \mathcal{P}_A(\mathbb{T})$, then:

$$[\![f]\!]_R \leqslant 8\,e \left[\!\!\left[ \sum_{k \in A} \widehat{P}(k)\widehat{f}(k)e_k \right]\!\!\right]_R \qquad (= 8\,e\,[\![P * f]\!]_R).$$

b) Let $f_\omega(t) = \sum_{k \in A} \varepsilon_k(\omega)\widehat{f}(k)\,e^{ikt}$, where $(\varepsilon_k)_{k \in A}$ is a Rademacher sequence. Show that, for $p \geqslant 2$, $\mathbb{E}_\omega \|f_\omega\|_p \leqslant \sqrt{p}\,\|f\|_2$.

c) Show that $\|P\|_{p^*} \leqslant 2^{5e\,q(A)/p}$, where $p^*$ is the conjugate exponent of $p$ (use $\|P\|_{p^*} \leqslant \|P\|_1^{1/p^*}\|P\|_\infty^{1/p}$).

d) With a proper choice of $p$, deduce that:

$$[\![f * P]\!]_R \leqslant \|P\|_{p^*}\mathbb{E}_\omega\|f\|_p \leqslant 16\sqrt{5}\,e^{3/2}\sqrt{q(A)}\,\|f\|_2.$$

e) Conclude.

**Part 2.**

Let $\{k_1, \ldots, k_n\}$ be a set of (distinct) integers, with $n \geqslant 2$, such that:

$$\int_{\mathbb{T}} \prod_{j=1}^n \left(1 + \cos(k_j t)\right) dt \leqslant \left(1 + \frac{1}{4e}\right)^n.$$

Let $\xi_1, \ldots, \xi_n$ be selectors such that:

$$\mathbb{P}(\xi_j = 1) = \frac{1}{2e} \quad \text{and} \quad \mathbb{P}(\xi_j = 0) = 1 - \frac{1}{2e}, \quad 1 \leqslant j \leqslant n.$$

Let $L = \left[\dfrac{n}{4e}\right] + 1$, and:

$$F_\omega(t) = \sum_{m \geqslant L} \sum_{\substack{S \subseteq \{1,\ldots,n\} \\ |S|=m}} \prod_{j \in S} \xi_j(\omega)\left(e^{ik_j t} + e^{-ik_j t}\right).$$

1) By noting that:

$$\int_{\mathbb{T}} \sum_{m=1}^n \sum_{\substack{S \subseteq \{1,\ldots,n\} \\ |S|=m}} \prod_{j \in S} \left[\frac{e^{ik_j t} + e^{-ik_j t}}{2}\right] dt$$

$$\leqslant \int_{\mathbb{T}} \prod_{j=1}^n \left(1 + \frac{e^{ik_j t} + e^{-ik_j t}}{2}\right) dt,$$

show that:

$$0 \leqslant \int_\Omega \int_{\mathbb{T}} F_\omega(t)\,dt\,d\mathbb{P}(\omega) < 1.$$

2) Deduce that:

$$\int_{\Omega} \left( \sum_{j=1}^{n} \xi_j(\omega) - \int_{\mathbb{T}} F_{\omega}(t) \, dt \right) d\mathbb{P}(\omega) > \frac{n}{2e} - 1.$$

3) Deduce that there exists $\omega_0 \in \Omega$ such that, with

$$D = \{ j \in \{1, \ldots, n\} \, ; \, \xi_j(\omega_0) = 1 \},$$

and with $N = \int_{\mathbb{T}} F_{\omega_0}(t) \, dt$, then: $|D| - N \geqslant (n/2e) - 1$.

4) Let $D_1 \subseteq D$ be a relation of maximal length among those of length $\leqslant L-1$. Show that every relation in $D \smallsetminus D_1$ forms with $D_1$ a relation of length $\geqslant L$. Deduce that there are at most $N$ such relations.

5) Show that the removal of a point from each of these relations in $D \smallsetminus D_1$ provides a quasi-independent set $B_1 \subseteq D \smallsetminus D_1$, such that $|B_1| \geqslant (n/2e) - L$.

6) Show that there exists a quasi-independent set $B \subseteq \{k_1, \ldots, k_n\}$ such that $|B| \geqslant n/5e$ (distinguish the cases $n \geqslant 20\,e$ and $n < 20\,e$).

*Part 3.*

1) Let $Q$ be a real trigonometric polynomial such that $\widehat{Q} \geqslant 0$ and, for $k \in \mathbb{Z}$, let:

$$R(t) = \big( 1 + \cos(kt) \big) \, Q(t).$$

Show that $\widehat{R}(k) \geqslant \widehat{R}(0)/2$.

2) Let $A$ be a finite subset of $\mathbb{Z}$ (with $|A| \geqslant 2$). First set $P_0 = \mathbb{I}$, and then construct trigonometric polynomials $P_1, \ldots, P_n$ of the form:

$$P_l(t) = \prod_{j=1}^{l} \big( 1 + \cos(k_j t) \big) \qquad 1 \leqslant l \leqslant n,$$

with $k_1, \ldots, k_l \in A$. Proceed as follows: if $P_1, \ldots, P_l$ have been constructed, we stop (and take $n = l$) if $\widehat{P}(k) \geqslant (1/4e)\widehat{P_l}(0)$ for every $k \in A$; and if there exists $k \in A$ such that $\widehat{P}(k) < (1/4e)\widehat{P_l}(0)$, denote by $k_{l+1}$ any one of such values, and set:

$$P_{l+1}(t) = P_l(t)\big( 1 + \cos(k_{l+1}t) \big).$$

a) Show, using 1), that $k_{l+1} \notin \{k_1, \ldots, k_l\}$, and hence that the procedure stops at $P_n$, with $n \leqslant |A|$.

b) Show that, for $l < n$: $\|P_{l+1}\|_1 \leqslant \left( 1 + \dfrac{1}{4e} \right) \|P_l\|_1$ (use $\|P_l\|_1 = \widehat{P_l}(0)$).

c) Show that $P = P_n / \|P_n\|_1$ satisfies the conditions (i), (ii), (iii) stated in 2) of Part 1.

**Exercise VI.2**

1) Let $d$ be an odd integer. With the Arkhipov–Oskolkov inequality:

$$\left|\sum_{k=1}^{n} \frac{\sin k^d x}{k}\right| \leqslant C_d,$$

stated in the Comments, use the method of selectors to show that, for any integer $N \geqslant 1$, there exists $\Lambda \subseteq (\mathbb{N}^*)^d = \{1^d, 2^d, 3^d, \ldots\}$, of cardinality $|\Lambda| = N$, satisfying

$$\left\|\sum_{k\in\Lambda} \sin kx\right\|_\infty \leqslant C_d' N^{2/3},$$

where $C_d' > 0$ depends only on $d$.

2) Let $d$ and $N$ be integers $\geqslant 1$. Show that $\left\|\sum_{k=1}^{N} \sin k^d x\right\|_\infty \gg N$ (take $x = \pi/(2N^d)$).

3) Show the same result for $\left\|\sum_{k=1}^{N} \sin a^k x\right\|_\infty$, with $a$ an integer $\geqslant 2$.

**Exercise VI.3**  Define:

$$\alpha_N = \sup \sup_{n\leqslant N}\{|a_1 + \cdots + a_n|\},$$

where the first supremum is taken over all the trigonometric polynomials

$$P(t) = a_1 e^{i\lambda_1 t} + \cdots + a_N e^{i\lambda_N t}$$

of "length" $N$ ($\lambda_1 < \cdots < \lambda_N$, $\lambda_k \in \mathbb{Z}$), with $\|P\|_\infty \leqslant 1$.

1) Show that $\alpha_N \leqslant \sqrt{N}$.
2) Using Theorem III.1, show that $\alpha_N \geqslant c N^{1/3}$, for some numerical constant $c > 0$ (*Problem of H. Bohr*).

# 6

## The Pisier Space of Almost Surely Continuous Functions. Applications

### I Introduction

The exponentials $e_n$, $n \in \mathbb{Z}$, are very far from being an unconditional sequence in $\mathcal{C}(\mathbb{T})$, as shown spectacularly by the Kahane–Katznelson–de Leeuw theorem already mentioned in Chapter 5 of Volume 1 (see Exercise VIII.16 in that chapter): for every sequence $(a_n)_{n \in \mathbb{Z}} \in \ell_2(\mathbb{Z})$, there is a continuous function $f \in \mathcal{C}(\mathbb{T})$ such that $|\widehat{f}(n)| \geqslant |a_n|$ for any $n$. Moreover, it is easy to see that if $\sum_{n \in \mathbb{Z}} \pm a_n e_n$ represents a continuous function for *every* choice of signs, then $\sum_{n \in \mathbb{Z}} |a_n| < +\infty$. P. Lefèvre [1999 b] even showed that a dense $G_\delta$-set of choices of signs was sufficient to reach the same conclusion. A natural question then arises: what happens with the requirement that $\sum_{n \in \mathbb{Z}} \pm a_n e_n \in \mathcal{C}(\mathbb{T})$ for *almost all* choices of signs? Necessarily $\sum_{n \in \mathbb{Z}} |a_n|^2 < +\infty$ (see Chapter 1 of Volume 1, Corollary III.6), and the Kahane–Katznelson–De Leeuw theorem indicates that little more can be said about the size of the coefficients $a_n$ (nonetheless $(a_n)_{n \in \mathbb{Z}} \in \ell_{2,1}$ is to be proved).

The space $\mathcal{C}^{as}$ of almost surely continuous Fourier series thus appears as the space of sequences $(a_n)_{n \in \mathbb{Z}} \in \ell_2$ such that $\sum_{n \in \mathbb{Z}} \pm a_n e_n \in \mathcal{C}(\mathbb{T})$ for almost all choices of signs (in a sense to be made precise!); or, if preferred, such that $\sum_{n \in \mathbb{Z}} \varepsilon_n(\omega) a_n e_n \in \mathcal{C}(\mathbb{T})$ for almost all $\omega$, where $(\varepsilon_n)_{n \in \mathbb{Z}}$ is an independent sequence of Bernoulli variables.

However, several difficulties appear: first, a deeper knowledge of Banach space valued variables is necessary; next, it is technically easier to replace the variables $\varepsilon_n$ by an (independent) complex standard Gaussian sequence $(Z_n)_{n \in \mathbb{Z}}$, for which additional tools are available: Slepian's lemma, and the theorems of Dudley and Fernique. Thus, even though less natural, the definition of the space $\mathcal{C}^{as}$ is given with Gaussian variables. A theorem of Marcus and Pisier shows *a posteriori* that these points of view are equivalent.

But first some complements on Banach-valued variables must be introduced.

# II Complements on Banach-Valued Variables

## II.1 The Itô–Nisio Theorem

Throughout this section, $(X_n)_{n \geqslant 1}$ denotes a sequence of independent random variables with values in a Banach space $E$, and

$$S_n = X_1 + \cdots + X_n$$

denotes its partial sums. In Chapter 4 of Volume 1, we saw that the convergence of $(S_n)_{n \geqslant 1}$ in probability, or even in distribution, implies its almost sure convergence. We give another statement of this sort, where the almost sure convergence of $S_n$ is automatic; indeed, for applications to random Fourier series, it is useful to know, for $S_n$, that the almost sure convergence for the weak topology ensures the almost sure convergence in norm. This is the object of the following two theorems.

The closed unit ball of $E^*$ is denoted $B_{E^*}$. A subset $\Delta$ of $B_{E^*}$ is said to be *norming* if

$$\|x\| = \sup_{\varphi \in \Delta} |\varphi(x)|$$

for every $x \in E$.

**Theorem II.1** (The Itô–Nisio Theorem)   *Let $E$ be a separable Banach space and $(X_n)_{n \geqslant 1}$ a sequence of independent* symmetric *variables with values in $E$, and $S_n = X_1 + \cdots + X_n$. Let $\Delta$ be a norming subset of $B_{E^*}$. Suppose that there exists a mapping $S \colon \Omega \to E$, such that, for every $\varphi \in \Delta$:*

$$\langle \varphi, S_n \rangle \xrightarrow[n \to +\infty]{a.s.} \langle \varphi, S \rangle.$$

*Then $S \in L^0(E)$ and $S_n \xrightarrow[n \to +\infty]{a.s.} S$, for the norm of $E$.*

We must point out that, in the hypothesis, a priori $S$ is not assumed measurable. Moreover, the convergence almost sure of $\langle \varphi, S_n \rangle$ takes place for each $\varphi$: there exists a negligible set $N_\varphi$ outside of which the convergence occurs.

*Proof*   Since $E$ is separable, the unit ball $B_{E^*}$ of its dual is $w^*$-compact and metrizable, hence separable. Therefore it is the same for its subset $\Delta$, which can thus be assumed countable. Then, on $E$, we consider the topology $\tau$ of pointwise convergence on the elements of $\Delta$. The hypothesis can be reformulated as:

$(*)$ $\qquad\qquad$ *almost surely:* $\qquad$ $S_n \xrightarrow[n \to +\infty]{\tau} S$;

here the "almost surely" no longer presents any ambiguity since the set $\Delta$ is countable: the convergence takes place outside of the negligible set $\bigcup_{\varphi \in \Delta} N_\varphi$. The proof is divided into five steps.

*Step 1. It suffices to present a proof when $X_n = \varepsilon_n x_n$, with $x_n \in E$ and $(\varepsilon_n)_{n \geqslant 1}$ an independent Bernoulli sequence.*

Indeed, suppose that this proof has been done. Let $(X_n)_{n \geqslant 1}$ be an arbitrary sequence of independent symmetric variables. For each fixed $\omega'$, the two sequences $(\varepsilon_n(\omega')X_n)_{n \geqslant 1}$ and $(X_n)_{n \geqslant 1}$ have the same distribution; hence $S_n^{\omega'} = \sum_{k=1}^n \varepsilon_k(\omega')X_k$ is almost surely $\tau$-convergent. Then, by the Fubinization principle, for almost all $\omega$: $S_n^{\omega'}(\omega) = S_n(\omega, \omega')$ $\tau$-converges for almost all $\omega'$, i.e. there exists a negligible set $N_0$ such that, for every $\omega \notin N_0$, there are negligible sets $N_\omega$ for which $S_n(\omega, \omega')$ converges for the topology $\tau$, for every $\omega' \notin N_\omega$.

Thanks to the statement assumed proved, there is convergence in norm. A second Fubinization shows that, for almost all $\omega'$, $S_n(\omega, \omega')$ converges in norm for almost all $\omega$. In particular, there exists an $\omega_0'$ such that $S_n(\omega, \omega_0')$ converges $\omega$-almost surely; by symmetry, $S_n(\omega)$ converges $\omega$-almost surely (in norm).

*Step 2. $S \in L^0(E)$.*

By the hypothesis (∗), $S$ is $\mathcal{B}_\tau$-measurable, with $\mathcal{B}_\tau$ the $\sigma$-algebra of Borel sets for the topology $\tau$. However, as $\Delta$ is norming, we have $B_E = \bigcap_{\varphi \in \Delta} \{x; \ |\varphi(x)| \leqslant 1\}$, thus $B_E$ is $\tau$-closed. All the closed balls are hence $\tau$-closed; therefore $\mathcal{B}_\tau$ is the Borel $\sigma$-algebra of $E$ (for the norm).

In fact, a general result states that every Borel $\sigma$-algebra of a Hausdorff topology coarser than a Polish topology (here the Polish topology of the *separable* Banach space $E$) is equal to the Borel $\sigma$-algebra of this Polish topology (Hoffmann-Jørgensen [1970]).

*Step 3. $S \in L^1(E)$ (in the special case of Step 1).*

This is the critical step. We write $\Delta = \{\varphi_1, \varphi_2, \ldots\}$, and fix $m \in \mathbb{N}^*$. Then:

$$\big(\varphi_1(S_n), \ldots, \varphi_m(S_n)\big) \xrightarrow[n \to +\infty]{a.s.} \big(\varphi_1(S), \ldots, \varphi_m(S)\big),$$

where the convergence is in $\mathbb{R}^m$, equipped with the norm of $\ell_\infty^m$. The Paul Lévy maximal inequality applied in this normed space gives, for any $t > 0$:

$$\mathbb{P}\left(\sup_{n \geqslant 1} \left\| \sum_{k=1}^n \varepsilon_k\big(\varphi_1(x_k), \ldots, \varphi_m(x_k)\big) \right\|_{\ell_\infty^m} > t \right)$$

$$= \mathbb{P}\left(\sup_{n \geqslant 1} \max_{j \leqslant m} |\varphi_j(S_n)| > t\right) \leqslant 2\mathbb{P}\left(\max_{j \leqslant m} |\varphi_j(S)| > t\right)$$

$$\leqslant 2\,\mathbb{P}\left(\|S\| > t\right).$$

As the events $A_m = \{\sup_{n \geqslant 1} \max_{j \leqslant m} |\varphi_j(S_n)| > t\}$ increase to the event $A = \{\sup_{n \geqslant 1} \|S_n\| > t\}$, we obtain:

$$\mathbb{P}(M > t) \leqslant 2\mathbb{P}(\|S\| > t),$$

where $M = \sup_{n \geqslant 1} \|S_n\|$. Thus we have $M < +\infty$ a.s., and by Theorem V.2 of Chapter 4 (Volume 1), $M \in L^1(\mathbb{R})$, hence $S \in L^1(E)$, since

$$\|S\| = \sup_{\varphi \in \Delta} |\varphi(S)| \leqslant \sup_{\varphi \in \Delta} \sup_{n \geqslant 1} |\varphi(S_n)| = \sup_{n \geqslant 1} \|S_n\| = M \quad a.s.$$

*Step 4.* $S_n = \mathbb{E}(S|\mathcal{A}_n)$, with $\mathcal{A}_n = \sigma(X_1, \ldots, X_n)$.

Let $A \in \mathcal{A}_n$. For $N > n$, we have $\mathbb{E}(S_N \mathbb{1}_A) = \mathbb{E}(S_n \mathbb{1}_A)$, since $S_N - S_n$ is centered and independent of $\mathcal{A}_n$. Then, if $\varphi \in \Delta$, we obtain:

$$
\begin{aligned}
|\langle \varphi, \mathbb{E}(\mathbb{1}_A S) \rangle - \langle \varphi, \mathbb{E}(\mathbb{1}_A S_n) \rangle| &= |\langle \varphi, \mathbb{E}(\mathbb{1}_A S) \rangle - \langle \varphi, \mathbb{E}(\mathbb{1}_A S_N) \rangle| \\
&\leqslant \mathbb{E}(|\varphi(S) - \varphi(S_N)| \mathbb{1}_A) \\
&\leqslant \mathbb{E}(|\varphi(S) - \varphi(S_N)|);
\end{aligned}
$$

this last term tends to zero when $N$ tends to infinity, by the dominated convergence theorem (since $M \in L^1$). Hence,

$$\langle \varphi, \mathbb{E}(\mathbb{1}_A S) \rangle = \langle \varphi, \mathbb{E}(\mathbb{1}_A S_n) \rangle$$

for every $\varphi \in \Delta$, and thus $\mathbb{E}(\mathbb{1}_A S) = \mathbb{E}(\mathbb{1}_A S_n)$. As this is true for every $A \in \mathcal{A}_n$, we finally obtain $S_n = \mathbb{E}(S|\mathcal{A}_n)$.

*Step 5. Conclusion.*

Steps 3 and 4 show that $(S_n)_{n \geqslant 1}$ is a closed martingale, hence it is convergent without any doubt, according to Theorem II.12 of Chapter 4 (Volume 1). It is in fact sufficient to use the convergence in $L^1(E)$, since this implies the convergence in probability, and hence the almost sure convergence, by the theorem of Paul Lévy (Chapter 4, Theorem III.3). □

**Remark** In the case of $\Delta = B_{E^*}$, a proof can be given based on Prokhorov's theorem instead of martingales (see LEDOUX–TALAGRAND, pages 48–50).

**Remark** The hypothesis of $E$ being separable is indispensable, as shown by the following example (indicated to us by F. Bayart):

$$E = \ell_\infty \quad \text{and} \quad \Delta = \{e_k \,;\, k \geqslant 1\},$$

where $(e_k)_{k \geqslant 1}$ is the canonical basis of $\ell_1$, and

$$X_j(\omega) = (0, \ldots, 0, \varepsilon_j(\omega), 0, \ldots),$$

where $(\varepsilon_j)_{j \geqslant 1}$ is a Rademacher sequence.

If $k \geqslant 1$ is given, we have $\lim_{n \to +\infty} \langle e_k, S_n \rangle = \langle e_k, S \rangle$, with

$$S(\omega) = \left( \varepsilon_j(\omega) \right)_{j \geqslant 1} \in \ell_\infty.$$

Nevertheless, we have:

$$\|S_n(\omega) - S(\omega)\|_\infty = 1 \quad \text{for all } n\,!$$

The reason is that, even though the variables $X_j$ take their values in the separable space $c_0$, the map $S$ is not separably valued.

For what follows, the most useful version of the abstract Itô–Nisio theorem is when $E = \mathcal{C}(K)$, the space of continuous scalar functions on a metrizable compact space $K$.

**Theorem II.2** (The Itô–Nisio Theorem for $\mathcal{C}(K)$)   *Let $K$ be a metrizable compact space, and $(X_n)_{n \geqslant 1}$ a sequence of independent symmetric random variables, with values in $\mathcal{C}(K)$. Suppose that:*

1) *for every $t \in K$, $S_n(t, \omega) = \sum_{j=1}^n X_j(t, \omega) \xrightarrow[n \to +\infty]{a.s.} X(t, \omega) = X_t(\omega)$ (i.e. for $\omega \notin N_t$, where $N_t$ is negligible);*
2) *the process $(X_t)_{t \in K}$ has a continuous version, i.e. there exists a process $(Y_t)_{t \in K}$ with the following properties:*
   - *there is a negligible set $N$ such that $t \in K \mapsto Y_t(\omega)$ is continuous for $\omega \notin N$;*
   - *for every $t \in K$: $\mathbb{P}(X_t = Y_t) = 1$, i.e. $X_t(\omega) = Y_t(\omega)$ if $\omega \notin N'_t$, where $N'_t$ is a negligible set depending on $t$.*

*Then $(S_n)_{n \geqslant 1}$ is almost surely uniformly convergent on $K$.*

*Proof*   If necessary replacing $Y_t(\omega)$ by $0$ when $\omega \in N$, we can suppose that all the paths of $Y = (Y_t)_{t \in K}$ are continuous. Let $\Delta$ be the subset of the unit ball of the dual $\mathcal{C}(K)^* = \mathcal{M}(K)$ formed by the Dirac masses $\delta_t$, for $t \in K$. The two hypotheses of the theorem can be reformulated as follows: for each $\delta_t \in \Delta$, we have:

$$\langle \delta_t, S_n(\omega) \rangle = S_n(t, \omega) \xrightarrow[n \to +\infty]{a.s.} \langle \delta_t, Y(\omega) \rangle = Y(t, \omega) = Y_t(\omega),$$

where the convergence takes place for every $\omega \notin N_t \cup N'_t$, and where $Y$ is a mapping from $\Omega$ into $E = \mathcal{C}(K)$. As $\Delta$ is evidently norming, the conditions of application of the abstract version of Theorem II.1 are exactly satisfied, and hence $S_n$ converges almost surely to $Y$ in $\mathcal{C}(K)$; in other words, uniformly on $K$.   $\square$

**Remark 1**   The Itô–Nisio theorem for $\mathcal{C}(K)$ brings to mind the Dini theorems, where, under certain hypotheses of monotonicity, the pointwise convergence of functions $f_n \in \mathcal{C}(K)$ to $f$ implies their uniform convergence, as soon as we assume $f \in \mathcal{C}(K)$. Here, the hypothesis of monotonicity is replaced by the stochastic and symmetric nature of the sequence, and the continuity of $f$ by the existence of a continuous version of the process.

**Remark 2** If $\tau$ is the topology of pointwise convergence on $\Delta$, as in Theorem II.1, the Hahn–Banach theorem implies that the hypothesis "$\Delta$ norming" is equivalent to "the unit ball $B_E$ is $\tau$-closed". The hypothesis of the theorem can thus be replaced by the following hypothesis: on $E$, there exists a Hausdorff locally convex topology $\tau$, coarser than the norm topology for which $B_E$ is closed, and such that $S_n \xrightarrow[n \to +\infty]{\tau} S$ almost surely.

## II.2 An Almost Sure "Tauberian" Theorem

Theorem III.5 of Chapter 4 (Volume 1) can usefully be generalized as follows:

**Definition II.3** Let $\Lambda$ be a set of parameters equipped with a filter $\mathcal{F}$ with a countable base, and let:

$$A = (a_{n,\lambda})_{(n,\lambda)}$$

be a matrix indexed by $\mathbb{N}^* \times \Lambda$, $\mathbb{N} \times \Lambda$, or $\mathbb{Z} \times \Lambda$. Suppose that, for any $n$:

$$\lim_{\lambda \in \Lambda} a_{n,\lambda} = 1.$$

Let $E$ be a Banach space; a series $\sum_{n \geqslant 1} x_n$ of elements of $E$ is said to be *A-convergent* (respectively *A-bounded*) if:

(i) $z_\lambda = \sum_n a_{n,\lambda} x_n$ exists for every $\lambda \in \Lambda$;
(ii) $\lim_{\lambda \in \Lambda} z_\lambda = z$ exists (respectively $(z_\lambda)_{\lambda \in \Lambda}$ is bounded).

The $\sum_{n \geqslant 1} x_n$ is also said to *converge to $z$* (respectively to *be bounded*) *for the summation procedure defined by $A$*.

**Examples**

1) $\Lambda = \mathbb{N}^*$ and $a_{n,\lambda} = \begin{cases} 1 & \text{if } n \leqslant \lambda \\ 0 & \text{if } n > \lambda. \end{cases}$

The condition $a_{n,\lambda} \xrightarrow[\lambda \to +\infty]{} 1$ is indeed satisfied, and $z_\lambda = \sum_{n=1}^{\lambda} x_n$: this is the usual partial sum. If $\Lambda$ were a subsequence of integers, we would have a subsequence of partial sums.

2) $\Lambda = ]0, 1[$ and $a_{n,\lambda} = \lambda^n$, $n \in \mathbb{N}$.

The condition $\lim_{\lambda \overset{<}{\to} 1} a_{n,\lambda} = 1$ is satisfied, and $z_\lambda = \sum_{n=0}^{+\infty} x_n \lambda^n$: this is the Abel–Poisson summation procedure.

3) $\Lambda = ]0, +\infty[$ and $a_{n,\lambda} = \left(\dfrac{\sin n\lambda}{n\lambda}\right)^2$.

The condition $\lim\limits_{\lambda \xrightarrow{>} 0} a_{n,\lambda} = 1$ is satisfied, and $z_\lambda = \sum\limits_{n=1}^{+\infty} x_n \left(\dfrac{\sin n\lambda}{n\lambda}\right)^2$ : this is the Riemann summation procedure, used to show that if a trigonometric series $\sum_{n\in\mathbb{Z}} c_n e^{int} = \sum_{n=0}^{+\infty}(a_n \cos nt + b_n \sin nt)$ converges everywhere to $0$, then necessarily all its coefficients are null.

4) $\Lambda = \mathbb{N}^*$ and $a_{n,\lambda} = \left(1 - \dfrac{n}{\lambda}\right)^+, n \in \mathbb{N}$.

We have $a_{n,\lambda} \xrightarrow[\lambda \to +\infty]{} 1$, and if $S_n = x_0 + \cdots + x_n$, then:

$$z_\lambda = \frac{S_0 + \cdots + S_{\lambda-1}}{\lambda} :$$

this is the Fejér–Cesàro summation procedure.

A *Tauberian theorem* for the summation procedure $A$ is a statement of the type: "if the series $\sum_n x_n$ is $A$-convergent (respectively $A$-bounded), and if the numbers $x_n$ satisfy an additional *Tauberian* condition, then the series $\sum_n x_n$ is convergent (respectively bounded) in the usual sense". The celebrated Hardy–Littlewood Tauberian theorem states that the Tauberian condition $nx_n = O(1)$, with $x_n \in \mathbb{C}$, is admissible in the Abel–Poisson summation procedure.

In the case of the almost sure convergence of *symmetric* random variables, a pleasant phenomenon occurs: the Tauberian theorems are automatically true, without the imposition of even the least additional condition. More precisely, we have the following result:

**Theorem II.4** (The Marcinkiewicz–Zygmund–Kahane Theorem)    *Let $\sum_{n\geqslant 1} X_n$ be a series of independent and symmetric variables with values in a Banach space $E$, and let $A = (a_{n,\lambda})_{n,\lambda}$ be a summation procedure. Then:*

1) *If $\sum_{n\geqslant 1} X_n$ is almost surely $A$-convergent, it converges almost surely.*
2) *If $\sum_{n\geqslant 1} X_n$ is almost surely $A$-bounded, it is almost surely bounded.*

*Proof*    We prove only 1); the method for 2) is entirely similar.

Note that Theorem III.5 of Chapter 4 (Volume 1) is a special case of this, with "a subsequence of partial sums" as the summation procedure. The general case will be reduced to this special case in two steps.

We set:

$$Y_\lambda = \sum_{n=1}^{+\infty} a_{n,\lambda} X_n.$$

*Step 1. We can replace $A = (a_{n,\lambda})_{(n,\lambda)\in\mathbb{N}^*\times\Lambda}$ by $B = (b_{n,p})_{(n,p)\in\mathbb{N}^*\times\mathbb{N}^*}$, with $b_{n,p} = 1$ if $n \leqslant p$ and $b_{n,p} = 0$ if $n > q_p$, where $q_p > p$.*

Indeed, as $\lim_{\lambda\in\Lambda} a_{n,\lambda} = 1$ for any $n \geqslant 1$, we have, for each $p \geqslant 1$: $\sum_{n=1}^{p}(1 - a_{n,\lambda})X_n \overset{a.s.}{\underset{\lambda\in\Lambda}{\to}} 0$; thus there exists $F_p \in \mathcal{F}$ such that:

$$(1) \qquad \mathbb{P}\left(\left\|\sum_{n=1}^{p}(1 - a_{n,\lambda})X_n\right\| > 2^{-p}\right) \leqslant \frac{1}{2^p} \quad \text{if } \lambda \in F_p.$$

We select $\lambda_p \in F_p$. The hypothesis of $A$-convergence states that the series $\sum_{n=1}^{+\infty} a_{n\lambda_p}X_n$ converges *a.s.*; thus, there exists $q_p > p$ such that:

$$(2) \qquad \mathbb{P}\left(\left\|\sum_{n>q} a_{n,\lambda_p}X_n\right\| > 2^{-p}\right) \leqslant \frac{1}{2^p} \quad \text{if } q \geqslant q_p.$$

We define $b_{n,p}$ by:

$$b_{n,p} = \begin{cases} 1 & \text{if } n \leqslant p \\ a_{n,\lambda_p} & \text{if } p < n \leqslant q_p \\ 0 & \text{if } n > q_p, \end{cases}$$

and set:

$$Z_p = \sum_{n=1}^{+\infty} b_{n,p}X_n = \sum_{n=1}^{q_p} b_{n,p}X_n.$$

Then:

$$Z_p - Y_{\lambda_p} = \sum_{n=1}^{p}(1 - a_{n,\lambda_p})X_n - \sum_{n>q_p} a_{n,\lambda_p}X_n.$$

Now (1) and (2) show that:

$$\mathbb{P}(\|Z_p - Y_{\lambda_p}\| > 2^{-p+1}) \leqslant \frac{1}{2^{p-1}}.$$

By the Borel–Cantelli lemma, there exists $\Omega_0 \subseteq \Omega$ such that $\mathbb{P}(\Omega_0) = 1$ and such that:

$$(3) \quad \|Z_p(\omega) - Y_{\lambda_p}(\omega)\| \leqslant \frac{1}{2^{p-1}} \quad \text{for } \omega \in \Omega_0 \quad \text{and } p \geqslant p_0 = p_0(\omega).$$

Moreover, as $Y_\lambda \overset{a.s.}{\underset{\lambda\in\Lambda}{\to}} Y$, and as $\mathcal{F}$ has a countable base, we can also assume that:

$$\mathbb{P}(\|Y_\lambda - Y\| > 2^{-p}) \leqslant \frac{1}{2^p} \quad \text{if } \lambda \in F_p.$$

Hence, diminishing $\Omega_0$ while keeping $\mathbb{P}(\Omega_0) = 1$, and augmenting $p_0 = p_0(\omega)$ if necessary, we can also assume that:

(4)    $\|Y_{\lambda_p}(\omega) - Y(\omega)\| \leqslant \dfrac{1}{2^{p-1}}$    if $\omega \in \Omega_0$ and if $p \geqslant p_0 = p_0(\omega)$.

Then, (3) and (4) show that $Z_p = \sum_{n=1}^{+\infty} b_{n,p} X_n$ converges almost surely to $Y$.

*Step 2.*

The matrix $B = (b_{n,p})_{n,p}$ would be associated with the convergence of a subsequence of partial sums were it not for the "no man's land" $b_{n,p} = a_{n,\lambda_p}$ for $p < n \leqslant q_p$. Step 2 consists of getting rid of this parasite range. For this, first, a sequence of integers $p_j$ is defined by induction: $p_1 = 1$ and $p_{j+1} = q_{p_j}$, where $q_p$ has been introduced in (2). Then, two new sequences of independent and symmetric variables $(X'_n)_{n \geqslant 1}$ and $(X''_n)_{n \geqslant 1}$ are defined as follows:

$$\begin{cases} X'_n = X_n \\ X''_n = 0 \end{cases} \quad \text{if } p_{2j-1} < n \leqslant p_{2j}$$

and

$$\begin{cases} X'_n = 0 \\ X''_n = X_n \end{cases} \quad \text{if } p_{2j} < n \leqslant p_{2j+1}.$$

They satisfy:

$$2X'_n - X_n = \begin{cases} X_n & \text{if } p_{2j-1} < n \leqslant p_{2j} \\ -X_n & \text{if } p_{2j} < n \leqslant p_{2j+1} \end{cases}$$

and

$$2X''_n - X_n = \begin{cases} -X_n & \text{if } p_{2j-1} < n \leqslant p_{2j} \\ X_n & \text{if } p_{2j} < n \leqslant p_{2j+1}. \end{cases}$$

Since the variables $X_n$ are symmetric, the sequences $(2X'_n - X_n)_{n \geqslant 1}$ and $(2X''_n - X_n)_{n \geqslant 1}$ have the same distribution as the sequence $(X_n)_{n \geqslant 1}$. The series $\sum_{n \geqslant 1}(2X'_n - X_n)$ and $\sum_{n \geqslant 1}(2X''_n - X_n)$ are thus almost surely $B$-convergent. Therefore, since $X'_n = [(2X'_n - X_n) + X_n]/2$ and $X''_n = [(2X''_n - X_n) + X_n]/2$, the series $\sum_{n \geqslant 1} X'_n$ and $\sum_{n \geqslant 1} X''_n$ are themselves almost surely $B$-convergent.

The gain is the disappearance in $(X'_n)_{n \geqslant 1}$ and $(X''_n)_{n \geqslant 1}$ of the parasite range in which $b_{n,p} \neq 0$ or $1$; therefore we can conclude. Indeed, since $X'_n = 0$ for $p_{2j} < n \leqslant p_{2j+1}$, keeping the notation of Step 1 but replacing $X_n$ by $X'_n$, we obtain:

$$Z'_{p_{2j}} = \sum_{n=1}^{q_{p_{2j}}} b_{n,p_{2j}} X'_n = \sum_{n=1}^{p_{2j+1}} b_{n,p_{2j}} X'_n = \sum_{n=1}^{p_{2j}} X'_n = S'_{p_{2j}}.$$

However, the $B$-convergence of $\sum_{n\geqslant 1} X'_n$ implies that $Z'_{p_{2j}} \xrightarrow[j\to+\infty]{a.s.} Y'$; hence $S'_{p_{2j}} \xrightarrow[j\to+\infty]{a.s.} Y'$. Thus, thanks to Theorem III.5 of Chapter 4 (Volume 1), we obtain $S'_n = \sum_{k=1}^{n} X'_k \xrightarrow[n\to+\infty]{a.s.} Y'$. Similarly, the series $\sum_{n\geqslant 1} X''_n$ converges almost surely, and thus so does $\sum_{n\geqslant 1} X_n$, since $X_n = X'_n + X''_n$. $\qquad\square$

**Remark** When $E$ is the scalar field, there is a much simpler proof, presented at the beginning of Section VI.

# III The $\mathcal{C}^{as}$ Space

## III.1 Equivalent Definitions

In what follows, $G$ is a compact metrizable Abelian group, and $\Gamma$ its (countable) dual group. An enumeration $\gamma_1, \gamma_2, \ldots$ of $\Gamma$ is set.

Let $(Z_n)_{n\geqslant 1}$ be a standard complex Gaussian sequence. For $(a_n)_{n\geqslant 1} \in \ell_2$, a complex centered Gaussian process can be defined by:

$$(GP) \qquad X_t(\omega) = \sum_{n=1}^{+\infty} a_n Z_n(\omega) \gamma_n(t), \quad \omega \in \Omega, \, t \in G.$$

If $m$ is the Haar measure of $G$, $\mathcal{C}(G)$ is denoted $\mathcal{C}$ and $L^p(G, m)$ is denoted $L^p$ ($1 \leqslant p \leqslant +\infty$). For $f \in L^1$, the Fourier coefficient of $f$ at $\gamma_n$ is denoted $\widehat{f}(n) = \widehat{f}(\gamma_n)$.

A (formal) series $\sum_{n=1}^{+\infty} c_n \gamma_n$ is said to be the Fourier series of the function $f$ if $c_n = \widehat{f}(n)$ for every $n \geqslant 1$.

The following theorem provides a number of equivalent properties for the process $(X_t)_{t\in G}$, or, alternatively, for the random series (in $\mathcal{C}$) $\sum_{n=1}^{+\infty} a_n Z_n \gamma_n$.

**Theorem III.1** (Equivalence Theorem) *For any sequence $(a_n)_{n\geqslant 1} \in \ell_2$, the following five properties are equivalent:*

1) $\mathbb{E} \left\| \sum_{n=1}^{N} a_n Z_n \gamma_n \right\|_\infty$ *is bounded for $N \geqslant 1$; in other words the partial sums $\sum_{n=1}^{N} a_n Z_n \gamma_n$ are bounded in $L^1(\Omega, \mathbb{P}; \mathcal{C})$.*
2) *Almost surely, (GP) is the Fourier series of a function $f^\omega \in L^\infty$, i.e. $\widehat{f^\omega}(n) = a_n Z_n(\omega)$ for all $n \geqslant 1$.*
3) *Almost surely, (GP) possesses a continuous version.*
4) *Almost surely, $\sum_{n=1}^{+\infty} a_n Z_n \gamma_n$ converges in the space $\mathcal{C}$.*
5) *Almost surely, (GP) is the Fourier series of a continuous function $f^\omega \in \mathcal{C}$.*

*Proof*

1) $\Rightarrow$ 2) Let:

$$\lambda = \sup_{N \geqslant 1} \mathbb{E} \left\| \sum_{n=1}^{N} a_n Z_n \gamma_n \right\|_\infty \quad \text{and} \quad M = \sup_{N \geqslant 1} \left\| \sum_{n=1}^{N} a_n Z_n \gamma_n \right\|_\infty .$$

As the variables $Z_n$ are symmetric, the Paul Lévy maximal inequality shows that $\mathbb{P}(M > t) \leqslant 2\lambda/t$ for $t > 0$, and hence $M < +\infty$ a.s. When $M(\omega) < +\infty$, the vectors $S_N(\omega) = \sum_{n=1}^{N} a_n Z_n(\omega) \gamma_n$ form a bounded sequence in $L^\infty$. As $L^\infty$ is a dual space, there exists a weak* cluster point $f^\omega$ of $(S_N(\omega))_{N \geqslant 1}$ in $L^\infty$. For any $n \geqslant 1$, and for $N \geqslant n$, we have $\widehat{S_N}(\omega)(n) = a_n Z_n(\omega)$; hence the cluster point $f^\omega$ satisfies $\widehat{f^\omega}(n) = a_n Z_n(\omega)$.

This shows the uniqueness of $f^\omega$ as a cluster point and hence the weak* convergence of the series $\sum_{n=1}^{+\infty} a_n Z_n(\omega) \gamma_n$ in $L^\infty$.

2) $\Rightarrow$ 3) Let $(K_N)_{N \geqslant 1}$ be a polynomial approximate identity:

$$K_N \in \mathcal{P}(G), \quad K_N \geqslant 0, \quad \|K_N\|_1 = 1 \quad \text{and} \quad \widehat{K}_N \xrightarrow[N \to +\infty]{} 1.$$

Let $(\alpha_{n,N})_{n,N} = (\widehat{K}_N(n))_{n,N}$ be the summation procedure defined by $(K_N)_{N \geqslant 1}$ (indeed, $\alpha_{n,N} \xrightarrow[N \to +\infty]{} 1$ for any $n \geqslant 1$). The series of symmetric independent variables $\sum_{n \geqslant 1} a_n Z_n \gamma_n$ is almost surely bounded for this summation procedure, since

$$\left\| \sum_{n=1}^{+\infty} a_n \alpha_{n,N} Z_n(\omega) \gamma_n \right\|_\infty = \|f^\omega * K_N\|_\infty \leqslant \|f^\omega\|_\infty.$$

Thanks to the Tauberian theorem (Theorem II.4), it is almost surely bounded; in other words, the process $(X_t)_{t \in G}$ almost surely has bounded trajectories. Moreover, this process is stationary, since, for $x \in G$, the Gaussian process $(X_{t+x})_{t \in G}$, with $X_{t+x} = \sum_{n=1}^{+\infty} a_n \gamma_n(x) Z_n \gamma(t)$, has the same covariance as $(X_t)_{t \in G}$. We can thus apply Fernique's minoration theorem (Theorem V.4 of Chapter 3, Volume 2); part 3) of this theorem states that $(X_t)_{t \in G}$ admits a continuous version as soon as it admits a bounded version.

3) $\Rightarrow$ 4) As the variables $a_n Z_n \gamma_n$ are symmetric and independent, this is an immediate application of the Itô–Nisio theorem for $\mathcal{C} = \mathcal{C}(G)$.

4) $\Rightarrow$ 5) This is trivial: if $S_N = \sum_{n=1}^{N} a_n Z_n(\omega) \gamma_n$ converges uniformly to $f^\omega \in \mathcal{C}$, a passage to the limit in $a_n Z_n(\omega) = \widehat{S_N}(n)$ for $N \geqslant n$ leads to $a_n Z_n(\omega) = \widehat{f^\omega}(n)$ for any $n \geqslant 1$.

5) $\Rightarrow$ 1) Again let $(K_N)_{N \geqslant 1}$ be a polynomial approximate identity and $\alpha_{n,N} = \widehat{K}_N(n)$. We have already stated that $(\alpha_{n,N})_{n,N}$ is a summation procedure. Now, the series of symmetric variables $\sum_{n=1}^{+\infty} a_n Z_n \gamma_n$ is almost surely convergent for

this procedure, since $\sum_{n=1}^{+\infty} \alpha_{n,N} a_n Z_n \gamma_n = f^\omega * K_N \xrightarrow[N \to +\infty]{} f^\omega$ in the space $\mathcal{C}$. By the Tauberian theorem (Theorem II.4), the series $\sum_{n=1}^{+\infty} a_n Z_n \gamma_n$ converges almost surely to $f^\omega$ in the space $\mathcal{C}$.

Fernique's theorem of automatic integrability (Theorem V.26 of Chapter 6, Volume 1) then shows that $f^\omega \in L^1(\Omega, \mathbb{P})$. Next, by the symmetry of the variables $Z_n$, we obtain $\mathbb{P}(\|S_N\|_\infty > t) \leqslant 2\mathbb{P}(\|f^\omega\|_\infty > t)$, and hence:

$$
\begin{aligned}
\mathbb{E}\|S_N\|_\infty &= \int_0^{+\infty} \mathbb{P}(\|S_N\|_\infty > t)\, dt \\
&\leqslant 2 \int_0^{+\infty} \mathbb{P}(\|f^\omega\|_\infty > t)\, dt = 2\,\|f\|_{L^1(\Omega,\mathbb{P};\mathcal{C})}
\end{aligned}
$$

where $F(\omega) = f^\omega$, which completes the proof. $\qquad\square$

**Remark** Clearly, the preceding equivalences are far from trivial; we must call to the rescue two theorems of Fernique, the Itô–Nisio theorem and the Marcinkiewicz-Zygmund-Kahane theorem. The implication 2) $\Rightarrow$ 5) by itself is known as *Billard's theorem* ; the initial proof of Billard [1965] (see also KAHANE 2) preceded Fernique's theorem (Fernique [1970]). Billard also showed the following result (see Billard [1965] or KAHANE 2):

*If $\sum_{n=1}^{+\infty} a_n Z_n \gamma_n(t)$ converges almost surely for every $t \in G$, then $\sum_{n=1}^{+\infty} a_n Z_n(\omega)\gamma_n$ is the Fourier series of a function $f^\omega \in \mathcal{C}$.*

In fact, Billard worked with series of Bernoulli variables, but the Marcus–Pisier theorem shows that it amounts to the same.

If $(a_n)_{n \geqslant 1} \in \ell_2$, here is an implication of the Fubinization principle:

*Almost surely, $\sum_{n=1}^{+\infty} a_n Z_n(\omega)\gamma_n(t)$ converges for almost every $t \in G$.*

Indeed, for every fixed $t \in G$, we have $\sum_{n=1}^{+\infty} |a_n \gamma_n(t)|^2 = \sum_{n=1}^{+\infty} |a_n|^2$; hence, by Kolmogorov's theorem (see Chapter 1 of Volume 1, Theorem III.2), the series $\sum_{n=1}^{+\infty} a_n Z_n(\omega)\gamma_n(t)$ converges $\omega$-almost surely.

Theorem III.1 leads to the definition of the space $\mathcal{C}^{as}$.

**Definition III.2** The space $\mathcal{C}^{as} = \mathcal{C}^{as}(G)$ is the space of functions $f \in L^2(G)$ for which the sequence $(a_n)_{n \geqslant 1} = (\widehat{f}(n))_{n \geqslant 1}$ satisfies one of the equivalent conditions of Theorem III.1.

It is equipped with the norm defined by:

$$
[\![f]\!] = \sup_{N \geqslant 1} \mathbb{E} \left\| \sum_{n=1}^{N} \widehat{f}(n) Z_n \gamma_n \right\|_\infty .
$$

Note that $[\![f]\!] < +\infty$ by 1) of Theorem III.1.

The indicator functions of arcs of $\mathbb{T}$ are typical examples of functions in $\mathcal{C}^{as}$ (see Exercise VII.3). The following two simple properties of this norm will be constantly used:

**Lemma III.3**

1) *For every $f \in \mathcal{C}^{as}$, we have $\|f\|_2 \leqslant [\![f]\!]$.*
2) *For every trigonometric polynomial $f = \sum_{n=1}^{N} a_n \gamma_n$, we have:*

$$\left[\!\!\left[ \sum_{n=1}^{N} a_n \gamma_n \right]\!\!\right] = \mathbb{E} \left\| \sum_{n=1}^{N} a_n Z_n \gamma_n \right\|_{\infty}.$$

*Proof*

1) This is trivial:

$$\|f\|_2 = \sup_{N \geqslant 1} \left( \sum_{n=1}^{N} |\widehat{f}(n)|^2 \right)^{1/2} = \sup_{N \geqslant 1} \mathbb{E} \left| \sum_{n=1}^{N} \widehat{f}(n) Z_n \gamma_n(0) \right|$$

$$\leqslant \sup_{N \geqslant 1} \mathbb{E} \left\| \sum_{n=1}^{N} \widehat{f}(n) Z_n \gamma_n \right\|_{\infty} = [\![f]\!].$$

2) Since the variables $a_1 Z_1 \gamma_1, \ldots, a_N Z_N \gamma_N$ are independent and centered in $\mathcal{C}$, then, for any $n \leqslant N$:

$$\mathbb{E} \left\| \sum_{k=1}^{n} a_k Z_k \gamma_k \right\|_{\infty} \leqslant \mathbb{E} \left\| \sum_{k=1}^{N} a_k Z_k \gamma_k \right\|_{\infty}$$

(see Chapter 4 of Volume 1, Theorem IV.3); hence:

$$\mathbb{E} \left\| \sum_{n=1}^{N} a_n Z_n \gamma_n \right\|_{\infty} = \left[\!\!\left[ \sum_{n=1}^{N} a_n \gamma_n \right]\!\!\right]. \quad \square$$

An important property of the space $\mathcal{C}^{as}$ is given by the following theorem:

**Theorem III.4**    *The space $\mathcal{C}^{as}$ equipped with the norm $[\![\,.\,]\!]$ is a Banach space, and the sequence $(\gamma_n)_{n \geqslant 1}$ of characters is an unconditional basis for it, with constant 1.*

*Proof*    As the norm $[\![\,.\,]\!]$ of $\mathcal{C}^{as}$ dominates the norm $L^2$, if $(f_p)_{p \geqslant 1}$ is a Cauchy sequence in $\mathcal{C}^{as} = \mathcal{C}^{as}(G)$, it is also Cauchy in $L^2$; hence it has a limit $f$ in $L^2$. Let us show that there is also convergence for the norm of $\mathcal{C}^{as}$. Indeed, let $\varepsilon > 0$, and let $p_0$ be an integer such that:

$$[\![f_p - f_q]\!] \leqslant \varepsilon \quad \text{if} \quad p_0 \leqslant p < q.$$

Fix $N \geqslant 1$; for $p_0 \leqslant p < q$, we have:

$$\mathbb{E} \left\| \sum_{n=1}^{N} [\widehat{f_p}(n) - \widehat{f_q}(n)] Z_n \gamma_n \right\|_{\infty} \leqslant [\![ f_p - f_q ]\!] \leqslant \varepsilon.$$

When $q$ tends to infinity, the dominated convergence theorem implies:

$$\mathbb{E} \left\| \sum_{n=1}^{N} [\widehat{f_p}(n) - \widehat{f}(n)] Z_n \gamma_n \right\|_{\infty} \leqslant \varepsilon \quad \text{for} \quad p \geqslant p_0.$$

As this occurs for every $N \geqslant 1$, we first obtain that $f_p - f \in \mathcal{C}^{as}$, hence $f \in \mathcal{C}^{as}$. Therefore the preceding inequality can be interpreted as:

$$[\![ f_p - f ]\!] \leqslant \varepsilon \quad \text{if} \quad p \geqslant p_0.$$

Hence, indeed $(f_p)_{p \geqslant 1}$ converges to $f$ for the norm of $\mathcal{C}^{as}$, and this space is a Banach space.

Now let $f \in \mathcal{C}^{as}$. Condition 4) of the equivalence theorem states that, almost surely, the series $\sum_{n=1}^{+\infty} \widehat{f}(n) Z_n \gamma_n$ converges in $\mathcal{C}$; hence, almost surely $M = \sup_{N \geqslant 1} \left\| \sum_{n=1}^{N} \widehat{f}(n) Z_n \gamma_n \right\|_{\infty} < +\infty$. Fernique's integrability theorem (Chapter 6 of Volume 1, Theorem V.26) implies that $M \in L^1(\Omega, \mathbb{P})$; hence we can use the dominated convergence theorem to obtain that the partial sums $\sum_{n=1}^{N} \widehat{f}(n) Z_n \gamma_n$ converge in the space $L^1(\Omega, \mathbb{P}; \mathcal{C}(G))$. In other words, since $\mathbb{E} \left\| \sum_{n=1}^{N} a_n Z_n \gamma_n \right\|_{\infty} = [\![ \sum_{n=1}^{N} a_n \gamma_n ]\!]$, we have:

$$\left[\!\!\left[ f - \sum_{n=1}^{N} \widehat{f}(n) \gamma_n \right]\!\!\right] \xrightarrow[N \to +\infty]{} 0.$$

Hence $(\gamma_n)_{n \geqslant 1}$ is a basis of $\mathcal{C}^{as}(G)$ (as the representation is clearly unique).

To conclude, let $u_1, \ldots, u_N$ be complex numbers of modulus 1. The invariance of $(Z_1, \ldots, Z_N)$ under the unitary group $\mathbb{U}(n)$ shows that $(u_1 Z_1, \ldots, u_N Z_N)$ has the same distribution as $(Z_1, \ldots, Z_N)$. Hence, when $a_1, \ldots, a_N \in \mathbb{C}$, by Lemma III.3 we have:

$$\left[\!\!\left[ \sum_{n=1}^{N} u_n a_n \gamma_n \right]\!\!\right] = \mathbb{E} \left\| \sum_{n=1}^{N} a_n (u_n Z_n) \gamma_n \right\|_{\infty}$$

$$= \mathbb{E} \left\| \sum_{n=1}^{N} a_n Z_n \gamma_n \right\|_{\infty} = \left[\!\!\left[ \sum_{n=1}^{N} a_n \gamma_n \right]\!\!\right].$$

This proves the unconditionality of the basis $(\gamma_n)_{n \geqslant 1}$ in $\mathcal{C}^{as}$, with an unconditional constant equal to 1. $\qquad\square$

**Remark**  Because the basis has an unconditional constant equal to 1, and hence is monotone, then, for $f \in \mathcal{C}^{as}$:

$$\mathbb{E}\|f^\omega\|_\infty = \lim_{N \to +\infty} \mathbb{E}\left\|\sum_{n=1}^N \widehat{f}(n)Z_n\gamma_n\right\|_\infty = \sup_{N \geqslant 1} \mathbb{E}\left\|\sum_{n=1}^N \widehat{f}(n)Z_n\gamma_n\right\|_\infty ;$$

this can be written:

$$\left[\!\left[\sum_{n=1}^{+\infty} \widehat{f}(n)\,\gamma_n\right]\!\right] = \mathbb{E}\left\|\sum_{n=1}^{+\infty} \widehat{f}(n)Z_n\gamma_n\right\|_\infty ,$$

providing a more agreeable form for the norm $[\![\,.\,]\!]$.

As we now know that $\mathcal{C}^{as}$ is a Banach space, the question of describing its dual arises spontaneously. This is the subject of Section III.3.

Prior to this, we show that, had we used Bernoulli variables (*a priori* more natural), we would have obtained the same space.

## III.2  The Marcus–Pisier Theorem

This theorem states that the space $\mathcal{C}^{as}$ remains unchanged when the complex Gaussian variables are replaced by Bernoulli variables.

**Theorem III.5** (The Marcus–Pisier Theorem)  *There exist numerical constants $C_1 > 0$ and $C_2 > 0$ such that, for every $a_1, \ldots, a_N \in \mathbb{C}$:*

$$C_1\mathbb{E}\left\|\sum_{n=1}^N a_n\varepsilon_n\gamma_n\right\|_\infty \leqslant \mathbb{E}\left\|\sum_{n=1}^N a_nZ_n\gamma_n\right\|_\infty \leqslant C_2\mathbb{E}\left\|\sum_{n=1}^N a_n\varepsilon_n\gamma_n\right\|_\infty ,$$

*where $(Z_n)_{n \geqslant 1}$ is a standard complex Gaussian sequence and $(\varepsilon_n)_{n \geqslant 1}$ a Bernoulli sequence.*

*Proof*

1) The left-hand inequality is valid in a general Banach framework, in the form:

$$(1) \qquad \mathbb{E}(|Z_1|)\,\mathbb{E}\left\|\sum_{n=1}^N \varepsilon_n x_n\right\| \leqslant \mathbb{E}\left\|\sum_{n=1}^N Z_n x_n\right\|,$$

with $x_n \in E$, for an arbitrary Banach space $E$, and $(Z_n)_{n \geqslant 1}$ complex random variables, *i.i.d.* integrable and symmetric in the complex sense: $uZ_n \sim Z_n$ for any complex number $u$ of modulus 1; this follows from the contraction principle (see Theorem IV.3, Chapter 4, Volume 1).

By specializing (1) to standard complex Gaussians $Z_n$, we obtain the left-hand inequality of the theorem with:

$$C_1 = \frac{1}{2\pi} \iint_{\mathbb{R}^2} \left| \frac{x+iy}{\sqrt{2}} \right| e^{-\frac{x^2+y^2}{2}} \, dxdy = \frac{1}{\sqrt{2}} \int_0^{+\infty} r^2 e^{-r^2/2} \, dr = \frac{\sqrt{\pi}}{2}.$$

2) The right-hand inequality is less evident, and is even more striking because when $G$ is infinite, $\mathcal{C}(G)$ contains $\ell_\infty^n$'s uniformly, and hence, by the easy part of a theorem due to Maurey and Pisier ([1976]), the Gaussian series $\sum_n Z_n f_n$ and the Rademacher series $\sum_n \varepsilon_n f_n$, where $f_n \in \mathcal{C}(G)$, do not have the same behavior: they are not equiconvergent (see Chapter 4 of Volume 1, Exercise VII.1). The Marcus–Pisier inequality states that, for the $f_n$'s proportional to the characters $\gamma_n$, the behavior in this case is the same. The key lies in the following theorem:

**Theorem III.6** (The Marcus–Pisier Inequality)   *There exists a numerical constant $\beta > 0$ such that, for every independent sequence $(\xi_n)_{n \geqslant 1}$ of symmetric variables with finite variance, we have*:

$$\mathbb{E} \left\| \sum_{n=1}^N a_n \xi_n \gamma_n \right\|_\infty \leqslant \beta \sigma \, \mathbb{E} \left\| \sum_{n=1}^N a_n Z_n \gamma_n \right\|_\infty ,$$

*where $\sigma = \sup_{n \geqslant 1} \|\xi_n\|_2$ and $(Z_n)_{n \geqslant 1}$ is a complex standard Gaussian sequence.*

Before proving this theorem, let us see how it almost immediately implies the preceding theorem. We truncate $Z_n$ by writing

$$Z_n = Z_n \mathbb{1}_{(|Z_n| \leqslant M)} + Z_n \mathbb{1}_{(|Z_n| > M)} = Z_n' + Z_n'',$$

with the numerical constant $M$ to be adjusted later.

If $X_t = X(t) = \sum_{n=1}^N a_n Z_n \gamma_n(t)$, we thus have $X_t = X_t' + X_t''$, with $X_t' = X'(t) = \sum_{n=1}^N a_n Z_n' \gamma_n(t)$ and $X_t'' = X''(t) = \sum_{n=1}^N a_n Z_n'' \gamma_n(t)$. Then we use the contraction principle and symmetry to bound $\mathbb{E}\|X'\|_\infty$, and Theorem III.6 to bound $\mathbb{E}\|X''\|_\infty$:

$$\mathbb{E}\|X\|_\infty \leqslant \mathbb{E}\|X'\|_\infty + \mathbb{E}\|X''\|_\infty$$

$$\leqslant 2M \, \mathbb{E} \left\| \sum_{n=1}^N \varepsilon_n a_n \gamma_n \right\|_\infty + \beta \, \|Z_1''\|_2 \mathbb{E} \left\| \sum_{n=1}^N a_n Z_n \gamma_n \right\|_\infty .$$

We now adjust $M$ large enough to have $\beta \|Z_1''\|_2 \leqslant 1/2$. Then the preceding inequality becomes:

$$\mathbb{E}\|X\|_\infty \leqslant 2M\,\mathbb{E}\left\|\sum_{n=1}^N \varepsilon_n a_n \gamma_n\right\|_\infty + \frac{1}{2}\mathbb{E}\|X\|_\infty;$$

thus:

$$\mathbb{E}\|X\|_\infty \leqslant 4M\,\mathbb{E}\left\|\sum_{n=1}^N \varepsilon_n a_n \gamma_n\right\|_\infty,$$

which proves Theorem III.5                                                    □

The proof of Theorem III.6 relies heavily on the Dudley–Fernique theorem of Chapter 3 of this volume (in the form of Proposition V.5), which brings into play a non-decreasing rearrangement. The following lemma is useful.

**Lemma III.7**  *Let G be an infinite group. Then:*

a) *Let $\overline{\psi}$ be the non-decreasing rearrangement of a positive integrable function $\psi : G \to \mathbb{R}^+$. Then, for $0 \leqslant h \leqslant 1$:*

$$\int_0^h \overline{\psi}(x)\,\mathrm{d}x = \inf_{m(A)=h}\left[\int_A \psi(t)\,\mathrm{d}m(t)\right].$$

b) *If $\psi_1, \psi_2$ are as in a), then, for $0 \leqslant h \leqslant 1$:*

$$\int_0^h \overline{\psi_1 + \psi_2}(x)\,\mathrm{d}x \geqslant \int_0^h \overline{\psi_1}(x)\,\mathrm{d}x + \int_0^h \overline{\psi_2}(x)\,\mathrm{d}x.$$

Recall the definition of the non-decreasing rearrangement $\overline{\psi} : [0,1] \to \mathbb{R}^+$ of the function $\psi : G \to \mathbb{R}^+$:

$$\overline{\psi}(x) = \sup\{y \in \mathbb{R}^+ \,;\, \mu(y) < x\},$$

where $\mu(y) = m(\{t \in G \,;\, \psi(t) < y\})$.

In the proof that follows, the elements of the group $G$ are denoted by the letter $t$; $x$ is a number in $[0,1]$, and $y \in \mathbb{R}^+$.

*Proof of the lemma*

a) We can assume $0 < h < 1$. Let $\lambda$ be the Lebesgue measure on $[0,1]$. The formula of integration by parts gives:

$$\int_0^h \overline{\psi}(x)\,\mathrm{d}x = \int_0^{+\infty} \lambda\big(\{x \,;\, \overline{\psi}(x) \geqslant y\} \cap [0,h]\big)\,\mathrm{d}y$$

$$= \int_0^{+\infty} \lambda\big(]\mu(y), 1] \cap [0,h]\big)\,\mathrm{d}y$$

$$= \int_0^\infty [h - \mu(y)]^+\,\mathrm{d}y.$$

Now, for $A \subseteq G$ such that $m(A) = h$, we have:

$$\int_A \psi(t)\, dm(t) = \int_0^{+\infty} m\big(\{t \in G\,;\ \psi(t) \geqslant y\} \cap A\big)\, dy$$

$$= \int_0^{+\infty} \big[m(A) - m(\{t \in G\,;\ \psi(t) < y\} \cap A)\big]\, dy$$

$$= \int_0^{+\infty} \big[h - m(\{t \in G\,;\ \psi(t) < y\} \cap A)\big]\, dy.$$

However, for any $y > 0$:

$$[h - \mu(y)]^+ = \big[h - m(\{t \in G\,;\ \psi(t) < y\})\big]^+$$
$$\leqslant h - m\big(\{t \in G\,;\ \psi(t) < y\} \cap A\big)\,;$$

therefore we obtain:

$$\int_0^h \overline{\psi}(x)\, dx \leqslant \inf_{m(A)=h} \int_A \psi(t)\, dm(t).$$

To show that it is in fact an equality, let $y_0 = \overline{\psi}(h)$. We distinguish two cases.

*Case 1:* $\mu(y_0) = h$

Then, with $A = \{t \in G\,;\ \psi(t) < y_0\}$, we have $m(A) = h$, by the definition of $\mu$. Moreover:

$$[h - \mu(y)]^+ = h - m(\{t \in G\,;\ \psi(t) < y\} \cap A)$$

for any $y \in \mathbb{R}^+$; indeed, if $y < y_0$, both sides are equal to $h - \mu(y)$; and if $y \geqslant y_0$, they are both null.

*Case 2:* $\mu(y_0) < h$

Denote $\delta = m(\{t \in G\,;\ \psi(t) = y_0\})$. Then $\delta \geqslant h - \mu(y_0)$. Indeed, if not, we would have $\delta + \mu(y_0) < h$, which would mean that

$$m(\{t \in G\,;\ \psi(t) \leqslant y_0\}) < h\,;$$

however:

$$m(\{t \in G\,;\ \psi(t) \leqslant y_0\}) = \lim_{n \to +\infty} m(\{t \in G\,;\ \psi(t) < y_0 + 1/n\})$$

$$= \lim_{n \to +\infty} \mu(y_0 + 1/n),$$

which contradicts the definition of $y_0 = \sup\{y > 0\,;\ \mu(y) < h\}$.

Now, as the compact group $G$ is infinite, its Haar measure $m$ is *diffuse* (see the Annex, Volume 1, Corollary III.7, and the Remark that follows), hence without atoms. Thus, thanks to Lyapounov's theorem (Chapter 7 of Volume 1, Theorem IV.10), it takes on all the values of the interval $[0, 1]$.

We can thus find $B \subseteq \{t \in G; \; \psi(t) = y_0\}$ such that $m(B) = h - \mu(y_0)$. Hence, the set $A = \{t \in G; \; \psi(t) < y_0\} \cup B$ has measure $m(A) = h$. Moreover, we also have $[h - \mu(y)]^+ = h - m(\{t \in G; \; \psi(t) < y\} \cap A)$ for any $y$, hence $\int_0^h \overline{\psi}(x) \, dx = \int_A \psi(t) \, dm(t)$.

b) If $m(A) = h$, then by a):

$$\int_A \big(\psi_1(t) + \psi_2(t)\big) \, dm(t) = \int_A \psi_1(t) \, dm(t) + \int_A \psi_2(t) \, dm(t)$$
$$\geqslant \int_0^h \overline{\psi}_1(x) \, dx + \int_0^h \overline{\psi}_2(x) \, dx.$$

Passing to the infimum over all possible sets $A$, and using a) again, we obtain the stated inequality. $\qquad\qquad\square$

Now let $\mathfrak{C}$ denote the convex cone consisting of the non-negative bounded measurable functions $\psi \colon G \to \mathbb{R}_+$. Let $I \colon \mathfrak{C} \to \overline{\mathbb{R}}_+$ be the functional introduced in Chapter 3 of this volume:

$$I(\psi) = \int_0^1 \frac{\overline{\psi}(x)}{x\sqrt{\log(e/x)}} \, dx = \int_0^1 \overline{\psi}(x) f(x) \, dx.$$

We will see that: *I is a concave functional on the convex cone* $\mathfrak{C}$.

As $I(\lambda\psi) = \lambda I(\psi)$ for $\lambda > 0$, since clearly $\overline{\lambda\psi} = \lambda\overline{\psi}$, it suffices to show that, for $\psi_1, \psi_2 \in \mathfrak{C}$:

$$I(\psi_1 + \psi_2) \geqslant I(\psi_1) + I(\psi_2).$$

For this, we use a lemma of Hardy (BENNETT –SHARPLEY, page 56):

**Lemma III.8** *Let* $\varphi \colon [0,1] \to \overline{\mathbb{R}}_+$ *be a non-negative non-increasing function, and also let* $a, b \colon [0,1] \to \mathbb{R}_+$ *be two non-negative functions such that:*

$$\int_0^x a(t) \, dt \leqslant \int_0^x b(t) \, dt, \qquad \forall x \in [0,1].$$

*Then:*

$$\int_0^1 a(t)\varphi(t) \, dt \leqslant \int_0^1 b(t)\varphi(t) \, dt.$$

*Proof of the lemma*   By approximation, and by the monotone convergence theorem, we can assume that $\varphi$ is a step function, and, since $\varphi$ is non-increasing, write:

$$\varphi = \sum_{j=1}^n c_j \mathbb{1}_{[0,x_j]},$$

where the scalars $c_j$ are positive, and where $0 < x_1 < \ldots < x_n \leqslant 1$. The hypothesis then gives:

$$\int_0^1 a(t)\varphi(t)\,\mathrm{d}t = \sum_{j=1}^n c_j \int_0^{x_j} a(t)\,\mathrm{d}t$$

$$\leqslant \sum_{j=1}^n c_j \int_0^{x_j} b(t)\,\mathrm{d}t = \int_0^1 b(t)\varphi(t)\,\mathrm{d}t. \qquad \square$$

To prove the concavity of $I$, it then suffices to apply this lemma to $a = \overline{\psi}_1 + \overline{\psi}_2$, $b = \overline{\psi_1 + \psi_2}$, and $\varphi = f$, which is possible, by Lemma III.7, and since $f$ is non-increasing.

We now return to:

*Proof of Theorem III.6* Set:

$$A = \mathbb{E}\left\|\sum_{n=1}^N a_n \xi_n \gamma_n\right\|_\infty.$$

Let $(\varepsilon_n)_{n\geqslant 1}$ be a Bernoulli sequence, independent of the sequence $(\xi_n)_{n\geqslant 1}$. Let $\mathbb{E}_\xi$ denote the expectation with respect to the $\xi_n$'s, and $\mathbb{E}_\varepsilon$ denote that with respect to the $\varepsilon_n$'s: then the symmetry of the variables $\xi_n$ leads to:

$$A = \mathbb{E}_\varepsilon \mathbb{E}_\xi \left\|\sum_{n=1}^N a_n \varepsilon_n(\omega_1)\xi_n(\omega_2)\gamma_n\right\|_\infty = \mathbb{E}_\xi \mathbb{E}_\varepsilon \left\|\sum_{n=1}^N a_n \varepsilon_n(\omega_1)\xi_n(\omega_2)\gamma_n\right\|_\infty.$$

For a fixed $\omega_2$, we use the left-hand inequality of Theorem III.5 (already proved!), with $a_n$ replaced by $a_n \xi_n(\omega_2)$, and then the Dudley majoration theorem (Theorem IV.3 of Chapter 3, this volume), to obtain:

$$\mathbb{E}_\varepsilon \left\|\sum_{n=1}^N a_n \xi_n(\omega_2)\varepsilon_n(\omega_1)\gamma_n\right\|_\infty \leqslant \alpha\left[I(\psi_{\omega_2}) + \left(\sum_{n=1}^N |a_n|^2 |\xi_n(\omega_2)|^2\right)^{1/2}\right],$$

with $\alpha > 0$ a numerical constant, and with:

$$\psi_{\omega_2}(t) = \left(\sum_{n=1}^N |a_n|^2 |\xi_n(\omega_2)|^2 |\gamma_n(t) - 1|^2\right)^{1/2}.$$

Next we integrate this inequality with respect to $\omega_2$. By using Jensen's inequality, allowed by the concavity of the functional $I$, we obtain (note that the mapping $t \mapsto [\mathbb{E}_\xi(\psi_{\omega_2})](t)$ is in $\mathfrak{C}$):

$$A \leqslant \alpha \left[ I\big(\mathbb{E}_\xi (\psi_{\omega_2})\big) + \mathbb{E}_\xi \left( \sum_{n=1}^{N} |a_n|^2 |\xi_n(\omega_2)|^2 \right)^{1/2} \right].$$

However:

$$\mathbb{E}_\xi \left( \sum_{n=1}^{N} |a_n|^2 |\xi_n(\omega_2)|^2 \right)^{1/2} \leqslant \left[ \mathbb{E}_\xi \left( \sum_{n=1}^{N} |a_n|^2 |\xi_n(\omega_2)|^2 \right) \right]^{1/2}$$

$$\leqslant \sigma \left( \sum_{n=1}^{N} |a_n|^2 \right)^{1/2}.$$

Now the functional $I$ is non-increasing on $\mathfrak{C}$; indeed, if $\psi_1, \psi_2 \in \mathfrak{C}$, and $\psi_1(t) \leqslant \psi_2(t)$ for every $t \in G$, we see that:

$$m(\{t \in G; \; \psi_1(t) < y\}) \geqslant m(\{t \in G; \; \psi_2(t) < y\}),$$

and hence:

$$\overline{\psi}_1(x) = \sup \{y > 0; \; m(\{t \in G; \; \psi_1(t) < y\}) < x\}$$
$$\leqslant \overline{\psi}_2(x) = \sup \{y > 0; \; m(\{t \in G; \; \psi_2(t) < y\}) < x\}.$$

For $t \in G$, if we set:

$$\psi(t) = \left( \sum_{n=1}^{N} |a_n|^2 |\gamma_n(t) - 1|^2 \right)^{1/2},$$

we have:

$$[\mathbb{E}_\xi (\psi_{\omega_2})](t) \leqslant \left[ \mathbb{E}_\xi \left( \sum_{n=1}^{N} |a_n|^2 |\xi_n(\omega_2)|^2 |\gamma_n(t) - 1|^2 \right) \right]^{1/2} \leqslant \sigma \, \psi(t);$$

hence:

$$A \leqslant \alpha \left[ I(\sigma \psi) + \sigma \left( \sum_{n=1}^{N} |a_n|^2 \right)^{1/2} \right] \leqslant \beta \sigma \, \mathbb{E} \left\| \sum_{n=1}^{N} a_n Z_n \gamma_n \right\|_\infty,$$

this time by a use of Fernique's inequality (Theorem V.4 of Chapter 3, this volume), where $\beta > 0$ is a new numerical constant.

The proof of Theorem III.6 is thus complete. $\qquad\square$

**Remark**  We have been working with finite sums, but there is an immediate extension to arbitrary sums; for example, if $\sum_{n=1}^{+\infty} a_n Z_n \gamma_n \in \mathcal{C}^{as}$, and if

$\sup_{n \geqslant 1} \|\xi_n\|_2 = \sigma < +\infty$, then the series $\sum_{n=1}^{+\infty} a_n \xi_n \gamma_n$ is almost surely uniformly convergent on $G$, and satisfies:

$$\mathbb{E} \left\| \sum_{n=1}^{+\infty} a_n \xi_n \gamma_n \right\|_\infty \leqslant C \sigma \, \mathbb{E} \left\| \sum_{n=1}^{+\infty} a_n Z_n \gamma_n \right\|_\infty.$$

### III.3 Duality Between $\mathcal{C}^{as}$ and $M_{2,\Psi_2}$

Let $G$ be a compact metrizable Abelian group, $\Gamma$ its dual and $\Psi_2$ the Orlicz function $\Psi_2(x) = e^{x^2} - 1$. A *multiplier* from $L^2$ to $L^{\Psi_2}$ is an operator $T \colon L^2 \to L^{\Psi_2}$ that commutes with the translations. Such an operator is determined by a sequence $\big(\widehat{T}(\gamma)\big)_{\gamma \in \Gamma}$ of "Fourier coefficients" such that:

$$T(\gamma) = \widehat{T}(\gamma)\,\gamma \qquad \text{for every } \gamma \in \Gamma,$$

and

$$\left\| \sum_{\gamma \in \Gamma} a_\gamma \widehat{T}(\gamma)\gamma \right\|_{\Psi_2} \leqslant C \left\| \sum_{\gamma \in \Gamma} a_\gamma \gamma \right\|_2$$

for every trigonometric polynomial $\sum_{\gamma \in \Gamma} a_\gamma \gamma$. The smallest possible constant $C > 0$ is denoted $\|T\|$.

The space $M_{2,\Psi_2}$ is the set of all multipliers from $L^2$ to $L^{\Psi_2}$. We show that it can be identified as the dual of $\mathcal{C}^{as}$. The proof of this duality is comparable in difficulty to the proof of the celebrated Fefferman–Stein $H^1 - BMO$ duality, and relies on the theorems of Dudley and Fernique. More precisely, we have the following theorem:

### Theorem III.9

1) *There exists a numerical constant $C_0 > 0$ such that:*

$$(*) \qquad \sum_{\gamma \in \Gamma} |\widehat{f}(\gamma)|\,|\widehat{T}(\gamma)| \leqslant C_0 \, [\![f]\!] \, \|T\|, \qquad \forall f \in \mathcal{C}^{as},\, \forall\, T \in M_{2,\Psi_2}.$$

2) *By this duality, the dual of $\mathcal{C}^{as}$ is isomorphically identified with $M_{2,\Psi_2}$.*

### Proof

1) If $u, v \in \mathcal{C}^{as}$, and $|\widehat{u}| \leqslant |\widehat{v}|$, then $[\![u]\!] \leqslant [\![v]\!]$, thanks to the unconditionality of characters in $\mathcal{C}^{as}$. Thus it suffices to prove $(*)$ when $f$ is a trigonometric polynomial. Moreover, we can assume $\|T\| = 1$. For $t, x \in G$, we set $g_t(x) = g(x+t)$. Then, the translation invariance of $T$ leads to $T(f_t) = (Tf)_t$.

We define:

$$X_t = (Tf)_t = \sum_{\gamma \in \Gamma} \widehat{f}(\gamma)\widehat{T}(\gamma)\gamma(t)\gamma.$$

This is a process $(X_t)_{t \in G}$ whose underlying probability space is the group $G$ with its Haar measure $m$. Then, with $d(s, t) = \|f_s - f_t\|_2$:

$$\|X_s - X_t\|_{\psi_2} = \|T(f_s - f_t)\|_{\psi_2} \leqslant \|f_s - f_t\|_2;$$

hence the process $(X_t)_{t \in G}$ satisfies the Lipschitz condition of Dudley's theorem, and therefore:

$$(\diamond) \qquad \mathbb{E}\left(\sup_{t \in G} |X_t|\right) \leqslant \alpha\left[J(d) + \|X_0\|_2\right],$$

where $\mathbb{E}$ is none other than the integral with respect to the Haar measure $m$. Furthermore, since $d(s, t) = \left(\sum_{\gamma \in \Gamma} |\widehat{f}(\gamma)|^2 |\gamma(s) - \gamma(t)|^2\right)^{1/2}$, $d$ is also the (pseudo)-metric associated with the Gaussian process:

$$Y_t = \sum_{\gamma \in \Gamma} \widehat{f}(\gamma)Z_\gamma \gamma(t),$$

where $(Z_\gamma)_{\gamma \in \Gamma}$ is a standard complex Gaussian sequence. The Fernique minoration theorem then shows that $J(d)$ is dominated by:

$$\mathbb{E}\left(\sup_{t \in G} |Y_t|\right) = [\![f]\!],$$

and via $(\diamond)$ we obtain:

$$\mathbb{E}\left(\sup_{t \in G} |X_t|\right) \leqslant C_0 [\![f]\!],$$

with $C_0 > 0$ a numerical constant. Note that, as $X_t$ is a trigonometric polynomial, its paths are continuous, and the upper bound does not present any difficulty of measurability.

Moreover, $|X_t(x)| = \left|\sum_{\gamma \in \Gamma} \widehat{f}(\gamma)\widehat{T}(\gamma)\gamma(t + x)\right|$; hence:

$$\sup_{t \in G} |X_t(x)| \geqslant |X_{-x}(x)| = \left|\sum_{\gamma \in \Gamma} \widehat{f}(\gamma)\widehat{T}(\gamma)\right|,$$

and by integrating with respect to $x$, we obtain:

$$\left|\sum_{\gamma \in \Gamma} \widehat{f}(\gamma)\widehat{T}(\gamma)\right| \leqslant \int_G \sup_{t \in G} |X_t(x)| \, dm(x) = \mathbb{E}\left(\sup_{t \in G} |X_t|\right) \leqslant C_0 [\![f]\!].$$

Finally, the unconditionality of the characters for the norm $\mathcal{C}^{as}$ implies (∗). Indeed, let $u = (u_\gamma)_{\gamma \in \Gamma}$ be a family of complex numbers of modulus 1 for which $\sum_{\gamma \in \Gamma} |\widehat{f}(\gamma)| |\widehat{T}(\gamma)| = \sum_{\gamma \in \Gamma} u_\gamma \widehat{f}(\gamma) \widehat{T}(\gamma)$, and let $f^u = \sum_{\gamma \in \Gamma} u_\gamma \widehat{f}_\gamma \gamma$. The preceding inequality gives:

$$\sum_{\gamma \in \Gamma} |\widehat{f}(\gamma)| |\widehat{T}(\gamma)| \leqslant C_0 [\![ f^u ]\!] = C_0 [\![ f ]\!].$$

2) If $T \in M_{2,\Psi_2}$, the linear functional $\widetilde{T}$ defined by:

$$\langle \widetilde{T}, f \rangle = \sum_{\gamma \in \Gamma} \widehat{f}(\gamma) \widehat{T}(\gamma)$$

is in the dual $[\mathcal{C}^{as}]^*$, and we have $\|\widetilde{T}\| \leqslant C_0 \|T\|$, according to (∗). Conversely, if $\Phi \in [\mathcal{C}^{as}]^*$, we must show that the sequence $(\Phi(\gamma))_{\gamma \in \Gamma}$ is the sequence of Fourier coefficients $\widehat{T}(\gamma)$ of a multiplier $T \in M_{2,\Psi_2}$; *a priori* this is far from evident, and requires the following proposition and lemmas.

Denote by $\varphi_2$ the Orlicz function $\varphi_2(x) = x\sqrt{1 + \log(1 + x)}$.

**Proposition III.10** *Let $A_{2,\varphi_2}$ be the space of functions $f : G \to \mathbb{C}$ that can be written $\sum_{n=1}^{+\infty} u_n * v_n$, with $\sum_{n=1}^{+\infty} \|u_n\|_2 \|v_n\|_{\varphi_2} < +\infty$, and normed by:*

$$\|f\|_* = \inf \left\{ \sum_{n=1}^{+\infty} \|u_n\|_2 \|v_n\|_{\varphi_2} \, ; \, f = \sum_{n=1}^{+\infty} u_n * v_n \right\}.$$

*Then $M_{2,\Psi_2}$ can be identified isomorphically as the dual of $A_{2,\varphi_2}$.*

*Proof* This result is classical and easy (see LARSEN, Chapter 5) when $\varphi_2(x)$ is replaced by $x^p$ and $\Psi_2(x)$ by $x^q$, with $p > 1$ and $q > 1$ conjugate exponents. The extension to the Orlicz spaces in duality $L^{\varphi_2}$ and $L^{\Psi_2} = [L^{\varphi_2}]^*$ is purely formal, and left as an exercise. □

**Lemma III.11** *Let $f = \sum_{\gamma \in \Gamma} a_\gamma \gamma$ be a trigonometric polynomial, and $(\varepsilon_\gamma)_{\gamma \in \Gamma}$ a Rademacher sequence. Then:*

$$\mathbb{E} \left\| \sum_{\gamma \in \Gamma} \varepsilon_\gamma a_\gamma \gamma \right\|_{\Psi_2} \leqslant C_1 \|f\|_2,$$

*with $C_1 > 0$ a numerical constant.*

*Proof* We can assume $\|f\|_2 = 1$. Set $f^\omega(t) = \sum_{\gamma \in \Gamma} \varepsilon_\gamma(\omega) a_\gamma \gamma(t)$, as above, and denote:

$$J(\omega) = \int_G \Psi_2 \left( \frac{|f^\omega(t)|}{C} \right) dm(t),$$

where $C > 0$ is the constant given by the Khintchine inequalities:

$$\int_\Omega \Psi_2 \left( \frac{\left| \sum_{\gamma \in \Gamma} \varepsilon_\gamma(\omega) b_\gamma \right|}{C} \right) d\mathbb{P}(\omega) \leqslant 1 \qquad \text{if} \quad \sum_{\gamma \in \Gamma} |b_\gamma|^2 = 1.$$

It ensues from this inequality and from Fubini's theorem that:

$$(1) \qquad\qquad \int_\Omega J(\omega) \, d\mathbb{P}(\omega) \leqslant 1.$$

We now show that:

$$(2) \qquad\qquad \|f^\omega\|_{\Psi_2} > C \implies \|f^\omega\|_{\Psi_2} \leqslant CJ(\omega).$$

For this, note that if $\|f^\omega\|_{\Psi_2} > C$, then, for $t \in G$:

$$\Psi_2 \left( \frac{|f^\omega(t)|}{C} \right) = \Psi_2 \left( \frac{|f^\omega(t)|}{\|f^\omega\|_{\Psi_2}} \frac{\|f^\omega\|_{\Psi_2}}{C} \right) \geqslant \frac{\|f^\omega\|_{\Psi_2}}{C} \Psi_2 \left( \frac{|f^\omega(t)|}{\|f^\omega\|_{\Psi_2}} \right),$$

because $\dfrac{\Psi_2(x)}{x}$ is increasing on $\mathbb{R}^+$ and hence $\Psi_2(\lambda x) \geqslant \lambda \Psi_2(x)$ if $\lambda \geqslant 1$. However, by definition we have:

$$\int_G \Psi_2 \left( \frac{|f^\omega(t)|}{\|f^\omega\|_{\Psi_2}} \right) dm(t) = 1.$$

Hence, an integration with respect to $t$ in the preceding inequality leads to $J(\omega) \geqslant \dfrac{\|f^\omega\|_{\Psi_2}}{C}$, which proves (2). Thus we always have:

$$\|f^\omega\|_{\Psi_2} \leqslant C + CJ(\omega);$$

which, by (1), implies:

$$\mathbb{E} \left\| \sum_{\gamma \in \Gamma} \varepsilon_\gamma a_\gamma \gamma \right\|_{\Psi_2} = \mathbb{E}\|f^\omega\|_{\Psi_2} \leqslant C\big(1 + \mathbb{E}(J)\big) \leqslant 2C :$$

the lemma is proved with $C_1 = 2C$. $\qquad\qquad\qquad\qquad\qquad\qquad\qquad$ □

**Lemma III.12** *For every $u \in L^{\Psi_2}$ and every $v \in L^{\varphi_2}$, we have $u * v \in \mathcal{C}(G)$, and there exists a numerical constant $C_2 > 0$ such that:*

$$\|u * v\|_\infty \leqslant C_2 \|u\|_{\Psi_2} \|v\|_{\varphi_2}.$$

*Proof* The continuity of $u * v$ ensues from the continuity of translation in $L^{\varphi_2}$. Moreover, if we set $u_x(t) = u(x - t)$, then:

$$|(u * v)(x)| \leqslant \int_G |u_x(t) v(t)| \, dm(t) \leqslant C_2 \|u_x\|_{\Psi_2} \|v\|_{\varphi_2} = C_2 \|u\|_{\Psi_2} \|v\|_{\varphi_2},$$

where $C_2$ is the constant of duality between $L^{\varphi_2}$ and $L^{\Psi_2}$, i.e.:

$$\left| \int_G fg \, dm \right| \leqslant C_2 \|f\|_{\Psi_2} \|g\|_{\varphi_2}$$

for $f \in L^{\Psi_2}$ and $g \in L^{\varphi_2}$. □

We return to the proof of Theorem III.9. By Proposition III.10, this is equivalent to showing that, if $\Phi \in [\mathcal{C}^{as}]^*$, the formula:

$$\langle \widetilde{\Phi}, w \rangle = \sum_{\gamma \in \Gamma} \Phi(\gamma)\widehat{w}(\gamma)$$

defines a continuous linear functional on $A_{2,\varphi_2}$. By convexity, it suffices to examine the case where $w = u * v$, with $u \in L^2$ and $v \in L^{\varphi_2}$. Furthermore, we can assume that $u$ and $v$ are trigonometric polynomials. Then Lemmas III.11 and III.12 imply:

$$\left| \sum_{\gamma \in \Gamma} \Phi(\gamma)\widehat{w}(\gamma) \right| = \left| \Phi\left( \sum_{\gamma \in \Gamma} \widehat{w}(\gamma)\gamma \right) \right|$$

$$\leqslant \|\Phi\| \, [\![w]\!] \leqslant C_3 \|\Phi\| \, \mathbb{E} \left\| \sum_{\gamma \in \Gamma} \varepsilon_\gamma \widehat{u}(\gamma)\widehat{v}(\gamma)\gamma \right\|_\infty,$$

according to the Marcus–Pisier theorem. Hence:

$$\left| \sum_{\gamma \in \Gamma} \Phi(\gamma)\widehat{w}(\gamma) \right| \leqslant C_2 C_3 \|\Phi\| \, \mathbb{E} \left\| \sum_{\gamma \in \Gamma} \varepsilon_\gamma \widehat{u}(\gamma)\gamma \right\|_{\Psi_2} \|v\|_{\varphi_2}$$

$$\leqslant C_4 \|\Phi\| \, \|u\|_2 \|v\|_{\varphi_2};$$

therefore:

$$|\Phi(w)| \leqslant C_4 \|\Phi\| \, \|w\|_{A_{2,\varphi_2}},$$

which completes the proof. □

An interesting use of the duality $\mathcal{C}^{as} - M_{2,\Psi_2}$ is the following inequality, which has a variety of applications in Harmonic Analysis (see also Queffélec and Saffari [1996], etc.):

**Proposition III.13** (The Salem–Zygmund Theorem) *There exist constants* $A, B > 0$ *such that, for arbitrary integers* $0 < \lambda_1 < \cdots < \lambda_n$ *($n \geqslant 2$), we have*:

$$A\sqrt{n \log n} \leqslant \left[\!\!\left[ \sum_{j=1}^{n} e_{\lambda_j} \right]\!\!\right] \leqslant B\sqrt{n \log \lambda_n},$$

*where* $e_k(t) = \mathrm{e}^{ikt}$.

*Proof*  The right-hand inequality is easy: $f(t) = \sum_{j=1}^{n} e^{i\lambda_j t}$ is a trigonometric polynomial of degree $\lambda_n$; hence Bernstein's inequality (Chapter 6 of Volume 1, Subsection V) implies:

$$\left\| \sum_{j=1}^{n} \varepsilon_j(\omega) e_{\lambda_j} \right\|_{\infty} \leqslant 5 \sup_{t \in R} \left| \sum_{j=1}^{n} \varepsilon_j(\omega) e^{i\lambda_j t} \right|,$$

where $R = \{k\pi/2\lambda_n ; \ k = 0, 1, \ldots, 4\lambda_n - 1\}$ is the set of $(4\lambda_n)$-th roots of unity; the majoration theorem (Theorem IV.5 of Chapter 1, Volume 1) implies:

$$\mathbb{E}\left( \sup_{t \in R} \left| \sum_{j=1}^{n} \varepsilon_j(\omega) e^{i\lambda_j t} \right| \right) \leqslant C\sqrt{\log(1 + |R|)}\sqrt{n}.$$

For the left-hand inequality, let $\Lambda = \{\lambda_1, \ldots, \lambda_n\}$, and let $T \in M_{2,\Psi_2}$ such that $\widehat{T} = \mathbb{1}_{\Lambda}$. We need to bound $\|T\|_{2,\Psi_2}$. For this, let $h \in L^2$ with norm 1, and let $2 \leqslant p < +\infty$. Denote $\theta = 2/p$, so that $\dfrac{1}{p} = \dfrac{1 - \theta}{\infty} + \dfrac{\theta}{2}$. Hölder's inequality provides:

$$\left\| \sum_{j=1}^{n} \widehat{T}(\lambda_j)\widehat{h}(\lambda_j) e_{\lambda_j} \right\|_{p} \leqslant \left\| \sum_{j=1}^{n} \widehat{T}(\lambda_j)\widehat{h}(\lambda_j) e_{\lambda_j} \right\|_{\infty}^{1-\theta} \left\| \sum_{j=1}^{n} \widehat{T}(\lambda_j)\widehat{h}(\lambda_j) e_{\lambda_j} \right\|_{2}^{\theta}.$$

However:

$$\left\| \sum_{j=1}^{n} \widehat{T}(\lambda_j)\widehat{h}(\lambda_j) e_{\lambda_j} \right\|_{\infty} \leqslant \sum_{j=1}^{n} |\widehat{T}(\lambda_j)||\widehat{h}(\lambda_j)| \leqslant \sum_{j=1}^{n} |\widehat{h}(\lambda_j)| \leqslant \sqrt{n}\|h\|_2,$$

and:

$$\left\| \sum_{j=1}^{n} \widehat{T}(\lambda_j)\widehat{h}(\lambda_j) e_{\lambda_j} \right\|_{2} \leqslant \|h\|_2;$$

hence:

$$\left\| \sum_{j=1}^{n} \widehat{T}(\lambda_j)\widehat{h}(\lambda_j) e_{\lambda_j} \right\|_{p} \leqslant \|h\|_2(\sqrt{n})^{1-\theta} = \|h\|_2 \, n^{\frac{1}{2} - \frac{\theta}{2}};$$

and thus:

$$\|T\|_{2,\Psi_2} \leqslant \alpha \sup_{p \geqslant 2} \frac{n^{1/2}}{\sqrt{p} \, n^{1/p}} \leqslant \beta \sqrt{\frac{n}{\log n}}$$

(the supremum is attained for $p = 2 \log n$).

The duality inequality of Theorem III.9 then reads:

$$n \leqslant C_0 [\![f]\!] \|T\|_{2,\Psi_2} \leqslant C_0 \beta \, [\![f]\!] \sqrt{\frac{n}{\log n}} \, ,$$

which provides the desired result.                                       □

# IV  Applications of the Space $\mathcal{C}^{as}$

## IV.1  Characterization of Sidon Sets

Let $G$ be a compact metrizable Abelian group, and $\Gamma$ its dual; Rudin's theorem (Chapter 4, this volume) states that if a subset $\Lambda$ of $\Gamma$ is a Sidon set, with constant $S(\Lambda)$, then:

$(\heartsuit)$ $\qquad\qquad \|f\|_p \leqslant S(\Lambda)\sqrt{p}\,\|f\|_2 \qquad \text{for } 2 \leqslant p < +\infty$

for every trigonometric polynomial $f \in \mathcal{P}_\Lambda$.

The converse, i.e. the "Sidon nature" of $\Lambda$ under the hypothesis $(\heartsuit)$, was known for particular groups, such as $G = (\mathbb{Z}/q\mathbb{Z})^{\mathbb{N}}$, where $q$ is a prime number (Malliavin and Malliavin-Brameret [1967]). Pisier [1978 b] proved it in full generality, with the introduction of the space $\mathcal{C}^{as}$ and the use of Gaussian processes in Harmonic Analysis. This proof is presented here: it is actually quite simple... once the space $\mathcal{C}^{as}$ is available and its duality with $M_{2,\Psi_2}$ highlighted, with Rider's theorem used throughout! An "elementary"(!) proof, i.e. one using selectors instead of Gaussian processes, was later given by Bourgain [1985 a] and [1985 b]; this was seen in Chapter 5 of this volume.

**Theorem IV.1** (Pisier's Theorem)  *Let $G$ be a compact metrizable Abelian group, and $\Lambda = \{\lambda_n\}_{n\geqslant 1}$ a subset of its dual $\Gamma$. If there exists a constant $C > 0$ such that:*

$$\|f\|_p \leqslant C\sqrt{p}\,\|f\|_2 \qquad \forall p \geqslant 2, \, \forall f \in L^2_\Lambda,$$

*then $\Lambda$ is a Sidon set.*

The Gaussian version of Rider's theorem is used here; this version follows immediately from the Marcus–Pisier theorem, but Chapter 6 (Volume 1), Proposition V.28, presented a direct proof as a simple consequence of Rider's theorem for the Rademacher functions.

*Proof of the theorem*   We know that:

$$\|\cdot\|_{\Psi_2} \approx \sup_{p\geqslant 2} \frac{\|\cdot\|_2}{\sqrt{p}}\,.$$

Hence, changing the constant $C$ if necessary, we can reformulate the hypothesis as follows:

$$\|f\|_{\Psi_2} \leqslant C\|f\|_2, \quad \forall f \in L^2_\Lambda.$$

Another interpretation is that $\mathbb{I}_\Lambda$ defines a multiplier of $L^2$ in $L^{\Psi_2}$: there exists $T \in M_{2,\Psi_2}$ such that $\widehat{T} = \mathbb{I}_\Lambda$. Indeed, for every $f \in L^2$, we can write:

$$\left\| \sum_{\gamma \in \Lambda} \widehat{f}(\gamma)\gamma \right\|_{\Psi_2} \leqslant C \left\| \sum_{\gamma \in \Lambda} \widehat{f}(\gamma)\gamma \right\|_2 \leqslant C\|f\|_2.$$

Now let $f \in \mathcal{P}_\Lambda$ be a trigonometric polynomial with spectrum in $\Lambda$. The duality inequality between $\mathcal{C}^{as}$ and $M_{2,\Psi_2}$ (Theorem III.9) provides:

$$\sum_{\gamma \in \Lambda} |\widehat{f}(\gamma)| = \sum_{\gamma \in \Gamma} |\widehat{f}(\gamma)| \mathbb{I}_\Lambda(\gamma) \leqslant C_0 [\![f]\!] \, \|\mathbb{I}_\Lambda\|_{2,\Psi_2}$$

$$\leqslant C_0 C [\![f]\!] = C_0 C \, \mathbb{E} \left\| \sum_{n=1}^{+\infty} \widehat{f}(\lambda_n) Z_n \lambda_n \right\|_\infty,$$

and hence, by Proposition V.28 of Chapter 6 (Volume 1), $\Lambda$ is a Sidon set.  □

**Remark**    This proof of Pisier only uses 1) of Theorem III.9, and does not require the entire theory of the duality developed in this theorem. For more clarity, we outline the proof as follows:

1)  The abstract Dudley theorem and Fernique's Theorem show the duality of $\mathcal{C}^{as}$ and $M_{2,\Psi_2}$, thanks to the inequality:

$$\sum_{\gamma \in \Gamma} |\widehat{f}(\gamma)| \, |\widehat{T}(\gamma)| \leqslant C_0 [\![f]\!] \, \|T\|, \quad \forall f \in \mathcal{C}^{as}, \, \forall T \in M_{2,\Psi_2}.$$

2)  Under the hypotheses of Theorem IV.1, the indicator function $\mathbb{I}_\Lambda$ of $\Lambda$ defines a multiplier from $L^2$ to $L^{\Psi_2}$, and thus 1) leads to:

$$\sum_{\gamma \in \Gamma} |\widehat{f}(\gamma)| \leqslant C [\![f]\!], \quad \forall f \in \mathcal{P}_\Lambda.$$

3)  Then the Gaussian Rider theorem (Chapter 6 of Volume 1, Proposition V.28) shows that $\Lambda$ is a Sidon set, according to 2).

## IV.2  The Katznelson Dichotomy Problem

This problem can be stated as follows: let $B$ be a complex Banach space of continuous functions on the circle $\mathbb{T}$; the function $F: [-1, 1] \to \mathbb{C}$ is said to *operate on B* if:

$$f \in B \text{ and } f(\mathbb{T}) \subseteq [-1, 1] \implies F \circ f \in B.$$

Two extreme examples are:

1) $B = C(\mathbb{T})$, with its uniform norm: all continuous functions $F$ operate on $C(\mathbb{T})$;
2) $B = A(\mathbb{T})$, the Wiener algebra i.e. the set of functions $f \in C(\mathbb{T})$ such that $\|f\|_{A(\mathbb{T})} = \sum_{n \in \mathbb{Z}} |\widehat{f}(n)| < +\infty$: only the (real-) analytic functions operate on $A(\mathbb{T})$ (Katznelson [1958]).

In particular, there exists $f \in A(\mathbb{T})$ such that $|f| \notin A(\mathbb{T})$ (Kahane [1956]), and provides a negative answer to a question of Kahane [1958].

Given these two extremes, the Katznelson dichotomy conjecture was the following: let $B$ be an intermediate Banach algebra between $A(\mathbb{T})$ and $C(\mathbb{T})$: $A(\mathbb{T}) \subseteq B \subseteq C(\mathbb{T})$, where the inclusions are continuous; if $B$ is "nice", in a sense to be made precise later, then there are two extreme cases:

1) either all the continuous functions operate on $B$;
2) or only the analytic functions operate on $B$.

It is easy to see that the analytic functions always operate on any Banach algebra. The additional properties that we assume for $B$ are the following:

– $B$ is *semi-simple*, i.e. its Gelfand transform is injective;
– $B$ is self-adjoint: $f \in B \Rightarrow \overline{f} \in B$;
– the spectrum $X(B)$ is equal to $\mathbb{T}$;
  Moreover, we assume that:
– $B$ is *strongly homogeneous*, i.e. on one hand, the translations $T_a$, defined by $(T_a f)(x) = f(x+a)$, are isometries from $B$ into $B$ and $\lim_{a \to 0} \|T_a f - f\|_B = 0$ for every $f \in B$ ($B$ is then said to be *homogeneous*), and, on the other hand, for any $k \in \mathbb{Z}^*$, the operator $D_k$ defined by $(D_k f)(x) = f(kx)$ sends $B$ into $B$, and is of norm 1.

A result of Katznelson [1960] shows that if "lots" of functions operate on $B$, then $B = C(\mathbb{T})$. More precisely, if there exists a function $F$ operating on $B$ and such that $\lim_{t \to 0} |f(t)|/|t| = +\infty$, then $B = C(\mathbb{T})$. Nothing more was known until 1978, when Zafran [1978] resolved the Katznelson conjecture in the negative, and then constructed an intermediate algebra, strictly contained in $C(\mathbb{T})$, on which all $C^3$ functions operate (Zafran [1979]). Pisier was able to go further, thanks to the space $C^{as}$.

**Theorem IV.2** (Pisier)    *Define*:

$$\mathfrak{P} = \mathfrak{P}(\mathbb{T}) = \mathcal{C}^{as} \cap C(\mathbb{T}),$$

*and equip $\mathfrak{P}$ with the norm $\|f\|_{\mathfrak{P}} = 8\,\|f\|_\infty + [\![f]\!]$. Then:*

1) $\mathfrak{P}$ *is a Banach algebra strictly contained in $C(\mathbb{T})$;*
2) *all Lipschitz functions operate on $\mathfrak{P}$.*

*The algebra $\mathfrak{P}$ is called the Pisier algebra.*

*Proof*   First, clearly the space $\mathfrak{P}$ thus normed is complete. Let us show that $\mathfrak{P} \neq C(\mathbb{T})$. If the two were equal, since $8\,\|f\|_\infty \leqslant \|f\|_{\mathfrak{P}}$, the Banach isomorphism theorem would imply the existence of a constant $C' > 8$ such that $\|f\|_{\mathfrak{P}} \leqslant C'\,\|f\|_\infty$, and then also of a constant $C > 0$ such that $[\![f]\!] \leqslant C\|f\|_\infty$ for every $f \in C(\mathbb{T})$. A test with $f(t) = \sum_{j=0}^{n} \delta_j e^{ijt}$, where $(\delta_j)_{j \geqslant 0}$ is the Rudin–Shapiro sequence:

$$\delta_0 = 1, \quad \delta_{2n} = \delta_n \quad \text{and} \quad \delta_{2n+1} = (-1)^n \delta_n,$$

leads to a contradiction. Indeed, the characters are unconditional with constant 1 and $|\delta_j| = 1$; hence, by the Salem–Zygmund theorem (Proposition III.13): $[\![f]\!] = \left[\!\left[ \sum_{j=0}^{n} e^{ijt} \right]\!\right] \geqslant A\sqrt{n \log n}$, which is not possible for every $n \geqslant 1$, since $\|f\|_\infty = \mathrm{O}(\sqrt{n})$.

We now show that $\mathfrak{P} \neq A(\mathbb{T})$. This would be evident *a posteriori* via the result of Katznelson, stating that only the analytic functions operate on $A(\mathbb{T})$. But this, as with $C(\mathbb{T})$, can be shown directly: if there were equality, we would have $\|f\|_{A(\mathbb{T})} \leqslant C(\|f\|_\infty + [\![f]\!])$, and the same polynomial as above leads to a contradiction, since $\|f\|_{A(\mathbb{T})} = n + 1$, and hence, as already said, $\|f\|_\infty = \mathrm{O}(\sqrt{n})$ and $[\![f]\!] = \mathrm{O}(\sqrt{n \log n})$.

Admitting for now that $\mathfrak{P}$ is indeed a Banach algebra, we verify that it has the "nice" properties required.

First, its spectrum is $\mathbb{T}$. Indeed, every element $a \in \mathbb{T}$ defines a character of $\mathfrak{P}$, by evaluation at the point $a$, $\chi_a \colon f \mapsto f(a)$, since $\mathfrak{P} \subseteq C(\mathbb{T})$. Conversely, let $\chi \in X(\mathfrak{P})$. Recall that every character is continuous and of norm 1; hence if we denote $e_n(t) = e^{int}$, then $|\chi(e_1)| \leqslant 1$. However, as $\chi(e_1)\chi(e_{-1}) = \chi(e_1 e_{-1}) = \chi(\mathbb{I}) = 1$, in fact $|\chi(e_1)| = 1$, and there exists $a \in \mathbb{R}$ such that $\chi(e_1) = e^{ia}$. Therefore $\chi(f) = f(a)$ for every trigonometric polynomial $f$, and thus for every $f \in \mathfrak{P}$, as the latter are dense in both $C(\mathbb{T})$ and $\mathcal{C}^{as}$, by Theorem III.4.

Clearly, then, $\mathfrak{P}$ is semi-simple since, modulo the identification of $X(\mathfrak{P})$ with $\mathbb{T}$, its Gelfand transform is none other than the canonical injection of $\mathfrak{P}$ into $C(\mathbb{T})$.

Next, the translations $T_a$ are isometries. Indeed, this is true for the norm $\|\cdot\|_\infty$; and for the norm $[\![\,.\,]\!]$ we have:

$$\mathbb{E}\left\|\sum_{n\in\mathbb{Z}}Z_n e^{ina}\widehat{f}(n)e_n\right\|_\infty = \mathbb{E}\left\|\sum_{n\in\mathbb{Z}}Z_n\widehat{f}(n)e_n\right\|_\infty$$

by the rotation-invariance of complex Gaussian variables, and hence $[\![T_a f]\!] = [\![f]\!]$.

It suffices to show the continuity of translations for the norm $[\![\,.\,]\!]$. However, if $f \in \mathcal{C}^{as}$, the unconditionality of characters implies that the function $\sum_{n\in\mathbb{Z}}Z_n|\widehat{f}(n)|e_n$ is also in $\mathcal{C}^{as} \subseteq L^1(\Omega,\mathbb{P};\mathcal{C})$.

Now denote:

$$X_a = \left\|\sum_{n\in\mathbb{Z}}Z_n(e^{ina}-1)\widehat{f}(n)\,e_n\right\|_\infty \quad \text{and} \quad X = \left\|\sum_{n\in\mathbb{Z}}Z_n|\widehat{f}(n)|e_n\right\|_\infty.$$

The 1-unconditionality (contraction principle) gives:

$$\mathbb{E}(|X_a|^2) \leqslant 4\,\mathbb{E}(|X|^2).$$

Thus, since $X \in L^2(\Omega,\mathbb{P})$, the set of functions $X_a$, for $a \in \mathbb{T}$, is *uniformly integrable*.

However, as $f_a^\omega = \sum_{n\in\mathbb{Z}}Z_n(\omega)e^{ina}\widehat{f}(n)e_n$ is almost surely continuous, $X_a \xrightarrow[a\to 0]{a.s.} 0$, and hence $[\![T_a f - f]\!] = \mathbb{E}(X_a) \xrightarrow[a\to 0]{} 0$. Finally, it is evident that the mappings $D_k$ are isometries from $\mathfrak{P}$ into itself.

We now show that all Lipschitz functions operate on $\mathfrak{P}$. Note that this allows us to conclude that $\mathfrak{P}$ is thus an *algebra*: the function $x \mapsto x^2$ is Lipschitz on $[-1,1]$; hence $f^2 \in \mathfrak{P}$ if $f \in \mathfrak{P}$.

This requires an intermediate result:

**Theorem IV.3** *Let $f$ and $g$ be two functions of $L^2(\mathbb{T})$ such that*:

$$\|g_s - g_t\|_2 \leqslant \|f_s - f_t\|_2$$

*for any $s,t \in \mathbb{T}$. Then*:

$$f \in \mathcal{C}^{as} \implies g \in \mathcal{C}^{as}.$$

*Moreover*:

$$[\![g]\!] \leqslant 4\sqrt{2}\,[\![f]\!] + \|g\|_2.$$

*Proof*  To the complex Gaussian process $X_t = \sum_{n\in\mathbb{Z}}Z_n\widehat{f}(n)e^{int}$ we associate the real Gaussian process:

$$X_t^r = \operatorname{Re}X_t + \operatorname{Im}X_t.$$

Note that if we set $\widehat{f}(n) = \rho_n e^{i\theta_n}$ and $Z_n = (g_n' + ig_n'')/\sqrt{2}$, then:

$$\operatorname{Re}X_t = \sum_{n\in\mathbb{Z}}\frac{\rho_n}{\sqrt{2}}\left[g_n'\cos(nt+\theta_n) - g_n''\sin(nt+\theta_n)\right]$$

and

$$\operatorname{Im} X_t = \sum_{n \in \mathbb{Z}} \frac{\rho_n}{\sqrt{2}} \left[ g_n' \sin(nt + \theta_n) + g_n'' \cos(nt + \theta_n) \right]$$

are orthogonal Gaussian processes, and hence independent. Similarly, we define $Y_t = \sum_{n \in \mathbb{Z}} Z_n \widehat{g}(n)\, e^{int}$ and $Y_t^r = \operatorname{Re} Y_t + \operatorname{Im} Y_t$. Then:

$$\mathbb{E}[(Y_s^r - Y_t^r)^2] = \mathbb{E}\left[ \left( \operatorname{Re}(Y_s - Y_t) \right)^2 + \left( \operatorname{Im}(Y_s - Y_t) \right)^2 \right]$$

$$= \mathbb{E}|Y_s - Y_t|^2 = \sum_{n \in \mathbb{Z}} |\widehat{g}(n)|^2 |e^{ins} - e^{int}|^2$$

$$= \|g_s - g_t\|_2^2 \leqslant \|f_s - f_t\|_2^2$$

$$= \mathbb{E}|X_s - X_t|^2 = \mathbb{E}[(X_s^r - X_t^r)^2].$$

According to the Marcus–Shepp theorem (Chapter 3 of this volume, Theorem II.2), $(Y_t^r)_{t \in \mathbb{T}}$ has a continuous version, satisfying

(1)                  $$\mathbb{E}\left( \sup_{t \in \mathbb{T}} Y_t^r \right) \leqslant \mathbb{E}\left( \sup_{t \in \mathbb{T}} X_t^r \right).$$

A similar comparison of the processes $(Y_t^i)_{t \in \mathbb{T}}$ and $(X_t^i)_{t \in \mathbb{T}}$, with

$$Y_t^i = \operatorname{Re} Y_t - \operatorname{Im} Y_t \quad \text{and} \quad X_t^i = \operatorname{Re} X_t - \operatorname{Im} X_t,$$

shows that $(Y_t)_{t \in \mathbb{T}}$ has a continuous version, and hence $g \in \mathcal{C}^{as}$.

It remains to compare the norms. However, since the processes

$$(\operatorname{Re} Y_t)_{t \in \mathbb{T}} \quad \text{and} \quad (\operatorname{Im} Y_t)_{t \in \mathbb{T}}$$

are independent and centered, we have (see Chapter 4 of Volume 1, Proposition II.13):

$$\mathbb{E}\| \operatorname{Re}(Y_t - Y_0) \|_\infty \leqslant \mathbb{E}\| \operatorname{Re}(Y_t - Y_0) + \operatorname{Im}(Y_t - Y_0) \|_\infty = \mathbb{E}\|(Y_t^r - Y_0^r)\|_\infty,$$

and similarly $\mathbb{E}\| \operatorname{Im}(Y_t - Y_0) \|_\infty \leqslant \mathbb{E}\|(Y_t^r - Y_0^r)\|_\infty$; thus:

$$\mathbb{E}\left( \sup_{t \in \mathbb{T}} |Y_t - Y_0| \right) \leqslant \mathbb{E}\left( \sup_{t \in \mathbb{T}} | \operatorname{Re}(Y_t - Y_0)| \right) + \mathbb{E}\left( \sup_{t \in \mathbb{T}} | \operatorname{Im}(Y_t - Y_0)| \right)$$

$$\leqslant 2\, \mathbb{E}\left( \sup_{t \in \mathbb{T}} |Y_t^r - Y_0^r| \right).$$

By (1), it ensues that:

$$\mathbb{E}\left( \sup_{t \in \mathbb{T}} |Y_t^r - Y_0^r| \right) \leqslant \mathbb{E}\left( \sup_{s,t \in \mathbb{T}} |Y_s^r - Y_t^r| \right) = \mathbb{E}\left( \sup_{s,t \in \mathbb{T}} (Y_s^r - Y_t^r) \right)$$

$$= 2\, \mathbb{E}\left( \sup_{t \in \mathbb{T}} Y_t^r \right) \leqslant 2\, \mathbb{E}\left( \sup_{t \in \mathbb{T}} X_t^r \right) \qquad \text{by (1)}$$

$$\leqslant 2\sqrt{2}\, \mathbb{E}\left( \sup_{t \in \mathbb{T}} |X_t| \right) \qquad \text{since } X_t^r \leqslant \sqrt{2}|X_t|$$

(note the importance of having $Y_t^r - Y_0^r$ instead of $Y_t^r$, in order to get rid of the absolute value). Finally:

$$\mathbb{E}\left(\sup_{t\in\mathbb{T}}|Y_t - Y_0|\right) \leqslant 4\sqrt{2}\,\mathbb{E}\left(\sup_{t\in\mathbb{T}}|X_t|\right) = 4\sqrt{2}\,[\![f]\!],$$

and hence:

$$[\![g]\!] = \mathbb{E}\left(\sup_{t\in\mathbb{T}}|Y_t|\right) \leqslant \mathbb{E}\left(\sup_{t\in\mathbb{T}}|Y_t - Y_0| + |Y_0|\right)$$
$$\leqslant 4\sqrt{2}\,[\![f]\!] + \left(\mathbb{E}|Y_0|^2\right)^{1/2} = 4\sqrt{2}\,[\![f]\!] + \|g\|_2,$$

which completes the proof of Theorem IV.3. $\qquad\square$

Back to the proof of Theorem IV.2: let $f \in \mathfrak{P}$, with values in $[-1, 1]$, and let $F\colon [-1,1] \to \mathbb{C}$ be a Lipschitz function. Multiplying $F$ by a constant if necessary, we can assume that $|F(x)-F(y)| \leqslant |x-y|$. Thus, with $g = F\circ f$, we have:

$$\|g_s - g_t\|_2^2 = \int_{\mathbb{T}} |F(f_s) - F(f_t)|^2\, dm \leqslant \int_{\mathbb{T}} |f_s - f_t|^2\, dm = \|f_s - f_t\|_2^2,$$

and $g \in \mathcal{C}^{as}$, by Theorem IV.3. Since clearly $g = F\circ f \in \mathcal{C}$, we indeed have $F\circ f \in \mathfrak{P}$ thus the Lipschitz functions indeed operate on $\mathfrak{P}$.

To complete the proof, it only remains to see that $\|uv\|_{\mathfrak{P}} \leqslant \|u\|_{\mathfrak{P}}\|v\|_{\mathfrak{P}}$ for $u, v \in \mathfrak{P}$. We have:

$$(uv)_s - (uv)_t = u_s v_s - u_t v_t = u_s(v_s - v_t) + v_t(u_s - u_t),$$

so that:

$$\|(uv)_s - (uv)_t\|_2 \leqslant \|u\|_\infty\|v_s - v_t\|_2 + \|v\|_\infty\|u_s - u_t\|_2$$
$$\leqslant \sqrt{2}\left(\|u\|_\infty^2\|v_s - v_t\|_2^2 + \|v\|_\infty^2\|u_s - u_t\|_2^2\right)^{1/2}$$
$$= \|w_s - w_t\|_2,$$

where $w \in L^2$ is defined by its Fourier coefficients:

$$\widehat{w}(n) = \sqrt{2}\left(\|u\|_\infty^2|\widehat{v}(n)|^2 + \|v\|_\infty^2|\widehat{u}(n)|^2\right)^{1/2}.$$

Since:

$$\widehat{w}(n) \leqslant \sqrt{2}\,\|u\|_\infty|\widehat{v}(n)| + \sqrt{2}\,\|v\|_\infty|\widehat{u}(n)|,$$

and since the characters are 1-unconditional in $\mathcal{C}^{as}$, then $w \in \mathcal{C}^{as}$ and:

$$[\![w]\!] \leqslant \sqrt{2}\,(\|u\|_\infty[\![v]\!] + \|v\|_\infty[\![u]\!]).$$

Next, by the inequality $\|(uv)_s - (uv)_t\|_2 \leqslant \|w_s - w_t\|_2$ and Theorem IV.3, we obtain $uv \in \mathcal{C}^{as}$, and

$$[\![uv]\!] \leqslant 4\sqrt{2}\,[\![w]\!] + \|uv\|_2 \leqslant 8\left(\|u\|_\infty[\![v]\!] + \|v\|_\infty[\![u]\!]\right) + \|u\|_\infty\|v\|_\infty,$$

which gives:

$$\|uv\|_{\mathfrak{B}} = 8\,\|uv\|_{\infty} + [\![uv]\!]$$
$$\leqslant 8\,\|u\|_{\infty}\|v\|_{\infty} + 8\left(\|u\|_{\infty}[\![v]\!] + \|v\|_{\infty}[\![u]\!]\right) + \|u\|_{\infty}\|v\|_{\infty}$$
$$\leqslant (8\,\|u\|_{\infty} + [\![u]\!])(8\,\|v\|_{\infty} + [\![v]\!]) = \|u\|_{\mathfrak{B}}\|v\|_{\mathfrak{B}}.$$

This completes the proof of Theorem IV.2.    □

**Remark**   The result of Katznelson cited at the beginning of this section indicates that Theorem IV.2 is in a way optimal: as soon as a function just a bit less stringent than a Lipschitz function (i.e. $\lim_{t\to 0} |f(t)|/|t| = +\infty$) operates on $B$, this algebra must be the whole of $\mathcal{C}(\mathbb{T})$.

## V  The Bourgain–Milman Theorem

This section does not make use of the space $\mathcal{C}^{as}$. We include it in this chapter because the notion of Sidon set, which it characterizes, is at the origin of the introduction of the space $\mathcal{C}^{as}$. It uses a large variety of tools and results seen throughout this book and hence provides a fitting conclusion. However, we thought that a separate chapter was not really necessary.

Back to the framework of a compact metrizable Abelian group $G$: in Chapter 6 of Volume 1 (Theorem V.24), we saw that a subset $\Lambda \subseteq \Gamma = \widehat{G}$ is a Sidon set as soon as the space $\mathcal{C}_{\Lambda}(G)$ has cotype 2. Pisier [1981 c] conjectured that in fact it was sufficient for $\mathcal{C}_{\Lambda}(G)$ to have a finite cotype in order to reach this conclusion. Bourgain and Milman [1985] showed that this is indeed the case.

**Theorem V.1** (The Bourgain–Milman Theorem)   *Let $G$ be a compact (metrizable) Abelian group. If the Banach space $\mathcal{C}_{\Lambda}(G)$ has a finite cotype, then $\Lambda$ is a Sidon set.*

In view of the result of Maurey and Pisier (see Chapter 5 of Volume 1, Subsection IV.2, Remark 3, stating that a Banach space has a finite cotype if and only if it does not contain $\ell_{\infty}^{n}$'s uniformly), this theorem expresses the following remarkable dichotomy:
*For every subset $\Lambda \subseteq \Gamma$:*

a) *either $\Lambda$ is a Sidon set, and $\mathcal{C}_{\Lambda}(G)$ is isomorphic to $\ell_1$;*
b) *or $\mathcal{C}_{\Lambda}(G)$ contains $\ell_{\infty}^{n}$'s uniformly.*

The latter condition means that a constant $C > 0$ can be found such that, for any $n \geqslant 1$, there exist "bump" functions $f_1^{(n)}, \ldots, f_n^{(n)} \in \mathcal{C}_{\Lambda}(G)$ satisfying:

$$\sup_{1\leqslant j\leqslant n}|a_j|\leqslant\left\|\sum_{j=1}^{n}a_jf_j^{(n)}\right\|_{\infty}\leqslant C\sup_{1\leqslant j\leqslant n}|a_j|$$

for every $a_1,\ldots,a_n\in\mathbb{C}$.

The proof of this theorem requires two notions of diameter, introduced in the next subsection.

## V.1 Banach and Arithmetic Diameters

**Definition V.2**  Let $E$ be a Banach space of dimension $n$. The *Banach diameter* of $E$, denoted $n(E)$, is the smallest integer $m$ such that there exist $\varphi_1,\ldots,\varphi_m\in B_{E^*}$ satisfying:

$$\|x\|\leqslant 2\sup_{1\leqslant k\leqslant m}|\varphi_k(x)|$$

for every $x\in E$.

Equivalently, $n(E)$ is the smallest integer $m$ such that $E$ is 2-isomorphic to a subspace of $\ell_\infty^m$, where the isomorphism is given by:

$$x\mapsto\big(\varphi_1(x),\ldots,\varphi_m(x)\big).$$

Note that the constant of isomorphism was taken equal to 2 as a clarifying example, but this 2 plays no particular role.

**Definition V.3**  Let $G$ be a compact (metrizable) Abelian group, and $A$ a finite subset of $\Gamma=\widehat{G}$. The *arithmetic diameter* of $A$ is defined as the entropy number $m=\overline{N_A}(1/2)$, where $\overline{N_A}$ is the function associated with the pseudo-metric $\overline{d_A}$ on $G$, defined by:

$$\overline{d_A}(s,t)=\sup_{\substack{f\in\mathcal{C}_A(G)\\\|f\|_\infty\leqslant 1}}|f(s)-f(t)|.$$

Here, $\overline{N_A}(\varepsilon)$ is the minimal number of *closed* $\overline{d_A}$-balls of radius $\varepsilon$ necessary to cover $G$. Note that in the theorems of Dudley, Fernique, minoration of Sudakov, etc., the entropy function $N_0$ used there was associated with coverings by *open* balls. This is of no importance, since $N(\varepsilon)\leqslant N_0(\varepsilon)\leqslant N(\varepsilon')$, for all $\varepsilon'<\varepsilon$, and the constants that appear can be slightly increased if necessary.

By the definition of the arithmetic diameter, there exist $t_1,\ldots,t_m\in G$ such that:

$$\|f\|_\infty\leqslant 2\sup_{1\leqslant k\leqslant m}|f(t_k)|$$

for every $f \in \mathcal{C}_A(G)$. Indeed, let $f \in \mathcal{C}_A$, and $t_1, \ldots, t_m \in G$ such that $G = \bigcup_{j=1}^{m} \overline{B}(t_j, 1/2)$. For every $t \in G$, we can find $j \leqslant m$ such that $\overline{d_A}(t, t_j) \leqslant 1/2$; hence $|f(t) - f(t_j)| \leqslant \|f\|_\infty / 2$, and then:

$$|f(t)| \leqslant |f(t) - f(t_j)| + |f(t_j)| \leqslant \frac{1}{2}\|f\|_\infty + \sup_{1 \leqslant k \leqslant m} |f(t_k)|.$$

By taking the supremum over $t \in G$, we obtain:

$$\|f\|_\infty \leqslant \frac{1}{2}\|f\|_\infty + \sup_{1 \leqslant k \leqslant m} |f(t_k)|,$$

which leads to the desired inequality.

Thus, whereas the Banach diameter is the minimal number of linear functionals of norm $\leqslant 1$ required to calculate the norm in $E$ (up to a factor of 2), the arithmetic diameter of $A$ provides a majoration of the number of points of $G$ necessary to calculate the norm in $\mathcal{C}_A(G)$ (up to a factor of 2). This arithmetic diameter thus corresponds to particular linear functionals $\delta_{t_1}, \ldots, \delta_{t_m}$, so that the arithmetic diameter of $A$ and the Banach diameter of $\mathcal{C}_A$ are linked by the inequality:

$$n(\mathcal{C}_A) \leqslant \overline{N_A}(1/2).$$

With these definitions, the two essential ingredients for the proof of the Bourgain–Milman theorem are the following:

*Step 1. If $\mathcal{C}_\Lambda$ has a finite cotype, then $\mathcal{C}_A$ has a large Banach diameter, and hence $A$ has a large arithmetic diameter, for every finite subset $A$ of $\Lambda$.*

*Step 2. If all the finite subsets of $\Lambda$ have a large arithmetic diameter, then $\Lambda$ is a Sidon set.*

These Steps 1 and 2 correspond more precisely to the two following theorems:

**Theorem V.4** (Maurey)   *Let $E$ be a finite-dimensional Banach space with* $\dim E \geqslant 2$, *and let* $2 \leqslant q < +\infty$. *Then, there exists a constant* $\beta = \beta(q, C_q(E)) = \alpha_q / (C_q(E))^q$, *with* $\alpha_q > 0$ *depending only on $q$, such that:*

$$\log(n(E)) \geqslant \beta \sup_{F \subseteq E} \left[ \frac{\dim F}{(\log(1 + d_F))^q} \right],$$

*where $F$ runs over the subspaces of $E$, and $d_F$ is the Banach–Mazur distance between $F$ and $\ell_2^{\dim F}$.*

**Theorem V.5** (Pisier)  *Let $G$ be a compact (metrizable) Abelian group, and let $\Lambda$ be a finite subset of $\Gamma = \widehat{G}$. If there exists $\delta > 0$ such that $\overline{N_A}(\delta) \geqslant e^{\delta|A|}$ for every subset $A \subseteq \Lambda$, then the inequality:*

$$S(\Lambda) \leqslant a \left( \frac{1}{\delta} \right)^b$$

*holds for the Sidon constant $S(\Lambda)$ of $\Lambda$, where $a, b > 0$ are two numerical constants.*

Theorem V.4 is due to Maurey (see Pisier [1981 c]), and Theorem V.5 essentially to Pisier, with a metric $d_A$ smaller than $\overline{d_A}$. However, Bourgain [1987] and Rodríguez-Piazza [1987] independently found that we have $\overline{d_A} \leqslant \pi \, d_A$. The proof of these two theorems is quite long; we thus show right away how Bourgain and Milman deduced Theorem V.1 from these two theorems.

*Proof of Theorem V.1*  To show that $\Lambda$ is a Sidon set, it suffices to show that the Sidon constant of all its finite subsets is uniformly bounded. We can thus assume that $\Lambda$ is finite; it then suffices to obtain a majoration of its Sidon constant by a quantity that does not depend on $\Lambda$, but only on the cotype-$q$ constant $C_q = C_q(\mathcal{C}_\Lambda)$ of $\mathcal{C}_\Lambda(G)$ (and on $q$).

For every subset $A \subseteq \Lambda$, because

$$\beta(q, C_q(\mathcal{C}_A)) \geqslant \beta(q, C_q) \quad \text{and} \quad \overline{N_A}(1/2) \geqslant n(\mathcal{C}_A),$$

Theorem V.4 provides:

$$(1) \qquad \log \overline{N_A}\left( \frac{1}{2} \right) \geqslant \beta(q, C_q) \sup_{F \subseteq \mathcal{C}_A} \left[ \frac{\dim F}{\left( \log(1 + d_F) \right)^q} \right].$$

To exploit this minoration, it is necessary to find a subspace $F$ of $\mathcal{C}_A$ for which $\dim F$ is large and $d_F$ small. For this, we consider the Fourier isomorphism $\phi \colon \mathcal{C}_A \to \ell_1^{|A|}$, defined by $\phi(f) = \big( \widehat{f}(\gamma) \big)_{\gamma \in A}$, for which $\|\phi\| = S(A)$ and $\|\phi^{-1}\| = 1$.

Given that $\ell_1$ has cotype 2, Dvoretzky's theorem for the cotype-2 spaces (Chapter 1 of this volume, Subsection IV.4, Example 2), Theorem IV.14) provides a subspace $G$ of $\ell_1^{|A|}$ such that $d_G \leqslant 2$ and $\dim G \geqslant \alpha \, |A|$, where $\alpha > 0$ is a numerical constant (see also Chapter 1, Exercise VII.6, which is sufficient here). Then, with $F = \phi^{-1}(G)$, we have $\dim F = \dim G \geqslant \alpha \, |A|$, and:

$$d_F \leqslant d_G \, d(F, G) \leqslant 2 \, \|\phi\| \, \|\phi^{-1}\| \leqslant 2 \, S(A) \leqslant 2 \, S(\Lambda).$$

The inequality (1) thus leads to:

$$(2) \qquad \log \overline{N_A}\left( \frac{1}{2} \right) \geqslant \frac{\beta(q, C_q) \, \alpha}{\left( \log(1 + 2 \, S(\Lambda)) \right)^q} \, |A|.$$

We set $\delta = \dfrac{\beta(q, C_q)\,\alpha}{\left(\log(1 + 2\,S(\Lambda))\right)^q}$, and distinguish two cases:

*Case 1*: $\delta \geqslant 1/2$.

Then, clearly, $S(\Lambda) \leqslant \varphi_1(q, C_q)$, where $\varphi_1$ is a function depending only on $q$ and $C_q = C_q(\mathcal{C}_\Lambda)$.

*Case 2*: $\delta < 1/2$.

In this case, (2) implies $\log \overline{N}_A(\delta) \geqslant \log \overline{N}_A(1/2) \geqslant \delta\,|A|$, and Theorem V.5 provides the inequality:

$$S(\Lambda) \leqslant a\,\left(\frac{1}{\delta}\right)^b = a\,\left[\frac{\left(\log(1 + 2\,S(\Lambda))\right)^q}{\beta(q, C_q)\,\alpha}\right]^b .$$

This inequality easily implies the existence of a function $\varphi_2$, depending only on $q$ and $C_q$, such that $S(\Lambda) \leqslant \varphi_2(q, C_q)$.

Finally, $S(\Lambda) \leqslant \max(\varphi_1, \varphi_2)(q, C_q)$, and this majoration completes the proof of Theorem V.1.    □

## V.2  Proof of Theorem V.4

The proof is based on the following crucial lemma. In it, $B_1^n$ is the unit ball of $\ell_1^n$ and $N(K, \varepsilon)$ the minimal number of closed balls of radius $\varepsilon$ necessary to cover the compact subset $K$ of the normed space $E$, i.e. $N(K, .)$ is the entropy of $K$.

**Lemma V.6** (Maurey)  *Let $E$ be a real finite-dimensional Banach space, and $1 < p \leqslant 2$. Then, for every linear mapping $u \colon \ell_1^n \to E$, and for any $k \geqslant 1$, we have:*

$$N\!\left(u(B_1^n), 2\,k^{-1/p^*}\,T_p(E)\,\|u\|\right) \leqslant (2n)^k,$$

*where $p^*$ is the conjugate exponent of $p$ and $T_p(E)$ is the type-$p$ constant of $E$.*

**Remark**  When the space $E$ is complex, and if $(B_1^n)_r$ denotes the unit ball of the real space $\ell_1^n$, we have: $B_1^n \subseteq (B_1^n)_r + i\,(B_1^n)_r$; hence $u(B_1^n) \subseteq u\!\left[(B_1^n)_r\right] + i\,u\!\left[(B_1^n)_r\right]$, thus Lemma V.6 holds, if the coefficient 2 in the left-hand side of the inequality is replaced by 4.

*Proof*  A main point in the proof is the fact that $B_1^n$ only has a few extreme points: the $2n$ points $\pm e_1, \ldots, \pm e_n$, where $(e_1, \ldots, e_n)$ is the canonical basis of $\ell_1^n$. Thus if $x_j = u(e_j)$, and $A = \{\pm x_1, \ldots, \pm x_n\}$, then $u(B_1^n) = \operatorname{conv} A$.

We select an arbitrary $x \in u(B_1^n)$. It can be written $x = \sum_{j=1}^{2n} \lambda_j\,a_j$, with $a_j \in A$, $\lambda_j \geqslant 0$, and $\sum_{j=1}^{2n} \lambda_j = 1$. In a more probabilistic style, this can be written $x = \mathbb{E}(Z)$, where $Z$ is a random variable with values in $A \subseteq E$, such that $\mathbb{P}(Z = a_j) = \lambda_j$, for $1 \leqslant j \leqslant 2n$. We now approximate $\mathbb{E}(Z)$, thanks to a quantitative form of the law of large numbers in the type-$p$ spaces. For each $k \geqslant 1$, let $Z_1, \ldots, Z_k$ be a sampling of $Z$. Note that:

$$
(1) \qquad \mathbb{E} \left\| \sum_{i=1}^{k} (Z_i - \mathbb{E}Z_i) \right\| \leqslant 2 T_p(E) \left( \sum_{i=1}^{k} \mathbb{E} \|Z_i\|^p \right)^{1/p}.
$$

Indeed, let us consider the symmetrization

$$
\overline{X}_i(\omega, \omega') = X_i(\omega) - X_i(\omega') = Z_i(\omega) - Z_i(\omega')
$$

of $X_i = Z_i - \mathbb{E}Z_i$; we use the results of Chapter 5 (Volume 1) and the symmetry of $\overline{X}_i$: denoting by $(\varepsilon_1, \ldots, \varepsilon_k)$ a Bernoulli sequence independent of the variables $\overline{X}_i$, we have:

$$
\mathbb{E} \left\| \sum_{i=1}^{k} (Z_i - \mathbb{E}Z_i) \right\| = \mathbb{E} \left\| \sum_{i=1}^{k} X_i \right\| \leqslant \mathbb{E} \left\| \sum_{i=1}^{k} \overline{X}_i \right\| = \mathbb{E} \left\| \sum_{i=1}^{k} \varepsilon_i \overline{X}_i \right\|
$$

$$
\leqslant T_p(E) \, \mathbb{E} \left[ \left( \sum_{i=1}^{k} \|\overline{X}_i\|^p \right)^{1/p} \right]
$$

$$
\leqslant T_p(E) \left[ \mathbb{E} \left( \sum_{i=1}^{k} \|\overline{X}_i\|^p \right) \right]^{1/p}
$$

$$
= T_p(E) \left( \sum_{i=1}^{k} \mathbb{E} \|\overline{X}_i\|^p \right)^{1/p}
$$

$$
\leqslant 2 T_p(E) \left( \sum_{i=1}^{k} \mathbb{E} \|Z_i\|^p \right)^{1/p},
$$

since $\|\overline{X}_i\|_p = \|Z_i(\omega) - Z_i(\omega')\|_p \leqslant 2 \|Z_i\|_p$.

The equidistribution of the variables $Z_i$ and a division by $k$ lead to this new form of (1):

$$
(2) \qquad \mathbb{E} \left\| \frac{1}{k} \sum_{i=1}^{k} Z_i - \mathbb{E}Z \right\| \leqslant 2 T_p(E) \, k^{-1/p^*} \left( \mathbb{E} \|Z\|^p \right)^{1/p}
$$

$$
\leqslant 2 T_p(E) \, k^{-1/p^*} \|u\|.
$$

Indeed, as $Z$ takes its values in $A \subseteq u(B_1^n)$, then $\|Z(\omega)\| \leqslant \|u\|$ for any $\omega$. Set $r_k = 2 T_p(E) \, k^{-1/p^*} \|u\|$. By the inequality (2), we can find an $\omega_0 \in \Omega$ (the underlying probability space) such that, setting $z_i = Z_i(\omega_0) \in A$, we obtain:

$$
\left\| x - \frac{z_1 + \cdots + z_k}{k} \right\| \leqslant r_k
$$

(note that we started with $x \in u(B_1^n)$, written $x = \mathbb{E}Z$).

Thus every $x \in u(B_1^n)$ has been approximated, up to an error $\leqslant r_k$, by an arithmetic mean of $k$ terms of $A$. Consequently the set of points of the form $\dfrac{z_1 + \cdots + z_k}{k}$, with $z_i \in A$, is a $r_k$-net of $u(B_1^n)$. As $|A| \leqslant 2n$, the cardinality of this $r_k$-net is $\leqslant (2n)^k$, hence $N\big(u(B_1^n), r_k\big) \leqslant (2n)^k$, as stated in the lemma. $\quad\square$

Back to the proof of Theorem V.4: let $F$ be a subspace of $E$. Increasing if necessary the constant $\beta = \beta(q, C_q)$, we can assume $\dim F \geqslant 2$, since $\dfrac{\dim F}{\big(\log(1 + d_F)\big)^{p^*}} = \dfrac{1}{(\log 2)^{p^*}}$ when $\dim F = 1$. By the definition of $n = n(F)$, there exists an injective linear mapping $j \colon F \to \ell_\infty^n$ such that

$$\|x\| \leqslant \|j(x)\|_\infty \leqslant 2\,\|x\|$$

for every $x \in F$. By duality, a surjective mapping $j^* \colon \ell_1^n \to F^*$ can be deduced such that:

$$B_{F^*} \subseteq j^*(B_1^n).$$

Indeed, by the Hahn–Banach theorem, every linear functional $\varphi \in B_{F^*}$ can be written $\varphi = \psi \circ j$, with $\psi \in \ell_1^n = (\ell_\infty^n)^*$, and the inequality $\|x\| \leqslant \|j(x)\|_\infty$ shows that $\psi_{|j(F)}$ is of norm $\leqslant 1$, and hence so is $\psi$.

Then by Lemma V.6 applied to $u = j^*$, since $B_{F^*} \subseteq j^*(B_1^n)$ and $\|j^*\| \leqslant 2$, we obtain:

$$N\big(B_{F^*}, 4\,k^{-1/p^*}T_p(F^*)\big) \leqslant (2n)^k.$$

Hence, if $k_0$ is the smallest integer $\geqslant 1$ such that $4\,k_0^{-1/p^*}T_p(F^*) \leqslant 1/2$, then:

$$N(B_{F^*}, 1/2) \leqslant (2n)^{k_0}.$$

Moreover, we always have:

$$N(B_{F^*}, 1/2) \geqslant 2^{\dim F^*}$$

(Chapter 1 of this volume, Lemma IV.12); consequently:

$$2^{\dim F} = 2^{\dim F^*} \leqslant (2n)^{k_0} \leqslant n^{2k_0}$$

(since $n = n(F) \geqslant \dim F \geqslant 2$), and, taking the logarithms, we obtain:

$$2\,k_0 \log n \geqslant (\log 2)\,\dim F.$$

Our choice of $k_0$ (note that $k_0 \geqslant \big(8T_p(F^*)\big)^{p^*} \geqslant 8^2 > 1$) leads to:

$$\log n \geqslant \alpha\,\frac{\dim F}{\big(T_p(F^*)\big)^{p^*}},$$

with $\alpha = \log 2/(4 \times 8^q)$ depending only on $q = p^*$. Then, since:

$$T_p(F^*) \leqslant K(F)\,C_q(F)$$

(Chapter 5 of Volume 1, Theorem IV.7), and since:

$$K(F) \leqslant C \log(1 + d_F)$$

for a numerical constant $C$ (Chapter 4 of this volume, Theorem V.3), we obtain:

$$\log n \geqslant \frac{\alpha_q}{\left(C_q(F)\right)^q} \frac{\dim F}{\left(\log(1 + d_F)\right)^q},$$

with $\alpha_q = \log 2 / 4(8C)^q$. Then $n(E) \geqslant n(F) = n$ and $C_q(F) \leqslant C_q(E)$; hence finally:

$$\log n(E) \geqslant \frac{\alpha_q}{\left(C_q(E)\right)^q} \frac{\dim F}{\left(\log(1 + d_F)\right)^q},$$

which completes the proof of Theorem V.4. $\qquad\square$

## V.3 Proof of Theorem V.5

This difficult proof makes a crucial use of the characterization of Sidon sets in terms of extraction of quasi-independent subsets, seen in Chapter 5 of this volume (Theorem II.3). It was stated for $\Gamma = \mathbb{Z}$, but remains valid without change for an arbitrary compact Abelian group $G$. For convenience, we recall:

**Theorem V.7** *A subset $\Lambda \subseteq \Gamma$ is a Sidon set, of constant $S(\Lambda)$ if, and only if, there exists a constant $\delta > 0$ such that every finite subset $A$ of $\Lambda$, not reduced to $\{0\}$, contains a quasi-independent subset $B$ of cardinality $|B| \geqslant \delta |A|$.*

*Moreover, when $\Lambda$ is a Sidon set, $\delta$ can be chosen $\geqslant c'/S(\Lambda)^2$, and, conversely, if the property of extraction holds, then $S(\Lambda) \leqslant c/\delta^2$, where $c$ and $c'$ are numerical constants.*

Here is the strategy to prove Theorem V.5, based on the preceding characterization: once and for all we assume that $\Lambda$ is *finite* and that $0 \notin \Lambda$ (clearly not restrictions); for $A \subseteq \Lambda$, three pseudo-metrics on $G$ are introduced:

$$\begin{cases} d_{2,A}(s,t) = \left(\displaystyle\sum_{\gamma \in A} |\gamma(s) - \gamma(t)|^2\right)^{1/2} \\[2ex] d_A(s,t) = \displaystyle\sup_{\gamma \in A} |\gamma(s) - \gamma(t)| \\[2ex] \overline{d_A}(s,t) = \sup_{\substack{f \in C_A \\ \|f\|_\infty \leqslant 1}} |f(s) - f(t)|. \end{cases}$$

As seen previously, the metric $\overline{d_A}$ is the best adapted to our problem; however, the metric $d_{2,A}$ is the best adapted to Gaussian processes and to the use of Rider's theorem; thus we begin with a study of the latter.

Let $N_{2,A}$, $N_A$ and $\overline{N_A}$ be the entropy functions associated to the three pseudo-metrics $d_{2,A}$, $d_A$ and $\overline{d_A}$ respectively. We successively show that:

*Step 1. if $N_{2,A}$ is large for every $A \subseteq \Lambda$, then the Sidon constant $S(\Lambda)$ of $\Lambda$ is controlled;*
*Step 2. if $N_A$ is large, then so is $N_{2,A}$;*
*Step 3. if $\overline{N_A}$ is large, then so is $N_A$.*

The combination of these three steps leads to Theorem V.5. First we make these three steps more precise. Step 1 corresponds to the following key proposition:

**Proposition V.8**  *Suppose that there exists $\delta > 0$ such that, for all $A \subseteq \Lambda$, we have:*

$$\sup_{\varepsilon > 0} \left( \varepsilon \sqrt{\log N_{2,A}(\varepsilon)} \right) \geqslant \delta \, |A|.$$

*Then $S(\Lambda) \leqslant a_1/\delta^{18}$, for a numerical constant $a_1 > 0$.*

*Proof*  Let $(Z_\gamma)_{\gamma \in A}$ be a complex standard Gaussian sequence. If $N_{2,A}$ denotes the entropy function associated with the Gaussian process:

$$t \in G \mapsto \sum_{\gamma \in A} Z_\gamma \, \gamma(t),$$

the Sudakov minoration (Chapter 1 of this volume, Corollary III.6) gives, for $P = \sum_{\gamma \in A} \gamma$:

$$[\![P]\!] \geqslant \alpha \sup_{\varepsilon > 0} \left( \varepsilon \sqrt{\log N_{2,A}(\varepsilon)} \right),$$

where $\alpha > 0$ is a numerical constant. In fact, with $K(\varepsilon) = K_{2,A}(\varepsilon)$ the maximal number of points $t_1, \ldots, t_k \in G$ such that $d_{2,A}(t_j, t_{j'}) > \varepsilon$ for $j \neq j'$, we have seen in Chapter 1, Sub-lemma IV.13, that $K(\varepsilon) \geqslant N_{2,A}(\varepsilon)$; hence if $t_1, \ldots, t_{K(\varepsilon)} \in G$ are $K(\varepsilon)$ such points, we have:

$$\inf_{j \neq j'} \left\| \sum_{\gamma \in A} Z_\gamma \, \gamma(t_j) - \sum_{\gamma \in A} Z_\gamma \, \gamma(t_{j'}) \right\|_{L^2(\mathbb{P})} = \inf_{j \neq j'} d_{2,A}(t_j, t_{j'}) > \varepsilon.$$

Hence the Sudakov minoration gives:

$$[\![P]\!] = \mathbb{E} \|P_\omega\|_\infty \geqslant \mathbb{E} \left( \sup_{1 \leqslant j \leqslant K} |P_\omega(t_j)| \right)$$

$$\geqslant \alpha \, \varepsilon \sqrt{\log K(\varepsilon)} \geqslant \alpha \, \varepsilon \sqrt{\log N_{2,A}(\varepsilon)}.$$

The hypothesis thus implies $[\![P]\!] \geqslant \alpha\delta \, |A|$.

We now seek a minoration of $\left[\!\left[\sum_{\gamma \in A} a_\gamma\, \gamma\right]\!\right]$ for arbitrary coefficients $a_\gamma \in \mathbb{C}$. This is done thanks to the following simple lemma:

**Lemma V.9** *Let $E$ be a complex normed space of finite dimension $n$, with a normalized basis $(e_1, \ldots, e_n)$ that is 1-unconditional in the complex sense, such that $\left\| \sum_{k=1}^n e_k \right\| \geqslant cn$, with $c > 0$ a fixed constant. Then there exists $B \subseteq \{1, \ldots, n\}$ such that:*

1) $|B| \geqslant \dfrac{c}{2}\, n$;

2) $\left\| \displaystyle\sum_{k \in B} a_k\, e_k \right\| \geqslant \dfrac{c}{2} \displaystyle\sum_{k \in B} |a_k|,\ \text{for every } a_k \in \mathbb{C}.$

Indeed, we can use this lemma with $E = \mathcal{C}_A(G)$, equipped with the norm $[\![\,.\,]\!]$, for which the elements of $A$ form a 1-unconditional basis; as $\left[\!\left[\sum_{\gamma \in A} \gamma\right]\!\right] \geqslant \alpha\delta\,|A|$, we obtain $B \subseteq A$ such that $|B| \geqslant \dfrac{\alpha\delta}{2}\,|A|$, and $\left[\!\left[\sum_{\gamma \in B} a_\gamma\, \gamma\right]\!\right] \geqslant \dfrac{\alpha\delta}{2} \sum_{\gamma \in B} |a_\gamma|$ for any $a_\gamma \in \mathbb{C}$.

The Gaussian Rider theorem (Chapter 6 of Volume 1, Proposition V.28) shows that $S(B) \leqslant a'/\delta^4$, where $a' > 0$ is a numerical constant.

From Theorem V.7 above, it ensues that $B$ contains a quasi-independent set $B'$ of cardinality:

$$|B'| \geqslant \frac{c'}{S(B)^2}\,|B| \geqslant \frac{c'}{a'^2}\,\delta^8\,|B| \geqslant a''\,\delta^9\,|A|.$$

Another application of Theorem V.7, only of its sufficient condition this time, shows that $S(\Lambda) \leqslant a_1/\delta^{18}$. $\qquad\square$

*Proof of Lemma V.9* Let $(e_1^*, \ldots, e_n^*)$ be the dual basis in $E^*$, and take $\xi = \sum_{j=1}^n \xi_j\, e_j^* \in E^*$ of norm 1 such that $\left\| \sum_{j=1}^n e_j \right\| = \langle \xi, \sum_{j=1}^n e_j \rangle$. As the dual basis $(e_1^*, \ldots, e_n^*)$ is also normalized and 1-unconditional, we have $|\xi_j| \leqslant \|\xi\| = 1$. With $B = \{j \leqslant n\, ;\, |\xi_j| \geqslant c/2\}$, then:

$$cn \leqslant \left\| \sum_{j=1}^n e_j \right\| = \left\langle \xi, \sum_{j=1}^n e_j \right\rangle = \left| \sum_{j=1}^n \xi_j \right| \leqslant \sum_{j=1}^n |\xi_j|$$

$$\leqslant \sum_{j \in B} |\xi_j| + \frac{c}{2}\,|B^c| \leqslant |B| + \frac{c}{2}\,n;$$

hence $|B| \geqslant cn/2$. Moreover, for any $a_j \in \mathbb{C}$, $j \in B$, the complex unconditionality shows that:

$$\left\| \sum_{j \in B} a_j e_j \right\| = \left\| \sum_{j \in B} |a_j| \operatorname{sign} \xi_j e_j \right\|$$

$$\geqslant \left\langle \xi, \sum_{j \in B} |a_j| (\operatorname{sign} \xi_j) e_j \right\rangle = \sum_{j \in B} |a_j| |\xi_j| \geqslant \frac{c}{2} \sum_{j \in B} |a_j|. \qquad \square$$

We now examine point 2), i.e. the link between the "round" balls of $\ell_2^n$ and the "square" balls of $\ell_\infty^n$. We begin with some notation:

**Notation** If $K_1$ and $K_2$ are two compact subsets of $\mathbb{R}^n$ or $\mathbb{C}^n$ with non-empty interiors, $N(K_1, K_2)$ denotes the minimal number of translates of $K_2$ necessary to cover $K_1$.

Note that $N(\lambda K_1, \lambda K_2) = N(K_1, K_2)$ for any $\lambda > 0$. Then:

**Proposition V.10**  *Let $B_2^n$ be the unit ball of $\ell_2^n$ and $B_\infty^n$ that of $\ell_\infty^n$. Let $A \subseteq \Gamma \smallsetminus \{0\}$, of cardinality $|A| = n \geqslant 1$. For any $\varepsilon, \varepsilon' > 0$, we have:*

a) $N_A(2\varepsilon) \leqslant N_{2,A}(\varepsilon \varepsilon' \sqrt{n}) \, N \left( B_2^n, \frac{1}{\varepsilon' \sqrt{n}} B_\infty^n \right)$;

b) $N \left( B_2^n, \frac{1}{\varepsilon' \sqrt{n}} B_\infty^n \right) \leqslant K \exp(K \sqrt{\varepsilon'} n)$, *if $\varepsilon' \leqslant 1$, where $K \geqslant 1$ is a numerical constant.*

Point a) poses no difficulties; it is the majoration b) that allows a) to be exploited. For the proof of b), the unit ball $B_1^n$ of $\ell_1^n$ is a useful intermediary:

**Lemma V.11**  *For every integer $p \geqslant 1$, and for balls of the real spaces $\ell_1^n$ and $\ell_\infty^n$, we have:*

$$N(p \, B_1^n, B_\infty^n) \leqslant 2^p \, C_{n+p}^p.$$

*Proof of Proposition V.10*

a) Set $m = N_{2,A}(\varepsilon \varepsilon' \sqrt{n})$, and, for $t \in G$, $M(t) = \big(\gamma(t)\big)_{\gamma \in A}$.
Let $\{t_1, \ldots, t_m\}$ be an $\varepsilon \varepsilon' \sqrt{n}$-lattice of $(G, d_{2,A})$. For every $t \in G$, there exists $j \leqslant m$ such that

$$d_{2,A}(t, t_j) = \| M(t) - M(t_j) \|_{\ell_2^n} \leqslant \varepsilon \varepsilon' \sqrt{n}.$$

Now we denote

$$l = N\big(\varepsilon \varepsilon' \sqrt{n} \, B_2^n, \varepsilon \, B_\infty^n\big) = N \left( B_2^n, \frac{1}{\varepsilon' \sqrt{n}} B_\infty^n \right).$$

Let $A_1, \ldots, A_l \in \mathbb{C}^n$ such that $\varepsilon \varepsilon' \sqrt{n} \, B_2^n \subseteq \bigcup_{k=1}^l (A_k + \varepsilon \, B_\infty^n)$. We can find $k \leqslant l$ such that $M(t) \in M(t_j) + A_k + \varepsilon \, B_\infty^n$. For $1 \leqslant j \leqslant m$ and $1 \leqslant k \leqslant l$, the points $M(t_j) + A_k$ hence form an $\varepsilon$-net of

$$M(G) = \{M(t)\,;\ t \in G\},$$

in $\ell_\infty^n$. To obtain a net inside $M(G)$, we fix a point $M(s_{j,k})$ in each of the sets $M(t_j) + A_k + \varepsilon\,B_\infty^n$ that intersect $M(G)$.

Then $M(t) \in M(t_j) + A_k + \varepsilon\,B_\infty^n$ implies

$$d_A(t, s_{j,k}) = \|M(t) - M(s_{j,k})\|_\infty \leqslant 2\varepsilon;$$

thus $N_A(2\varepsilon)$ is less than or equal to to the number of points $s_{j,k}$, i.e. $N_A(2\varepsilon) \leqslant ml$, and hence a) is proved.

b) We use Lemma V.11. Note that it implies, in the complex case as well as in the real case:

$$(*) \qquad N(\varepsilon' n\, B_1^n, B_\infty^n) \leqslant K \exp\left(K\,(\sqrt{\varepsilon'} + \varepsilon')\,n\right)$$

for a numerical constant $K \geqslant 1$. Indeed, $(*)$ is evident when $\varepsilon' n \leqslant 1$, since $B_1^n \subseteq B_\infty^n$, and thus

$$N(\varepsilon' n\, B_1^n, B_\infty^n) = 1 \leqslant K \leqslant K \exp\left(K\,(\sqrt{\varepsilon'} + \varepsilon')\,n\right).$$

We can hence assume $\varepsilon' n \geqslant 1$, and apply Lemma V.11 with $p = [\varepsilon' n]+1$. For this value of $p$, we have $\varepsilon' n \leqslant p \leqslant 2\,\varepsilon' n$, and obtain, in the real case:

$$N(\varepsilon' n\, B_1^n, B_\infty^n) \leqslant N(p\, B_1^n, B_\infty^n) \leqslant 2^p\,\frac{(n+p)!}{n!\,p!} \leqslant C\,2^p\,\frac{(n+p)^{n+p}}{n^n\,p^p}\,,$$

by Stirling's formula, $C > 0$ being a numerical constant

$$= C\,2^p \left(1 + \frac{p}{n}\right)^n \left(1 + \frac{n}{p}\right)^p \leqslant C\,2^p\,e^p \left(1 + \frac{1}{\varepsilon'}\right)^{2\varepsilon' n}$$

$$\leqslant \alpha \exp\left[\beta\, n\varepsilon' \left(1 + \log\left(1 + \frac{1}{\varepsilon'}\right)\right)\right],$$

with $\alpha, \beta > 0$ numerical constants. As $\log\left(1 + 1/\varepsilon'\right) \leqslant \beta'/\sqrt{\varepsilon'}$, this indeed leads to $N(\varepsilon' n\, B_1^n, B_\infty^n) \leqslant K \exp\left(K\,(\sqrt{\varepsilon'} + \varepsilon')\,n\right)$ for a numerical constant $K \geqslant 1$.

Up to an increase of the constant $K$, clearly this estimation remains valid in the complex case.

Finally, in the case $\varepsilon' \leqslant 1$, we have $\varepsilon' + \sqrt{\varepsilon'} \leqslant 2\sqrt{\varepsilon'}$; thus we obtain point b) of the proposition, by replacing $K$ by $2K$ and noting that $B_2^n \subseteq \sqrt{n}\, B_1^n$ implies:

$$N\left(B_2^n, \frac{1}{\varepsilon'\,\sqrt{n}}\, B_\infty^n\right) = N(\varepsilon'\,\sqrt{n}\, B_2^n, B_\infty^n) \leqslant N(\varepsilon' n\, B_1^n, B_\infty^n). \qquad \square$$

*Proof of Lemma V.11*    Here, $[x]$ denotes the integer part of $x$, if $x \geqslant 0$, and that of $x + 1$, if $x < 0$. Then $|[x]| \leqslant |x|$ and $|x - [x]| \leqslant 1$. For $x = (x_1, \ldots, x_n) \in p B_1^n$, set $\bar{x} = ([x_1], \ldots, [x_n])$. Then $\bar{x} \in p B_1^n$, since $|[x_j]| \leqslant |x_j|$, and $x - \bar{x} \in B_\infty^n$. Next, $N(p B_1^n, B_\infty^n)$ is smaller than the number $N'$ of points with integer coordinates in $p B_1^n$, i.e.:

$$N' = \sharp\{(k_1, \ldots, k_n) \in \mathbb{Z}^n; \ |k_1| + \cdots + |k_n| \leqslant p\}.$$

Now, the estimation $N' \leqslant 2^p \, C_{n+p}^p$ was already obtained in Chapter 5 of this volume (see the proof of Sub-Lemma II.12). $\qquad\square$

Proposition V.10 is now exploited as follows:

**Proposition V.12**    *Let $\Lambda$ be a finite subset of $\Gamma \setminus \{0\}$. Assume that there exists $\delta > 0$ such that $N_A(\delta) \geqslant e^{\delta|A|}$ for every $A \subseteq \Lambda$. Then:*

$$S(\Lambda) \leqslant a_2/\delta^{b_2},$$

*where $a_2 > 0$ and $b_2$ are numerical constants.*

*Proof*    Let $\varepsilon = \delta/2$ and $\varepsilon' = \delta^2/4K^2$, where $K \geqslant 1$ is the constant appearing in Proposition V.10, b), and let $A \subseteq \Lambda$, with cardinality $|A| = n$. First note that $\delta < 2$, thus $\varepsilon' < 1$. Indeed, we can embed $G$ isometrically in $\ell_\infty^n$ with the mapping $M : t \mapsto (\gamma(t))_{\gamma \in A}$; therefore the entropy function $N_A$ can be identified, on $M(G)$, with the entropy function of $\ell_\infty^n$, hence the estimation:

$$N_A(\delta) \leqslant \left(1 + \frac{2}{\delta}\right)^{2n}$$

(Chapter 1 of this volume, Lemma IV.12). It ensues that:

$$e^{\delta n} \leqslant N_A(\delta) \leqslant \left(1 + \frac{2}{\delta}\right)^{2n} < e^{4n/\delta},$$

and hence $\delta < 2$. Then we use Proposition V.10, taking into account the hypothesis:

$$e^{\delta n} \leqslant N_A(\delta) = N_A(2\varepsilon) \leqslant N_{2,A}(\varepsilon \varepsilon' \sqrt{n}) \, N\left(B_2^n, \frac{1}{\varepsilon' \sqrt{n}} B_\infty^n\right)$$

$$\leqslant N_{2,A}(\varepsilon \varepsilon' \sqrt{n}) \, K \exp\left(K \sqrt{\varepsilon' n}\right)$$

$$= N_{2,A}\left(\frac{\delta^3}{8K^2} \sqrt{n}\right) K \exp\left(\frac{\delta}{2} n\right).$$

Increasing the Sidon constant of $\Lambda$ by a fixed constant if necessary, we can assume that $n = |A|$ is large enough so that $K \leqslant \exp(\delta n/4)$; hence:

$$e^{\delta n} \leqslant N_{2,A}\left(\frac{\delta^3}{8K^2}\sqrt{n}\right)\exp\left(\frac{\delta}{4}n\right)\exp\left(\frac{\delta}{2}n\right);$$

thus:

$$N_{2,A}\left(\frac{\delta^3}{8K^2}\sqrt{n}\right) \geqslant e^{\delta n/4}.$$

Setting $t = \delta^3\sqrt{n}/8K^2$, we obtain:

$$t\sqrt{\log N_{2,A}(t)} \geqslant t\sqrt{\frac{\delta n}{4}} = \frac{\delta^3\sqrt{n}}{8K^2}\sqrt{\frac{\delta n}{4}} = c\,\delta^{7/2}n,$$

and, *a fortiori*:

$$\sup_{s>0} s\sqrt{\log N_{2,A}(s)} \geqslant c\,\delta^{7/2}n.$$

Then by Proposition V.8, $S(\Lambda) \leqslant a_2/\delta^{63}$.                    □

Finally, step 3 of the proof of Theorem V.5 corresponds to the following unexpected proposition:

**Proposition V.13**   *For every finite subset $A \subseteq \Gamma = \widehat{G}$, we have:*

$$\overline{d_A} \leqslant \pi\, d_A,$$

*and hence $N_A(\varepsilon) \geqslant \overline{N_A}(\pi\varepsilon)$ for any $\varepsilon > 0$.*

*Proof* (L. Rodríguez-Piazza)   The proof is based on the following inequality, in which $P(t) = \sum_{j=1}^{n} a_j\, e^{i\lambda_j t}$, with $\lambda_j \in \mathbb{R}$ and $|\lambda_j| \leqslant \alpha < 2$, is an almost periodic polynomial with low frequencies:

$$(\diamond) \qquad |P(0) - P(1)| \leqslant \frac{\alpha}{1 - \alpha/2} \sup_{k\in\mathbb{Z}} |P(k)|.$$

Indeed, the generalized Bernstein's inequality (Chapter 6 of Volume 1, Proposition V.13) shows that $\|P'\|_\infty \leqslant \alpha\,\|P\|_\infty$; hence if $t \in \mathbb{R}$ and if we select $k \in \mathbb{Z}$ such that $|t - k| \leqslant 1/2$, we obtain:

$$|P(t) - P(k)| \leqslant |t - k|\,\|P'\|_\infty \leqslant \frac{\alpha}{2}\,\|P\|_\infty;$$

thus $\|P\|_\infty \leqslant \dfrac{1}{1 - \alpha/2} \sup_{k\in\mathbb{Z}} |P(k)|$. Another application of Bernstein's inequality then gives:

$$|P(0) - P(1)| \leqslant \|P'\|_\infty \leqslant \alpha\,\|P\|_\infty \leqslant \frac{\alpha}{1 - \alpha/2} \sup_{k\in\mathbb{Z}} |P(k)|,$$

as stated.

To prove that $\overline{d_A}(s,t) \leqslant \pi\, d_A(s,t)$, we can assume $t = 0$, since the metrics $d_A$ and $\overline{d_A}$ are translation-invariant. We set $d = d_A(s,0)$, and consider two cases:

*Case 1: $d > 2/\pi$.*
  The result is trivial: $\overline{d_A}(s,0) \leqslant 2 \leqslant \pi d$.

*Case 2: $d \leqslant 2/\pi$.*
  We denote $A = \{\gamma_1, \ldots, \gamma_n\}$ and $\gamma_j(s) = e^{i\lambda_j}$, with $|\lambda_j| \leqslant \pi$. By hypothesis: $|e^{i\lambda_j} - 1| \leqslant d$. However:

$$|e^{i\lambda_j} - 1| = 2 \left|\sin \frac{\lambda_j}{2}\right| \geqslant 2\, \frac{2}{\pi}\, \frac{|\lambda_j|}{2};$$

hence $|\lambda_j| \leqslant \pi d/2 \leqslant 1$.

  Now let $f = \sum_{j=1}^{n} a_j \gamma_j \in \mathcal{C}_A$, with $\|f\|_\infty \leqslant 1$. To $f$ we associate the almost periodic polynomial $P(u) = \sum_{j=1}^{n} a_j\, e^{i\lambda_j u}$, $u \in \mathbb{R}$. Then for $k \in \mathbb{Z}$:

$$P(k) = \sum_{j=1}^{n} a_j [\gamma_j(s)]^k = \sum_{j=1}^{n} a_j \gamma_j(ks) = f(ks),$$

and thus $|P(k)| \leqslant 1$. The inequality ($\diamond$) hence leads to:

$$|f(s) - f(0)| = |P(1) - P(0)| \leqslant \frac{\pi d/2}{1 - \pi d/4} \leqslant \pi d,$$

since $1 - \pi d/4 \geqslant 1 - 1/2 = 1/2$. By taking the upper bound over all $f \in \mathcal{C}_A$ of norm $\leqslant 1$, we obtain $\overline{d_A}(s,0) \leqslant \pi d = \pi d_A(s,0)$.   $\square$

*Proof of Theorem V.5*   When $\overline{N_A}(\delta) \geqslant e^{\delta|A|}$, then, by Proposition V.13, $N_A(\delta/\pi) \geqslant e^{\delta|A|} \geqslant e^{(\delta/\pi)|A|}$. Proposition V.12 then implies $S(\Lambda) \leqslant a_2/(\delta/\pi)^{b_2} = a_3/\delta^{b_2}$, which completes the proof of Theorem V.5, and hence also of the Bourgain–Milman theorem.   $\square$

## VI  Comments

1) The Tauberian theorem (Theorem II.4) was first shown by Marcinkiewicz and Zygmund [1938] in the scalar case; in this case the proof is much simpler.
  For example, for the Abel–Poisson procedure, let us show that if

$$\varlimsup_{n \to +\infty} |a_n|^{1/n} \leqslant 1 \quad \text{and} \quad \sum_{n=0}^{+\infty} |a_n|^2 = +\infty,$$

then $M_r = \left| \sum_{n=0}^{+\infty} \varepsilon_n a_n r^n \right|$ is almost surely unbounded when $r \overset{<}{\to} 1$. For this, let, $\sigma_r = \left( \sum_{n=0}^{+\infty} |a_n|^2 r^{2n} \right)^{1/2}$ and $M = \sup_{0 < r < 1} M_r$; the Paley–Zygmund inequality implies $\mathbb{P}(M_r \geqslant \delta \sigma_r) \geqslant \delta$, for a numerical constant $\delta > 0$; then Fatou's lemma provides:

$$\delta \leqslant \varlimsup_{r \overset{<}{\to} 1} \mathbb{P}(M_r \geqslant \delta \sigma_r) \leqslant \mathbb{P}\left( \varlimsup_{r \overset{<}{\to} 1} (M_r \geqslant \delta \sigma_r) \right) \leqslant \mathbb{P}(M = +\infty),$$

since $\sigma_r \to +\infty$ when $r \overset{<}{\to} 1$. Thus $\mathbb{P}(M = +\infty) = 1$, by the zero–one law.

The vectorial case is due to Kahane (KAHANE 2). As we have seen, it is very useful in the equivalent characterizations of the space $\mathcal{C}^{as}$.

As examples, two typical applications of the scalar case of Theorem II.4 are presented.

A) The proof by Duren [1969 a] of the non-validity of the Bloch–Nevanlinna conjecture: "if $f \in \mathcal{N}$, then $f' \in \mathcal{N}$", where $\mathcal{N}$ is the *Nevanlinna class*, i.e. the set of functions in the open unit disk $\mathbb{D}$ of the complex plan that satisfy the growth condition:

$$\sup_{0 \leqslant r < 1} \frac{1}{2\pi} \int_0^{2\pi} \log^+ |f(re^{it})| \, dt < +\infty.$$

Before stating the result, we also introduce the *analytic Zygmund class* $\Lambda_*$: the set of functions $f$ analytic in $\mathbb{D}$ and continuous on $\overline{\mathbb{D}}$, such that $g(t) = f(e^{it})$ satisfies the *Zygmund condition*: $g(t + h) + g(t - h) - 2g(t) = O(h)$. For a sequence $(a_n)_{n \geqslant 0}$ of complex numbers and a Bernoulli sequence $(\varepsilon_n)_{n \geqslant 0}$, consider the random analytic function $f_\omega$ defined by $f_\omega(z) = \sum_{n=0}^{+\infty} \varepsilon_n(\omega) a_n z^n$. Then, if $H^p = H^p(\mathbb{D})$ $(0 < p \leqslant +\infty)$ denotes the usual Hardy space:

**Theorem VI.1** *Let $(a_n)_{n \geqslant 0}$ be a bounded sequence of complex numbers such that $\varlimsup_{n \to +\infty} |a_n|^{1/n} \leqslant 1$.*

1) *If $\sum_{n=0}^{+\infty} |a_n|^2 < +\infty$, then $f_\omega \in \bigcap_{p < +\infty} H^p$ almost surely.*

2) *If $\sum_{n=0}^{+\infty} |a_n|^2 = +\infty$, then $f_\omega \notin \mathcal{N}$ almost surely.*

3) *There exists $f \in \Lambda_*$ such that $f' \notin \mathcal{N}$. In particular, there exists $f \in \mathcal{N}$ such that $f' \notin \mathcal{N}$.*

*Proof*

1) follows easily from Khintchine's inequalities and the Lévy symmetry principle.

2) Fix $t \in \mathbb{R}$. The random series $\sum_{n=0}^{+\infty} \varepsilon_n(\omega) a_n e^{int}$ diverges almost surely ($\omega \notin N_t$) according to the hypothesis and the three-series theorem.

Therefore $f_\omega(re^{it}) = \sum_{n=0}^{+\infty} \varepsilon_n(\omega) a_n r^n e^{int}$ lacks radial limits when $r \nearrow 1$, with probability $\alpha > 0$. Indeed, $\alpha = 0$ would imply that the series of symmetric variables is almost surely summable by the Abel–Poisson procedure, hence almost surely convergent by the Tauberian theorem (Theorem II.4). Thus, by the zero–one law, $\alpha = 1$. Now, letting $t$ vary, and applying the Fubinization principle, we see that: almost surely, $f_\omega(re^{it})$ lacks radial limits, and this for almost all $t$. Consequently $f_\omega \notin \mathcal{N}$ almost surely, since the functions of $\mathcal{N}$ notably have radial limits almost everywhere (RUDIN 2, § 17.19).

3) We use the following criterion of membership in $\Lambda_*$ (see Exercise VII.6): if $g(t) = \sum_{n=0}^{+\infty} c_n e^{int}$ is a uniformly convergent trigonometric series and if $\left\| \sum_{n=N+1}^{+\infty} c_n e^{int} \right\|_\infty = O(1/N)$, then $g \in \Lambda_*$. By this criterion, $f_\omega(z) = \sum_{n=0}^{+\infty} \varepsilon_n(\omega) 2^{-n} z^{2^n} \in \Lambda_*$ for every $\omega \in \Omega$, whereas $f'_\omega(z) = \sum_{n=0}^{+\infty} \varepsilon_n(\omega) z^{2^n-1}$ is not in $\mathcal{N}$ for (almost) any $\omega$, by 2). The result ensues, and is close to optimal: if $f \in \Lambda_1$, the *analytic Lipschitz class*, i.e. $f(t+h) - f(t) = O(h)$, then $f' \in H^\infty \subseteq \mathcal{N}$.    □

Interestingly, the first refutation of the Bloch–Nevanlinna conjecture was based on lacunary Taylor series, which in many ways have the same behavior as series of independent random variables (in fact, it was the properties of lacunary series that led Kolmogorov to his theorems about series of independent random variables). A more difficult result, that seemingly escapes the use of random methods, is due to Hayman [1964]: there exists $f \in \mathcal{N}$ with primitives $F \notin \mathcal{N}$.

B) Another interesting application of Theorem II.4 concerns random Dirichlet series $\sum_{n=1}^{+\infty} X_n e^{-\lambda_n z}$, where $(\lambda_n)_{n \geqslant 1}$ is a sequence of non-negative real numbers increasing to infinity. This result is due to Marczinkiewicz and Zygmund [1938].

**Theorem VI.2** *Let $(X_n)_{n \geqslant 1}$ be a sequence of independent and symmetric complex variables. Assume that the random Dirichlet series*

$$\sum_{n=1}^{+\infty} X_n e^{-\lambda_n z}$$

*has an abscissa of convergence $\sigma \in \mathbb{R}$ (which is almost surely constant). Then, for the sum $f_\omega$ of the series, the line $\mathrm{Re}\, z = \sigma$ is almost surely a natural boundary: i.e. $f_\omega$ cannot be analytically extended in any neighborhood of any point of this line.*

*Proof* The proof is based on the following general property: let $A_0 \subsetneqq A$ be two open subsets of $\mathbb{C}$, and let $(\varphi_n)_{n \geqslant 1}$ be a sequence of analytic functions in $A$, without zeros in $A \setminus A_0$. Suppose that the series $\sum_{n=1}^{+\infty} c_n \varphi_n(z)$ ($c_n \in \mathbb{C}$) converges uniformly to $f$ on every compact subset of $A_0$. Let $a \in A_0$, $D(a, r) \subseteq A$ and $z \in D(a, r) \setminus A_0$. Then:

(*)     *If $f$ has an analytic extension $F$ in $A$, there exists a summation procedure $M = (a_{n,j})_{n,j}$ such that $\sum_{n=1}^{+\infty} c_n \varphi_n(z)$ is M-convergent to $F(z)$.*

Indeed, for such a $z$, we have:

$$F(z) = \sum_{k=0}^{+\infty} \frac{(z-a)^k}{k!} f^{(k)}(a) = \sum_{k=0}^{+\infty} \frac{(z-a)^k}{k!} \left( \sum_{n=1}^{+\infty} c_n \varphi_n^{(k)}(a) \right);$$

hence, if we could interchange the sums, then:

$$F(z) = \sum_{n=1}^{+\infty} c_n \left( \sum_{k=0}^{+\infty} \frac{(z-a)^k}{k!} \varphi_n^{(k)}(a) \right) = \sum_{n=1}^{+\infty} c_n \varphi_n(z) \qquad !$$

However, as this interchange is in general forbidden, we truncate the series at $k$ by setting:

$$a_{n,j} = \frac{1}{\varphi_n(z)} \sum_{k=0}^{j} \frac{(z-a)^k}{k!} \varphi_n^{(k)}(a).$$

For each $n$, we have $\lim_{j \to +\infty} a_{n,j} = \frac{\varphi_n(z)}{\varphi_n(z)} = 1$. Moreover:

$$\begin{aligned}
\sum_{n=1}^{+\infty} a_{n,j} c_n \varphi_n(z) &= \sum_{n=1}^{+\infty} c_n \sum_{k=0}^{j} \frac{(z-a)^k}{k!} \varphi_n^{(k)}(a) \\
&= \sum_{k=0}^{j} \frac{(z-a)^k}{k!} \left( \sum_{n=1}^{+\infty} c_n \varphi_n^{(k)}(a) \right) = \sum_{k=0}^{j} \frac{(z-a)^k}{k!} f^{(k)}(a) \\
&= \sum_{k=0}^{j} \frac{(z-a)^k}{k!} F^{(k)}(a) \xrightarrow[j \to +\infty]{} F(z),
\end{aligned}$$

which proves (*).

Now set $A_0 = \{\operatorname{Re} z > \sigma\}$. If $\alpha \in \partial A_0$ and if $f_\omega$ can be extended in $D(\alpha, 2r) \cup A_0 = A$, by setting $a = \alpha + (r/2)$, we see that $D(a, r)$ intersects the half-plane $\operatorname{Re} z < \sigma$ at $z \in D(a, r) \subseteq A$, and hence that $\sum_{n=1}^{+\infty} X_n(\omega) e^{-\lambda_n z}$ is M-convergent according to (*), but diverges in the usual sense. Then, by Theorem II.4, $f_\omega$ is almost surely not analytically extendable in a neighborhood of $\alpha$. If we take a countable dense set of numbers $\alpha$ on $\partial A_0 = \{\operatorname{Re} z = \sigma\}$, the stated result ensues. $\qquad \square$

Note that the result and its proof are valid for random power series $\sum_{n=0}^{+\infty} X_n z^n$, but in this case we can give a much simpler proof (KAHANE 2).

2) Theorem II.2 was first shown by Itô and Nisio [1968] for Brownian motion on [0, 1]. The formulation given here is due to MARCUS–PISIER. The abstract version (Theorem II.1) was then obtained by Hoffmann-Jørgensen [1974].

3) The initial proof of Billard's theorem was for Bernoulli or Steinhaus variables; it is presented in KAHANE 2. We can now see it as a consequence of the Fernique minoration theorem (Chapter 3 of this volume, Theorem V.4) and the Marcus–Pisier equivalence theorem (Theorem III.5).

4) The space $\mathcal{C}^{as}$ was introduced by Pisier [1978 b], in relation to the Gaussian processes and the Dudley–Fernique theorems, initially to resolve the problem of Rudin. Pisier then made a systematic study of the Banach space $\mathcal{C}^{as}$. In the book MARCUS–PISIER, numerous properties are presented in a rather condensed style: it has cotype 2, and is a 2-concave lattice; it can be identified isomorphically to $A_{2,\varphi_2}$, as we have essentially seen in Proposition III.10 (hence, by a general result, it has a natural predual: the space $K_{2,\Psi_2}$ of compact multipliers of $L^2$ in $L^{\Psi_2}$).

Extensions to the case of locally compact groups, and to compact non-Abelian groups, are also given.

Halász [1973] specified the Salem–Zygmund theorem (Proposition III.13) for consecutive integers:

$$\left[\!\!\left[ \sum_{j=1}^{n} e_j \right]\!\!\right] \sim \sqrt{n \log n},$$

when $n$ tends to infinity.

5) L. Rodríguez-Piazza ([1987]; see also [1991]) defined a class of sets that extend naturally the notion of Sidon sets, in the form given by Rider's theorem. These are the sets $\Lambda \subseteq \Gamma$ for which $\|\widehat{f}\|_p \leqslant C[\![f]\!]$ for every trigonometric polynomial $f \in \mathcal{P}_\Lambda$ ($1 \leqslant p < 2$). The above subtle properties of $\mathcal{C}^{as}$ were used to give their complete characterization in terms of extraction of quasi-independent subsets, generalizing the result of Pisier for Sidon sets. These sets, first called *almost sure p-Sidon*, are now called *p-Rider*. This characterization was used by Li, Queffélec and Rodríguez-Piazza [2002] to construct sets $\Lambda \subseteq \mathbb{N}$ that are $p$-Sidon for every $p > 1$, but for which $\mathcal{C}_\Lambda$ contains $c_0$ (and hence $L_\Lambda^\infty$ is not separable). For $p = 1$, by Rider's theorem, this notion coincides with that of a Sidon set, but for $p > 1$ it is *a priori* weaker, even though Lefèvre and Rodríguez-Piazza [2003] showed that any $p$-Rider set with $p < 4/3$ is $q$-Sidon for every $q > p/(2 - p)$; in particular, $p$-Rider for every $p > 1$ implies $p$-Sidon for every $p > 1$.

6) Pisier introduced the notion of a *stationary set*: this is a set $\Lambda$ of $\Gamma$ satisfying $[\![f]\!] \leqslant C\|f\|_\infty$ for every $f \in \mathcal{P}_\Lambda$ (recall that the reverse inequality $\|f\|_\infty \leqslant C[\![f]\!]$ characterizes the Sidon sets). Notably, using Slepian's lemma, he showed that every finite product of Sidon sets is stationary in the product group. An extensive study of these sets can be found in Lefèvre [1998], where in particular it is shown that

$$\Lambda_N = \{3^{k_1} + \cdots + 3^{k_N} \, ; \, 1 \leqslant k_1 < \cdots < k_N\}$$

is stationary, whereas the set of prime numbers is not.

7) Pisier [1979] showed that the algebra $\mathfrak{P} = \mathcal{C}^{as} \cap \mathcal{C}$ is not the only one to satisfy the conclusions of Theorem IV.2. For $2 \leqslant q < +\infty$, consider the Orlicz function $\varphi_q(x) = x\big(1 + \log(1 + x)\big)^{1/q}$, and denote by $A_{2,\varphi_q}$ the subspace of functions $f \in L^2$ that can be written $f = \sum_{n=1}^{+\infty} f_n * g_n$, with $f_n \in L^2$, $g_n \in L^{\varphi_q}$ and $\sum_{n=1}^{+\infty} \|f_n\|_2 \|g_n\|_{\varphi_q} < +\infty$. Denote $\mathfrak{P}_q = A_{2,\varphi_q} \cap \mathcal{C}(\mathbb{T})$ and equip it with a norm analogous to that of $\mathfrak{P} = \mathfrak{P}_2$, by setting:

$$\|f\|_{2,\varphi_q} = \inf\left\{ \sum_{n=1}^{+\infty} \|f_n\|_2 \|g_n\|_{\varphi_q} \, ; \, f = \sum_{n=1}^{+\infty} f_n * g_n \right\}$$

and

$$\|f\|_{\mathfrak{P}_q} = \alpha_q \|f\|_\infty + \|f\|_{2,\varphi_q},$$

where $\alpha_q > 0$ is a suitable constant.

Then $\mathfrak{P}_q$ has the same properties as $\mathfrak{P}$: it is an algebra, strictly intermediate between $A(\mathbb{T})$ and $\mathcal{C}(\mathbb{T})$, and all the Lipschitz functions operate on $\mathfrak{P}_q$. All these algebras are thus distinct counterexamples to the Katznelson dichotomy problem, initially resolved by Zafran.

Other properties of the Pisier algebra $\mathfrak{P}$ were given by Pedersen [2000].

8) A detailed proof of the Bourgain–Milman theorem was presented by Prignot [1987].

# VII Exercises

Most of the time, for $\mathcal{C}^{as}$, the use of the equivalent norm given by:

$$[\![f]\!]_R = \mathbb{E} \left\| \sum_{n \in \mathbb{Z}} \varepsilon_n \widehat{f}(n) e_n \right\|_\infty$$

is more convenient.

**Exercise VII.1**    Let $(a_n)_{n\geqslant 1}$ be a non-increasing sequence of non-negative real numbers, such that $\sum_{n=1}^{+\infty} a_n e_n \in C^{as}$. Using the contraction principle and the inequality $\left[\!\!\left[\sum_{n=1}^{N} e_n\right]\!\!\right] \geqslant \delta\sqrt{N\log N}$, show that:

$$\sum_{n=2}^{+\infty} \left(\frac{1}{n^2 \log n} \sum_{m\geqslant n} a_m^2\right)^{1/2} < +\infty.$$

**Exercise VII.2**    The following result (see MARCUS–PISIER) is admitted: if $f \in C^{as}(G)$, and if $(a_n)_{n\geqslant 1}$ is the non-increasing rearrangement of $\left(|\widehat{f}(\gamma)|\right)_{\gamma\in\widehat{G}}$, then $\sum_{n=1}^{+\infty} a_n e_n \in C^{as}(\mathbb{T})$.

1) Show that:

$$\sum_{n=2}^{+\infty} \frac{a_n}{\sqrt{n}} \leqslant \sum_{n=2}^{+\infty} \frac{a_n}{\sqrt{n}\log n} \cdot 2 \sum_{m=2}^{n} \frac{1}{m};$$

and deduce that $\displaystyle\sum_{n=2}^{+\infty} \frac{a_n}{\sqrt{n}} < +\infty$.

2) Show that if $f \in C^{as}(G)$, then $\widehat{f} \in \ell_{2,1}$, the Lorentz space.

**Exercise VII.3**    Let $I$ be an arc of the unit circle $\mathbb{T}$. Show that its indicator function $\mathbb{I}_I$ is in $C^{as}$. Does this hold for the indicator function of an arbitrary Borel set?

**Exercise VII.4**    Let $f(t) = \sum_{n=1}^{+\infty} a_n e^{int}$, with

$$\sum_{n=3}^{+\infty} |a_n| \log n (\log\log n)^\beta < +\infty,$$

where $\beta > 1$ is fixed. Show that there exist $g$ and $h \in C(\mathbb{T})$, with $\widehat{g}(n) = \widehat{h}(n) = 0$ for $n < 0$, such that $f = g * h$.

Show the same result with $\displaystyle\sum_{n=2}^{+\infty} \frac{1}{\sqrt{n}(\log n)^{1/2+\varepsilon}} e^{int}$, where $\varepsilon > 0$ is fixed. Does this still hold with just the hypothesis $\sum_{n=1}^{+\infty} |a_n| < +\infty$?

**Exercise VII.5**    Explain in detail point 1) of Theorem VI.1.

**Exercise VII.6**    Let $f : \mathbb{R} \to \mathbb{C}$ be a continuous periodic function with period $2\pi$. Let $E_n(f)$ be the distance between $f$ and the trigonometric polynomials of degree at most $n$, i.e.:

$$E_n(f) = \inf \left\{ \|f - P\|_\infty \; ; \; P(t) = \sum_{k=-n}^{n} a_k e^{ikt} \right\}.$$

1) Suppose that $E_n(f) = O(1/n)$ when $n$ tends to infinity.

   a) Show that $f$ can be written $f = \sum_{n=0}^{+\infty} P_n$ with $\deg P_n \leqslant 2^n$ and $\|P_n\|_\infty \leqslant C2^{-n}$.

   b) Show that:

   $$|f(t+h) + f(t-h) - 2f(t)| \leqslant C h^2 \sum_{n=0}^{N} 2^n + 4 \sum_{n=N+1}^{+\infty} 2^{-n}.$$

   c) Show that $f \in \Lambda_*$, the Zygmund class.

2) Suppose that $f \in \Lambda_*$. By approximating $f$ by $f * J_{\lambda_n}$, where $J_n$ is the Jackson kernel and $\lambda_n$ the integer part of $n/2$, show that $E_n(f) = O(1/n)$.

**Exercise VII.7**   Let $(a_{n,\lambda})_{n,\lambda}$ be a summation matrix, and let $(b_n)_{n \geqslant 1}$ be a sequence of complex numbers such that $\sum_{n=1}^{+\infty} |b_n|^2 = +\infty$ and $\sum_{n=1}^{+\infty} |a_{n,\lambda} b_n| < +\infty$. Let:

$$M_\lambda(\omega) = \left| \sum_{n=1}^{+\infty} \varepsilon_n(\omega) a_{n,\lambda} b_n \right|,$$

where $(\varepsilon_n)_{n \geqslant 1}$ is a Rademacher sequence, and $M = \sup_\lambda M_\lambda$.

1) Show that:

$$\mathbb{P}\left( M_\lambda \geqslant \delta \left( \sum_{n=1}^{+\infty} |a_{n,\lambda} b_n|^2 \right)^{1/2} \right) \geqslant \delta,$$

   for a numerical constant $\delta > 0$.

2) Show that $\mathbb{P}(M = +\infty) \geqslant \delta$, and then that $\mathbb{P}(M = +\infty) = 1$.

Hence, if $\sum_{n=1}^{+\infty} |b_n|^2 = +\infty$, not only does $\sum_{n=1}^{+\infty} \varepsilon_n b_n$ diverge almost surely, but even $\sum_{n=1}^{+\infty} \varepsilon_n a_{n,\lambda} b_n$ is *a.s.* unbounded for every summation procedure.

# Appendix A

## News in the Theory of Infinite-Dimensional Banach Spaces in the Past 20 Years

Gilles Godefroy

Université Pierre et Marie Curie, Paris, France

### Hereditarily Indecomposable Spaces

During the summer of 1991, Timothy Gowers and Bernard Maurey independently solved an important problem on the structure of Banach spaces: they constructed a Banach space containing no subspace with an unconditional basis. It turned out that they actually constructed the same space, say $GM$, which therefore contains no Banach lattice. In particular, the quite satisfactory classification results which are available for lattices are powerless on such a space. The Gowers–Maurey construction [8] relied in part on a space $S$ constructed earlier by Thomas Schlumprecht, which could be renormed in such a way that all basic sequences had a large unconditionality constant. Schlumprecht's space, as with Tsirelson's space before it (and Gowers–Maurey after it), is constructed inductively; in other words, with a definition which appears to be circular at first glance. Its advantage on Tsirelson's space relies on its distortion properties, and actually it plays an important part in the Odell–Schlumprecht solution to the distortion problem ([13]): given $M > 0$, there is an equivalent renorming of the Hilbert space $l_2$ such that any subspace of the renormed space is at distance at least $M$ from the usual $l_2$.

The space $GM$ turned out to have a quite amazing geometrical property: if $U$ and $V$ are infinite-dimensional linear subspaces of $GM$, then for every $\epsilon > 0$, one can find $u \in U$ and $v \in V$ of norm one such that $\|u - v\| < \epsilon$. Therefore, no closed subspace of $GM$ is the direct sum of two infinite-dimensional closed subspaces. Such spaces are now called hereditarily indecomposable (in short, $HI$), and a proper use of spectral theory shows that they admit "few" operators: if $X$ is $HI$, every bounded linear operator on $X$ is a strictly singular perturbation of a scalar operator. It follows that an $HI$ space is not isomorphic to any of its proper subspaces, and in particular is not isomorphic to its hyperplanes. This solves Banach's hyperplane problem (whose first solution, due to Tim Gowers, used an unconditional variant of the space *GM*). Among other remarkable constructions due to Gowers, one should single out his example of a space $X$ which is isomorphic to its cube $X^3$ but not to its square $X^2$. Hence each one of the spaces $X$ and $X^2$ is isomorphic to a complemented subspace of the other but they are not isomorphic, and this solves the so-called Schroeder–Bernstein problem for Banach spaces.

An easy gliding hump argument shows that no subspace of a space with an unconditional basis can be *HI*. But this easy obstruction is somehow the only one, where constructing *HI* spaces is concerned. Indeed, Tim Gowers [7] discovered the following dichotomy (proved in 1993, but fully published in 2002; see the end of Chapter 3 in Vol. 1 of this book): every Banach space contains an unconditional basic sequence or an *HI* subspace. The beautiful proof relies on a topological game, played by *I* and *II*: while *I* tries to construct an unconditional basic sequence, *II* tries to construct an *HI* space, and one of them has to win. An important application of Gowers' dichotomy is his solution of the homogeneous space problem: if a Banach space $X$ is isomorphic to every closed infinite-dimensional subspace of $X$, then $X$ is isomorphic to $l_2$. Indeed, since an *HI* space is very far from being homogeneous, Gowers' dichotomy reduces the homogeneous space problem to the case where $X$ has an unconditional basis, and then a result of Ryszard Komorowski and Nicole Tomczak-Jaegermann provides the conclusion. Gowers' dichotomy led him to formulate what is now called the *Gowers program*: find a list of classes $(C_i)$ of Banach spaces which satisfy the following conditions:

(i) each class is hereditary, at least for block bases if the definition of the class is associated with a basis;

(ii) the classes $(C_i)$ are unavoidable; that is, every Banach space contains a subspace in one of the classes;

(iii) the classes are disjoint, i.e. no infinite-dimensional space belongs to more than one class;

(iv) belonging to a class provides a lot of information on the operators which can be defined on the space and its subspaces.

Gowers' dichotomy provides two such classes, but of course a larger list provides a finer classification. This program was investigated by Gowers himself, who proved a second dichotomy, and then by Valentin Ferenczi and Christian Rosendal [5], who proved three more dichotomies, in such a way that the list of unavoidable classes now contains 19 items.

Let us conclude this section with a mention of Piotr Koszmider's constructions of indecomposable $C(K)$ spaces [10]: he showed, for instance (in 2004), that there exists a (necessarily non-metrizable) compact space $K$ such that every operator from the Banach space $C(K)$ to itself is a weakly compact perturbation of a multiplication operator. It follows in particular that this space $C(K)$ is indecomposable, not isomorphic to its hyperplanes and not isomorphic to a $C(L)$ space with $L$ a totally discontinuous compact space. It should be stressed that Koszmider's arguments are totally independent of the Gowers–Maurey techniques.

The last important result (to this day) which has been obtained through *HI* spaces is the solution (published in 2011) by Spiros Argyros and Richard Haydon of the "scalar plus compact" problem [2]: there exists an infinite-dimensional Banach space $E$ such that every operator from $E$ to itself is a compact perturbation of a scalar operator. This space $E$ is actually an *HI* isomorphic predual of $l_1$, and its construction requests a combination of the Bourgain–Delbaen constructions of exotic preduals of $l_1$ with the full use of *HI* techniques. We are now quite close to the ultimate Banach space with few operators, namely a space $Z$ such that every operator on $Z$ is a nuclear perturbation of a scalar operator. It is plausible that constructing such a space, if it exists, will require a visit to the realm of Banach spaces which fail Grothendieck's approximation property.

## Descriptive Set Theory and Banach Spaces

Descriptive set theory is present, at least implicitly, when index theory is used: in other words, when separable Banach spaces are classified by a countable ordinal. Deepening this remark led Jean Bourgain to realize, as early as 1979, that the Lusin–Suslin theory of analytic sets and the related tree-techniques were directly applicable to universality results, for showing more precisely that universal spaces fail to exist for certain families. The proper frame for the descriptive theory of Banach spaces was subsequently designed by Benoît Bossard in 1994 (and mostly published in 2002 [3]), through the Effros–Borel structure on the set of closed subspaces of a universal space. This structure turns the collection of separable Banach spaces into a standard Borel space, and it is now recognized that the construction is canonical for all practical purposes. Within this frame, it turns out that the isomorphism equivalence relation is analytic non-Borel, and actually that this relation has maximal complexity among all analytic relations: this gives a precise meaning to the statement that Banach spaces cannot be fully classified. Actually, a rule of thumb is that local properties of Banach spaces (that is, properties which can be checked on the collection of finite-dimensional subspaces, such as super-reflexivity) lead to

Borel families, while infinite-dimensional conditions (e.g. reflexivity) usually lead to non-Borel families, in that case to a coanalytic family.

For instance, the Szlenk index of a separable space $X$, which measures how close the weak-star and norm topologies are on the dual unit ball $B_{X^*}$, has been intensively used in the last 20 years. This index is a coanalytic rank on the (coanalytic) set $Asp$ of Banach spaces with separable dual, and it is therefore uniformly bounded by some countable ordinal on any analytic subset of $Asp$. Similar situations occur for a bunch of coanalytic families $C$. It is easy to show that a coanalytic non-Borel class admits no universal space within the class. But it is by no means trivial that, on the other hand, analytic classes do admit, in great generality, non-trivial universal spaces. Such converse theorems have been shown by Spiros Argyros and Pandelis Dodos [1] in their work on amalgamation of classes of Banach spaces (published in 2007). For instance, the Borel character of super-reflexivity directly shows the existence of a separable reflexive space which contains an isomorphic copy of every uniformly convex space. Very recently (2016), Ondrej Kurka developed an isometric version of Argyros–Dodos amalgamation theory, and established similar results on the existence on non-trivial universal spaces for analytic families – although isomorphic and isometric universalities are quite different notions. An example among many: there exists a separable strictly convex Banach space which contains isometrically every separable uniformly convex space.

As mentioned above, Tim Gowers showed that a Banach space $X$ which is not isomorphic to $l_2$ contains an infinite-dimensional subspace $Y$ which is not isomorphic to $X$. What can be said, more generally, of the quotient of the set of infinite-dimensional subspaces of $X$ by the isomorphism equivalence relation? Gowers' theorem asserts that the cardinality of this quotient is at least 2 if $X$ is not $l_2$, but quite surprisingly it is still unknown whether this quotient is necessarily infinite, or even if there is a continuum of mutually non-isomorphic subspaces. Descriptive set theory and the classification of equivalence relations is naturally useful in this context. Following Valentin Ferenczi and Christian Rosendal [4], let us say that a Banach space $X$ is *ergodic* if the equivalence relation $E_0$ is Borel reducible to the isomorphism equivalence relation of the subspaces of $X$: it follows constructively that there is a continuum of mutually non-isomorphic subspaces of $X$. It is conjectured that $l_2$ is the only non-ergodic Banach space. A very recent result of Wilson Cuellar (2016) asserts that a non-ergodic Banach space is near Hilbert, that is, has type $2 - \epsilon$ and cotype $2 + \epsilon$ for all $\epsilon > 0$. This is a significant step towards a positive solution to this conjecture.

## Nonlinear Geometry of Banach Spaces

A Banach space $X$ is in particular a metric space. It is quite natural to investigate which properties of $X$ are invariant under isomorphisms which preserve all or at least part of the metric structure, for instance bi-Lipschitz or bi-uniform isomorphisms (see [9]). It turns out that Lipschitz isomorphisms remain somewhat close to linear isomorphisms, since some kind of differentiation is usually available. It follows that many properties which are stable under linear isomorphisms can be shown, more or less easily, to be stable under Lipschitz isomorphisms. On the other hand, a remarkable theorem shown by Ribe in 1976 asserts that the local structure of Banach spaces is invariant under uniform isomorphisms, while infinite-dimensional properties (such as reflexivity) are not stable under such maps. Ribe's theorem led to what is now called the Ribe program: given a local property $(p)$ of Banach spaces, find a property $(P)$ of metric spaces $M$ which reduces, when $M$ happens to be a Banach space, to $(p)$. Therefore the Ribe program intends to transfer properties from the well-structured field of Banach spaces to the wider realm of metric spaces, and this program has been successfully fulfilled, in particular by Assaf Naor. This approach has allowed the discovery of new phenomena about metric spaces, which could have remained unnoticed otherwise: for instance, Dvoretzky's theorem on Euclidean sections in normed spaces of large finite dimensions has a metric analogue shown by Manor Mendel and Assaf Naor [11], which reads as follows: for any $\epsilon > 0$ and any integer $n$, every metric space $M$ of cardinality $n$ contains a subset $S$ of cardinality at least $n^{1-\epsilon}$, which embeds into an ultrametric space with distortion at most $(9/\epsilon)$. Note that Dvoretzky's theorem and this ultrametric skeleton result are related by the fact that ultrametric sets embed isometrically into the Hilbert space. The Mendel–Naor theorem, and several related metric results, have applications in theoretical computer science: indeed, when a subset $E$ of a given metric space $M$ (thought of as a weighted graph) is essentially Euclidean, then the numerous algorithms from linear algebra operate with full power on the set $E$, and they can provide information on the original set $M$. For instance, Mendel and Naor show in this manner that one can solve the approximate distance oracle problem, which consists in keeping in stock a minimal number of mutual distances between points in a metric space $M$ in order to recover, in constant query time, the distance between any two points up to a fixed distortion. Another important technique consists in relaxing an optimization problem to a larger frame where semi-definite programming is available, to obtain the result up to a certain constant – which happens to be frequently related to Grothendieck's constant $K_G$. For instance, in order to solve the sparsest cut problem for graphs with $n$ vertices, in polynomial time and with a precision of $\sqrt{\log n}$, Assaf Naor [12] follows a similar approach, which relies on bi-Lipschitz embeddings of finite metric spaces of negative type into $L_1$.

What are the Banach spaces $X$ which are determined by their Lipschitz structure, in the sense that the existence of a bi-Lipschitz bijective map between $X$ and a Banach space $Y$ implies that $Y$ is linearly isomorphic to $X$? This is known to fail for some (non-separable and non-reflexive) Banach spaces: for instance, elaborating on a construction of Israel Aharoni and Joram Lindenstrauss proves that if a compact set $K$ has a finite Cantor index, then the space $C(K)$ is Lipschitz-isomorphic to a $c_0(\Gamma)$-space. However, a major open problem asks whether two *separable* Banach spaces which are Lipschitz-isomorphic are linearly isomorphic. On this latter problem, relatively few positive results are available: for instance, it is not known if the spaces $C(I)$, $L_1(I)$ (where $I$ is the unit interval) or $l_1(\mathbb{N})$ are determined by their Lipschitz structure.

If $M$ is a pointed metric space equipped with some distinguished point 0, the space of real-valued Lipschitz functions which vanish at 0 is isometric to a dual space, and its predual $\mathcal{F}(M)$ is what is now called the Lipschitz-free space over $M$. The properties of the Banach space $\mathcal{F}(M)$ reflect to some extent those of the metric space $M$: for instance, a Banach space $X$ has the bounded approximation property *(BAP)* if and only if $\mathcal{F}(X)$ enjoys it. Since the free spaces of Lipschitz-isomorphic spaces are linearly isomorphic, it follows that the *BAP* is Lipschitz invariant. Free spaces are also used in the proof by Nigel Kalton and Gilles Godefroy [6] that, if a separable Banach space $X$ isometrically embeds into a Banach space $Y$, there exists a linear isometric embedding from $X$ into $Y$. However, this statement fails for every non-separable reflexive space. Free spaces provide a class of Banach spaces whose structure is not yet well understood, and this class has been thoroughly investigated in the last 15 years. We refer in particular to Nigel Kalton's posthumous papers for sharp results on this class and its Hölder counterpart. Nigel Kalton asked, for instance, whether the free space over any uniformly separated metric space has the bounded approximation property. This important problem remains open.

## References

[1] S. Argyros & P. Dodos, Genericity and amalgamation of classes of Banach spaces, *Adv. Math.* **209**, 2 (2007), 666–748.

[2] S. Argyros & R. G. Haydon, A hereditary indecomposable $\mathcal{L}_\infty$-space that solves the scalar-plus-compact problem, *Acta Math.* **206**, 1 (2011), 1–54.

[3] B. Bossard, A coding of separable Banach spaces: Analytic and coanalytic families of Banach spaces, *Fund. Math.* **172**, 2 (2002), 117–152.

[4]  V. Ferenczi & C. Rosendal, Erogodic Banach spaces, *Adv. Math.* **195**, 1 (2005), 259–282.

[5]  V. Ferenczi & C. Rosendal, Banach spaces without minimal subspaces, *J Funct. Anal.* **257** (2009), 149–193.

[6]  G. Godefroy & N. J. Kalton, Lipschitz-free Banach spaces, *Studia Math.* **159**, 1 (2003), 121–141.

[7]  W. T. Gowers, An infinite Ramsey theorem and some Banach-space dichotomies, *Ann. of Math. (2)* **156**, 3 (2002), 797–833.

[8]  W. T. Gowers & B. Maurey, The unconditional basic sequence problem, *J. Amer. Math. Soc.* **6**, 4 (1993), 851–874.

[9]  N. J. Kalton, The nonlinear geometry of Banach spaces, *Rev. Mat. Complut.* **21** (2008), 7–60.

[10]  P. Koszmider, Banach spaces of continuous functions with few operators, *Math. Ann.* **330**, 1 (2004), 151–183.

[11]  M. Mendel & A. Naor, Ultrametric skeletons, *Proc. Nat. Acad. Sci. USA* **110**, 48 (2013), 19256–19262.

[12]  A. Naor, $L_1$ embeddings of the Heisenberg group and fast estimation of graph isoperimetry, in *Proceedings of the International Congress of Mathematicians: Hyderabad, August 19–27, 2010, Vol. III*, Hindustan Book Agency (2010), 1549–1575.

[13]  E. Odell & T. Schlumprecht, The distorsion problem, *Acta Math.* **173** (1994), 259–281.

# Appendix B

## An Update on Some Problems in High-Dimensional Convex Geometry and Related Probabilistic Results

Olivier Guédon
Université Paris-Est Marne-la-Vallée, France

### Concentration Phenomena in High-Dimensional Convex Geometry

In Chapter 3 of Vol. 2, Lemma VI.7, the authors prove the Prékopa–Leindler inequality which tells in particular that if $f : \mathbb{R}^n \to \mathbb{R}_+$ satisfies $\forall x, y \in \mathbb{R}^n, \forall \theta \in [0, 1]$,

$$f((1 - \theta)x + \theta y) \geqslant f(x)^{1-\theta} f(y)^{\theta},$$

then the measure $\mu$ with density $f \in L_1^{\mathrm{loc}}$ is log-concave, i.e. for every $\theta \in [0, 1]$, for all compact sets $A, B \subset \mathbb{R}^n$,

$$\mu((1 - \theta)A + \theta B) \geqslant \mu(A)^{1-\theta} \mu(B)^{\theta}. \tag{B.1}$$

Classical examples are the exponential distribution $f(x) = \frac{1}{2^n} \exp(-|x|_1)$, the Gaussian density $f(x) = \frac{1}{(2\pi)^{n/2}} \exp(-|x|_2^2/2)$, the density of the uniform measure on a convex body $K$, $f(x) = \frac{1}{|K|} 1_K(x)$, and any integrable density of the form $f(x) = Z^{-1} e^{-V(x)}$, where $V : \mathbb{R}^n \to \mathbb{R} \cup \{+\infty\}$ is convex and $Z$ is a normalization factor. The class of log-concave measures is stable under linear transformations, and another consequence of the Prékopa–Leindler inequality is that it is stable under convolution, while the class of uniform distributions on a convex body is stable under linear transformations but not under convolution. This is one of several reasons why it is more pleasant to work with log-concave measures. Concentration phenomena for log-concave measures like $(1 - \mu(uC)) \leqslant (1 - \mu(C))^{(u+1)/2}$ (where $C \subset \mathbb{R}^n$ is a convex body containing 0 in its interior and $u > 0$) are a powerful tool to prove Kahane–Khinchine type inequalities comparing the $L_p(\mu)$-norms of $\| \sum_{i=1}^{n} x_i v_i \|$ for any $p > -1$, where $v_i$ are vectors in a Banach space and $X = (x_1, \ldots, x_n)$ is a random vector with log-concave measure $\mu$ [40, 24]. However, the weakness of such an inequality is that the right-hand side does not decrease when $n$ grows to infinity. At the end of the 1990s, Keith Ball pointed out that, in the simple case of the Euclidean norm, we were not aware of any concentration inequality better than the trivial inequality $\text{Var}|X|_2 \leqslant \mathbb{E}|X|_2^2$ when $X$ was a random log-concave vector. He motivated his question, proving with Anttila and Perissinaki [4] that any non-trivial better bound would imply a central limit theorem for convex bodies. In 2005, Paouris [42] made a first breakthrough in this direction, proving that, for any isotropic probability $\mu$ on $\mathbb{R}^n$ with log-concave density, one has

$$\mu(|X|_2 \geqslant t\sqrt{n}) \leqslant e^{-ct\sqrt{n}}, \quad \text{for any } t \geqslant 10, \tag{B.2}$$

where $c > 0$ is a universal constant. In other words, it tells us that, in isotropic position, for any $p \leqslant \sqrt{n}$,

$$\left(\mathbb{E}|X|_2^p\right)^{1/p} \leqslant C\mathbb{E}|X|_2 \approx \sqrt{n},$$

where $C > 1$ is a universal constant. This concentration inequality is very different from all the previous ones in that the right-hand side depends strongly on $n$. Actually, this inequality is optimal when $X$ is uniformly distributed on the $\ell_1^n$ unit ball. Simultaneously, Klartag [30] proved that, for any $\varepsilon > 0$ and any convex body $K$, one can associate a convex body $T$ as close as you wish in Banach–Mazur distance, with an isotropic constant bounded by $C/\sqrt{\varepsilon}$. Combining this with the result of Paouris, he proved the best up-to-date general bound on the isotropic constant $L_K \leqslant cn^{1/4}$. In 2006, Klartag [31] proved the central limit theorem for convex bodies, showing that any isotropic log-concave measure is concentrated in a thin shell, which means that, for any $t > 0$, $\mu(|\,|X|_2 - \sqrt{n}\,| \geqslant t\sqrt{n})$ decreases to 0 when the dimension $n$ goes to infinity. This was also done independently using a different method by Fleury, Guédon and Paouris [19], showing that, for any $p \leqslant (\log n)^{1/3}$,

$$\left(\mathbb{E}|X|_2^p\right)^{1/p} \leqslant \left(1 + c\,\frac{p}{(\log n)^{1/3}}\right)\mathbb{E}|X|_2.$$

Once a thin-shell estimate has been proved, it is natural to study the rate of convergence. The results have been improved to polynomial estimates in the dimension $n$ by Klartag [32], Fleury [17] and Guédon and E. Milman [26], showing that

$$\forall t \geqslant 0, \quad \mu\left(\big|\,|X|_2 - \sqrt{n}\,\big| \geqslant t\sqrt{n}\right) \leqslant C\exp(-c\sqrt{n}\,\min(t^3, t)).$$

The isoperimetric problem for convex bodies is closely related to this. Let $K$ be a convex body in $\mathbb{R}^n$, and $\mu_K$ be the uniform measure on $K$. Let $S$ be a subset of $K$ and define the boundary measure of $S$ as

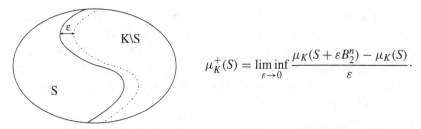

$$\mu_K^+(S) = \liminf_{\varepsilon \to 0} \frac{\mu_K(S + \varepsilon B_2^n) - \mu_K(S)}{\varepsilon}.$$

This definition is also valid for any measure $\mu$ with log-concave density on $\mathbb{R}^n$. The problem is to evaluate the largest possible $h$ such that

$$\forall\ S \subset K,\ \mu^+(S) \geqslant h\,\mu(S)(1 - \mu(S)). \tag{B.3}$$

Without any assumptions on the measure, you can easily imagine a situation where $h$ may be as close to 0 as you wish. In our situation, we assume that $\mu$ is isotropic and log-concave. This avoids a lot of non-regular situations. Kannan, Lovász and Simonovits [28] conjectured that, up to a universal constant in the inequality (B.3) the worst set should be a half-space with the same measure as $S$.

**Theorem** **The KLS conjecture** ([28]) *There exists $c > 0$ such that, for any dimension n and any isotropic log-concave probability on $\mathbb{R}^n$,*

$$\forall\ S \subset \mathbb{R}^n,\ \mu^+(S) \geqslant c\,\mu(S)(1 - \mu(S)).$$

This conjecture is supported by the Gaussian setting, where it is known that half-spaces are the exact solutions of the isoperimetric problem. Inequality (B.3) is called a Cheeger type inequality and $h$ is usually referred to as the *Cheeger constant* of the measure $\mu$. From the work of Buser [10] and Ledoux [36], we know that, for a log-concave probability, the Cheeger constant is related to the best constant in the Poincaré inequality. Let $X$ be the random vector distributed according to $\mu$, and let $D_2$ be the largest constant such that, for every regular function $F : \mathbb{R}^n \to \mathbb{R}$, one has

$$D_2\,\mathrm{Var}F(X) \leqslant \mathbb{E}|\nabla F(X)|_2^2; \tag{B.4}$$

then $h^2 \approx D_2$. E. Milman [39] proved the more surprising result that $h^2 \approx D_\infty$, where $D_\infty$ is the largest constant such that, for every 1-Lipschitz function $F : \mathbb{R}^n \to \mathbb{R}$,

$$D_\infty\,\mathrm{Var}F(X) \leqslant 1.$$

Both inequalities are easy consequences of the KLS conjecture. The difficult part of the proofs concerns the reverse statement. A paper of Gozlan, Roberto and Samson [23] completes the picture of the different equivalent formulations of the question. Few positive answers are known. It was proved only for some classes of convex bodies like the unit balls of $\ell_p^n$ [48, 35], hyperplane projections of these unit balls [3] and generalized Orlicz balls [33], and a weaker form is proved for random Gaussian polytopes in [18]. We emphasize the fact that the KLS conjecture implies a very strong concentration inequality of the Euclidean norm:

**Theorem    The thin-shell conjecture** *There exists $c > 0$ such that, for any log-concave isotropic probability $\mu$ on $\mathbb{R}^n$, for any $t > 0$,*

$$\mu\left(\big|\,|X|_2 - \sqrt{n}\,\big| \geqslant t\sqrt{n}\right) \leqslant 2e^{-ct\sqrt{n}}.$$

Eldan [16] developed tools from stochastic probabilities to prove strong connection between the thin-shell estimate and the Cheeger constant. Pursuing this stochastic approach, Lee and Vempala [37] very recently posted on ArXiv a paper proving that the Cheeger constant in the isoperimetric problem for convex bodies is bounded below by $cn^{-1/4}$. This would be the best up-to-date bound and would imply all the known general results about these conjectures, like the bound $n^{1/4}$ of Klartag [30] for the isotropic constant.

### Approximation of the Inertia Matrix

Kannan, Lovász and Simonovits published two papers [28, 29] where they asked questions related to high-dimensional geometry of convex bodies. We have already discussed the first one, concerning isoperimetric inequalities for convex bodies. The second one concerns a rounding procedure. It consists in finding algorithmically an Euclidean structure such that $B_2^n \subset K \subset d\,B_2^n$, where $d$ depends polynomially on the dimension $n$. Instead of using the classical ellipsoid algorithm, which achieved $d = O(n^{3/2})$, they considered the inertia ellipsoid $\mathcal{E}$ associated with $K$, and asked how to approximate it. We refer to [52] for new developments on the algorithmic aspects of convex geometry.

Let $X$ be a random vector uniformly distributed on a convex body in $\mathbb{R}^n$. The inertia matrix is given by $\mathbb{E}(X \otimes X)$. The simplest procedure is to understand how many samples are needed to approximate it. Given $\varepsilon \in (0, 1)$, the problem is to estimate the smallest number $N$ such that

$$\left\|\frac{1}{N}\sum_{i=1}^{N} X_i \otimes X_i - \mathbb{E}\,X \otimes X\right\| \leqslant \varepsilon\|\mathbb{E}\,X \otimes X\|,$$

where $\|\cdot\|$ denotes the operator norm from $\ell_2^n$ to $\ell_2^n$. Since the procedure is random, we hope to have such a result with large positive probability. Without loss of generality, we may assume that $X$ is isotropic; that is, $\mathbb{E}X = 0$ and

$\mathbb{E} X \otimes X = \text{Id}$. In terms of random processes, the problem is to evaluate $N$ such that, with the highest possible probability,

$$\sup_{y \in S^{n-1}} \left| \frac{1}{N} \sum_{i=1}^{N} \langle X_j, y \rangle^2 - 1 \right| \leqslant \varepsilon. \tag{B.5}$$

Using the language of random matrices, this is nothing else that evaluating $N$ such that all the singular values of the random matrix $A$, with rows $(\frac{X_i}{\sqrt{N}}, 1 \leqslant i \leqslant N)$, are in the interval $[1 - \varepsilon, 1 + \varepsilon]$. In [29] Kannan, Lovász and Simonovits proved that if $N \approx \frac{n^2}{\varepsilon^2 \eta^2}$, then (B.5) holds with probability larger than $1 - \eta$. Shortly afterwards, Bourgain [8] improved this estimate to $N \approx \frac{n \log^3 n}{\varepsilon^2 \eta^2}$. Over more than 15 years, several people [43, 22, 42] have proposed different strategies to improve the result. A breakthrough has been made by Adamczak, Litvak, Pajor and Tomczak-Jaegermann [1, 2], who proved that if $N \approx \frac{n}{\varepsilon^2}$, then (B.5) holds with probability at least $1 - e^{-c\sqrt{n}}$. The achievement in the result is that $N$ is taken of the order of the dimension $n$ and that the probability of the event is not only large but increases extremely fast with respect to $n$. An important step in the proof is to correctly evaluate the operator norm of a random matrix with independent log-concave isotropic rows. A trivial lower bound is the Euclidean norm of one row, and you can easily see why it was important to have in hand the concentration result of Paouris [42]; see (B.2). It is now of interest to understand which other random vector probability distributions than the log-concave ones satisfy such types of estimates. The subject has developed a lot. What can be said when the linear forms are heavy-tailed? This has been solved recently by Tikhomirov [49]. Pursuing the results of [38, 25], he proved that, to approximate the covariance matrix with $N$ samples, $N \approx n$, it is enough that uniformly on $\theta \in S^{n-1}$, all the linear forms $\langle X_i, \theta \rangle$ have a finite moment of order strictly greater than 2.

### Harmonic Analysis and Compressed Sensing

Deep relations connect empirical processes, random matrices and Harmonic Analysis. For example, Bourgain [7] proved the existence of true $\Lambda(p)$ sets, $p > 2$, using probabilistic methods. More precisely, he proved that there exist sets $\Lambda \subset \{1, \ldots, N\}$ of cardinality greater than $N^{2/p}$ such that, for every $(\alpha_j) \in \mathbb{C}^\Lambda$,

$$\left\| \sum_{j \in \Lambda} \alpha_j e^{2i\pi jt} \right\|_p \leqslant C(p) \left\| \sum_{j \in \Lambda} \alpha_j e^{2i\pi jt} \right\|_2,$$

where $C(p)$ depends only on $p$. In a slightly different direction, we can look at another problem of Harmonic Analysis: find a subset of a bounded orthonormal system (typically the Fourier or the Walsh system) on which the $L_1$ and $L_2$-norms will be comparable on the subspace that they span. This means, if $\varphi_1, \ldots, \varphi_N$ are $N$ orthonormal vectors in $L_2$, uniformly bounded in $L_\infty$ norm, can we extract a subset $(\varphi_i)_{i \in I}$ with $\#I = N - k$ such that

$$\left\| \sum_{i \in I} a_i \varphi_i \right\|_{L_1} \leqslant \left\| \sum_{i \in I} a_i \varphi_i \right\|_{L_2} \leqslant C(N, k) \left\| \sum_{i \in I} a_i \varphi_i \right\|_{L_1}$$

with the best possible value of $C(N, k)$? This problem is related to the study of the radius of $(\ker \Phi \cap B_1^N)$, where $\Phi$ is obtained by selecting correctly some rows of the original full orthogonal matrix. A comparable problem was completely understood in the 1980s in the local theory of Banach spaces when studying the Gelfand width of the $\ell_1^N$ unit ball, where the matrix $\Phi$ was supposed to be a Gaussian matrix. It is now of importance to understand what happens when the matrix $\Phi$ is built differently, coming for example from an extraction of the Fourier or Walsh system. While this problem could be seen as a purely theoretical one, it turns out to have strong connections with compressed sensing. This was done after some work of Donoho [15], Candès, Romberg and Tao [11] and Rudelson and Vershynin [45]. Candès and Tao [12] proposed studying a restricted isometry property for a matrix (deterministic or random) to ensure the existence and uniqueness of a solution to the basis pursuit algorithm

$$(\mathcal{P}) \qquad \min_{t \in \mathbb{R}^N} \{ |t|_1, \quad \Phi U = \Phi t \},$$

where $\Phi U$ is the received compressed signal and $U$ is the unknown signal which is supposed to have sparse coordinates. We say that a $n \times N$ matrix $A$ satisfies the restricted isometry property for $s$-sparse vectors of order $\delta \in (0, 1)$ if, for any vectors with at most $s$ non-zero coordinates, we have

$$(1 - \delta)|x|_2 \leqslant |Ax|_2 \leqslant (1 + \delta)|x|_2.$$

The study of matrices with this property and good dependence in the parameters $s$, $n$ and $N$ has led to a lot of problems. It was quickly observed that random Gaussian matrices satisfy this property with the optimal choice of parameters. However, the problem came with a computational point of view and an Harmonic Analysis flavor. And the question is still open for discrete Fourier matrices or Walsh matrices. We refer to the books [13, 20] for more information. The problem is also to describe a way to select the $n$ rows of this matrix. Until now, it was done using the method of selectors. Some progress is due to [9, 27]. It is also of interest to find more deterministic or algorithmic procedures. Other possible choices of matrices have been described with other types of restricted isometry properties. This has led to results in graph theory [6], local theory of Banach spaces [21] and learning theory [14].

### Random Matrices: A Non-Asymptotic Point of View

In this book, you can find a beautiful proof of Gordon's inequalities. A consequence of these inequalities is the following. Let $T$ be a subset of the unit sphere $S^{n-1}$. Then you can find a random subspace $E$, generated by the kernel of a Gaussian matrix, such that $E \cap T = \emptyset$. The condition on $k$, the codimension of $E$, is:

$$\mathbb{E} \sup_{t \in T} \sum_{i=1}^{n} t_i g_i > \mathbb{E} \left( \sum_{j=1}^{k} g_j^2 \right)^{1/2} \sim \sqrt{k},$$

where $g_1, \ldots, g_n$ are independent standard Gaussian random variables. This is known as Gordon's escape theorem. To my knowledge, the proof has not been reproduced in any previous book. This may be due to its high level of sophistication. However, in Chapter 1 of Vol. 2, you can find a very clear and clever explanation of how Gordon's inequalities extend the classical comparison inequalities of Slepian or Fernique for Gaussian processes. Recently, Gordon's inequalities have received a lot of attention in various areas. Rudelson and Vershynin [45] studied the restricted isometry property and sparse reconstruction from Fourier and Gaussian measurements. Tikhomirov and Youssef [50] studied the smallest number $N$ of steps needed for a discrete random walk $W$ in $\mathbb{R}^n$ to be such that the origin is included in the convex hull of $\{W(i)\}_{i \leq N}$, and related this problem to Gordon's escape theorem. Oymak and Tropp [41] extended the result to a universal statement when the entries of the matrix are independent, symmetric and centered, with variance 1 and finite moments of order $p$, with $p > 4$. A weakness of Gordon's inequalities was that, for square matrices, it was not possible to get any non-asymptotic estimate of the smallest singular value of a Gaussian matrix. This was done by Rudelson and Vershynin [44, 46], for not only Gaussian matrices but any matrix with independent, identically distributed sub-Gaussian entries, like Rademacher entries.

Sometimes, it is a difficult task to compute the operator norm of a random operator. A main step in [1] is to get such estimates when the rows of the matrix are independent, identically distributed log-concave vectors. When the entries of $A = (X_{ij})$ are independent identically distributed, Seginer [47] proved that

$\mathbb{E}\|A\|$ is equivalent, up to a universal constant, to $\mathbb{E}\max |R_i|_2 + \mathbb{E}\max |C_j|_2$, where $R_i$ and $C_j$, are the rows and columns of the matrix $A$. However, he proved that when the entries are not identically distributed, the situation is very different. Very little was known in that direction even when $G = (a_{ij}g_{ij})$, with $g_{ij}$ independent standard Gaussian random variables. A first result was given by Latała [34]. Recently a breakthrough has been made by Bandeira and Van Handel [5]:

$$\mathbb{E}\left\|G : \ell_2^m \longrightarrow \ell_2^n\right\|$$

$$\leqslant C\left(\max_{i\leqslant m} |(a_{ij})_{j=1}^n|_2 + \max_{j\leqslant n} |(a_{ij})_{i=1}^m|_2 + \sqrt{\log(n \vee m)}\max_{\substack{i\leqslant m \\ j\leqslant n}} |a_{ij}|\right),$$

where $C \geqslant 1$ is a universal constant. This was further developed in [51].

# References

[1] Adamczak, R., Litvak, A. E., Pajor, A. & Tomczak-Jaegermann, N. Quantitative estimates of the convergence of the empirical covariance matrix in log-concave ensembles, *J. Amer. Math. Soc.* **23**, 2 (2010), 535–561.

[2] Adamczak, R., Litvak, A. E., Pajor, A. & Tomczak-Jaegermann, N. Sharp bounds on the rate of convergence of the empirical covariance matrix, *C. R. Math. Acad. Sci. Paris* **349**, 3-4 (2011), 195–200.

[3] Alonso-Guttiérez, D. & Bastero, J. The variance conjecture on hyperplane projections of $\ell_p^n$ balls, *https://arxiv.org/abs/1610.04023* (preprint).

[4] Anttila, M., Ball, K. & Perissinaki, I. The central limit problem for convex bodies, *Trans. Amer. Math. Soc.* **355**, 12 (2003), 4723–4735 (electronic).

[5] Bandeira, A. S. & van Handel, R. Sharp nonasymptotic bounds on the norm of random matrices with independent entries, *Ann. Probab.* **44**, 4 (2016), 2479–2506.

[6] Berinde, R., Gilbert, A., Indyk, P., H., K. & Strauss, M. Combining geometry and combinatorics: a unified approach to sparse signal recovery, in *Communication, Control, and Computing, 2008 46th Annual Allerton Conference on*, IEEE (2008), 798–805.

[7] Bourgain, J. Bounded orthogonal systems and the $\Lambda(p)$-set problem, *Acta Math.* **162**, 3-4 (1989), 227–245.

[8] Bourgain, J. Random points in isotropic convex sets, in *Convex Geometric Analysis (Berkeley, CA, 1996)*, Mathematical Sciences Research Institute Publications **34**, Cambridge University Press (1999), 53–58.

[9] Bourgain, J. An improved estimate in the restricted isometry problem, in *Geometric Aspects of Functional Analysis, Lecture Notes in Mathematics* **2116**, Springer (2014), 65–70.

[10] Buser, P. A note on the isoperimetric constant, *Ann. Sci. Éc. Norm. Super. (4)* **15**, 2 (1982), 213–230.

[11] Candès, E. J., Romberg, J. K. & Tao, T. Stable signal recovery from incomplete and inaccurate measurements, *Comm. Pure Appl. Math.* **59**, 8 (2006), 1207–1223.

[12] Candès, E. J. & Tao, T. Decoding by linear programming, *IEEE Trans. Inform. Theory* **51**, 12 (2005), 4203–4215.

[13] Chafaï, D., Guédon, O., Lecué, G. & Pajor, A. *Interactions Between Compressed Sensing Random Matrices and High Dimensional Geometry*, Panoramas et Synthèses **37**, Société Mathématique de France (2012).

[14] Dirksen, S., Lecué, G. & Rauhut, H. On the gap between restricted isometry properties and sparse recovery conditions, *https:doi.org/10.1109/TIT.2016.2570244*, to appear in *IEEE Trans. Inform. Theory*.

[15] Donoho, D. L. Compressed sensing, *IEEE Trans. Inform. Theory* **52**, 4 (2006), 1289–1306.

[16] Eldan, R. Thin shell implies spectral gap up to polylog via a stochastic localization scheme, *Geom. Funct. Anal.* **23**, 2 (2013), 532–569.

[17] Fleury, B. Concentration in a thin Euclidean shell for log-concave measures, *J. Funct. Anal.* **259**, 4 (2010), 832–841.

[18] Fleury, B. Poincaré inequality in mean value for Gaussian polytopes, *Probab. Theory Related Fields* **152**, 1-2 (2012), 141–178.

[19] Fleury, B., Guédon, O. & Paouris, G. A stability result for mean width of $L_p$-centroid bodies, *Adv. Math.* **214**, 2 (2007), 865–877.

[20] Foucart, S. & Rauhut, H. *A Mathematical Introduction to Compressive Sensing*, Applied and Numerical Harmonic Analysis. Birkhäuser/Springer (2013).

[21] Friedland, O. & Guédon, O. Sparsity and non-Euclidean embeddings, *Israel J. Math.* **197**, 1 (2013), 329–345.

[22] Giannopoulos, A. A. & Milman, V. D. Concentration property on probability spaces, *Adv. Math.* **156**, 1 (2000), 77–106.

[23] Gozlan, N., Roberto, C. & Samson, P.-M. From dimension free concentration to the Poincaré inequality, *Calc. Var. Partial Differential Equations* **52**, 3-4 (2015), 899–925.

[24] Guédon, O. Kahane–Khinchine type inequalities for negative exponent, *Mathematika* **46**, 1 (1999), 165–173.

[25] Guédon, O., Litvak, A. E., Pajor, A. & Tomczak-Jaegermann, N. On the interval of fluctuation of the singular values of random matrices, *https://arxiv.org/abs/1509.02322* (preprint), to appear in *J. Eur. Math. Soc. (JEMS)*.

[26] Guédon, O. & Milman, E. Interpolating thin-shell and sharp large-deviation estimates for isotropic log-concave measures, *Geom. Funct. Anal.* **21**, 5 (2011), 1043–1068.

[27] Haviv, I. & Regev, O. The restricted isometry property of subsampled Fourier matrices, *https://arxiv.org/abs/1507.01768* (preprint).

[28] Kannan, R., Lovász, L. & Simonovits, M. Isoperimetric problems for convex bodies and a localization lemma, *Discrete Comput. Geom.* **13**, 3-4 (1995), 541–559.

[29] Kannan, R., Lovász, L. & Simonovits, M. Random walks and an $O^*(n^5)$ volume algorithm for convex bodies, *Random Structures Algorithms* **11**, 1 (1997), 1–50.

[30] Klartag, B. On convex perturbations with a bounded isotropic constant, *Geom. Funct. Anal.* **16**, 6 (2006), 1274–1290.

[31] Klartag, B. A central limit theorem for convex sets, *Invent. Math.* **168**, 1 (2007), 91–131.

[32] Klartag, B. Power-law estimates for the central limit theorem for convex sets, *J. Funct. Anal.* **245**, 1 (2007), 284–310.

[33] Kolesnikov, A. V. & Milman, E. The KLS isoperimetric conjecture for generalized Orlicz balls, *https://arxiv.org/abs/1610.06336* (preprint).

[34] Latała, R. Some estimates of norms of random matrices, *Proc. Amer. Math. Soc.* **133**, 5 (2005), 1273–1282.

[35] Latała, R. & Wojtaszczyk, J. O. On the infimum convolution inequality, *Studia Math.* **189**, 2 (2008), 147–187.

[36] Ledoux, M. A simple analytic proof of an inequality by P. Buser, *Proc. Amer. Math. Soc.* **121**, 3 (1994), 951–959.

[37] Lee, Y. T. & Vempala, S. S. Eldan's stochastic localization and the KLS hyperplane conjecture: an improved lower bound for expansion, *https://arxiv.org/abs/1612.01507* (preprint).

[38] Mendelson, S. & Paouris, G. On the singular values of random matrices, *J. Eur. Math. Soc. (JEMS)* **16**, 4 (2014), 823–834.

[39] Milman, E. On the role of convexity in isoperimetry, spectral gap and concentration, *Invent. Math.* **177**, 1 (2009), 1–43.

[40] Milman, V. D. & Schechtman, G. *Asymptotic Theory of Finite-Dimensional Normed Spaces*, Lecture Notes in Mathematics **1200**, Springer (1986), With an appendix by M. Gromov.

[41] Oymak, S. & Tropp, J. A. Universality laws for randomized dimension reduction, with applications, *https://arxiv.org/abs/1511.09433* (preprint).

[42] Paouris, G. Concentration of mass on convex bodies, *Geom. Funct. Anal.* **16**, 5 (2006), 1021–1049.

[43] Rudelson, M. Random vectors in the isotropic position, *J. Funct. Anal.* **164**, 1 (1999), 60–72.

[44] Rudelson, M. & Vershynin, R. The Littlewood–Offord problem and invertibility of random matrices, *Adv. Math.* **218**, 2 (2008), 600–633.

[45] Rudelson, M. & Vershynin, R. On sparse reconstruction from Fourier and Gaussian measurements, *Comm. Pure Appl. Math.* **61**, 8 (2008), 1025–1045.

[46] Rudelson, M. & Vershynin, R. Smallest singular value of a random rectangular matrix, *Comm. Pure Appl. Math.* **62**, 12 (2009), 1707–1739.

[47] Seginer, Y. The expected norm of random matrices, *Combin. Probab. Comput.* **9**, 2 (2000), 149–166.

[48] Sodin, S. An isoperimetric inequality on the $l_p$ balls, *Ann. Inst. Henri Poincaré Probab. Stat.* **44**, 2 (2008), 362–373.

[49] Tikhomirov, K. Sample covariance matrices of heavy-tailed distributions, *https://arxiv.org/abs/1606.03557* (preprint), to appear in *Int. Math. Res. Not. IMRN*.

[50] Tikhomirov, K. & Youssef, P. When does a discrete-time random walk in $\mathbb{R}^n$ absorb the origin into its convex hull? *https://arxiv.org/abs/1410.0458* (preprint), to appear in *Ann. Probab.*

[51] van Handel, R. On the spectral norm of Gaussian random matrices, *https://arxiv.org/abs/1502.05003* (preprint), to appear in *Trans. Amer. Math. Soc.*

[52] Vempala, S. S. Recent progress and open problems in algorithmic convex geometry, in *30th International Conference on Foundations of Software Technology and Theoretical Computer Science*, LIPIcs: Leibniz International Proceedings in Informatics **8**, Schloss Dagstuhl – Leibniz-Zentrum für Informatik (2010), pp. 42–64.

# Appendix C

## A Few Updates and Pointers

Gilles Pisier

Texas A&M University, College Station, Texas; and
Université Pierre et Marie Curie, Paris, France

### Sidon sets

Very recently Bourgain and Lewko [2] reopened the study of Sidon sets and
their connection with Rudin's $\Lambda(p)$-sets. Their goal was to extend the known
equivalences between Sidon, randomly Sidon and $\Lambda(p)$ with constant $O(\sqrt{p})$
to uniformly bounded orthonormal systems. More precisely, they considered
an orthonormal system $(\varphi_n)$ in $L_2$ over an arbitrary probability space $(T, m)$
with the mere assumption that there is a bound $b < \infty$ such that

$$\forall n \quad \|\varphi_n\|_\infty \leqslant b.$$

Let $\Lambda = \{\varphi_n \mid n \geqslant 1\}$ be such a sequence. We say that $\Lambda$ is Sidon if there is
a constant $C$ such that, for any finitely supported scalar sequence $(a_n)$, we have

$$\sum |a_n| \leqslant C \left\| \sum a_n \varphi_n \right\|_\infty.$$

We say that $(\varphi_n)$ is sub-Gaussian with constant $C$ if, for any $n$ and any complex
sequence $(a_k)$, we have

$$\left\| \sum_1^n a_k \varphi_k \right\|_{L_{\psi_2}} \leqslant C \left( \sum_1^n |a_k|^2 \right)^{1/2}.$$

Here $\psi_2$ is the function $\psi_2(x) = \exp(x^2) - 1$ on $\mathbb{R}_+$ and $L_{\psi_2}$ is the associated
Orlicz space. These notions are modeled on the familiar ones for sequences
of continuous characters on a compact Abelian group. However, the next one,
introduced in [2] is new.

We say that $(\varphi_n)$ is $\otimes^k$-Sidon with constant $C$ if the system $\{\varphi_n(t_1) \cdots \varphi_n(t_k)\}$
(or equivalently $\{\varphi_n^{\otimes k}\}$) is Sidon with constant $C$ in $L_\infty(T^k, m^{\otimes k})$. Bourgain
and Lewko [2] proved that sub-Gaussian $\Rightarrow \otimes^k$-Sidon for $k = 5$, but *not* for
$k = 1$ (in other words, sub-Gaussian $\Rightarrow$ Sidon fails in this generality). In the

case of characters, this remarkable result extends the implication sub-Gaussian $\Rightarrow$ Sidon (see Theorem IV.1 in Chapter 6 of this volume), since, obviously, when the orthonormal sequence is composed of characters, the notions of $\otimes^k$-Sidon and Sidon are identical for any $k$. Bourgain and Lewko asked whether sub-Gaussian $\Rightarrow$ $\otimes^k$-Sidon for $k = 2$, which was proved in [19]. The proof relies heavily on Talagrand's solution in [25] of Fernique's majorizing measure conjecture, or more precisely on a generalization of Slepian's lemma that follows as a corollary. In [19], the notion of randomly Sidon set is also considered, in analogy with Rider's results from [23].

We say that $(\varphi_n)$ is randomly Sidon if there is a constant $C$ such that, for any finitely supported scalar sequence $(a_n)$, we have

$$\sum |a_n| \leqslant C\mathbb{E} \left\| \sum \varepsilon_n a_n \varphi_n \right\|_\infty.$$

Here $(\varepsilon_n)$ is an *i.i.d.* sequence of independent choices of sign taking the values $\pm 1$ with probability $1/2$. It is proved in [19] that randomly Sidon is equivalent to $\otimes^4$-Sidon or to $\otimes^k$-Sidon for some (or equivalently for all) $k \geqslant 4$.

In particular, this yields as a corollary an extension of Drury's solution of the union problem: if a system $(\varphi_n)$ is the union of two systems spanning mutually orthogonal subspaces, and if each of the two subsystems is Sidon, then $(\varphi_n)$ is $\otimes^4$-Sidon. Indeed, this follows since the union will automatically be randomly Sidon.

Concerning the implication Sidon $\Rightarrow$ sub-Gaussian due to Rudin [24] for characters, the situation is less satisfactory. A simple observation shows that it cannot hold for bounded orthonormal systems, because if $(T, m)$ is split into two parts $T_1, T_2$, say of measure $1/2$, and if $(\varphi_n^1)$ and $(\varphi_n^2)$ are the restrictions of $(\varphi_n)$ respectively to $T_1$ and $T_2$, then for $(\varphi_n)$ to be Sidon it suffices that $(\varphi_n^1)$ be Sidon over $T_1$ with respect to $m(T_1)^{-1}m_{|T_1}$. In sharp contrast, for $(\varphi_n)$ to be sub-Gaussian it is necessary that *both* $(\varphi_n^1)$ and $(\varphi_n^2)$ be sub-Gaussian. So if the other part $(\varphi_n^2)$ is chosen "bad" enough, it will prevent $(\varphi_n)$ from being sub-Gaussian.

Let us say that a sequence $(\varphi_n)$ in $L_1(m)$ is $C$-dominated by another one $(g_n)$ in $L_1(\mathbb{P})$ if there are $C \geqslant 0$ and $u : L_1(\mathbb{P}) \to L_1(m)$ with $\|u\| \leqslant C$ such that

$$\forall n \ u(g_n) = \varphi_n.$$

Let $(g_n)$ be an *i.i.d.* sequence of standard $N(0, 1)$ Gaussian random variables. The key fact (from [25]) used in [19] is that there is a numerical constant $\tau$ such that any $C$-sub-Gaussian sequence is $\tau C$-dominated by $(g_n)$. For characters, the converse also holds.

The paper [19] also includes similar results for the case of Sidon sets in duals of *non-Abelian* compact groups.

Among the many interesting open questions raised by Bourgain and Lewko in [2], we should mention this one: is any sub-Gaussian uniformly bounded orthonormal system the union of finitely many Sidon systems?

## Grothendieck's Inequality

First we should mention some important progress on the Grothendieck constant $K_G$ in the real case. It is now known that Krivine's upper bound is not the best constant. Krivine proved that

$$K_G \leqslant \pi/(2 \operatorname{Log}(1 + \sqrt{2})) = 1.782\ldots,$$

and conjectured that this is the exact value. This remained open until the recent paper [4] that proved that his bound is not optimal. Curiously the latter paper does not produce a new upper bound.

In recent years, Grothendieck's inequality has popped up in various areas where it was totally unexpected – in particular, in computer science and in theoretical physics in connection with Bell's inequality. The computer science connection can be described roughly like this: let us write Grothendieck's inequality as saying that, for any $n \times n$ matrix $[a_{ij}]$ with real entries, we have (here $S_{\ell_2^n}$ denotes the unit sphere of $\ell_2^n$, which for $n = 1$ is $\{-1, 1\}$)

$$\sup \left\{ \left| \sum a_{ij}\langle x_i, y_j \rangle \right| \ \Big| \ x_i, y_j \in S_{\ell_2^n} \right\} \leqslant K_G \sup \left\{ \left| \sum a_{ij} x_i y_j \right| \ \Big| \ x_i, y_j \in \{-1, 1\} \right\}.$$

Then the inequality says that $q_1 \leqslant K_G q_2$ (and $q_2 \leqslant q_1$ is trivial), where we view $q_1, q_2$ as quantities that we want to compute. It turns out that, although $q_2$ cannot be computed "fast," in sharp contrast $q_1$ can. Here "fast" means in polynomial "time" (here we remain deliberately vague; see the references for the precise meaning). More precisely, computing $q_2$ in polynomial time is an NP-hard problem (i.e. one that would imply $P = NP$), while computing $q_1$ can be done using semi-definite programming, and hence in polynomial time. The surprising connection with $K_G$ then appears in the papers [21, 22]: they show that, for any $0 < K < K_G$, assuming a strengthening of $P \neq NP$ called the UGC (short for "unique games conjecture"), it is NP-hard to compute *any* quantity $q$ such that $q \leqslant Kq_2$. Thus, if one accepts the UGC, $K = K_G$ appears as the critical value: for $K \geqslant K_G$ there is such a $q$ computable in polynomial time (since $q = q_1$ does the job) but for $K < K_G$ there is no such $q$.

In [1], the Grothendieck constant of a (finite) graph $\mathcal{G} = (V, E)$ is introduced, as the smallest constant $K$ such that, for every $a \colon E \to \mathbb{R}$, we have

$$\sup_{f:\, V \to S} \sum_{\{s,t\} \in E} a(s,t) \langle f(s), f(t) \rangle \leqslant K \sup_{f:\, V \to \{-1,1\}} \sum_{\{s,t\} \in E} a(s,t) f(s) f(t),$$

where $S$ is the unit sphere of $H = \ell_2$. Note that we may replace $H$ by the span of the range of $f$ and hence we may always assume $\dim(H) \leqslant |V|$. The constant $K_G$ appears as the best upper bound for all bipartite graphs. However, for general non-bipartite graphs, the constant is unbounded and a growth of order $\log n$ is proved in [1] for the complete graph on $n$ vertices. See the survey [10] by S. Khot and A. Naor for more on this computer science connection.

The connection with the mathematics of quantum mechanics can be described roughly like this: Tsirelson discovered a close relationship between Grothendieck's inequality and the famous Bell's inequality in quantum physics. The latter was crucial to test the Einstein–Podolsky–Rosen (EPR) framework of "hidden variables" proposed as a sort of substitute to quantum mechanics. In 1964, J.S. Bell proposed a way to test the hidden variables theory. He introduced a specific inequality (now called "Bell's inequality"), roughly of the form $P_{qm} \leqslant K P_{hv}$ between the respective predictions of quantum mechanics and of the hidden variables theory for the result of a certain kind of experiment involving the spin of two particules of a very special type. Since the best constant $K$ is $> 1$, he suggested that experimentation might be able to detect which one is correct. Using Bell's ideas, experiments were made and the experts' consensus seems to be now that the measurements agree with the predictions of quantum mechanics. In 1980, Tsirelson [26] observed that Grothendieck's inequality could be interpreted as giving an upper bound for the best constant $K$ in a generalized Bell's inequality. Thus Bell's whole approach seems closely related to the assertion that $K_G > 1$.

More precisely, consider an $n \times n$ matrix $[a_{ij}]$ with *real* entries. On one hand let

$$P_{qm}(a) = \sup \left\| \sum a_{ij} u_i v_j \right\|,$$

where the sup runs over all $N$ and all self-adjoint unitary matrices of size $N \times N$ such that $u_i v_j = v_j u_i$ for all $i, j$. On the other hand, let

$$P_{hv}(a) = \sup \left\| \sum a_{ij} u_i v_j \right\|,$$

where the sup is now restricted to self-adjoint unitary matrices such that $\{u_1, \cdots, u_n, v_1, \cdots, v_n\}$ all mutually commute.

Tsirelson proved that $P_{qm}(a) = \sup | \sum a_{ij} \langle x_i, y_j \rangle |$, where the sup runs over all $x_i, y_j$ in the unit ball of $\ell_2$. Moreover it is easy to see that $P_{hv}(a) = \sup | \sum a_{ij} x_i y_j |$, where the sup runs over all $x_i, y_j \in [0, 1]$. Thus the

Grothendieck inequality (in the real case), as stated in Theorem III.3 in Chapter 5 (Vol. 1) coincides precisely with $P_{qm}(a) \leqslant K_G P_{hv}(a)$.

In quantum mechanics, systems with two observers correspond to tensor products with two factors. It is quite natural to wonder, as Tsirelson did in [26], what happens in the case of three observers, corresponding to a triple tensor product. The natural question of whether the constant in the associated inequality is still bounded was answered negatively by Marius Junge with Perez-Garcia, Wolf, Palazuelos and Villanueva [15].

In another direction, a non-commutative version of Grothendieck's theorem (in short, GT) has been given by the author and U. Haagerup. The original GT gives a special factorization for all the bounded linear maps $u : L_\infty \to L_1$ (see Theorem II.8 in Chapter 5 (Vol. 1)). The non-commutative version describes an analogous factorization for bounded linear maps $u : M \to N_*$, where $M, N$ are von Neumann algebras and $N_*$ is the predual of $N$. In other words, $M$ (resp. $N_*$) can be thought of as a non-commutative $L_\infty$-space (resp. $L_1$-space). More generally, the result applies to maps $u : A \to B^*$, where $A, B$ are arbitrary $C^*$-algebras. One surprising feature is that the value of the non-commutative analogue of $K_G$ is known: it is equal to 2. Some more general kinds of factorization theorems for maps on $C^*$-algebras appear in [12, 14]. We refer to our survey [18] for more references on all this.

### Non-Commutative Banach Spaces: Operator Spaces

The last three decades have seen the appearance of a notion of "non-commutative" Banach space, which has led to a new theory called "operator space theory," somewhat intermediate between the operator algebra and Banach space theories. The objects, called operator spaces, are simply Banach spaces given together with an isometric embedding into the space $B(H)$ of bounded operators on a Hilbert space $H$. Since $B(H)$ contains $\ell_\infty(I)$ if $H = \ell_2(I)$, any Banach space embeds isometrically into some $B(H)$. Thus any Banach space can appear. But the novelty of operator space theory lies in its morphisms: the bounded linear maps are replaced by the completely bounded ones. A linear map $u : E \to F$ between operator spaces $E \subset B(H)$ and $F \subset B(K)$ is called completely bounded (c.b. in short) if the mappings $u_n : M_n(E) \to M_n(F)$ are bounded uniformly over $n$, and we define

$$\|u\|_{cb} = \sup_{n \geqslant 1} \|u_n : M_n(E) \longrightarrow M_n(F)\|.$$

Here $M_n(E)$ denotes the space of $n \times n$ matrices with entries in $E$, equipped with the norm induced by the natural one on $M_n(B(H)) = B(H \oplus \cdots \oplus H)$. By the "operator space structure" (in short o.s.s.) on $E$ we mean the sequence of the norms on $M_n(E)$, indexed by $n$.

The map $u$ is called completely isometric if $u_n$ is isometric for all $n$.

The definition of $\|u\|_{cb}$ requires only knowledge of the o.s.s. on $E$ and $F$. But actually, it is crucial to have a way to recognize which sequences of norms truly come from embeddings $E \subset B(H)$ and $F \subset B(K)$. This is provided by a theorem due to Ruan, often described as the starting point of operator space theory (see [5, 17]), after which the foundations of the theory were laid by Effros and Ruan and also by Blecher and Paulsen.

**Theorem** (Ruan's Theorem)  *Let $V$ be a vector space. Consider, for each $n$, a norm $\alpha_n$ on the vector space $M_n(V)$. Then the sequence of norms $(\alpha_n)$ comes from a linear embedding of $V$ into some $B(H)$ iff the following two properties hold:*

- $\forall n, \forall x \in M_n(V) \qquad \forall a, b \in M_n; \qquad \alpha_n(a.x.b) \leqslant \|a\|_{M_n} \alpha_n(x) \|b\|_{M_n}.$
- $\forall n, m \,\forall x \in M_n(V) \,\forall y \in M_m(V) \qquad \alpha_{n+m}(x \oplus y) = \max\{\alpha_n(x), \alpha_m(y)\},$
  *where we denote by $x \oplus y$ the $(n+m) \times (n+m)$ matrix defined by*
  $$x \oplus y = \begin{pmatrix} x & 0 \\ 0 & y \end{pmatrix}.$$

After completion of $V$ with respect to the norm $\alpha_1$, one obtains an operator space $E$, which is isometric to the closure of $V$ in $B(H)$. Of course, if $E$ is a $C^*$-algebra then there is a canonical norm on $E$, namely the *unique* norm (the unicity requires $E$ to be complete) such that

$$\forall x, y \in E \quad \|xy\| \leqslant \|x\| \|y\| \quad \|x\| = \|x^*\| \quad \|x^*x\| = \|x\|^2.$$

Such norms are called $C^*$-norms. Since $M_n(E)$ is then a $C^*$-algebra for the natural matricial $*$-algebra structure, we also have a unique norm on $M_n(E)$. By Gelfand's classical theorem, these norms come from a realization of $E$ as a $C^*$-subalgebra of $B(H)$ for some $H$. Thus there is a *canonical* o.s.s. on a $C^*$-algebra $E$.

The space of completely bounded maps $u \colon E \to F$ is denoted by $CB(E, F)$. It becomes a Banach space when equipped with the cb-norm. But the flavor of the new theory is to always consider operator spaces rather than just Banach spaces. Thus, for instance, by Ruan's theorem the space $CB(E, F)$ can be equipped with a *distinguished* operator space structure: there is a Hilbert space $\mathcal{H}$ and an isometric embedding $CB(E, F) \subset B(\mathcal{H})$ such that, for each $n$, the norm induced on $M_n(CB(E, F))$ by $M_n(B(\mathcal{H}))$ coincides with the norm of the space $CB(E, M_n(F))$. In particular, when $F = \mathbb{C}$, this gives us a distinguished operator space structure on the dual $E^*$ of $E$.

Note that in particular this gives us an operator space structure on the dual of any $C^*$-algebra.

The dual structure just defined is closely related to the minimal tensor product, defined as follows. Given operator spaces $G \subset B(\mathcal{H})$ and $F \subset B(K)$, there is a natural embedding

$$J: G \otimes F \longrightarrow B(\mathcal{H} \otimes_2 K)$$

of the *algebraic* tensor product $G \otimes F$ into $B(\mathcal{H} \otimes_2 K)$. The closure of $J(G \otimes F)$ in $B(\mathcal{H} \otimes_2 K)$ is called the minimal tensor product. The resulting operator space is denoted by

$$G \otimes_{\min} F \subset B(\mathcal{H} \otimes_2 K).$$

The distinguished operator space structures on $E^*$ and on $CB(E, F)$ are such that, if $\dim(F) < \infty$, we have

$$E^* \otimes_{\min} F = CB(E, F)$$

completely isometrically, and in general we have a completely isometric embedding

$$E^* \otimes_{\min} F \subset CB(E, F).$$

These identities show that $G \otimes_{\min} F$ is analogous to the injective tensor product of Banach space theory. Similarly there is an analogue of the projective tensor product and of Grothendieck's approximation property in operator space theory (see [5]).

Using Ruan's theorem, one extends the complex interpolation method (see [3]) from Banach spaces to operator spaces. Applied to the pair $(\ell_1, \ell_\infty)$, this produces an operator space structure on $\ell_p$ for any $1 < p < \infty$. Here $\ell_\infty$ is equipped with its canonical operator space structure as a $C^*$-algebra, while $\ell_1$ is equipped with its distinguished structure as the dual of a $C^*$-algebra.

Let $E$ be any operator space. Using complex interpolation, one can define an operator space structure on the space $\ell_p(E)$. This leads to an analogue of $p$-summing mappings ([16]): a map $u: E \to F$ is called completely $p$-summing if the associated mapping $\ell_p \otimes_{\min} E \to \ell_p(F)$ is completely bounded. This is analogous to the notion considered in Chapter 5 of Vol. 1.

More generally, the space $\ell_p$ can be replaced by the Schatten $p$-class $S_p$ or by any of the so-called non-commutative $L_p$-spaces. Again, by complex interpolation, one can define an operator space structure on the space $L_p(E)$ when $L_p$ is a non-commutative $L_p$-space. Assuming that the latter is associated to an injective von Neumann algebra, one can reproduce most of the known properties of the ordinary Banach space valued $L_p$-spaces (see [16]).

Since operator spaces admit a nice duality theory parallel to that of Banach spaces, it is natural to wonder whether there is an analogue of Hilbert space. Indeed, this turns out to be the case: for any Hilbert space $H$ there is a unique o.s.s. on $H$ such that the canonical isometry $\overline{H} = H^*$ becomes a *complete* isometry. See [17, Chapter 7] for more on this.

One then wonders whether there is an analogue of Dvoretzky's theorem, and again there is, although the resulting statement seems less powerful (see [17], § 9.11 and references there).

In yet another direction, analogues of Rosenthal's theorem and Maurey's factorization, considered in Chapter 4 of this volume, are developed for operator spaces in Junge and Parcet's work [6, 7, 8]. Extensions of type, cotype and $K$-convexity are discussed in [9].

All these results were made possible by the remarkable non-commutative generalization of Khintchine's inequality to non-commutative $L_p$-spaces ($1 < p < \infty$) proved by Lust-Piquard [11] in 1986. The case $p = 1$ was later obtained in [13], while the case $0 < p < 1$ resisted until the very recent paper [20].

## References

[1] N. Alon, K. Makarychev, Y. Makarychev & A. Naor, Quadratic forms on graphs, *Invent. Math.* **163**, 3 (2006), 499–522.

[2] J. Bourgain & M. Lewko, Sidonicity and variants of Kaczmarz's problem *https:// arxiv.org/abs/1504.05290v3* (preprint), to appear in *Ann. Inst. Fourier*.

[3] J. Bergh & J. Löfström, *Interpolation Spaces: An Introduction*, Springer (1976).

[4] M. Braverman, K. Makarychev, Y. Makarychev & A. Naor, The Grothendieck constant is strictly smaller than Krivine's bound, *Forum Math. Pi* **1** (2013).

[5] E.G. Effros & Z.J. Ruan, *Operator Spaces*, Clarendon Press, (2000).

[6] M. Junge & J. Parcet, Rosenthal's theorem for subspaces of noncommutative $L_p$, *Duke Math. J.* **141**, 1 (2008), 75–122.

[7] M. Junge & J. Parcet, Maurey's factorization theory for operator spaces, *Math. Ann.* **347**, 2 (2010), 299–338.

[8] M. Junge & J. Parcet, Mixed-norm inequalities and operator space $L_p$ embedding theory, *Memoirs of the American Mathematical Society* **203**, 953 (2010).

[9] M. Junge & J. Parcet, The norm of sums of independent noncommutative random variables in $Lp(\ell_1)$, *J. Funct. Anal.* **221** (2005), 366–406.

[10] S. Khot & A. Naor, Grothendieck-type inequalities in combinatorial optimization, *Comm. Pure Appl. Math.* **65**, 7 (2012), 992–1035.

[11] F. Lust-Piquard, Inégalités de Khintchine dans $C_p$ ($1 < p < \infty$) (in French) [Khinchine inequalities in $C_p$ ($1 < p < \infty$)], *C.R.A.S. Paris Sér. I Math.* **303**, 7 (1986), 289–292.

[12] F. Lust-Piquard, A Grothendieck factorization theorem on 2-convex Schatten spaces, *Israel J. Math.* **79**, 2–3 (1992), 331–365.

[13] F. Lust-Piquard & G. Pisier, Noncommutative Khintchine and Paley inequalities, *Ark. Mat.* **29**, 2 (1991), 241–260.

[14] F. Lust-Piquard & Q. Xu, The little Grothendieck theorem and Khintchine inequalities for symmetric spaces of measurable operators, *J. Funct. Anal.* **244**, 2 (2007), 488–503.

[15] D. Pérez-García, M.M. Wolf, C. Palazuelos, I. Villanueva & M. Junge, Unbounded violation of tripartite Bell inequalities, *Comm. Math. Phys.* **279**, 2 (2008), 455–486.

[16] G. Pisier, Non-commutative vector valued $L_p$-spaces and completely $p$-summing maps, *Astérisque* **247** (1998).

[17] G. Pisier, *Introduction to Operator Space Theory*, Cambridge University Press (2003).

[18] G. Pisier, Grothendieck's theorem, past and present. *Bull. Amer. Math. Soc.* **49** (2012), 237–323.

[19] G. Pisier, On uniformly bounded orthonormal Sidon systems, *https:arxiv.org/abs/1602.02430v6* (preprint), to appear in *Math. Res. Lett.*

[20] G. Pisier & É. Ricard, The non-commutative Khintchine inequalities for $0 < p < 1$, *https:doi.org/10.1017/S1474748015000353*, *J. Inst. Math. Jussieu* (2017).

[21] P. Raghavendra, Optimal algorithms and inapproximability results for every CSP?, in *Proceedings of the Fortieth Annual ACM Symposium on the Theory of Computing*, ACM (2008), 245–254.

[22] P. Raghavendra & D. Steurer, Towards computing the Grothendieck constant, in *Proceedings of the Twentieth Annual ACM-SIAM Symposium on Discrete Algorithms*, ACM (2009), 525–534.

[23] D. Rider, Randomly continuous functions and Sidon sets, *Duke Math. J.* **42** (1975), 752–764.

[24] W. Rudin, Trigonometric series with gaps, *J. Math. Mech.* **9** (1960), 203–227.

[25] M. Talagrand, Regularity of Gaussian processes, *Acta Math.* **159** (1987), 99–149.

[26] B.S. Tsirelson, Quantum generalizations of Bell's inequality, *Lett. Math. Phys.* **4**, 2 (1980), 93–100.

# Appendix D

## On the Mesh Condition for Sidon Sets

Luis Rodríguez-Piazza

Universidad de Sevilla, Sevilla, Spain

### Introduction

It is known that if $\Lambda$ is a Sidon set of a discrete Abelian group $\Gamma$, then $\Lambda$ has to be lacunary in an arithmetic sense. For example (see [3], Chapter 6), if $\Gamma = \mathbb{Z}$, $\Lambda$ cannot contain arbitrarily large arithmetic progressions, or even a big proportion of such progressions. This can be generalized to a so-called "mesh condition" (to be described here). Whether this mesh condition is sufficient for implying Sidonicity is an open problem.

Here, we prove that the slightest weakening of the mesh condition is not sufficient to imply this Sidonicity. We first prove this result for some group $\Gamma$, relying on a random construction of Blei and Körner [1] on a notion of fractional dimension (Theorem 4 to follow). Then, we are able to transfer it to the (most interesting) case of $\mathbb{Z}$, this time using arguments of combinatorial type (Theorem 5).

Let us begin by describing our framework and the mesh condition alluded to above.

### Notation

$G$ will be an Abelian compact group, $\Gamma$ its dual group (where we will use additive notation for the operation). Let $A$ be a finite subset of $\Gamma$, and $s \in \mathbb{N}$:

- $|A|$ will denote the cardinal number of $A$.
- $\mathrm{gr}(A)$ will denote the subgroup of $\Gamma$ generated by $A$.
- $M_s(A)$ will denote the "mesh" of length $s$ over $A$:

$$M_s(A) = \left\{ \sum_{\gamma \in A} m_\gamma \gamma : (m_\gamma)_\gamma \in \mathbb{Z}^A, \quad \sum_{\gamma \in A} |m_\gamma| \leqslant s \right\}.$$

316

- [A] will denote the set

$$[A] = \left\{ \sum_{\gamma \in A} m_\gamma \gamma : (m_\gamma)_\gamma \in \{-1, 0, 1\}^A \right\}.$$

Observe that we have $[A] \subset M_{|A|}(A)$ for every finite set $A \subset \Gamma$.

## Mesh Condition

The mesh condition is an arithmetic condition that every Sidon set must satisfy. This is the statement of the following theorem (see [3], Corollary 6.4 or [2], Theorem 5, page 71).

**Theorem 1** *Suppose $\Lambda$ is a Sidon set included in $\Gamma$. Then there exists a constant $C > 0$ such that*

$$|\Lambda \cap M_s(A)| \leqslant C|A| \log(1 + s), \quad \text{for every } s \geqslant 1, \text{ and every finite set } A \subset \Gamma.$$
$$(*)$$

It is unknown whether every set $\Lambda$ satisfying $(*)$ is a Sidon set. The following result is an easy corollary:

**Corollary 2** *Suppose $\Lambda$ is a Sidon set included in $\Gamma$. Then there exists a constant $C > 0$ such that*

$$|\Lambda \cap [A]| \leqslant C|A| \log(1 + |A|), \quad \text{for every finite set } A \subset \Gamma. \qquad (**)$$

Thanks to an example of Pisier ([4], Corollary 7.3), one can see that $(**)$ cannot be improved for general groups. Indeed, if an increasing function $\varphi [1, +\infty) \to [1, +\infty)$ satisfies that for every Sidon set $\Lambda$ in any group $\Gamma$ there exists a constant $C > 0$ such that

$$|\Lambda \cap [A]| \leqslant C\varphi(|A|), \quad \text{for every finite set } A \subset \Gamma,$$

then we have $x \log x = O(\varphi(x))$, when $x \to \infty$. Otherwise, if $\Gamma$ is a group of bounded order (there exists an integer $M \geqslant 1$ such that $M\gamma = 0$, for every $\gamma \in \Gamma$), then $(**)$ can be improved, and in fact we have (see [4], Section 3):

**Theorem 3** *Let $\Gamma$ be a group of bounded order, and $\Lambda \subset \Gamma$. The following are equivalent:*

*(a) $\Lambda$ is a Sidon set.*
*(b) There exists $C > 0$ such that $|\Lambda \cap \mathrm{gr}(A)| \leqslant C|A|$, for every finite set $A \subset \Gamma$.*
*(c) There exists $C > 0$ such that $|\Lambda \cap [A]| \leqslant C|A|$, for every finite set $A \subset \Gamma$.*

In fact, in every group, condition $(c)$ is sufficient for $\Lambda$ to be a Sidon set. We see that this cannot be improved, in the sense that there is no weaker condition than $|\Lambda \cap [A]| \leqslant C|A|$ implying $\Lambda$ is a Sidon set. We present two different constructions based on the same idea. In the first one (Theorem 4) we use the group $\mathbb{Z}^{(\mathbb{N})}$, which is a $\mathbb{Z}$-module of infinite dimension, containing an infinite $\mathbb{Z}$-independent set. In the second one (Theorem 5) this construction is carried out in the group of integers $\mathbb{Z}$, where such an infinite $\mathbb{Z}$-independent set does not exist.

**Theorem 4** *Let* $\varphi \, (0, +\infty) \to (0, +\infty)$ *be an increasing function such that*

$$\lim_{x \to \infty} \frac{\varphi(x)}{x} = \infty.$$

*There exist* $C > 0$, *a group* $\Gamma$ *and a set* $\Lambda \subset \Gamma$ *such that*:

*(1)* $\Lambda$ *is not a Sidon set;*
*(2)* *for every finite set* $A \subset \Gamma$ *we have* $|\Lambda \cap \mathrm{gr}(A)| \leqslant C\varphi(|A|)$; *in particular,*

$$|\Lambda \cap [A]|, |\Lambda \cap M_s(A)| \leqslant C\varphi(|A|), \quad \textit{for every finite set } A \subset \Gamma \textit{ and every } s \geqslant 1.$$

**Theorem 5** *Let* $\varphi \, (0, +\infty) \to (0, +\infty)$ *be an increasing function such that*

$$\lim_{x \to \infty} \frac{\varphi(x)}{x} = \infty.$$

*There exist* $C > 0$ *and a set* $\Lambda \subset \mathbb{Z}$ *such that*:

*(1)* $\Lambda$ *is not a Sidon set;*
*(2)* *for every finite set* $A \subset \mathbb{Z}$, *if* $s = |A|$, *we have*

$$|\Lambda \cap M_s(A)| \leqslant C\varphi(|A|), \qquad \textit{and so} \qquad |\Lambda \cap [A]| \leqslant C\varphi(|A|).$$

*Proof of Theorem 4*   Put $h(x) = \min\{x^{3/2}, \varphi(x)\}$. Taking the pointwise supremum of all the convex positive functions under $h$, it is easy to see that there exists a continuous convex increasing function $\varphi_1 [0, +\infty) \to [0, +\infty)$ such that $\varphi_1(0) = 0$, $\min\{x^{3/2}, \varphi(x)\} \geqslant \varphi_1(x) > 0$, for every $x > 0$, and

$$\lim_{x \to \infty} \frac{\varphi_1(x)}{x} = \infty.$$

So we can and will assume that $\varphi$ in the statement of Theorem 4 is a continuous convex increasing function and $0 < \varphi(x) \leqslant x^{3/2}$, for every $x > 0$.

For a set $S \subset \mathbb{N} \times \mathbb{N}$, and for every integer $k \geqslant 1$, let us define

$$\rho_S(k) = \max\{|S \cap D \times E| : D, E \subset \mathbb{N} \, ; |D|, |E| \leqslant k\}.$$

We will use the following result due to Blei and Körner ([1], Proposition C).

**Proposition 6** *Let $\psi\ [0, +\infty)\ \to\ [0, +\infty)$ be an increasing function such that:*

*(a)* $\lim_{x\to\infty} \psi(x)/x = \infty$;
*(b)* *there exists $\delta > 0$, so that $\lim_{x\to\infty} \psi(x)/x^{2-\delta} = 0$;*
*(c)* *there exists $C > 0$ such that, whenever $x_1, x_2, \ldots, x_l$ are greater than 1, we have $\psi(x_1) + \psi(x_2) + \cdots + \psi(x_l) \leqslant C\psi(x_1 + x_2 + \cdots + x_l)$.*

*Then there exist a set $S \subset \mathbb{N} \times \mathbb{N}$ and $C_1, C_2 > 0$ such that:*

*(1)* $\rho_S(k) \leqslant C_1\psi(k)$, *for every $k \geqslant 1$, and*
*(2)* $\rho_S(k_j) \geqslant C_2\psi(k_j)$ *for some sequence $(k_j)$ tending to $\infty$.*

In fact Blei and Körner give the proof of this proposition only for the case $\psi(x) = x^\alpha$, with $1 < \alpha < 2$ (Proposition B in [1]). Then they give the statement of Proposition 6 without condition (c). I think this extra condition should be added and then one can get the proof of Proposition 6 as a straightforward generalization of the proof of Proposition B in [1].

We can apply Proposition 6 with $\psi = \varphi$ (observe that $\varphi(0) = 0$ and convexity together yield condition (c) for $C = 1$). Let $S \subset \mathbb{N} \times \mathbb{N}$ satisfy (1) and (2) in Proposition 6 with $\varphi$ instead of $\psi$. Let $\mathbb{Z}^{(\mathbb{N})}$ be the dual group of $\mathbb{T}^\mathbb{N}$, and denote by $\{e_n\}_{n\in\mathbb{N}}$ the natural basis of $\mathbb{Z}^{(\mathbb{N})}$ as a $\mathbb{Z}$-module. $\mathbb{Z}^{(\mathbb{N})}$ can be viewed as the set of sequences in $\mathbb{Z}$ which are eventually 0, and then each $e_n$ is the sequence with entry 1 in the $n$-th position, and 0 otherwise.

The group $\Gamma$ will be $\Gamma = \mathbb{Z}^{(\mathbb{N})} \times \mathbb{Z}^{(\mathbb{N})}$, and $\Lambda$ its subset

$$\Lambda = \{(e_m, e_n) : (m, n) \in S\}.$$

Let $A$ be a finite subset of $\Gamma$. Let $\pi_1 \colon \Gamma \to \mathbb{Z}^{(\mathbb{N})}$ be the first projection, $\pi_1(x, y) = x$. Then

$$\pi_1\big(\mathrm{gr}(A) \cap \Lambda\big) \subset \mathrm{gr}\big(\pi_1(A)\big) \cap \{e_m : m \in \mathbb{N}\}.$$

The basis $\{e_m : m \in \mathbb{N}\}$ is a $\mathbb{Z}$-independent set and $\mathrm{gr}\big(\pi_1(A)\big)$ is generated by less than $|A|$ elements. So there exists a set $D \subset \mathbb{N}$ such that $|D| \leqslant |A|$, and $\pi_1\big(\mathrm{gr}(A) \cap \Lambda\big) \subset \{e_m : m \in D\}$. Analogously, there exists a set $E \subset \mathbb{N}$ such that $|E| \leqslant |A|$, and $\pi_2\big(\mathrm{gr}(A) \cap \Lambda\big) \subset \{e_n : n \in E\}$.

We have $\{(m, n) \in S : (e_m, e_n) \in \Lambda \cap \mathrm{gr}(A)\} \subset D \times E$, and

$$|\mathrm{gr}(A) \cap \Lambda| = |\{(m, n) \in S : (e_m, e_n) \in \Lambda \cap \mathrm{gr}(A)\}| \leqslant |S \cap D \times E| \leqslant \rho_S(|A|).$$

By condition (2) in Proposition 6, we obtain $|\mathrm{gr}(A) \cap \Lambda| \leqslant C_2\varphi(|A|)$, proving (2) in the statement of Theorem 4.

If $\Lambda$ were a Sidon set, by Theorem 1, there would exist a constant $C > 0$ such that

$$|\Lambda \cap M_2(A)| \leqslant C|A|, \qquad \text{for every finite set } A \subset \Gamma. \qquad (\clubsuit)$$

Take an integer $k \geqslant 1$, and $D, E \subset \mathbb{N}$, with $|D|, |E| \leqslant k$. Put

$$A = \{(e_m, 0) : m \in D\} \cup \{(0, e_n) : n \in E\}.$$

Then $|A| \leqslant 2k$, and $\{(e_m, e_n) : m \in D, n \in E\}$ is included in $M_2(A)$. So we have, thanks to $(\clubsuit)$,

$$|S \cap D \times E| = |\{(e_m, e_n) \in \Lambda : m \in D, n \in E\}| \leqslant |M_2(A) \cap \Lambda| \leqslant 2Ck.$$

Taking the supremum in $D$ and $E$, we get $\rho_S(k) \leqslant 2Ck$, for every $k \geqslant 1$. This is a contradiction with (2) in Proposition 6, and so $\Lambda$ is not a Sidon set. $\qquad \square$

*Proof of Theorem 5*    In the proof of Theorem 4, we used the fact that, for every set $S \subset \mathbb{N} \times \mathbb{N}$, taking the set $\Lambda = \{(e_j, e_k) : (j, k) \in S\}$, we have

$$|M_s(A) \cap \Lambda| \leqslant \rho_S(|A|), \quad \text{for all finite set } A \text{ and all } s \geqslant 1. \qquad (\diamondsuit)$$

We will use the same ideas to prove Theorem 5. Now we will construct two sequences $(\sigma_j)_{j \geqslant 1}$, $(\tau_j)_{j \geqslant 1}$ of positive integers and we will consider the set

$$\Lambda = \{\sigma_j + \tau_k : (j, k) \in S\} \subset \mathbb{Z}.$$

We need to have something similar to $(\diamondsuit)$, but now this is impossible for all $s \geqslant 1$, and we will need it only for $1 \leqslant s \leqslant |A|$. Actually we only have to check the case $|A| = s$. So Theorem 5 is a direct consequence of the following Proposition 7, once we use the same set $S$ given by Proposition 6, and the same ideas of the proof of Theorem 4. $\qquad \square$

**Proposition 7**    *There exist two sequences of positive integers $(\sigma_j)_{j \geqslant 1}$, $(\tau_j)_{j \geqslant 1}$, and a constant $c_0 > 0$ such that:*

*(1) $\sigma_j + \tau_k \neq \sigma_{j'} + \tau_{k'}$ whenever we have $(j, k) \neq (j', k')$;*
*(2) for every set $S \subset \mathbb{N} \times \mathbb{N}$, if we put $\Lambda = \{\sigma_j + \tau_k : (j, k) \in S\} \subset \mathbb{Z}$, we have*

$$|M_s(A) \cap \Lambda| \leqslant c_0 \rho_S(s), \quad \text{for all finite set } A \subset \mathbb{Z} \text{ and } s = |A|.$$

We will finish once we have proved Proposition 7. We need some definitions and results. Let $s$ be a positive integer. We will say that a finite subset $B$ of an Abelian group $\Gamma$ is $s$-independent if the relation $\sum_{\gamma \in B} m_\gamma \gamma = 0$, with $m_\gamma \in \mathbb{Z}$, $|m_\gamma| \leqslant s$, for all $\gamma \in B$, implies $m_\gamma = 0$, for all $\gamma \in B$. This is equivalent to saying that the set

$$[B]_s^+ = \left\{ \sum_{\gamma \in B} m_\gamma \gamma : (m_\gamma)_\gamma \in \mathbb{Z}^B, \quad 0 \leqslant m_\gamma \leqslant s \right\}$$

has exactly $(1 + s)^{|B|}$ different elements, because $s$-independence of $B$ is equivalent to the fact that distinct $B$-tuples $(m_\gamma)_\gamma$, with $0 \leqslant m_\gamma \leqslant s$, produce distinct elements in $\Gamma$ when taking $\sum_{\gamma \in B} m_\gamma \gamma$.

We will need the following lemma:

**Lemma 8** *There exists a positive integer constant $c_1$ (actually $c_1 = 8$ suffices) such that, if $B$ is an $s$-independent set in $\Gamma$ and $B \subset M_s(A)$, for certain set $A$ with $|A| = s$, then*

$$|B| \leqslant c_1 s.$$

*Proof* For $s = 1$ it is clear that $|B| \leqslant 1$. So we can assume that $s \geqslant 2$. Observe that $[B]_s^+$ is included in $M_{s|B|}(B)$. Using the fact that $B \subset M_s(A)$ implies $M_t(B) \subset M_{st}(A)$, we have $[B]_s^+ \subset M_{s^2|B|}(A)$. The following estimate about the cardinality of a mesh can be found, for instance, in Sub-lemma II.12, Chapter 5 of this volume:

$$|M_r(A)| \leqslant \left( 4e \frac{r}{|A|} \right)^{|A|}, \quad \text{if } |A| \leqslant r.$$

Using this estimate, and the fact that $|[B]_s^+| = (1 + s)^{|B|}$, we have

$$(1 + s)^{|B|} \leqslant |M_{s^2|B|}(A)| \leqslant \left( 4e \frac{s^2|B|}{|A|} \right)^{|A|} = (4es|B|)^s,$$

and then

$$|B| \log(1 + s) \leqslant s \log(4es) + s \log |B|.$$

This implies that either $|B| \log(1 + s) \leqslant 2s \log(4es)$ or $|B| \log(1 + s) \leqslant 2s \log |B|$. In the first case we have

$$|B| \leqslant \frac{2s \log(4es)}{\log(1 + s)} \leqslant 8s,$$

for every $s \geqslant 2$. In the second case, putting $y = |B|/s$ and taking into account that $\log(1 + s) \geqslant 1$ (since $s \geqslant 2$), we have

$$y \leqslant 2 \frac{\log(sy)}{\log(1 + s)} = 2 \frac{\log(s)}{\log(1 + s)} + 2 \frac{\log(y)}{\log(1 + s)} \leqslant 2 + 2 \log y.$$

As the function $g(t) = \frac{1 + \log t}{t}$ is decreasing in $(1, +\infty)$, and $g(8) < 1/2$, we have $y \leqslant 8$, and so $|B| \leqslant 8s$ in this case too. $\qquad \square$

In order to prove Proposition 7, we first construct by induction an increasing sequence $(\lambda_j)_{j \geqslant 1}$ of integers satisfying $\lambda_1 = 1$, and

$$\lambda_{k+1} > (c_1 + 1)(k + 1)^2 \sum_{j=1}^{k} \lambda_j, \qquad \text{for all } k \geqslant 1, \qquad (\bullet)$$

where $c_1$ is the integer constant in Lemma 8. Then we define:

$$\sigma_j = \lambda_{2j}, \qquad \tau_j = \lambda_{2j+1}, \qquad \text{for all } j \geqslant 1.$$

In the following lemma we set the main property of the sequence $(\lambda_j)_j$ that we will need to make our construction; in fact this construction, and consequently the statement of Theorem 5, can be carried out in any group where there exists a sequence with the property of the next lemma; that is, in any group of not bounded order.

**Lemma 9**  *Let $r$, $s$ be two positive integers, and consider $r$ integer numbers $m_1, m_2, \ldots, m_r$. If we have*

$$\sum_{k=1}^{r} m_k \lambda_k = 0, \qquad and \qquad |m_k| \leqslant (c_1 + 1)s^2, \quad \text{for all } k,$$

*then $m_k = 0$, for every $k \geqslant s$.*

*Proof*  Suppose the result is false and take $k_0 = \max\{k : m_k \neq 0\}$. We know $k_0 \geqslant s$, and we have

$$m_{k_0} \lambda_{k_0} = - \sum_{j=1}^{k_0-1} m_j \lambda_j.$$

Then we arrive at a contradiction, thanks to $(\bullet)$:

$$\lambda_{k_0} \leqslant |m_{k_0} \lambda_{k_0}| \leqslant \sum_{j=1}^{k_0-1} |m_j| \lambda_j \leqslant (c_1 + 1)s^2 \sum_{j=1}^{k_0-1} \lambda_j \leqslant (c_1 + 1)k_0^2 \sum_{j=1}^{k_0-1} \lambda_j < \lambda_{k_0}.$$

$\square$

*Proof of Proposition 7.*  Taking $s = 1$ in Lemma 9, we see that every finite subsequence of the sequence $(\lambda_j)_j$ is $(c_1 + 1)$-independent. Part (1) of the statement follows easily from this fact.

So we have to prove part (2). Take $S \subset \mathbb{N} \times \mathbb{N}$, $\Lambda = \{\sigma_j + \tau_k : (j,k) \in S\}$ and $A \subset \mathbb{Z}$, with $|A| = s$. Let us define

$$D = \{j \in \mathbb{N} : \exists k \in \mathbb{N}, \sigma_j + \tau_k \in M_s(A) \cap \Lambda \},$$
$$E = \{k \in \mathbb{N} : \exists j \in \mathbb{N}, \sigma_j + \tau_k \in M_s(A) \cap \Lambda \}.$$

We only have to prove $|D| \leqslant (c_1 + 1)s$ and $|E| \leqslant (c_1 + 1)s$, since then

$$|M_s(A) \cap \Lambda| \leqslant |S \cap (D \times E)| \leqslant \rho_S\big((c_1 + 1)s\big) \leqslant (c_1 + 1)^2 \rho_S(s),$$

and part (2) follows with $c_0 = (c_1 + 1)^2$. We have used the inequality $\rho_S(ms) \leqslant m^2 \rho_S(s)$, a consequence of the fact that a set of the form $D \times E$, with $|E|$, $|D| \leqslant ms$, can be covered by less than $m^2$ sets of the form $E' \times F'$ with $|E'|$, $|D'| \leqslant s$.

Let us look at $|D| \leqslant (c_1 + 1)s$; the proof for $E$ is the same. Suppose that $|D| > (c_1 + 1)s$; then $|D \cap [s, +\infty)| > c_1 s$. This implies that, for a certain integer $d$ with $c_1 s < d \leqslant (c_1 + 1)s$, there exist $j_1 < j_2 < \cdots < j_d$ in $D$ with $j_1 \geqslant s$. There exist $k_1, k_2, \ldots, k_d$ in $\mathbb{N}$, so that

$$\Lambda_0 = \{\sigma_{j_1} + \tau_{k_1}, \sigma_{j_2} + \tau_{k_2}, \ldots, \sigma_{j_d} + \tau_{k_d}\}$$

is included in $M_s(A) \cap \Lambda$. We will arrive to a contradiction with Lemma 8, proving that $\Lambda_0$ is $s$-independent, since then $M_s(A)$ would contain an $s$-independent set of cardinality $|\Lambda_0| = d > c_1 s$.

Take $(m_i)_{i=1}^d$ in $\mathbb{Z}^d$ such that $|m_i| \leqslant s$, for all $i$, and

$$0 = m_1(\sigma_{j_1} + \tau_{k_1}) + m_2(\sigma_{j_2} + \tau_{k_2}) + \cdots + m_d(\sigma_{j_d} + \tau_{k_d}).$$

Remembering the definition of the $\sigma$ and $\tau$ sequences, and regrouping the coefficients of the same $\tau_k$, this can be written in terms of the sequence $(\lambda_i)_i$, for certain $r$, as

$$0 = \sum_{i=1}^r n_i \lambda_i,$$

where $n_{j_i} = m_i$, for $1 \leqslant i \leqslant d$, and each $n_i$ is the sum of at most $d$ of the $m_i$'s. We then have $|n_i| \leqslant ds \leqslant (c_1 + 1)s^2$, for every $i$; and, using Lemma 9, $n_i = 0$, for every $i \geqslant s$. This implies that $m_i = n_{k_i} = 0$ for $1 \leqslant i \leqslant d$. We have proved that $\Lambda_0$ is $s$-independent and Proposition 7 follows. $\qquad\square$

## References

[1] R. C. Blei & T. W. Körner Combinatorial dimension and random sets. *Israel J. Math.* **47** (1984), 65–74.

[2] J.-P. Kahane, *Some Random Series of Functions*, 2nd edn., Cambridge University Press, (1985).

[3] J. M. López & K. A. Ross, *Sidon Sets*, Marcel Dekker (1975).

[4] G. Pisier, De nouvelles caractérisations des ensembles de Sidon, in *Mathematical Analysis and Applications, Part B*, Advances in Mathematics Supplementary Studies, **7B**, Academic Press (1981), 685–726.

# References

## Books

The following reference is a two-volume, multi-author overview of Banach spaces:
HANDBOOK OF THE GEOMETRY OF BANACH SPACES, 2 VOLS., *W.B. Johnson & J. Lindenstrauss*, eds., North-Holland, Vol. I (2001), Vol. II (2003).

E.M. ALFSEN. *Compact Convex Sets and Boundary Integrals*, Ergebnisse der Mathematik und ihrer Grenzgebiete **57**, Springer (1971).

K. BALL. An elementary introduction to modern convex geometry, in *Flavors of Geometry*, S. Levy, ed., Mathematical Sciences Research Institute Publications, Cambridge University Press (1997).

S. BANACH. *Théorie des Opérations Linéaires*, (first published 1932), Éditions Jacques Gabay (1993).

P. BARBE & M. LEDOUX. *Probabilité*, Belin (1998).

B. BEAUZAMY. *Introduction to Banach Spaces and Their Geometry*, Notas de Matemática **68**, North-Holland (1982).

B. BEAUZAMY & T. LAPRESTÉ. *Modèles Étalés des Espaces de Banach*, Travaux en Cours, Hermann (1984).

C. BENNETT & R. SHARPLEY. *Interpolation of Operators*, Pure and Applied Mathematics **129**, Academic Press (1988).

Y. BENYAMINI & J. LINDENSTRAUSS. *Geometric Nonlinear Functional Analysis*, Vol. 1, A.M.S. Colloquium Publications **48**, American Mathematical Society (2000).

J. BERGH & J. LÖFSTRÖM. *Interpolation Spaces: An Introduction*, Grundlehren der Mathematischen Wissenschaften **223**, Springer (1976).

P. BILLINGSLEY. *Probability and Measure*, 3rd edn, John Wiley & Sons (1995).

V.S. BORKAR. *Probability Theory: An Advanced Course*, Springer (1995).

P.G. CASAZZA & T.J. SHURA. *Tsirelson's Space: With an Appendix by J. Baker, O. Slotterbeck and R. Aron*, Lecture Notes in Mathematics **1363**, Springer (1989).

J.P.R. CHRISTENSEN. *Topology and Borel Structure*, North-Holland Mathematics Studies **10**, Elsevier (1974).

324

M.M. DAY. *Normed Linear Spaces*, Ergebnisse der Mathematik und ihrer Grenzgebiete **21**, Springer (1958), 3rd edn. (1973).

R. DEVILLE, G. GODEFROY & V. ZIZLER. *Smoothness and Renormings in Banach Spaces*, Pitman Monographs and Surveys in Pure and Applied Mathematics **64**, Longman (1993).

J. DIESTEL. *Sequences and Series in Banach Spaces*, Graduate Texts in Mathematics **92**, Springer (1984).

J. DIESTEL, H. JARCHOW & A. TONGE. *Absolutely Summing Operators*, Cambridge Studies in Advanced Mathematics **43**, Cambridge University Press (1995).

J. DIESTEL & J.J. UHL, JR. *Vector Measures*, Mathematical Surveys **13**, American Mathematical Society (1977).

D. VAN DULST. *Characterization of Banach Spaces Not Containing $\ell^1$*, CWI Tracts **59**, Centrum voor Wiskunde en Informatica (1989).

N. DUNFORD & J.T. SCHWARTZ. *Linear Operators: Part I, General Theory, with the Assistance of William G. Bade and Robert G. Bartle*, Interscience (1958), re-issue Wiley Classics Library, John Wiley & Sons (1988).

P. DUREN. *Theory of $H^p$-Spaces*, Academic Press (1970), 2nd edn., Dover (2000).

R.E. EDWARDS. *Fourier Series, A Modern Introduction,* 2 vols, Holt, Rinehart and Winston (1967).

M. FABIAN, P. HABALA, P. HÁJEK, V. MONTESINOS-SANTALUCÍA, J. PELANT & V. ZIZLER. *Functional Analysis and Infinite-Dimensional Geometry*, CMS Books in Mathematics **8**, Springer (2001).

H. FETTER & B. GAMBOA. *The James Forest*, London Mathematical Society Lecture Notes Series **236**, Cambridge University Press (1997).

M. FRÉCHET. *Les Espaces Abstraits*, Gauthiers-Villars, Paris (1928); re-issue Les Grands Classiques Gauthier-Villars, Éditions Jacques Gabay (1989).

N. GHOUSSOUB, G. GODEFROY, B. MAUREY & W. SCHACHERMAYER. *Some Topological and Geometrical Structures in Banach Spaces*, Memoirs of the American Mathematical Society **378** (1987).

C. GRAHAM & O. McGEHEE. *Essays in Commutative Harmonic Analysis*, Grundlehren der Mathematischen Wissenschaften **238**, Springer (1979).

S. GUERRE-DELABRIÈRE. *Classical Sequences in Banach Spaces*, Monographs and Textbooks in Pure and Applied Mathematics **166**, Marcel Dekker (1992).

P. HABALA, P. HÁJEK & V. ZIZLER. *Introduction to Banach Spaces*, 2 vols, Matfyz-Press, Charles University in Prague (1996).

P.R. HALMOS. *A Hilbert Space Problem Book*, Springer (1974), 2nd edn. (1982).

G.H. HARDY & E.M. WRIGHT. *An Introduction to the Theory of Numbers*, 5th edn., Clarendon Press (1979, reprinted 1996).

P. HARMAND, D. WERNER & W. WERNER. *M-Ideals in Banach Spaces and Banach Algebras*, Lecture Notes in Mathematics **1547**, Springer (1993).

V. HAVIN & B. JÖRICKE. *The Uncertainty Principle in Harmonic Analysis*, Ergebnisse der Mathematik und ihrer Grenzgebiete **28**, Springer (1994).

B. Host, J.-F. Méla & F. Parreau. *Analyse Harmonique des Mesures*, Astérisque 135–136, Société Mathématique de France (1986).

J.-P. Kahane 1. *Séries de Fourier Absolument Convergentes*, Ergebnisse der Mathematik und ihrer Grenzgebiete **50**, Springer (1970).

J.-P. Kahane 2. *Some Random Series of Functions*, D. C. Heath and Co. (1968), 2nd edn. Cambridge University Press (1985).

J.-P. Kahane & R. Salem. *Ensembles Parfaits et Séries Trigonométriques*, Hermann, 2nd edn. (1994).

B. Kashin & A.A. Saakyan. *Orthogonal Series*, Translations of Mathematical Monographs, American Mathematical Society (1989).

Y. Katznelson. *An Introduction to Harmonic Analysis*, Dover Books on Advanced Mathematics, Dover (1976).

J.-L. Krivine. *Théorie Axiomatique des Ensembles*, Collection SUP: Le Mathématicien, Presses Universitaires de France (1972).

H.E. Lacey. *The Isometric Theory of Classical Banach Spaces*, Grundlehren der Mathematischen Wissenschaften **208**, Springer (1974).

R. Larsen. *An Introduction to the Theory of Multipliers*, Springer (1971).

R. Lasser. *Introduction to Fourier Series*, Monographs and Textbooks in Pure and Applied Mathematics **199**, Marcel Dekker (1996).

M. Ledoux & M. Talagrand. *Probability in Banach Spaces*, Ergebnisse der Mathematik und ihrer Grenzgebiete **23**, Springer (1991).

M. Lifshits. *Gaussian Random Functions*, Kluwer Academic Publishers (1995).

L.A. Lindahl & F. Poulsen. *Thin Sets in Harmonic Analysis*, Marcel Dekker (1971).

J. Lindenstrauss & L. Tzafriri. *Classical Banach Spaces* 2 vols, Classics in Mathematics, Springer (1997).

M. Loève. *Probability Theory*, 2nd edn., Springer (1977).

L. Loomis. *An Introduction to Abstract Harmonic Analysis*, Van Nostrand (1953).

J.-M. Lopez & K.A. Ross. *Sidon Sets*, Lecture Notes in Pure and Applied Mathematics **13**, Marcel Dekker (1975).

E. Lukacs. *Characteristic Functions*, 2nd edn., Griffin (1970).

P. Malliavin. *Intégration et Probabilités, Analyse de Fourier et Analyse Spectrale*, Masson (1982).

M. Marcus & G. Pisier. *Random Fourier Series with Applications to Harmonic Analysis*, Annals of Mathematics Studies **101**, Princeton University Press (1981).

P. Mattila. *Geometry of Sets and Measures in Euclidean Spaces*, Cambridge Studies in Advanced Mathematics **44**, Cambridge University Press (1995).

B. Maurey. *Théorèmes de Factorisation pour les Opérateurs Linéaires à Valeurs dans les Espaces $L_p$*, Astérisque **11**, Société Mathématique de France (1974).

R.E. Megginson. *An Introduction to Banach Space Theory*, Graduate Texts in Mathematics **183**, Springer (1998).

V.D. MILMAN & G. SCHECHTMAN. *Asymptotic Theory of Finite-Dimensional Normed Spaces; With an Appendix by M. Gromov*, Lecture Notes in Mathematics **1200**, Springer (1986).

T.J. MORRISON. *Functional Analysis: An Introduction to Banach Space Theory*, Pure and Applied Mathematics, Wiley-Interscience (2001).

J. NEVEU 1. *Bases Mathématiques du Calcul des Probabilités*, Masson (1964).

J. NEVEU 2. *Martingales à Temps Discret*, Masson (1972).

J. NEVEU 3. *Processus Aléatoires Gaussiens*, Les Presses de l'Université de Montréal (1968).

A. PAJOR. *Sous-Espaces $l_1^n$ des Espaces de Banach*, Travaux en Cours **16**, Hermann, Paris (1985).

K.R. PARTHASARATY. *Probability Measures on Metric Spaces*, Academic Press (1967).

A. PEŁCZYŃSKI. *Banach Spaces of Analytic Functions and Absolutely Summing Operators*, CBMS Regional Conference Series in Mathematics **30**, American Mathematical Society (1977).

R.R. PHELPS. *Lectures on Choquet's Theorem*, Van Nostrand Mathematical Studies **7**, Van Nostrand (1966).

A. PIETSCH & J. WENZEL. *Orthonormal Systems and Banach Space Geometry*, Encyclopedia of Mathematics and its Applications **70**, Cambridge University Press (1998).

G. PISIER 1. *Factorization of Linear Operators and Geometry of Banach Spaces*, CBMS Regional Conference Series in Mathematics **60**, American Mathematical Society (1986).

G. PISIER 2. *The Volume of Convex Bodies and Banach Space Geometry*, Cambridge Tracts in Mathematics **94**, Cambridge University Press (1989).

G. PÓLYA & G. SZEGÖ. *Problems and Theorems in Analysis II*, Springer (1976).

H. QUEFFÉLEC & C. ZUILY. *Eléments d'Analyse pour l'Agrégation*, Dunod (2013), 4th edn., revised and updated.

D. REVUZ. *Probabilités*, Collection Méthodes, Hermann (1997).

A. RÉNYI. *Calcul des Probabilités*, Dunod (1966).

W. RUDIN 1. *Fourier Analysis on Groups*, Wiley Classics Library, John Wiley & Sons (1990).

W. RUDIN 2. *Analyse Réelle et Complexe*, 3rd edn., Dunod (1998).

W. RUDIN 3. *Functional Analysis*, International Series in Pure and Applied Mathematics, McGraw-Hill (1991).

Y. SAMORODNITZKY & G. TAQQU. *Stable Non-Gaussian Random Processes*, Chapman-Hall (1994).

Z. SEMADENI. *Banach Spaces of Continuous Functions*, Vol. I, Monografie Matematyczne **55**, PWN – Polish Scientific Publishers (1971).

A.N. SHIRYAEV. *Probability*, 2nd edn., Graduate Texts in Mathematics **95**, Springer (1996).

I. SINGER. *Bases in Banach Spaces I*, Grundlehren der Mathematischen Wissenschaften **154**, Springer (1970).

D.W. STROOCK. *Probability Theory: An Analytic View*, Cambridge University Press (1994).

V.N. SUDAKOV. *Geometric Problems of the Theory of Infinite-Dimensional Probability Distributions* (in Russian), Trudy Matematicheskogo Instituta Imemi V. A. Steklova **141** (1976).

M. TALAGRAND. *Pettis Integral and Measure Theory*, Memoirs of the American Mathematical Society **307** (1984).

N. TOMCZAK-JAEGERMANN. *Banach–Mazur Distances and Finite-Dimensional Operator Ideals*, Pitman Monographs and Surveys in Pure and Applied Mathematics **38**, Longman (1989).

P. TURÁN. *On a New Method of Analysis and its Applications*, Wiley Interscience (1984).

A. WEIL. *L'Intégration dans les Groupes Topologiques et ses Applications*, Gauthiers-Villars (1938).

P. WOJTASZCZYK. *Banach Spaces for Analysts*, Cambridge Studies in Advanced Mathematics **25**, Cambridge University Press (1991).

V.M. ZOLOTAREV. *One-Dimensional Stable Distributions*, Translations of Mathematical Monographs **65**, American Mathematical Society (1986).

A. ZYGMUND. *Trigonometric Series*, 2 vols., Cambridge University Press (1993).

Since this book appeared in French, two other books on Banach spaces deserve attention:

F. ALBIAC & N.J. KALTON. *Topics in Banach Space Theory*, Graduate Texts in Mathematics **233**, Springer (2006).

M. FABIAN, P. HABALA, P. HÁJEK, V. MONTESINOS & V. ZIZLER. *Banach Space Theory: The Basis for Linear and Nonlinear Analysis*, CMS Books in Mathematics, Springer (2011).

# Articles

L. Alaoglu
    1940    Weak topologies of normed linear spaces, *Ann. of Math.* **41**, 252–267.

D.J. Aldous
    1981    Subspaces of $L^1$, via random measures, *Trans. Amer. Math. Soc.* **267**, 445–463.

D.E. Alspach
    1983    Small into isomorphisms on $L_p$ spaces, *Illinois J. Math.* **27**, 300–314.

D.E. Alspach & E. Odell
    2001    $L_p$ spaces, in *Handbook of the Geometry of Banach Spaces I*, Elsevier, 123–160.

D. Amir
1965    On isomorphisms of continuous function spaces, *Israel J. Math.* **3**, 205–210.

S.A. Argyros & V. Felouzis
2000    Interpolating hereditarily indecomposable Banach spaces, *J. Amer. Math. Soc.* **13**, 243–294.

S.A. Argyros & I. Gasparis
2001    Unconditional structures of weakly null sequences, *Trans. Amer. Math. Soc.* **353**, 2019–2058.

G.I. Arkhipov & K.I. Oskolkov
1987    On a special trigonometric series and its applications, *Mat. Sb.* **134**, 145–155.

N.H. Asmar & S. Montgomery-Smith
1993    On the distribution of Sidon series, *Ark. Mat.* **31**, 13–26.

K. Azuma
1967    Weighted sums of certain dependent variables, *Tohoku Math. J.* **19**, 357–367.

G.F. Bachelis & S.F. Ebenstein
1974    On $\Lambda(p)$-sets, *Pacific J. Math.* **54**, 35–38.

K. Ball & F. Nazarov
1994    Zero Khinchin's theorem, unpublished work.

F. Barthe
1998    Optimal Young's inequality and its converse: A simple proof, *Geom. Funct. Anal.* **8**, 234–242.

B. Beauzamy
1973    Le théorème de Dvoretzky, *Séminaire Maurey–Schwartz 1972–1973*, Exposés XXVI et XXVII, École Polytechnique, Paris.

A. Beck
1962    A convexity condition in Banach spaces and the strong law of large numbers, *Proc. Amer. Math. Soc.* **13**, 329–334.

W. Beckner
1975    Inequalities in Fourier analysis, *Ann. of Math.* **102**, 159–182.

G. Bennett
1976    Unconditional convergence and almost everywhere convergence, *Z. Wahrscheinlichkeitstheorie und Verw. Gebiete* **34**, 135–155.

G. Bennett, L.E. Dor, V. Goodman, W.B. Johnson & C.M. Newman
1977    On uncomplemented subspaces of $L_p$, $1 < p < 2$, *Israel J. Math.* **26**, 178–187.

Y. Benyamini
1981    Small into-isomorphisms between spaces of continuous functions, *Proc. Amer. Math. Soc.* **83**, 479–485.

C. Bessaga & A. Pełczyński
1958 a  On bases and unconditional convergence of series in Banach spaces, *Studia Math.* **17**, 151–164.

1958 b  A generalization of results of R. C. James concerning absolute bases in Banach spaces, *Studia Math.* **17**, 165–174.

1960    Spaces of continuous functions IV, *Studia Math.* **19**, 53–60.

P. Billard
1965        Séries de Fourier aléatoirement bornées, continues, uniformément conver-
            gentes, *Ann. Sci. Éc. Norm. Supér* **82**, 131–179.

A. Biró
2000        An upper estimate in Turán's pure power sum problem, *Indag. Math.* **11**,
            499–508.

S. Bochner & A.E. Taylor
1938        Linear functionals on certain spaces of abstractly-valued functions, *Ann.
            of Math.* **39**, 913–944.

J. Boclé
1960        Sur la théorie ergodique, *Ann. Inst. Fourier* **10**, 1–45.

B. Bojanov & N. Naidenov
1999        An extension of the Landau–Kolmogorov inequality: Solution of a
            problem of Erdös, *J. Anal. Math.* **78**, 263–280.

A. Bonami
1970        Etude des coefficients de Fourier des fonctions de $L^p(G)$, *Ann. Inst.
            Fourier* **20**, 335–402.

C. Borell
1979        On the integrability of Banach space valued Walsh polynomials, in *Sémi-
            naire de Probabilités XIII*, Lecture Notes in Mathematics **721**, Springer,
            1–3.

N. Bourbaki
1938        Sur les espaces de Banach, *C.R.A.S. Paris* **206**, 1701–1704.

J. Bourgain
1979 a      La propriété de Radon–Nikodym, Cours de 3ème cycle, Publications
            Mathématiques de l'Université Pierre et Marie Curie **36**.

1979 b      An averaging result for $l_1$-sequences and applications to weakly condi-
            tionally compact sets in $L_X^1$, *Israel J. Math.* **32**, 289–298.

1981        A counterexample to a complementation problem, *Compos. Math.* **43**,
            133–144.

1982        A remark on finite-dimensional $P_\lambda$-spaces, *Studia Math.* **72** (3), 285–289.

1983 a      Some remarks on Banach spaces in which martingale differences are
            unconditional, *Ark. Mat.* **21**, 163–168.

1983 b      Propriétés de décomposition pour les ensembles de Sidon, *Bull. Soc. Math.
            France* **111**, 421–428.

1983 c      Sur les sommes de sinus, Publications Mathématiques d'Orsay 83-01,
            Exposé 3.

1984 a      New Banach space properties of the disk algebra and $H^\infty$, *Acta Math.* **152**,
            1–48.

1984 b      On martingale transforms in finite-dimensional lattices with an appendix
            on the $K$-convexity constant, *Math. Nachr.* **119**, 41–53.

1985 a      Sidon sets and Riesz products, *Ann. Inst. Fourier* **35**, 137–148.

1985 b      Subspaces of $L_N^\infty$, arithmetical diameter and Sidon sets, in *Probability in
            Banach Spaces V*, Lecture Notes in Mathematics **1153**, Springer, 96–127.

1986    Sur le minimum d'une somme de cosinus, *Acta Arith.* **45**, 381–389.

1987    A remark on entropy of abelian groups and the invariant uniform approximation property, *Studia Math.* **86**, 79–84.

1989    Bounded orthogonal sets and the $\Lambda(p)$-set problem, *Acta Math.* **162**, 227–246.

J. Bourgain, D. Fremlin & M. Talagrand

1978    Pointwise compact sets of Baire-measurable functions, *Amer. J. Math.* **100**, 845–886.

J. Bourgain, J. Lindenstrauss & V. Milman

1989    Approximation of zonoids by zonotopes, *Acta Math.* **162**, 73–141.

J. Bourgain & V. Milman

1985    Dichotomie du cotype pour les espaces invariants, *C.R.A.S. Paris* **300**, 263–266.

1987    New volume ratio properties for convex symmetric bodies in $\mathbf{R}^n$, *Invent. Math.* **88**, 319–340.

J. Bourgain & H.P. Rosenthal

1980    Martingales valued in certain subspaces of $L^1$, *Israel J. Math.* **37**, 54–75.

J. Bourgain & S. Szarek

1988    The Banach–Mazur distance to the cube and the Dvoretzky–Rogers factorization, *Israel J. Math.* **62**, 169–180.

H.J. Brascamp & E.H. Lieb

1976 a    On extensions of the Brunn–Minkowski and Prékopa–Leindler theorems, including inequalities for log concave functions, and with an application to the diffusion equation, *J. Funct. Anal.* **22**, 366–389.

1976 b    Best constants in Young's inequality, its converse and its generalization to more than three functions, *Adv. Math.* **20**, 151–173.

J. Bretagnolle, D. Dacunha-Castelle & J.-L. Krivine

1966    Lois stables et espaces $L^p$, *Ann. Inst. H. Poincaré. B. (N.S.)* **2**, 231–259.

J. Brillhart & L. Carlitz

1970    Note on the Shapiro polynomials, *Proc. Amer. Math. Soc.* **25**, 114–118.

A.V. Bukhvalov & G. Lozanovskiĭ [Lozanovski]

1978    On sets closed in measure, *Trans. Moscow Math. Soc.* **2**, 127–148.

D. L. Burkholder

1981    A geometrical characterization of Banach spaces in which martingale differences are unconditional, *Ann. Probab.* **9**, 997–1011.

1983    A geometric condition that implies the existence of certain singular integrals of Banach-space valued functions, in *Proceedings of the Conference in Harmonic Analysis in Honor of Antoni Zygmund, University of Chicago, 1981*, Wadsworth Mathematical Series, Wadsworth, 270–286.

1984    Boundary value problems and sharp inequalities for martingale transforms, *Ann. Probab.* **12**, 647–702.

1985    An elementary proof of an inequality of R.E.A.C. Paley, *Bull. London Math. Soc.* **17**, 474–478.

1988    A proof of Pełczyński's conjecture for the Haar system, *Studia Math.* **91**, 79–83.

D.L. Burkholder & R.F. Gundy
1970      Extrapolation and interpolation of quasi-linear operators on martingales, *Acta Math.* **124**, 249–304.

M. Cambern
1967      On isomorphisms with small bound, *Proc. Amer. Math. Soc.* **18**, 1062–1066.
1968      On mappings of sequence spaces, *Studia Math.* **30**, 73–77.

L. Carleson
1980      An explicit unconditional basis of $H^1$, *Bull. Sci. Math.* **104**, 405–416.

P. Casazza
1986      Finite-dimensional decompositions in Banach spaces, in *Geometry of Normed Linear Spaces: Proceedings of a Conference held June 9–12, 1983 in Honor of Mahlon Marsh Day* (Urbana–Champaign, IL) Contemporary Mathematics **52**, American Mathematical Society, 1–31.
1989      The commuting B.A.P. for Banach spaces, in *Analysis at Urbana II*, E. Berkson, N.T. Peck & J.J. Uhl, eds., London Mathematical Society Lecture Note Series **138**, 108–127.
2001      Approximation properties, in *Handbook of The Geometry of Banach Spaces I*, Elsevier, 271–316.

P.G. Casazza, W.B. Johnson & L. Tzafriri
1984      On Tsirelson's space, *Israel J. Math.* **47**, 81–98.

P. Casazza & N. Kalton
1990      Notes on approximation properties in separable Banach spaces, in *Geometry of Banach Spaces*, P.F.X. Müller and W. Schachermayer, eds., London Mathematical Society Lecture Note Series **158**, Cambridge University Press, 49–63.

J. Caveny
1966      Bounded Hadamard products of $H^p$ functions, *Duke Math. J.* **33**, 389–394.

S.D. Chatterji
1968      Martingale convergence and the Radon–Nikodym theorem in Banach spaces, *Math. Scand.* **22**, 21–41.

G. Choquet
1962      Remarques à propos de la démonstration de l'unicité de P. A. Meyer, in *Séminaire Brelot–Choquet–Deny (Théorie du Potentiel)* **6**, 2, Exposé 8, Secrétariat Mathématique.

F. Cobos
1983      On the type of interpolation spaces and $S_{p,q}$, *Math. Nachr.* **113**, 59–64.

M. Cotlar
1955      A unified theory of Hilbert transforms and ergodic theory, *Rev. Mat. Cuyana* **I**, 105–167.

H. Cramér
1935      Prime numbers and probability, *Skand. Mat.-Kongr.* **8**, 107–115.
1937      On the order of magnitude of the difference between consecutive prime numbers, *Acta Arith.* **2**, 23–46.

J. Creekmore
1981      Type and cotype in Lorentz $L_{p,q}$ spaces, *Indag. Math.* **43**, 145–152.

D. Dacunha-Castelle & J.-L. Krivine
1972    Applications des ultraproduits à l'étude des espaces et algèbres de Banach, *Studia Math.* **41**, 315–334.

A.M. Davie
1973    The approximation problem for Banach spaces, *Bull. London Math. Soc.* **5**, 261–266.
1975    The Banach approximation problem, *J. Approx. Theory* **13**, 392–394.

W.J. Davis, D.W. Dean & I. Singer
1968    Complemented subspaces and $\Lambda$ systems in Banach spaces, *Israel J. Math.* **6**, 303–309.

G. Debs
1987    Effective properties in compact sets of Borel functions, *Mathematika* **34**, 64–68.

M. Déchamps-Gondim [Déchamps]
1972    Ensembles de Sidon topologiques, *Ann. Inst. Fourier* **22**, 51–79.
1984    Sur les compacts associés aux ensembles lacunaires, les ensembles de Sidon et quelques problèmes ouverts, Publications Mathématiques d'Orsay 84-01.
1987    Densité harmonique et espaces de Banach invariants par translation ne contenant pas $c_0$, *Colloq. Math.* **51**, 67–84.

J. Diestel
1980    A survey of results related to the Dunford–Pettis property, *Contemp. Math.* **2**, 15–60.

S.J. Dilworth, M. Girardi & J. Hagler
2000    Dual Banach spaces which contain an isometric copy of $L_1$, *Bull. Pol. Acad. Sci. Math.* **48**, 1–12.

S.J. Dilworth & J.P. Patterson
2003    An extension of Elton's $\ell_1^n$ theorem to complex Banach spaces, *Proc. Amer. Math. Soc.* **131**, 1489–1500.

L. Dor
1975 a    On projections in $L_1$, *Ann. of Math.* **102**, 463–474.
1975 b    On sequences spanning a complex $l_1$ space, *Proc. Amer. Math. Soc.* **47**, 515–516.

S. Drury
1970    Sur les ensembles de Sidon, *C.R.A.S. Paris* **271**, 162–163.

R.M. Dudley
1967    The size of compact subsets of Hilbert space and continuity of Gaussian processes, *J. Funct. Anal.* **1**, 290–330.
1973    Sample functions of the Gaussian process, *Ann. Probab.* **1**, 66–103.

N. Dunford
1939    A mean ergodic theorem, *Duke Math. J.* **5**, 635–646.

N. Dunford & A.P. Morse
1936    Remarks on the preceding paper of James A. Clarkson: "Uniformly convex spaces," *Trans. Amer. Math. Soc.* **40**, 415–420.

N. Dunford & B.J. Pettis
1940    Linear operations on summable functions, *Trans. Amer. Math. Soc.* **47**, 323–392.

P. L. Duren
1969 a    On the Bloch–Nevanlinna conjecture, *Colloq. Math.* **20**, 295–297.
1969 b    On the multipliers of $H^p$ spaces, *Proc. Amer. Math. Soc.* **22**, 24–27.
1985      Random series and bounded mean oscillation, *Michigan Math. J.* **32**, 81–86.

A. Dvoretzky
1959      A theorem on convex bodies and applications to Banach spaces, *Proc. Nat. Acad. Sci. USA* **45**, 223–226; erratum, 1554.
1961      Some results on convex bodies and Banach spaces, in *Proceedings of the International Symposium on Linear Spaces (Jerusalem, 1960)*, Jerusalem Academic Press, Pergamon Press, 123–160.

A. Dvoretzky & C.A. Rogers
1950      Absolute and unconditional convergence in normed linear spaces, *Proc. Nat. Acad. Sci. USA* **36**, 192–197.

W.F. Eberlein
1947      Weak compactness in Banach spaces, *Proc. Nat. Acad. Sci. USA* **33**, 51–53.

J. Elton
1983      Sign-embeddings of $\ell_1^n$, *Trans. Amer. Math. Soc.* **279**, 113–124.

P. Enflo
1973      A counterexample to the approximation property in Banach spaces, *Acta Math.* **130**, 309–317.

P. Erdös
1955      Problems and results in additive number theory, in *Colloque sur la Théorie des Nombres, held in Bruxelles (December 1955)*, Librairie Universitaire 127–137.

P. Erdös & A. Rényi
1960      Additive properties of random sequences of positive integers, *Acta Arith.* **6**, 83–110.

T. Fack
1987      Type and cotype inequalities for noncommutative $L^p$-spaces, *J. Operator Theory* **17**, 255–279.

H. Fakhoury
1977      Sur les espaces de Banach ne contenant pas $\ell^1(\mathbb{N})$, *Math. Scand.* **41**, 277–289.

J. Farahat
1974      Espaces de Banach contenant $l^1$, d'après H.P. Rosenthal, in *Séminaire Maurey–Schwartz 1973–1974*, École Polytechnique.

J. Faraut & K. Harzallah
1974      Distances hilbertiennes invariantes sur un espace homogène, *Ann. Inst. Fourier* **24**, 171–217.

V. Farmaki
2002      Ordinal indices and Ramsey dichotomies measuring $c_0$-content and semi-bounded completeness, *Fund. Math.* **172**, 153–179.

S. Yu. Favorov
1998      A generalized Kahane–Khinchin inequality, *Studia Math.* **130**, 101–107.

V. Ferenczi
1995    Un espace de Banach uniformément convexe et héréditairement indécomposable, *C.R.A.S. Paris Sér. I Math.* **320**, 49–54.
1997 a  A uniformly convex hereditarily indecomposable Banach space, *Israel J. Math.* **102**, 199–225.
1997 b  Operators on subspaces of hereditary indecomposable Banach spaces, *Bull. London Math. Soc.* **29**, 338–344.

X. Fernique
1970    Intégrabilité des vecteurs gaussiens, *C.R.A.S. Paris Sér. A* **270**, 1698–1699.
1971    Régularité des processus gaussiens, *Invent. Math.* **12**, 304–320.
1975    Regularité des trajectoires des fonctions aléatoires gaussiennes, in *École d'Été de Probabilités de Saint-Flour IV-1974*, Lecture Notes in Mathematics **480**, Springer, 1–96.

T. Figiel
1976    A short proof of Dvoretzky's theorem, *Compos. Math.* **33**, 297–301.

T. Figiel & W.B. Johnson
1973    The approximation property does not imply the bounded approximation property, *Proc. Amer. Math. Soc.* **4**, 197–200.
1974    A uniformly convex Banach space which contains no $\ell_p$, *Compos. Math.* **29**, 179–190.

T. Figiel, J. Lindenstrauss & V. Milman
1977    The dimension of almost spherical sections of convex bodies, *Acta Math.* **139**, 53–94.

T. Figiel & N. Tomczak-Jaegermann
1979    Projections onto Hilbertian subspaces of Banach spaces, *Israel J. Math.* **33**, 155–171.

T. Figiel & P. Wojtaszczyk
2001    Special bases in function spaces, in *Handbook of the Geometry of Banach Spaces I*, Elsevier, 561–597.

D.J.H. Garling & Y. Gordon
1971    Relations between some constants associated with finite dimensional Banach spaces, *Israel J. Math.* **9**, 346–361.

B. Gelbaum
1958    Notes on Banach spaces and bases, *An. Acad. Brasil. Ciênc.* **30**, 29–36.

I.M. Gelfand
1938    Abstrakte funktionen und lineare operatoren, *Mat. Sb.* **4** (46), 235–286.

N. Ghoussoub, B. Maurey & W. Schachermayer
1992    Slicings, selections and their applications, *Canad. J. Math.* **44**, 483–504.

D.P. Giesy & R.C. James
1973    Uniformly non-$l^{(1)}$ and $B$-convex Banach spaces, *Studia Math.* **48**, 61–69.

E. Giné, M. Marcus & J. Zinn
1985    A version of Chevet's theorem for stable processes, *J. Funct. Anal.* **63**, 47–73.

E. Gluskin, M. Meyer & A. Pajor
1994    Zeros of analytic functions and norms of inverse matrices, *Israel J. Math.* **87**, 225–242.

G. Godefroy

1978    Espaces de Banach : Existence et unicité de certains préduaux, *Ann. Inst. Fourier* **28**, 87–105.

1979    Étude des projections de norme 1 de $E''$ sur $E$: Unicité de certains préduaux. Applications, *Ann. Inst. Fourier* **29**, 53–70.

1981    Points de Namioka, espaces normants: Applications à la théorie isométrique de la dualité, *Israel J. Math.* **38**, 209–220.

1983    Parties admissibles d'un espace de Banach: Applications, *Ann. Sci. Éc. Norm. Supér.* **16**, 109–122.

1984 a   Quelques remarques sur l'unicité des préduaux, *Quart. J. Math. Oxford* **35**, 147–152.

1984 b   Sous-espaces bien disposés de $L^1$: Applications, *Trans. Amer. Math. Soc.* **286**, 227–249.

1987    Boundaries of a convex set and interpolation sets, *Math. Ann.* **277**, 173–184.

1988    On Riesz subsets of abelian discrete groups, *Israel J. Math.* **61**, 301–331.

1989 a   Metric characterization of first Baire class linear forms and octahedral norms, *Studia Math.* **95**, 1–15.

1989 b   Existence and uniqueness of isometric preduals: A survey, Contemp. Math. **85**, 131–193.

G. Godefroy & N.J. Kalton

1989    The ball topology and its applications, in *Banach Space Theory: Proceedings of a Research Workshop held July 5–25, 1987* (Iowa City, IA), Contemporary Mathematics **85**, American Mathematical Society, 195-237.

1997    Approximating sequences and bidual projections, *Quart. J. Math. Oxford (2)* **48**, 179–202.

G. Godefroy, N. Kalton & D. Li

1995    Propriété d'approximation métrique inconditionnelle et sous-espaces de $L^1$ dont la boule est compacte en mesure, *C.R.A.S. Paris* **320**, 1069–1073.

1996    On subspaces of $L^1$ which embed in $\ell_1$, *J. Reine Angew. Math.* **471**, 43–75.

2000    Operators between subspaces and quotients of $L^1$, *Indiana Univ. Math. J.* **49**, 245–286.

G.B. Godefroy, N.J. Kalton & P.D. Saphar

1993    Unconditional ideals in Banach spaces, *Studia Math.* **104**, 13–59.

G. Godefroy & D. Li

1989    Banach spaces which are $M$-ideals in their bidual have property $(u)$, *Ann. Inst. Fourier* **39**, 361–371.

1998    Strictly convex functions on compact convex sets and their use, in *Functional Analysis: Selected Topics*, P.K. Jain, ed., Narosa Publishing House.

G. Godefroy & P.D. Saphar

1989    Three-space problems for the approximation properties, *Proc. Amer. Math. Soc.* **105**, 70–75.

H.H. Goldstine

1938    Weakly complete Banach spaces, *Duke Math. J.* **4**, 125–131.

Y. Gordon
1985    Some inequalities for Gaussian processes and applications, *Israel J. Math.* **50**, 265–289.
1992    Majorization of Gaussian processes and geometric applications, *Probab. Theory Related Fields* **91**, 251–267.

W.T. Gowers
1994 a    A Banach space not containing $c_0$, $l_1$ or a reflexive subspace, *Trans. Amer. Math. Soc.* **344**, 407–420.
1994 b    Analytic sets and games in Banach spaces, Preprint IHES M/94/42.
1995    Recent results in the theory of infinite-dimensional Banach spaces, in *Proceedings of the International Congress of Mathematicians, Vols 1 and 2 (held in Zürich, 1994)*, Birkhäuser, 933–942.
1996 a    A solution to the Schroeder–Bernstein problem for Banach spaces, *Bull. London Math. Soc.* **28**, 297–304.
1996 b    A new dichotomy for Banach spaces, *Geom. Funct. Anal.* **6**, 1083–1093.
2002    An infinite Ramsey theorem and some Banach-space dichotomies, *Ann. of Math.* **156**, 797–833.

W.T. Gowers & B. Maurey
1993    The unconditional basic sequence problem, *J. Amer. Math. Soc.* **6**, 851–874.
1997    Banach spaces with small spaces of operators, *Math. Ann.* **307**, 543–568.

C.C. Graham, B. Host & F. Parreau
1981    Sur les supports des transformées de Fourier–Stieltjes, *Colloq. Math.* **44**, 145–146.

L. Gross
1975    Logarithmic Sobolev inequalities, *Amer. J. Math.* **97**, 1061–1083.

A. Grothendieck
1956    Résumé de la théorie métrique des produits tensoriels topologiques, *Bol. Soc. Mat. São Paulo* **8**, 1–79.

O. Guédon
1997    Gaussian version of a theorem of Milman and Schechtman, *Positivity* **1**, 1–5.
1998    Sections euclidiennes des corps convexes et inégalités de concentration volumiques, Doctoral thesis, Université de Marne-la-Vallée.
1999    Kahane–Khinchine type inequalities for negative exponent, *Mathematika* **46**, 165–173.

S. Guerre [Guerre-Delabrière] & J.-T. Lapresté
1981    Quelques propriétés des espaces de Banach stables, *Israel J. Math.* **39**, 247–254.

S. Guerre [Guerre-Delabrière] & M. Lévy
1983    Espaces $\ell^p$ dans les sous-espaces de $L^1$, *Trans. Amer. Math. Soc.* **279**, 611–616.

U. Haagerup
1978    Les meilleures constantes de l'inégalité de Khintchine, *C.R.A.S. Paris Sér. A* **286**, 259–262.
1982    The best constants in the Khintchine inequality, *Studia Math.* **70** (1981), 231–283.

1987 A new upper bound for the complex Grothendieck constant, *Israel J. Math.* **60**, 199–224.

J. Hagler

1973 Some more Banach spaces which contain $l_1$, *Studia Math.* **46**, 35–42.

1977 A counterexample to several questions about Banach spaces, *Studia Math.* **60** (1977), 289–308.

J. Hagler & C. Stegall

1973 Banach spaces whose duals contain complemented subspaces isomorphic to $C[0, 1]^*$, *J. Funct. Anal.* **13**, 233–251.

G. Halász

1973 On a result of Salem and Zygmund concerning random polynomials, *Studia Sci. Math. Hungar.* **8**, 369–377.

A. Harcharras

1999 Fourier analysis, Schur multipliers on $S^p$ and non-commutative $\Lambda(p)$-sets, *Studia Math.* **137**, 203–260.

C.D. Hardin, Jr.

1981 Isometries on subspaces of $L^p$, *Indiana Univ. Math. J.* **30**, 449–465.

G.H. Hardy & J.E. Littlewood

1932 Some properties of fractional integrals II, *Math. Z.* **34**, 403–439.

K. Hare

1988 An elementary proof of a result on $\Lambda(p)$-sets, *Proc. Amer. Math. Soc.* **104**, 829–834.

V.P. Havin

1973 Weak completeness of the space $L^1/H_0^1$, *Vestnik Leningrad Univ.* **13**, 77–81 (in Russian).

R. Haydon

1976 Some more characterizations of Banach spaces containing $l^1$, *Math. Proceed. Cambridge Phil. Soc.* **80**, 269–276.

W.K. Hayman

1964 On the characteristic of functions meromorphic in the unit disk and of their integrals, *Acta Math.* **112**, 181–214.

E. Hewitt & K. Yosida

1952 Finitely additive measures, *Trans. Amer. Math. Soc.* **72**, 46–66.

J. Hoffmann-Jørgensen

1970 The theory of analytic sets, Matematisk Institut, Aarhus Universitet, Varions Publications Series **10**.

1973 Sums of independent Banach space random variables, Aarhus Universitet, Preprint Series **15**.

1974 Sums of independent Banach space random variables, *Studia Math.* **52**, 159–186.

B. Host & F. Parreau

1979 Sur les mesures dont la transformée de Fourier–Stieltjes ne tend pas vers 0 à l'infini, *Colloq. Math.* **41**, 285–289.

I.A. Ibragimov, V.N. Sudakov & B.S. Tsirelson

1976 Norms of Gaussian sample functions, in *Proceedings of the Third Japan-USSR Symposium on Probability Theory*, Lecture Notes in Mathematics **550**, Springer, 20–41.

K. Itô & M. Nisio
  1968      On the convergence of sums of independent Banach space valued random variables, *Osaka J. Math.* **5**, 35–48.
R.C. James
  1950      Bases and reflexivity of Banach spaces, *Ann. of Math.* **52**, 518–527.

  1951      A non-reflexive Banach space isometric with its second conjugate, *Proc. Nat. Acad. Sci. USA* **37**, 174–177.

  1957      Reflexivity and the supremum of linear functionals, *Ann. of Math.* **66**, 159–169.

  1964 a    Characterizations of reflexivity, *Studia Math.* **23**, 205–216.

  1964 b    Uniformly non square Banach spaces, *Ann. of Math.* **80**, 542–550.

  1972      Reflexivity and the sup of linear functionals, *Israel J. Math.* **13**, 289–300.

  1974      A separable somewhat reflexive Banach space with non-separable dual, *Bull. Amer. Math. Soc.* **80**, 738–743.
L. Janicka
  1979      Some measure-theoretic characterizations of Banach spaces not containing $\ell^1$, *Bull. Acad. Polon. Sci. Sér. Sci. Math.* **27**, 561–565.
F. John
  1948      Extremum problems with inequalities as subsidiary conditions, in *Studies and Essays: Presented to R. Courant on His 60th Birthday, January 8, 1948*, Interscience 187–204.
W.B. Johnson
  1970      Finite-dimensional Schauder decompositions in $\pi_\lambda$ and dual $\pi_\lambda$-spaces, *Illinois J. Math.* **14**, 642–647.

  1972      A complementary universal conjugate Banach space and its relation to the approximation problem, *Israel J. Math.* **13**, 301–310.

  1980      Banach spaces all of whose subspaces have the approximation property, in *Special Topics of Applied Mathematics: Proceedings of the Seminar Held at the GMD, Bonn 8–10 October, 1979*, North-Holland, 15–26.
W.B. Johnson & E. Odell
  1974      Subspaces of $L_p$ which embed into $\ell_p$, *Compos. Math.* **28**, 37–49.
W.B. Johnson & H.P. Rosenthal
  1972      On $w^*$-basic sequences and their applications to the study of Banach spaces, *Studia Math.* **43**, 77–92.
W.B. Johnson, H.P. Rosenthal & M. Zippin
  1971      On bases, finite-dimensional decompositions and weaker structures in Banach spaces, *Israel J. Math.* **9**, 488–506.
W.B. Johnson & G. Schechtman
  2001      Finite-dimensional subspaces of $L_p$, in *Handbook of the Geometry of Banach Spaces I*, Elsevier, 837–870.
W.B. Johnson & M. Zippin
  1974      Subspaces and quotient spaces of $(\sum G_n)_{\ell_p}$ and $(\sum G_n)_{c_0}$, *Israel J. Math.* **17**, 50–55.
M. Junge
  2002      Doob's inequality for non-commutative martingales, *J. Reine Angew. Math.* **549**, 149–190.

M.I. Kadeč
1958      On linear dimension of the spaces $L_p$, *Uspekhi Mat. Nauk* **13**, 95–98.

M.I. Kadeč & A. Pełczyński
1962      Bases, lacunary sequences and complemented subspaces in the spaces $L_p$,
          *Studia Math.* **21**, 161–176.

M.I. Kadeč & M.G. Snobar
1971      Some functionals over a compact Minkowski space, *Math. Notes* **10**,
          694–696.

V.M. Kadets, R.V. Shvidkoy, G. G. Sirotkin & D. Werner
2000      Banach spaces with the Daugavet property, *Trans. Amer. Math. Soc.* **352**,
          855–873.

J.-P. Kahane
1956      Sur certaines classes de séries de Fourier absolument convergentes,
          *J. Math. Pures Appl.* **35**, 249–259.

1958      Sur un théorème de Wiener–Lévy, *C.R.A.S. Paris* **246**, 1949–1951.

1980      Sur les polynômes à coefficients unimodulaires, *Bull. London Math. Soc.*
          **12**, 321–342.

1986      Une inégalité du type de Slepian et Gordon sur les processus gaussiens,
          *Israel J. Math.* **55**, 109–110.

J.-P. Kahane, Y. Katznelson & K. de Leeuw
1977      Sur les coefficients de Fourier des fonctions continues, *C.R.A.S. Paris Sér.
          A* **285**, 1001–1003.

N. Kalton
1995      The basic sequence problem, *Studia Math.* **116**, 167–187.

N. Kalton & A. Pełczyński
1997      Kernels of surjections from $\mathcal{L}_1$-spaces with an application to Sidon sets,
          *Math. Ann.* **309**, 135–158.

Y. Katznelson
1958      Sur les fonctions opérant sur l'algèbre des séries de Fourier absolument
          convergentes, *C.R.A.S. Paris Sér. A* **247**, 404–406.

1960      A characterization of the algebra of all continuous functions on a compact
          Hausdorff space, *Bull. Amer. Math. Soc.* **66**, 313–315.

1973      Suites aléatoires d'entiers, in *L'Analyse Harmonique dans le Domaine
          Complexe: Actes de la Table Ronde Internationale du Centre National de
          la Recherche Scientifique tenue à Montpelier du 11 au 15 Septembre 1972*,
          Lecture Notes in Mathematics **336**, Springer, 148–152.

Y. Katznelson & P. Malliavin
1966      Vérification statistique de la conjecture de la dichotomie sur une classe
          d'algèbres de restriction, *C.R.A.S. Paris* **262**, 490–492.

S. Kisliakov
1978      On spaces with "small" annihilators, *Sem. Inst. Steklov (LOMI)* **73**,
          91–101.

1991      Absolutely summing operators on the disk algebra, *Algebra i Analiz* **3**,
          1–79; translated in *St. Petersburg Math. J.* **3** (1992), 705–774.

H. Knaust & E. Odell
1989      On $c_0$ sequences in Banach spaces, *Israel J. Math.* **67**, 153–169.

A.L. Koldobsky
1992 The Schoenberg problem on positive-definite functions, *St. Petersburg Math. J.* **3**, 563–570.
A.L. Koldobsky & Y. Lonke
1999 A short proof of Schoenberg's conjecture on positive definite functions, *Bull. London Math. Soc.* **31**, 693–699.
J. Komlós
1967 A generalization of a problem of Steinhaus, *Acta Math. Acad. Sci. Hung.* **18**, 217–229.
R. Komorowski & N. Tomczak-Jaegermann
1995 Banach spaces without unconditional structure, *Israel J. Math.* **89**, 205–226.
1998 Erratum, *Israel J. Math.* **105**, 85–92.
H. König
1990 On the complex Grothendieck constant in the $n$-dimensional case, in *Geometry of Banach Spaces: Proceedings of the Conference held in Strobl, Austria, 1989*, London Mathematical Society Lecture Notes Series **158**, Cambridge University Press, 181–198.
S.V. Konyagin
1997 Estimates of maxima of sine sums, *East J. Approx.* **3**, 301–308.
M. Krein & D. Milman
1940 On extreme points of regular convex sets, *Studia Math.* **9**, 133-138.
J.-L. Krivine
1975 Sur les espaces isomorphes à $l_p$, *C.R.A.S. Paris Sér. A* **280**, 713–715.
1976 Sous-espaces de dimension finie des espaces de Banach réticulés, *Ann. of Math.* **104**, 1–29.
1977 Sur la constante de Grothendieck, *C.R.A.S. Paris Sér. A* **284**, 445–446.
1979 Constantes de Grothendieck et fonctions de type positif sur les sphères, *Adv. Math.* **31**, 16–30.
S. Kwapień
1970 a On a theorem of L. Schwartz and its applications to absolutely summing operators, *Studia Math.* **38**, 193–201.
1970 b A linear topological characterization of inner product spaces, *Studia Math.* **38**, 277–278.
1972 Isomorphic characterizations of inner product spaces by orthogonal series with vector valued coefficients, *Studia Math.* **44**, 583–595.
1974 On Banach spaces containing $c_0$, *Studia Math.* **52**, 187–188.
1976 A theorem on the Rademacher series with vector valued coefficients, in *Probability in Banach spaces: Proceedings of the First International Conference on Probability in Banach Space, 20–26 July 1975*, Oberwolfach, Lecture Notes in Mathematics **526**, Springer, 157–158.
S. Kwapień & A. Pełczyński
1970 The main triangle projection in matrix spaces and its applications, *Studia Math.* **34**, 43–68.
1980 Absolutely summing operators and translation invariant spaces of functions on compact abelian groups, *Math. Nachr.* **94**, 303–340.

R. Latała
1997    Sudakov minoration principle and supremum of some processes, *Geom. Funct. Anal.* **7**, 936–953.

R. Latała & K. Oleszkiewicz
1994    On the best constant in the Khinchin–Kahane inequality, *Studia Math.* **109**, 101–104.

P. Lefèvre
1998    On some properties of the class of stationary sets, *Colloq. Math.* **76**, 1–18.

1999 a    Measures and lacunary sets, *Studia Math.* **133**, 145–161.

1999 b    Topological dichotomy and unconditional convergence, *Serdica Math. J.* **25**, 297–310.

P. Lefèvre, D. Li, H. Queffélec & L. Rodríguez-Piazza
2002    Lacunary sets and function spaces with finite cotype, *J. Funct. Anal.* **188**, 272–291.

P. Lefèvre & L. Rodríguez-Piazza
2003    *p*-Rider sets are *q*-Sidon sets, *Proc. Amer. Math. Soc.* **131**, 1829–1838.

L. Leindler
1972    On a certain converse of Hölder's inequality II, *Acta Sci. Math.* **33**, 217–223.

D. Lewis
1978    Finite dimensional subspaces of $L_p$, *Studia Math.* **63**, 207–212.

1979    Ellipsoids defined by Banach ideal norms, *Mathematika* **26**, 18–29.

D. Li
1987    Espaces *L*-facteurs de leurs biduaux bonne disposition, meilleure approximation et propriété de Radon–Nikodym, *Quart. J. Math. Oxford (2)* **38**, 229–243.

1988    Lifting properties for some quotients of $L^1$-spaces and other spaces *L*-summand in their bidual, *Math. Z.* **199**, 321–329.

1991    Propriété d'approximation par des opérateurs qui commutent, d'après Casazza et Kalton, in *Séminaire d'Initiation à l'Analyse, 1990-1991*, Exposé 4, Publications Mathématiques de l'université Pierre et Marie Curie **104**.

1996    Complex unconditional metric approximation property for $\mathcal{C}_{(\mathbb{T})}$ spaces, *Studia Math.* **121**, 231–247.

1998    A remark about $\Lambda(p)$-sets and Rosenthal sets, *Proc. Amer. Math. Soc.* **126**, 3329–3333.

D. Li, H. Queffélec & L. Rodríguez-Piazza
2002    Some new thin sets of integers in harmonic analysis, *J. Anal. Math.* **86**, 105–138.

J. Lindenstrauss
1966 a    A short proof of Liapounoff's convexity theorem, *J. Math. Mech.* **15**, 971–972.

1966 b    Notes on Klee's paper "Polyhedral sections of convex bodies," *Israel J. Math.* **4**, 235–242.

1967    On complemented subspaces of *m*, *Israel J. Math.* **5**, 153–156.

1971    On James' paper "separable conjugate spaces," *Israel J. Math.* **9**, 263–269.

J. Lindenstrauss & A. Pełczyński
1968    Absolutely summing operators in $\mathcal{L}_p$ spaces and their applications, *Studia Math.* **29**, 275–326.
1971    Contributions to the theory of the classical Banach spaces, *J. Funct. Anal.* **8**, 225–249.

J. Lindenstrauss & H.P. Rosenthal
1969    The $\mathcal{L}_p$ spaces, *Israel J. Math.* **7**, 325–349.

J. Lindenstrauss & C. Stegall
1975    Examples of separable spaces which do not contain $l_1$ and whose duals are non-separable, *Studia Math.* **54**, 81–105.

J. Lindenstrauss & L. Tzafriri
1971    On the complemented subspaces problem, *Israel J. Math.* **9**, 263–269.

J. Lindenstrauss & M. Zippin
1969    Banach spaces with a unique unconditional basis, *J. Funct. Anal.* **3**, 115–125.

J.-L. Lions & J. Peetre
1964    Sur une classe d'espaces d'interpolation, *Publ. Math. Inst. Hautes Études Sci.* **19**, 5–68.

R.H. Lohman
1976    A note on Banach spaces containing $\ell_1$, *Canad. Math. Bull.* **19**, 365–367.

G. Lorentz & N. Tomczak-Jaegermann
1984    Projections of minimal norm, *The University of Texas Functional Analysis Seminar 1983–1984*, Longhorn Notes, University of Texas at Austin, 167–176.

W. Lusky
1978    Some consequences of Rudin's paper "$L_p$-Isometries and equimeasurability," *Indiana Univ. Math. J.* **27**, 859–866.

F. Lust (F. Lust-Piquard)
1975    Produits tensoriels injectifs d'espaces faiblement séquentiellement complets, *Colloq. Math.* **33**, 289–290.

F. Lust-Piquard
1976    Ensembles de Rosenthal et ensembles de Riesz, *C.R.A.S. Paris Sér. A* **282**, 833-835.
1978    Propriétés géométriques des sous-espaces invariants par translation de $L^1(G)$ et $C(G)$, *Séminaire sur la Géométrie des Espaces de Banach 1977–1978*, Exposé 26, École Polytechnique, Palaiseau.
1979    L'espace des fonctions presque-périodiques dont le spectre est contenu dans un ensemble compact dénombrable a la propriété de Schur, *Colloq. Math.* **41**, 273–284.
1989    Bohr local properties of $C_\Lambda(\mathbb{T})$, *Colloq. Math.* **58**, 29-38.
1997    On the coefficient problem: A version of the Kahane–Katznelson–de Leeuw theorem for spaces of matrices, *J. Funct. Anal.* **149**, 352–376.

P. Malliavin & M.-P. Malliavin-Brameret
1967    Caractérisation arithmétique d'une classe d'ensembles de Helson, *C.R.A.S. Paris Sér. A* **264**, 192–193.

J. Marcinkiewicz & A. Zygmund
  1938    Quelques théorèmes sur les fonctions indépendantes, *Studia Math.* **7**, 104–120.
M.B. Marcus & G. Pisier
  1984    Characterizations of almost surely continuous $p$-stable random Fourier series and strongly stationary processes, *Acta Math.* **152**, 245–301.
M.B. Marcus & L. Shepp
  1972    Sample behavior of Gaussian processes, in *Proceedings of the Sixth Berkeley Symposium on Mathematical Statistics and Probability, University of California, Berkeley, CA, 1970–1971, Vol. II: Probability Theory*, University of California Press, 423–441.
B. Maurey
  1972 a  Théorèmes de factorisation pour les applications linéaires à valeurs dans un espace $L_p$, *C.R.A.S. Paris Sér. A* **274**, 1825–1828.
  1972 b  Espaces de type $(p, q)$, théorèmes de factorisation et de plongement, *C.R.A.S. Paris Sér. A* **274**, 1939–1941.
  1972 c  Espaces de cotype $(p, q)$ et théorèmes de relèvement, *C.R.A.S. Paris Sér. A* **275**, 785–788.
  1973 a  Espaces de cotype $p$, $0 < p \leqslant 2$, *Séminaire Maurey–Schwartz 1972–1973*, Exposé VII, École Polytechnique, Paris.
  1973 b  Théorèmes de Nikishin: Théorèmes de factorisation pour les applications linéaires à valeurs dans un espace $L^0(\Omega, \mu)$, *Séminaire Maurey–Schwartz 1972–1973*, Exposés X et XI, École Polytechnique, Paris.
  1973 c  Théorèmes de Nikishin: Théorèmes de factorisation pour les applications linéaires à valeurs dans un espace $L^0(\Omega, \mu)$ (suite et fin), *Séminaire Maurey–Schwartz 1972–1973*, Exposé XII, École Polytechnique, Paris.
  1973 d  Théorèmes de factorisation pour les opérateurs à valeurs dans un espace $L^p(\Omega, \mu)$, $0 < p \leqslant +\infty$, *Séminaire Maurey–Schwartz 1972–1973*, Exposé XV, École Polytechnique, Paris.
  1975 a  Système de Haar, *Séminaire Maurey–Schwartz 1974–1975*, Exposés I et II, École Polytechnique, Paris.
  1975 b  Projections dans $L^1$, d'après L. Dor, *Séminaire Maurey–Schwartz 1974–1975*, Exposé 21, École Polytechnique, Paris.
  1980 a  Isomorphismes entre espaces $H_1$, *Acta Math.* **145**, 79–120.
  1980 b  Tout sous-espace de $L^1$ contient un $\ell_p$ (d'après D. Aldous), *Séminaire d'Analyse Fonctionnelle 1979–1980*, Exposés 1 et 2, École Polytechnique, Palaiseau.
  1983    Types and $\ell_1$-subspaces, *The University of Texas Functional Analysis Seminar 1982–1983*, Longhorn Notes, University of Texas at Austin, 123–137.
  1990    Lemme de Slepian, Exposés des 27 avril et 4 mai 1990 à l'Université Pierre et Marie Curie (Paris VI), unpublished seminar.
  1991    Some deviation inequalities, *Geom. Funct. Anal.* **1**, 188–197.
  1994    Quelques progrès dans la compréhension de la dimension infinie, *Journée Annuelle*, Société Mathématique de France, 1–29.

1995 a     A remark about distortion, in *Geometric Aspects of Functional Analysis: Israel Seminar (1992–1994)*, Operator Theory: Advances and Applications **77**, Birkhäuser, 131–142.

1995 b     Symmetric distortion in $\ell_2$, *Geometric Aspects of Functional Analysis: Seminar Israel (1992–1994)*, Operator Theory: Advances and Applications **77**, Birkhäuser, 143–147.

1998        A note on Gowers' dichotomy theorem, Convex Geometric Analysis **34**, MSRI Publications, 149–157.

2003 a     Type, cotype and $K$-convexity, in *Handbook of the Geometry of Banach Spaces II*, Elsevier, 1299–1332.

2003 b     Banach spaces with few operators, in *Handbook of the Geometry of Banach Spaces II*, Elsevier, 1247–1297.

B. Maurey & A. Nahoum

1973        Applications radonifiantes dans l'espace des séries convergentes, *C.R.A.S. Paris Sér. A-B* **276**, A751–A754.

B. Maurey & G. Pisier

1973        Caractérisation d'une classe d'espaces de Banach par des propriétés de séries aléatoires vectorielles, *C.R.A.S. Paris Sér. A* **277**, 687–690.

1976        Séries de variables aléatoires vectorielles indépendantes et propriétés géométriques des espaces de Banach, *Studia Math.* **58**, 45-90.

Y. Meyer

1968 a     Spectres des mesures et mesures absolument continues, *Studia Math.* **30**, 87–99.

1968 b     Endomorphismes des idéaux fermés de $L^1(G)$, classes de Hardy et séries de Fourier lacunaires, *Ann. Sci. Éc. Norm. Supér.* **1**, 499–580.

V.D. Milman

1971 a     The geometric theory of Banach spaces, Part II, *Russian Math. Surveys* **26**, 79–163.

1971 b     A new proof of A. Dvoretzky's theorem on cross-sections of convex bodies, *Funct. Anal. Appl.* **5**, 288–295.

1982        Some remarks about embeddings of $\ell_1^k$ in finite dimensional spaces, *Israel J. Math.* **43**, 129–138.

1992        Dvoretzky's theorem: Thirty years later, *Geom. Funct. Anal.* **2**, 455–479.

V.D. Milman & G. Schechtman

1995        An "isomorphic" version of Dvoretzky's theorem, *C.R.A.S. Paris Sér. I Math.* **321**, 541–544.

V.D. Milman & N. Tomczak-Jaegermann

1993        Asymptotic $\ell_p$ spaces and bounded distortions, *Contemporary Mathematics* **44**, 173–195.

V.D. Milman & H. Wolfson

1978        Minkowski spaces with extremal distances from the Euclidean space, *Israel J. Math.* **29**, 113–131.

A.A. Miljutin [Milyutin]

1966        Isomorphisms of spaces of continuous functions on compacts of the power continuum, in *Teoria Funcktsii*, Funktsionalnyi Analiz i ego Prilozheniya. **2**, 150–156 (in Russian).

S.J. Montgomery-Smith
1990      The distribution of Rademacher sums, *Proc. Amer. Math. Soc.* **109**, 517–522.

M.C. Mooney
1972      A theorem on bounded analytic functions, *Pacific J. Math.* **43**, 457–463.

K. Musial
1979      The weak Radon–Nikodym property in Banach spaces, *Studia Math.* **64**, 151–173.

F.L. Nazarov
1996      Summability of large powers of logarithm of classic lacunary series and its simplest consequences, unpublished work.

1998      The Bang solution of the coefficient problem, *Algebra i Analiz* **9** (1997), 272–287 (in Russian); translation in *St. Petersburg Math. J.* **9**, 407–419.

F.L. Nazarov & A. N. Podkorytov
2000      Ball, Haagerup, and distribution functions, in *Complex Analysis, Operators, and Related Topics: The S.A. Vinogradov Memorial Volume*, Operator Theory: Advances and Applications **113**, Birkhäuser, 247–267.

S. Neuwirth
1998      Metric unconditionality and Fourier analysis, *Studia Math.* **131**, 19–62.

1999      Two random constructions inside lacunary sets, *Ann. Inst. Fourier* **49**, 1853–1867.

E.M. Nikishin
1970      Resonance theorems and superlinear operators, *Uspekhi Mat. Nauk* **25**, 129–191 (in Rissian).

E. Odell
2002      Stability in Banach spaces, *Extracta Math.* **17**, 385–425.

E. Odell & H.P. Rosenthal
1975      A double-dual characterization of separable Banach spaces containing $l^1$, *Israel J. Math.* **20**, 375–384.

E. Odell & T. Schlumprecht
1994      The distortion problem, *Acta Math.* **173**, 259–281.

A.M. Olevskiĭ
1967      Fourier series and Lebesgue functions, *Uspekhi Mat. Nauk* **22**, 236–239 (in Russian).

W. Orlicz
1929      Beiträge zu theorie der orthogonalenwicklungen, II, *Studia Math.* **1**, 241–255.

1933 a    Über unbedingte konvergenz in funktionenräumen. I, *Studia Math.* **4**, 33–37.

1933 b    Über unbedingte konvergenz in funktionenräumen. II, *Studia Math.* **4**, 41–47.

P. Ørno
1976      A note on unconditionally converging series in $L_p$, *Proc. Amer. Math. Soc.* **59**, 252–254.

A. Pajor
1983      Plongement de $\ell_1^k$ dans les espaces de Banach complexes, *C.R.A.S. Paris* **296**, 741–743.

R.E.A.C. Paley
1932     A remarkable series of orthogonal functions I, *Proc. London Math. Soc.*
         **34**, 241–264.
1933     On the lacunary coefficients of power series, *Ann. of Math.* **34**, 615–616.

R.E.A.C. Paley & A. Zygmund
1932     A note on analytic functions in the unit circle, *Proc. Cambridge Philos.*
         *Soc.* **28**, 266–272.

S.F. Papadopoulos
1998     Minima of trigonometric polynomials, *Bull. London Math. Soc.* **30**,
         291–294.

T. Pedersen
2000     Some properties of the Pisier algebra, *Math. Proc. Cambridge Philos. Soc.*
         **128**, 343–354.

A.M. Pelczar
2001     Remarks on Gowers' dichotomy, in *Recent Progress in Functional Anal-*
         *ysis: Proceedings of the International Functional Analysis Meeting on the*
         *Occasion of the 70th Birthday of Professor Manual Valdivia, Valencia,*
         *Spain, July 3–7, 2000*, North-Holland Mathematics Studies **189**, Elsevier,
         201–213.

A. Pełczyński
1958     A connection between weakly unconditional convergence and weakly
         completeness of Banach spaces, *Bull. Acad. Polon. Sci. Sér. Math.* **6**,
         251–253.
1960     Projections in certain Banach spaces, *Studia Math.* **19**, 209–228.
1961     On the impossibility of embedding of the space $L$ in certain Banach spaces,
         *Colloq. Math.* **8**, 199–203.
1962     Banach spaces on which every unconditionally converging operator is
         weakly compact, *Bull. Acad. Polon. Sci. Sér. Math.* **10**, 641–648.
1968     On Banach spaces containing $L_1(\mu)$, *Studia Math.* **30**, 231–246.
1971     Any separable Banach space with the bounded approximation property is
         a complemented subspace of a Banach space with a basis, *Studia Math.*
         **40**, 239–243.
1988     Commensurate sequences of characters, *Proc. Amer. Math. Soc.* **104**,
         525–531.

A. Pełczyński & C. Bessaga
1979     Some aspects of the present theory of Banach spaces, in *S. Banach,*
         *Oeuvres, Vol. II: Travaux sur l'Analyse Fonctionnelle*, C. Bessaga, S.
         Mazur, W. Orlicz, A. Pełczyński, S. Rolewicz & W. Żelazko, eds., with
         an article by A. Pełczyński & C. Bessaga, PWN – Éditions Scientifiques
         de Pologne, Warsaw (1979). Also see S. BANACH, *Theory of Linear*
         *Operations*, translated from French by F. Jellett, with commentaries by
         A. Pełczyński and C. Bessaga, North-Holland Mathematical Library **38**,
         North-Holland (1987).

A. Pełczyński & W. Szlenk
1965     An example of a non-shrinking basis, *Rev. Roumaine Math. Pures Appl.*
         **10**, 961–966.

A. Pełczyński & P. Wojtaszczyk
1971        Banach spaces with finite dimensional expansions of identity and universal bases of finite dimensional subspaces, *Studia Math.* **40**, 91–108.

B.J. Pettis
1938 a      On integration in vector spaces, *Trans. Amer. Math. Soc.* **44**, 277–304.
1938 b      Linear functionals and completely additive set functions, *Duke Math. J.* **4**, 552–565.

H. Pfitzner
1993        *L*-summands in their biduals have property ($V^*$), *Studia Math.* **104**, 91–98.
1994        Weak compactness in $C^*$-algebras is determined commutatively, *Math. Ann.* **298**, 349–371.

R.S. Phillips
1940        On linear transformations, *Trans. Amer. Math. Soc.* **48**, 516–541.

A. Pietsch
1967        Absolut *p*-summierende abbildungen in normierten räumen, *Studia Math.* **28**, 333–353.

G. Pisier
1973 a      Type des espaces normés, *C.R.A.S. Paris Sér. A* **276**, 1673–1676.
1973 b      Sur les espaces de Banach qui ne contiennent pas uniformément de $\ell_n^1$, *C.R.A.S. Paris Sér. A* **277**, 991–994.
1973 c      Bases, suites lacunaires dans les espaces $L^p$ d'après Kadeč et Pełczyński, *Séminaire Maurey–Schwartz 1972–1973*, Exposés XVIII et XIX, École Polytechnique, Paris.
1973 d      Sur les espaces qui ne contiennent pas de $l_n^\infty$ uniformément, *Séminaire Maurey–Schwartz, 1972–1973*, Annexe, École Polytechnique, Paris.
1974 a      "Type" des espaces normés, *Séminaire Maurey–Schwartz 1973–1974*, Exposé III, École Polytechnique, Paris.
1974 b      Sur les espaces qui ne contiennent pas de $\ell_n^1$ uniformément, *Séminaire Maurey–Schwartz 1973–1974*, Exposé VII, École Polytechnique, Paris.
1974 c      Une propriété du type *p*-stable, *Séminaire Maurey–Schwartz 1973–1974*, Exposé No. VIII (errata, page E.1), École Polytechnique, Paris.
1978 a      Ensembles de Sidon et espaces de cotype 2, *Séminaire sur la Géométrie des Espaces de Banach 1977–1978*, Exposé 14, École Polytechnique, Paris.
1978 b      Sur l'espace de Banach des séries de Fourier aléatoires presque sûrement continues, *Séminaire sur la Géométrie des Espaces de Banach 1977–1978*, Éxposes 17–18, École Polytechnique, Paris.
1978 c      Grothendieck's theorem for non-commutative $C^*$-algebras with an appendix on Grothendieck's constants, *J. Funct. Anal.* **29**, 397–415.
1978 d      Une nouvelle classe d'espaces vérifiant le théorème de Grothendieck, *Ann. Inst. Fourier* **28**, 69–90.
1978 e      Les inégalités de Khintchine–Kahane, d'après C. Borell, *Séminaire sur la Géométrie des Espaces de Banach 1977–1978*, Exposé 7, École Polytechnique, Palaiseau.
1978 f      Une propriété de stabilité de la classe des espaces ne contenant pas $l_1$, *C.R.A.S. Paris Sér. A-B* **286**, A747–A749.

1979    A remarkable homogeneous Banach algebra, *Israel J. Math.* **34**, 38–44.

1980 a   Un théorème de factorisation pour les opérateurs linéaires entre espaces de Banach, *Ann. Sci. Éc. Norm. Supér.* **13**, 23–44.

1980 b   Conditions d'entropie assurant la continuité de certains processus et applications à l'analyse harmonique, *Séminaire d'Analyse Fonctionnelle 1979–1980* Exposés 13–14, École Polytechnique, Palaiseau.

1981 a   De nouvelles caractérisations des ensembles de Sidon, in *Mathematical Analysis and Applications, Part B*, Advances in Mathematics Supplementary Studies **7B**, Academic Press, 685–726.

1981 b   Semi-groupes holomorphes et espaces de Banach $K$-convexes, *Séminaire d'Analyse Fonctionnelle 1980–1981* Exposé II, École Polytechnique, Palaiseau.

1981 c   Remarques sur un résultat non publié de B. Maurey, *Séminaire d'Analyse Fonctionnelle 1980–1981*, Exposé V, École Polytechnique, Paris.

1982    Holomorphic semi-groups and the geometry of Banach spaces, *Ann. of Math.* **115**, 375–392.

1983 a   Arithmetical characterizations of Sidon sets, *Bull. Amer. Math. Soc.* **8**, 87–89.

1983 b   Conditions d'entropie et caractérisation arithmétiques des ensembles de Sidon, in *Topics in Modern Harmonic Analysis: Proceedings of a Seminar Held in Torino and Milano, May–June 1982*, Vol. II, Instituto Nazionale di Alta Matematica, 911–944..

1983 c   Some applications of the metric entropy condition to harmonic analysis, in *Banach Spaces, Harmonic Analysis, and Probability Theory: Proceedings of the Special Year in Analysis, Held at the University of Connecticut 1980–1981*, Lecture Notes in Mathematics **995**, Springer, 123–154.

1984    Remarques sur les classes de Vapnik et Červonenkis, *Ann. Inst. H. Poincaré B. Probab. Stat.* **20**, 287–298.

1986 a   Factorization of operators through $L_{p\infty}$ and $L_{p1}$ and non-commutative generalizations, *Math. Ann.* **276**, 105–136.

1986 b   Probabilistic methods in the geometry of Banach spaces, in *Probability and Analysis: Lectures Given at the 1st 1985 Session of the Centro Internazionale Matematico Estiro (CIME) held in Varenna (Como), Italy, May 31–June 8, 1985*, Lecture Notes in Mathematics **1206**, Springer, 167–241.

1988    The dual $J^*$ of the James space has cotype 2 and the Gordon–Lewis property, *Math. Proc. Cambridge Philos. Soc.* **103**, 323–331.

1996    Dvoretzky's theorem for operator spaces, *Houston J. Math.* **22**, 399–416.

G. Pisier & Q. Xu
1997    Non-commutative martingale inequalities, *Comm. Math. Phys.* **189**, 667–698.

H.R. Pitt
1936    A note on bilinear forms, *J. London Math. Soc. (1)* **11**, 174–180.

A. I. Plotkin
1974    Continuation of $L^p$ isometries, *J. Soviet Math.* **2**, 143–165.

A. Prékopa

1973    On logarithmically concave measures and functions, *Acta Sci. Math.* **34**, 335–343.

C. Preston

1971    Banach spaces arising from some integral inequalities, *Indiana Univ. Math. J.* **20**, 997–1015.

P. Prignot

1987    Dichotomie du cotype pour les espaces invariants, Publications Mathématiques d'Orsay 87–02, 1–50.

H. Queffélec

1993    Sur un théorème de Gluskin–Meyer–Pajor, *C.R.A.S. Paris Sér. I Math.* **317**, 155–158.

1995    Norm of the inverse of a matrix: Solution to a problem of Schäffer, in *Harmonic Analysis from the Pichorides Viewpoint: Recueil d'Articles Réunis à l'Occasion du Colloque tenu à Anogia (Crète, 24–28 Juillet 1995) en l'Honneur de Stylianos Pichorides*, Publications Mathématiques d'Orsay 96-01, 69–87.

H. Queffélec & B. Saffari

1996    On Bernstein's inequality and Kahane's ultraflat polynomials, *J. Fourier Anal. Appl.* **2**, 519–582.

Y. Raynaud

1981 a    Espaces de Banach superstables, *C.R.A.S. Paris Sér. I Math.* **292**, 671–673.

1981 b    Deux nouveaux exemples d'espaces de Banach stables, *C.R.A.S. Paris Sér. I Math.* **292**, 715–717.

1983    Espaces de Banach superstables, distances stables et homéomorphismes uniformes, *Israel J. Math.* **44**, 33–52.

C.J. Read

Different forms of the approximation property, unpublished work.

É. Ricard

2000    L'espace $H^1$ n'a pas de base complètement inconditionnelle, *C.R.A.S. Paris Sér. I Math.* **331**, 625–628.

D. Rider

1975    Randomly continuous functions and Sidon sets, *Duke Math. J.* **42**, 759–764.

M. Riesz

1926    Sur les maxima de formes bilinéaires et sur les fonctionnelles linéaires, *Acta Math.* **49**, 465–497.

1927    Sur les fonctions conjuguées, *Math. Z.* **27**, 218–244.

H. Robbins

1955    A remark on Stirling's formula, *Amer. Math. Monthly* **62**, 26–29.

L. Rodríguez–Piazza

1987    Caractérisation des ensembles $p$-Sidon p.s., *C.R.A.S. Paris Sér. I Math.* **305**, 237–240.

1991    Rango y propiedades de medidas vectoriales: Conjuntos $p$-Sidon p.s., Thesis, Universidad de Sevilla.

M. Rogalski
1968     Opérateurs de Lion, projecteurs boréliens, et simplexes analytiques, *J. Funct. Anal.* **2**, 458–488.

H.P. Rosenthal
1967     On trigonometric series associated with weak* closed subspaces of continuous functions, *J. Math. Mech.* **17**, 485–490.

1970     On the subspaces of $L^p$ $(p > 2)$ spanned by sequences of independent random variables, *Israel J. Math.* **8**, 273–303.

1972     On factors of $C([0, 1])$ with non-separable dual, *Israel J. Math.* **13**, 361–378.

1973     On subspaces of $L^p$, *Ann. of Math.* **97**, 344–373.

1974 a   Pointwise compact subsets of the first Baire class, *Amer. J. Math.* **99**, 362–378.

1974 b   A characterization of Banach spaces containing $\ell^1$, *Proc. Nat. Acad. Sci. USA* **71**, 2411–2413.

1979     Sous-espaces de $L^1$, Cours de 3ème cycle, Université Paris VI, unpublished manuscript.

1984     Double dual types and the Maurey characterization of Banach spaces containing $\ell^1$, *The University of Texas Functional Analysis Seminar 1983–1984*, Longhorn Notes, University of Texas at Austin, 1–37.

1994     A characterization of Banach spaces containing $c_0$, *J. Amer. Math. Soc.* **7**, 707–748.

W. Rudin
1959     Some theorems on Fourier coefficients, *Proc. Amer. Math. Soc.* **10**, 855–859.

1960     Trigonometric series with gaps, *J. Math. Mech.* **9**, 203–227.

1976     $L_p$-isometries and equimeasurability, *Indiana Univ. Math. J.* **25**, 215–228.

R. Salem & A. Zygmund
1954     Some properties of trigonometric series whose terms have random signs, *Acta Math.* **91**, 245–301.

N. Sauer
1972     On the density of families of sets, *J. Combin. Theory Ser. A* **13**, 145–147.

J. Schauder
1927     Zur theorie stetiger abbildungen in funktionalraümen, *Math. Z.* **26**, 46–65.

1928     Eine eigenschaft des haarschen orthogonalsystems, *Math. Z.* **28**, 317–320.

G. Schechtman
1979     Almost isometric $L_p$ subspaces of $L_p(0, 1)$, *J. London Math. Soc. (2)* **20**, 516–528.

1981     Random embeddings of Euclidean spaces in sequence spaces, *Israel J. Math.* **40**, 187–192.

1987     More on embedding subspaces of $L_p$ in $l_r^n$, *Compos. Math.* **61**, 159–170.

T. Schlumprecht
1991     An arbitrarily distortable Banach space, *Israel J. Math.* **76**, 81–95.

S. Shelah
1972     A combinatorial problem, *Pacific J. Math.* **41**, 247–261.

S. Sidon
  1927       Verallgemeinerung eines satzes über die absolute Konvergenz von Fourier reihen mit lücken, *Math. Ann.* **97**, 675–676.

S. Simons
  1972 a     A convergence theorem with boundary, *Pacific J. Math.* **40**, 703–708.
  1972 b     Maximinimax, minimax, and antiminimax theorems and a result of R.C. James, *Pacific J. Math.* **40**, 709–718.

W.T. Sledd
  1981       Random series which are BMO or Bloch, *Michigan Math. J.* **28**, 259–266.

D. Slepian
  1962       The one-sided barrier problem for Gaussian noise, *Bull. System Tech. J.* **41**, 463–501.

V.L. Šmulian
  1940       Über linear topologische räume, *Mat. Sb. N.S.* **7**, 425–448.

A. Sobczyk
  1941       Projection of the space $m$ on its subspace $c_0$, *Bull. Amer. Math. Soc.* **47**, 938–947.

S.B. Stečkin
  1956       On absolute convergence of Fourier series, *Izv. Akad. Nauk SSSR, Ser. Matem.* **20**, 385–412.

C. Stegall
  1975       The Radon–Nikodym property in conjugate Banach spaces, *Trans. Amer. Math. Soc.* **206**, 213–223.
  1980       A proof of the principle of local reflexivity, *Proc. Amer. Math. Soc.* **78**, 154–156.

H. Steinhaus
  1919       Additive und stetige funktionaloperationen, *Math. Z.* **5**, 186–221.

V.N. Sudakov
  1969       Gaussian measures, Cauchy measures and $\varepsilon$-entropy, *Soviet Math. Dokl.* **10**, 310–313.
  1971       Gaussian random processes, and measures of solid angles in Hilbert space, *Dokl. Akad. Nauk SSSR* **197**, 43–45 (in Russian); translation in *Soviet Math. Dokl.* **12**, 412–415.
  1973       A remark on the criterion of continuity of Gaussian sample function, in *Proceedings of the Second Japan-USSR Symposium on Probability Theory*, Lecture Notes in Mathematics **330**, Springer, 444–454.

A. Szankowski
  1974       On Dvoretzky's Theorem on almost spherical sections of convex bodies, *Israel J. Math.* **17**, 325–338.
  1976       A Banach lattice without the approximation property, *Israel J. Math.* **24**, 329–337.
  1978       Subspaces without the approximation property, *Israel J. Math.* **30**, 123–129.

S. Szarek
  1976       On the best constants in the Khinchin inequality, *Studia Math.* **58**, 197–208.
  1983       The finite dimensional basis problem with an appendix on nets of Grassmann manifolds, *Acta Math.* **151**, 153–179.

1987       A Banach space without a basis which has the bounded approximation property, *Acta Math.* **159**, 81–98.

W. Szlenk

1965       Sur les suites faiblement convergentes dans l'espace *L*, *Studia Math.* **25**, 337–341.

1968       The non-existence of a separable reflexive Banach space universal for all separable reflexive Banach spaces, *Studia Math.* **30**, 53–61.

M. Talagrand

1984 a     Weak Cauchy sequences in $L^1(E)$, *Amer. J. Math.* **106**, 703–724.

1984 b     A new type of affine Borel function, *Math. Scand.* **54**, 183–188.

1987       Regularity of Gaussian processes, *Acta Math.* **159**, 99–149.

1990 a     Embedding subspaces of $L_1$ into $l_1^N$, *Proc. Amer. Math. Soc.* **108**, 363–369.

1990 b     The three-space problem for $L^1$, *J. Amer. Math. Soc.* **3**, 9–29.

1992 a     Cotype of operators from $C(K)$, *Invent. Math.* **107**, 1–40.

1992 b     Type, infratype and the Elton–Pajor theorem, *Invent. Math.* **107**, 41–59.

1992 c     Cotype and $(q, 1)$-summing norm in a Banach space, *Invent. Math.* **110**, 545–556.

1992 d     A simple proof of the majorizing measure theorem, *Geom. Funct. Anal.* **2**, 118–125.

1994       Orlicz property and cotype in symmetric sequence spaces, *Israel J. Math.* **87**, 181–192.

1995       Sections of smooth convex bodies via majorizing measures, *Acta Math.* **175**, 273–300.

2001       Majorizing measures without measures, *Ann. Probab.* **29**, 411–417.

G.O. Thorin

1939       An extension of a convexity theorem due to M. Riesz, *Fysiogr. Sllsk. Lund Frh.* **8**, 166–170.

B. Tomaszewski

1982       Two remarks on the Khintchine–Kahane inequality, *Colloq. Math.* **46**, 283–288.

1987       A simple and elementary proof of the Khintchine inequality with the best constant, *Bull. Sci. Math.* (2) **111**, 103–109.

N. Tomczak-Jaegermann

1974       The moduli of smoothness and convexity and the Rademacher averages of trace classes $S_p$ $(1 \leqslant p < \infty)$, *Studia Math.* **50**, 163–182.

B.S. Tsirelson

1974       Not every Banach space contains $\ell_p$ or $c_0$, *Funct. Anal. Appl.* **8**, 139–141.

P. Turpin

1973       Un critère de compacité dans les espaces vectoriels topologiques, *Studia Math.* **46**, 141–148.

L. Tzafriri

1969       Reflexivity of cyclic Banach spaces, *Proc. Amer. Math. Soc.* **22**, 61–68.

1972       Reflexivity in Banach lattices and their subspaces, *J. Funct. Anal.* **10**, 1–18.

J.J. Uhl, Jr.

1969       The range of a vector-valued measure, *Proc. Amer. Math. Soc.* **23**, 158–163.

D. Ullrich
1988 a    An extension of the Kahane–Khinchine inequality, *Bull. Amer. Math. Soc.
          (N.S.)* **18**, 52–54.
1988 b    An extension of the Kahane–Khinchine inequality in a Banach space,
          *Israel J. Math.* **62**, 56–62.
1988 c    Khinchin's inequality and the zeroes of Bloch functions, *Duke Math. J.*
          **57**, 519–535.

V.N. Vapnik & A.Y. Červonenkis
1971      On the uniform convergence of relative frequencies of events to their
          probabilities, *Theory Probab. Appl.* **16**, 264–280.

N. Varopoulos
1976      Une remarque sur les ensembles de Helson, *Duke Math. J.* **43**, 387–390.

W.A. Veech
1971      Short proof of Sobczyk's theorem, *Proc. Amer. Math. Soc.* **28**, 627–628.

B. Virot
1981      Extensions vectorielles d'opérateurs linéaires bornés sur $L^p$, Publications
          Mathématiques d'Orsay 81-08, Exposé 7.

F. Weissler
1980      Logarithmic Sobolev inequalities and hypercontractive estimates on the
          circle, *J. Funct. Anal.* **37**, 218–234.

D. Werner
2001      Recent progress on the Daugavet property, *Irish Math. Soc. Bull.* **46**,
          77–97.

R.J. Whitley
1967      An elementary proof of the Eberlein–Šmulian theorem, *Math. Ann.* **172**,
          116–118.

G. Willis
1992      The compact approximation property does not imply the approximation
          property, *Studia Math.* **103**, 99–108.

P. Wojtaszczyk
1982      The Franklin system is an unconditional basis in $H_1$, *Ark. Mat.* **20**,
          293–300.
1999      Wavelets as unconditional bases in $L_p(\mathbb{R})$, *J. Fourier Anal. Appl.* **5**, 73–85.

V.V. Yurinskiĭ
1976      Exponential inequalities for sums of random vectors, *J. Multivariate Anal.*
          **6**, 473–499.

M. Zafran
1978      The dichotomy problem for homogeneous Banach algebras, *Ann. of Math.
          (2)* **108**, 97–105.
1979      On the symbolic calculus in homogeneous Banach algebras, *Israel J.
          Math.* **32**, 183–192.

M. Zippin
1966      On perfectly homogeneous bases in Banach spaces, *Israel J. Math.* **4**,
          265–272.
1977      The separable extension problem, *Israel J. Math.* **26**, 372–387.
1988      Banach spaces with separable duals, *Trans. Amer. Math. Soc.* **310**,
          371–379.

# Notation Index for Volume 2

# Author Index for Volume 2

# Subject Index for Volume 2

# Notation Index for Volume 1

# Author Index for Volume 1

# Subject Index for Volume 1